HANDBOOK ON THE DIGITAL ECONOMY

Handbook on the Digital Creative Economy

Edited by

Ruth Towse

Professor of Economics of Creative Industries, CIPPM, Bournemouth University, UK

Christian Handke

Assistant Professor of Cultural Economics, ESHCC, Erasmus University Rotterdam; Senior Researcher, IViR, University of Amsterdam, The Netherlands

Edward Elgar

Cheltenham, UK • Northampton, MA, USA

Published by
Edward Elgar Publishing Limited
The Lypiatts
15 Lansdown Road
Cheltenham
Glos GL50 2JA
UK

Edward Elgar Publishing, Inc.
William Pratt House
9 Dewey Court
Northampton
Massachusetts 01060
USA

A catalogue record for this book
is available from the British Library

Library of Congress Control Number: 2013943223

This book is available electronically in the ElgarOnline.com
Economics Subject Collection, E-ISBN 978 1 78100 487 6

ISBN 978 1 78100 486 9 (cased)
ISBN 978 1 78100 488 3 (paperback)

Typeset by Servis Filmsetting Ltd, Stockport, Cheshire
Printed and bound in Great Britain by T.J. International Ltd, Padstow

Contents

Figures

Tables

Contributors

Payal Arora, Erasmus University Rotterdam, The Netherlands

Kristín Atladottír, University of Iceland

Piet Bakker, Hogeschool Utrecht, The Netherlands

John Banks, Queensland University of Technology in Brisbane, Australia

William J. Baumol, Professor and Academic Director, the Berkley Center for Entrepreneurship and Innovation, Stern School of Business at New York University; and Professor Emeritus, Princeton University, USA

Cliff Bekar, Lewis and Clark College, USA

Axel Bruns, ARC Centre of Excellence for Creative Industries and Innovation, Queensland University of Technology, Australia

Stuart Cunningham, ARC Centre of Excellence for Creative Industries and Innovation, Queensland University of Technology, Australia

Peter DiCola, Northwestern University School of Law, USA

Gillian Doyle, Professor of Media Economics, Director of Centre for Cultural Policy Research, University of Glasgow, UK

Koen van Eijck, Erasmus University Rotterdam, The Netherlands

Joëlle Farchy, University of Paris 1, France

Marcella Favale, Research Fellow, Centre for Intellectual Property Policy and Management, Bournemouth University, UK

Terry Flew, Queensland University of Technology, Brisbane, Australia

Mathilde Gansemer, University of Paris 1, France

Peter Goodridge, Imperial College Business School, University of London, UK

Christian Handke, Assistant Professor of Cultural Economics, ESHCC, Erasmus University Rotterdam, and Senior Researcher, IViR, University of Amsterdam, The Netherlands

Erin Haswell, Lewis and Clark College, USA

Anders Henten, CMI, Aalborg University Copenhagen, Denmark

Reto M. Hilty, Director, Max Planck Institute for Intellectual Property and Competition Law, Munich and Professor at Zurich University, Switzerland and Ludwig Maximilians University, Munich, Germany

Fabian Homberg, Senior Lecturer, Business School, Bournemouth University, UK

Ronald Inglehart, Lowenstein Professor of Political Science and Research Professor at the Institute for Social Research at the University of Michigan, USA

Anette Johansson, Jönköping International Business School, Jönköping University, Sweden

Ariel Katz, Associate Professor, Innovation Chair – Electronic Commerce, Faculty of Law, University of Toronto, Canada

Hans van Kranenburg, Radboud University Nijmegen, Institute for Management Research, Nijmegen School of Management, The Netherlands

Martin Kretschmer, Professor of Law and Director, CREATe, University of Glasgow, UK

Michael Latzer, Professor of Media Change and Innovation at the Institute of Mass Communication and Media Research (IPMZ), University of Zurich, Switzerland

Stan J. Liebowitz, Ashbel Smith Professor of Economics and Director of the Center for the Analysis of Property Rights and Innovation, University of Texas at Dallas, USA

Max Majorana, Erasmus University Rotterdam, The Netherlands

Dinusha Mendis, Senior Lecturer in Law, Co-Director Centre for Intellectual Property Policy and Management, Bournemouth University, UK

Frank Mueller-Langer, Senior Research Fellow, Munich Center for Innovation and Entrepreneurship Research (MCIER), Max Planck Institute for Intellectual Property and Competition Law, Munich, Germany

Trilce Navarrete, University of Amsterdam, The Netherlands

Sylvie Nérisson, Senior Research Fellow, Max Planck Institute for Intellectual Property and Competition Law, Munich, Germany

Pippa Norris, Paul F. McGuire Lecturer in Comparative Politics, Kennedy School of Government, Harvard University, USA and ARC Laureate Fellow and Professor of Government at the University of Sydney, Australia

Jessica Petrou, University of Paris 1, France

Joost Poort, Institute for Information Law, University of Amsterdam, The Netherlands

Jason Potts, School of Economics, Finance and Marketing, RMIT University, Melbourne, Australia

Andy C. Pratt, Professor of Cultural Economy, City University London, UK

Marc Scheufen, DFG Graduate School in Law and Economics (GRK 1597/1), Institute of Law and Economics, University of Hamburg, Germany

Nicola Searle, Intellectual Property Office, UK

Davide Secchi, Senior Lecturer, Business School, Bournemouth University, UK

Paul Stepan, FOKUS Society for cultural economics and policy studies, Austria

Adam Swift, Queensland University of Technology, Brisbane, Australia

Reza Tadayoni, CMI, Aalborg University Copenhagen, Denmark

Ruth Towse, Professor of Economics of Creative Industries, CIPPM, Bournemouth University, UK and Professor Emerita, Erasmus University Rotterdam, The Netherlands

Peter Tschmuck, Professor of Culture Institutions Studies, University of Music and Performing Arts Vienna, Austria

Filip Vermeylen, Erasmus University Rotterdam, The Netherlands

Patrick Waelbroeck, Telecom ParisTech, France

Richard Watt, University of Canterbury, New Zealand

Gregor White, Institute for Arts, Media and Computer Games, University of Abertay Dundee, UK

Patrik Wikström, ARC Centre of Excellence for Creative Industries and Innovation, Queensland University of Technology, Australia

Glenn Withers, Australian National University, Australia

Richard van der Wurff, Amsterdam School of Communication Research (ASCoR), University of Amsterdam, The Netherlands

Gerrit Willem Ziggers, Radboud University Nijmegen, Institute for Management Research, Nijmegen School of Management, The Netherlands

Abbreviations

ACE	Arts Council of England
ATSC	Advanced Television System Committee
CRMO	collective rights management organization
DCMS	Department for Culture, Media and Sport (UK)
DMCA	Digital Millennium Copyright Act of 1998 (US)
DRM	digital rights management
DTT	digital terrestrial television
EC	European Commission
ECJ	European Court of Justice
EU	European Union
DVB	Digital Video Broadcasting
GPT	general purpose technology
GVA	gross value added
HBBTV	Hybrid Broadcast Broadband TV
HDTV	high-definition television
ICT	information and communication technology
IFPI	International Federation of the Phonographic Industry
IP	Internet Protocol; also intellectual property
IPR	intellectual property right
IPTV	Internet Protocol TV
ISDB	Integrated Services Digital Broadcasting
ISDN	Integrated Services Digital Network
IT	information technology
ITU	International Telecommunication Union
NEA	National Endowment for the Arts
NT	National Theatre
OA	open access
OECD	Organisation for Economic Co-operation and Development
OTT TV	Over-the-Top TV
PAL	Phase Alternating Line
SIC	Standard Industrial Classification
TPMs	technological protection measures
UHF	ultra-high frequency
UNESCO	United Nations Educational, Scientific and Cultural Organization
VHF	very high frequency
WIPO	World Intellectual Property Organization
wtp	willingness to pay

Introduction

Christian Handke and Ruth Towse

Digital information and communication devices have developed rapidly, and they are applied in increasingly sophisticated ways for many tasks and activities. This process of digitization is already having a considerable impact on the creative industries. An increasingly important aspect of creative works is the element of information goods and services that can be captured in digital bits. Digitization transforms the way creative works are generated, disseminated and used. Digitization has also enabled the development of new types of creative goods and services, such as video games, and new ways of financing creativity, such as crowd-funding, and it is blurring the boundary between producers and consumers. The effects of digitization are also felt in the traditional, non-reproducible arts, where it affects back-office tasks and brings up new related goods and services, for example. What is more, digital technologies generate plenty of data on activities related to creative works, which provides all types of stakeholders and researchers with new opportunities. This handbook contains 37 chapters that discuss the implications of digitization in the creative economy.

It has often been proclaimed that there will be substantial changes in the arts, cultural heritage and media industries in the course of digitization. It has also turned out to be difficult to make adequate predictions about the types of change, their timing, or their normative implications. Visions of perfectly competitive markets, an age of access and do-it-yourself culture contrast with concerns for excessive commercialization, market power of transnational conglomerates, or the erosion of the established, professional cultural practices and excellence. In so far as concepts and predictions are precise and specific enough to allow for comparisons, many seem incompatible with each other. This lack of understanding makes it hard to develop practical insights for stakeholders or generally acceptable public policy. The handbook seeks to present the results of sober academic work so far. The aim is to inform the public debate and those seeking to develop new insights.

The term 'creative economy' has taken hold in the last decade or so along with the adoption of the notion of creative industries encompassing the arts, heritage and cultural and media industries. This new paradigm is based on several shared features of the performing arts, the art market, museums, publishing of literary and academic texts, recorded music, film, broadcasting, fashion, advertising, games and so on. They all deal with novel ideas of cultural and social value that are produced by creative people. Valuable new ideas can be exploited to produce wealth, income and economic growth, though the goods and services that embody them are not perfectly excludable and they give rise to substantial external benefits. What is more, demand conditions for what are often thought of as experience goods are notoriously uncertain.

The idea of a creative economy, that is, an economy in which creativity and human capital rather than innovation in physical capital are the drivers of productivity and growth, seems to be relatively recent. Other nomenclature has been adopted previously for similar conceptions, such as the knowledge economy, the Internet economy, the digital economy, the information society and so on (Carlaw et al., 2006). Early on, it was the Internet rather than digital technology that was regarded as the force behind changes hitting the 'mass media', the preferred term among cultural sociologists and media scholars (Picard et al., 2008). Similarly, there was no consensus on what to call these 'cultural' industries, which overlapped with media industries and were not the same as the arts but related to them. Cultural economics, developed in the 1960s as the economics of the arts, acquired its title as a parallel to cultural sociology (Towse, 1997). These changes in terminology, taxonomy and conceptualization do not appear to have any methodological significance in the philosophical sense. However, agreement on terms is necessary for clarification of the scope of public policy and also for academic enquiry. A review of these matters was part of an important milestone in the creative industries paradigm, the publication of *Creative Economy Report 2008* by the United Nations Conference for Trade and Development (UNCTAD), which worked with other UN agencies, the UN Educational, Scientific and Cultural Organization (UNESCO) and the World Intellectual Property Organization (WIPO), which effectively endorsed the term 'creative economy' (UNCTAD, 2008). The accepted term for the arts and cultural enterprises in the creative economy now seems to be established in the international sphere as 'cultural and creative industries' or simply 'creative industries'; however, authors writing in this area still use somewhat differing terms, and that is also the case in this book.

Do terminology and defining boundaries matter? If we stick to talking in general about these developments, it hardly does. That is different when it comes to measuring the contribution to the national economy of the creative industries and their growth, which has been one of the main reasons for the promotion of the notion of the creative economy. To establish those features, it is necessary to have a listing of industries with consistent industrial classification that can be measured over time. The UK's list – advertising, architecture, art and antiques, craft, design, designer fashion, film and video, interactive leisure software, music, performing arts, publishing, software, television and radio – has been adopted elsewhere, but it is not definitive. If we want to measure the contribution of the creative industries to the national income and growth – and these figures precede many a government document – there have to be consistency and the avoidance of double-counting. Although there is some controversy as to how they should be measured, it is generally the case that they make a significant contribution to gross domestic product (GDP) and are growing faster than the average, and that is the case in a range of economies, not just in developed ones (WIPO, 2012).

Digitization has spread very quickly and in some ways has had unprecedented effects, both intended and unintended, on all aspects of life; there is, however, a 'digital divide', with the wealthy and citizens of developed countries being more 'connected' than the poor in less developed countries. Digitization is a general purpose technology (GPT) in the sense that it has spread universally through disparate aspects of production and consumption in the economy; Lipsey et al. (2005) argue that digitization has produced a technological revolution in the same way that steam and electricity did in previous industrial revolutions. Moreover, there have been other GPTs in the past that have impacted on the arts and effectively created cultural industries: the printing press, engraving and photography, sound recording and film, radio and television, and transistorization, to name some. All have increased access to the arts and culture to ever more consumers – readers, visitors, listeners and viewers. They have led to the production of both 'software' and 'hardware' that enable the widespread distribution of creative content in printed books, reproductions of works of art, sound recordings, films and the like, as well as the facilities needed to access them – record players, cinemas, radio and TV sets, DVD players and so on. Some developments have allowed people to produce their own work, such as cameras and film, camcorders and sound recording equipment. Digital technologies have taken all these activities a stage further, and the Internet has vastly extended their distribution. Indeed, the Internet has created its own possibilities for creators. Many technologies have fallen by the wayside and have not been heard of again (Picard et al., 2008). Others did not deliver on the initial promises associated with them. A significant question in cultural economics has been the ability of arts organizations, particularly in the performing arts with their labour-intensive production process, to adopt technologies that enable productivity to rise in tandem with that of the rest of the economy, thus reducing costs of production and holding down price rises. It is an interesting question whether the well-known 'cost disease' (Baumol and Bowen, 1966) can finally be defeated by digital technologies; according to Baumol's chapter in this volume (Chapter 2), it cannot.

This book includes all these aspects of the digital creative economy: some chapters deal with the transition to the present and others with the new and foreseeable. Other economists have placed emphasis on the Internet's impact on the creative industries; we prefer to consider digitization as the basis of change, viewing the Internet as a means of delivering digital goods and services. Both are valid approaches and no doubt offer somewhat different insights. Furthermore, any analysis of the creative economy has to take into account telecommunications, which both provide platforms for distribution of creative content and are responsible for accessibility to that content. That has not so far been part of the study of the creative industries.

Another vital dimension to considering the impact of digitization on the creative economy is the role played by copyright law, which is probably its chief means of regulation. Copyright law and the creative economy are so closely connected that the creative industries are also called 'copyright industries' or 'copyright-based industries' by some authors (UNCTAD, 2008; WIPO, 2012). One of the leading 'problems' in the creative economy has been the impact that digitization has had on the efficacy of copyright law. That has been confronted in two ways both in practice and in the literature: first, by the enhancement of copyright law to cope with digital content and distribution; second, by the development of new business models that enable creators and the creative industries to make money in the digital world. The economics of copyright is a whole field in itself, with theoretical and empirical literatures (Towse et al., 2008). Several aspects of digitization and copyright are included in the book: how copyright law has adapted to digitalization; the implications for the administration of copyright by collecting societies; and the connection between copyright law and competition policy. The impact of digitization and the Internet on unauthorized usage of copyright material has been a major topic in the economics of copyright and is represented here too.

Most chapters in the book have been specially commissioned. A few are based on existing longer articles. Authors were asked to keep their contributions to around 5000 words (in order to allow a wide range of topics to be covered) and to offer guidance for further reading. The focus of many chapters is on the impact of digitization on a particular industry or sector, but several chapters offer an overall perspective on the evolution of technologies in the arts and culture that provides a perspective on the impact of digital technology. Though there is some degree of overlap in the chapters, each author clearly has an individual outlook and contribution to make. As the field is new, there are as yet no established schools of thought or a single disciplinary basis. The outlook of many of the authors is interdisciplinary or multi-disciplinary, while the unifying bond is expertise in aspects of the creative economy. It is interesting to consider whether analysis of the digital creative economy will eventually emerge as a discipline in its own right. We should also consider whether new concepts are needed for the task or whether existing ones in the various disciplines are sufficient.

The book is divided into five thematic parts: perspectives on digitization in the creative economy; development of technologies in the creative economy; policy and copyright issues in the digital creative economy; copyright and digitization; and creative industry studies, in which the impact of digitization in specific industries is analysed in depth. Chapters have been allocated to these parts according to the main thrust of the authors' argument.

In Part I, perspectives on digitization in the creative economy include Bekar and Haswell's overview of digitization as a GPT (Chapter 1), Baumol on technological development in a range of labour-intensive services, one being the performing arts and its impact and costs (Chapter 2), Potts's evolutionary economics approach to the digital economy (Chapter 3), and Pratt on the spatial aspects of the digital creative economy (Chapter 4). Searle and White look at business models, which have been seen by some economists as the countermand to unauthorized and unremunerated use of copyright material, commonly called piracy (Chapter 5). van Kranenburg and Ziggers analyse industrial organization, and how firms can cope with perpetual change (Chapter 6). Bruns discusses the development from 'prosumption' to 'produsage', by which users adopt an increasingly important role in the production and reproduction of creative material – undoubtedly a fundamental change in the distinction normally made in economics between supply and demand (Chapter 7). Changing consumption patterns are traced by van Eijck and Majorana (Chapter 8). Finally, Norris and Inglehart discuss the global issue of the digital divide between those with and those without access to the Internet and the benefits of digital products and production methods (Chapter 9).

Part II, on development of technologies in the creative economy, moves into more specific treatment of the impact of successive technological progress in the arts and culture, which generally speaking has broadened the supply of new creative products through the ages, made them copiable and increased access for consumers. Bekar provides a historical analysis of copying technologies (Chapter 10), and Tschmuck traces the role of technological change in cultural development (Chapter 11). For the present time, Latzer shows the effects of media convergence (Chapter 12), while Tadayoni and Henten offer a detailed picture of the successes and failures of digitization in the context of digital television broadcasting, demonstrating that technological development is not smooth and its promises are not always fulfilled (Chapter 13).

Part III, on policy and copyright issues in the digital creative economy, considers the effect of digitization on various aspects of cultural policy, in particular copyright law. Perhaps a word of explanation is due here: copyright law clearly has a significant economic dimension to it, and the various aspects to the law establish incentives and rewards for the creation and the distribution of protected works. The changes being made to copyright law to cope with digitization are strongly related to creative industry policy and also to cultural policy, since they determine rights of access to cultural material, much of which is in copyright. Starting with cultural policy, Flew and Swift consider the development of cultural policy over the previous 50 or so years (Chapter 14). The chapter by Goodridge provides a detailed discussion of the problems of measuring the creative economy and recommends a new way of treating copyright content in national income accounts, a topic that has considerable policy implications given the emphasis in government circles on the growth and sustainability of the creative industries (Chapter 15). Doyle analyses a neglected area of the creative economy, namely international trade in audiovisual products, which has an important cultural policy dimension (Chapter 16). Turning to copyright law as such, DiCola gives an account of the changes to copyright law due to digitization (Chapter 17). Watt analyses the relation between copyright and contracts, a little-explored aspect of copyright as an incentive to create (Chapter 18), and Katz considers the competition policy aspects of copyright (Chapter 19). Hilty and Nérisson turn to policy on collective rights management, a topic of considerable political debate, as digitization impacts on remuneration of copyright holders (Chapter 20). Poort's chapter on copyright levies, an administrative solution to the problems of enforcing copyright, rounds off this part (Chapter 21).

Part IV focuses on empirical studies of the economic effects of digitization on copyright. Handke's chapter surveys the evidence on copyright's role, both positive and negative, in the creative economy (Chapter 22), and Liebowitz estimates the impact of Internet piracy on the sales and revenues of copyright owners (Chapter 23). Atladottír, Kretschmer and Towse report on evidence on artists' earnings from copyright and the incentives that copyright and authors' rights have for primary creators (Chapter 24). Another perspective is provided by Farchy, Gansemer and Petrou, who consider whether the new opportunities the Internet and digitization offer to authors and other creators alter their remuneration (Chapter 25). Homberg, Favale, Kretschmer, Mendis and Secchi report on their study of a problem that many governments are currently tackling – how to treat access to orphan works, that is, works that may be supposed to be in copyright but whose owners cannot be traced; this is holding up the process of digitizing archives in libraries of books, films, music and so on that fully exploits the potential of digital ICT for preservation and for dissemination of creative works (Chapter 26).

Part V consists of chapters on the impact of digitization on specific creative industries. First off, Towse summarizes the little work that has so far been done in cultural economics on the use of digital technologies in the performing arts that increase access and offer new sources of revenue (Chapter 27). Arora and Vermeylen discuss art markets, which may become much more transparent as digital databases become widely available (Chapter 28). Navarrete's research shows how far digital technologies have already pervaded various aspects of the work of museums (Chapter 29). Wikström and Johansson analyse the economic consequences of digitization on the markets for consumer magazines and trade books (Chapter 30), while Farchy, Gansemer and Petrou present their research on ebooks, drawing conclusions about the implications for cultural diversity in the book market (Chapter 31). Mueller-Langer and Scheufen discuss academic publishing – a topic many readers will have first-hand experience with – and the trend towards open access publishing online in particular (Chapter 32). Bakker and van der

Wurff report on the challenges for news publishers with digitization (Chapter 33). Waelbroeck on the music industry illustrates how the market for music has now become more complex and uncertain with digitization, while better digital data may help to drive back uncertainty in the future (Chapter 34). Stepan concentrates on the impact of digitization on film distribution (Chapter 35), and Withers discusses digitization in broadcasting and the wider implications for media policy (Chapter 36). Banks and Cunningham conclude this part with an account of games and entertainment software, the 'digital native' among the creative industries, arguing that innovative research is needed to better explain the dynamic development of this sector (Chapter 37).

Overall, this handbook covers most of the topics that have been researched on the implications of digitization in the creative economy. As with all research, some areas have received more attention than others. While we have striven to present as complete an overview as possible, there are gaps in the research and in our own ability to commission papers. It is still very early days for research in some areas. It is our hope that the book will inform researchers, foster an exchange of ideas and help to identify relevant gaps in the literature to work on. We also intend that this book will cause students, policy makers and all types of stakeholders to take inspiration from the academic work already done.

REFERENCES

Baumol, W. and W. Bowen (1966), *Performing Arts: The Economic Dilemma*, Hartford, CT: Twentieth Century Fund.

Carlaw, K., L. Oxley, L. Walker, D. Thorns and M. Nuth (2006), 'Beyond the hype: intellectual property and the knowledge economy/knowledge society', *Journal of Economic Surveys*, **20** (4), 633–90.

Lipsey, R., K. Carlaw and C. Bekar (2005), *Economic Transformations: General Purpose Technologies and Economic Growth*, Oxford: Oxford University Press.

Picard, R., L. Kueng and R. Towse (2008), *The Internet and the Mass Media*, London: Sage.

Towse, R. (ed.) (1997), *Cultural Economics: The Arts, the Heritage and the Media Industries*, Cheltenham, UK and Lyme, NH, USA: Edward Elgar Publishing.

Towse, R., C. Handke and P. Stepan (2008), 'The economics of copyright law: a stocktake of the literature', *Review of Economic Research in Copyright Issues*, **5** (1), 1–22.

UNCTAD (2008), *Creative Economy Report 2008*, Geneva: UNCTAD.

WIPO (2012), *WIPO Studies on the Economic Contribution of the Copyright Industries*, Geneva: WIPO, available at: http://www.wipo.int/ip-development/en/creative_industry/pdf/economic_contribution_analysis_2012.pdf.

PART I

PERSPECTIVES ON DIGITIZATION IN THE CREATIVE ECONOMY

1. General purpose technologies

Cliff Bekar and Erin Haswell

In this chapter we employ the framework of general purpose technologies (GPTs) to analyze the impact of digitization on the production, transmission, and consumption of goods and services in the creative sector of the economy. Digital technologies are in the very early stages of their development. Their long run impact on the creative sector will ultimately be complex, with a diverse range of unpredictable outcomes. We therefore do not engage in any form of 'sectoral forecasting'. Nevertheless, we argue some broad principles concerning the impacts of digitization suggested by the nature of the creative sector, the historical impact of innovation on the sector, and the impact of past information communication technologies (ICTs) on other sectors.

Digitization is one dimension of modern computer ICTs. Because the literature on modern ICTs is vast, and growing quickly, a thorough review of that literature is beyond the scope of this chapter. We instead start with a working definition of the creative sector. Next, we define GPTs, distinguishing our definition from others in the literature and deriving a simple framework to analyze the impact of ICTs on the creative sector. Finally we turn to both historical and contemporary applications of the model, developing some illustrative case studies.

THE CREATIVE SECTOR

The creative sector of the economy (*C*) is composed of a set of industries producing 'goods and services that we broadly associate with cultural, artistic, or simply entertainment value' (Caves, 2000: 1). Cultural goods and services 'comprise tangible products and intangible intellectual or artistic services with creative content, economic value and market objectives' (UNCTAD, 2008: 13). This includes, but is not limited to, the visual and performing arts, museums, cultural sites, literature, film, music, architecture, graphic design, and videogames. Throsby (2001: 7) argues that cultural goods: (1) are experience goods; (2) display aspects of public goods; (3) contain symbolic messages; (4) yield non-pecuniary value; (5) contain intellectual property; and (6) result from a production process relying on human creativity. For our analysis the key properties of cultural goods relate to their production. They tend to disproportionately employ intellectual and creative capital as primary inputs, constituting 'a set of knowledge-based activities, focused on but not limited to arts, potentially generating revenues from trade and intellectual property rights' (UNCTAD, 2008: 13).

Cultural goods are heterogeneous in their composition, as are the creative industries that make up the creative sector. Throsby (2010: 26) points out that 'the distinction between economic value and cultural value can be used to inform an approach to modeling the cultural industries . . . [and] different goods have different degrees of cultural content relative to their commercial value'. Caves (2000: 5) argues that producing creative goods requires 'humdrum' inputs (labor, capital, human capital, etc.) and 'creative' inputs. Creative inputs are relatively ephemeral and difficult to quantify, and possess few, if any, good substitutes. Since Baumol and Bowen (1966), scholars have noted that the core cultural industries employ relatively more creative inputs – that is, display more 'artisanal' production processes – than commercial industries. The artisanal nature of staging an opera is one reason why it requires more public support (from nonprofits, charitable contributions, government programs, and so on) than do videogames. Following this, we assume that the industries within the creative sector $\{C = c_0, c_1, \ldots, c_n\}$ may be ranked by the relative content of creative inputs so that core industries $\{c_0, c_1\}$ (high cultural content, for example painting, opera) display more artisanal production techniques than do more commercial industries $\{c_{n-1}, c_n\}$ (low cultural content, such as movies, videogames). We make no aesthetic claims in such a ranking.

While our ranking is a simplification – any real world ordering would be complex and multidimensional – it serves as a useful heuristic device and is consistent with the literature (for example, Throsby's concentric circles model, UNCTAD's distinction between upstream and downstream sectors, Baumol's work on the distribution of cost disease, and so on).

IS DIGITIZATION A GPT?

The idea that modern ICTs are best modeled as GPTs is widely accepted – 'For analytical purposes, ICTs are treated by economists as "general purpose technologies"' (Mansell et al., 2007: 7). For the most part, this conclusion is based on a definition of GPTs focusing on their disruptive effects, that is, adopting ICTs in production involves 'accommodations and adjustments that may be time-consuming and expensive to make. This observation has led to the claim that ICTs constitute a GPT' (Steinmueller, 2007: 202).

Our definition instead stresses a GPT's technological characteristics. We define a GPT as 'a single generic technology or technology system, recognizable as such over its whole lifetime, that initially has much scope for improvement and eventually comes to be widely used, to have many uses, and to have many spillover effects' (Lipsey et al., 2005: 98). GPTs fall into six classes: (1) materials; (2) energy; (3) ICTs; (4) tools; (5) transportation; and (6) organization. Historical examples of GPTs include writing, the steam engine, bronze, railways, and the waterwheel.

We define 'digital ICT' as the collection of components embodying electronic binary logic. The definition includes PCs, tablets, phones and other smart devices, and the software running on those devices. It includes the Internet and devices connected to the Internet. Digital ICTs are often characterized by clusters of computing systems embodied in a range of components with high degrees of connectivity. The core technology of digital ICT is the electronic computer: 'Like all transforming GPTs, the computer started as a crude, specific-purpose technology and took decades to be improved and diffused through the whole economy' (Lipsey et al., 2005: 114; see also Lipsey et al., 1998). In 1953, the total amount of digital memory on the planet – about 5 kilobytes – was less than that used today to store an address on a modern smartphone. That memory was also incredibly expensive despite massive government subsidies. Processing speeds were glacial by modern standards. Networking options were extremely slow or, in most cases, nonexistent.

Since those earliest implementations, the power of computers has evolved at an exponential rate, with processor speeds doubling every few years or so. This increase in processing power is mirrored in the evolution of other elements in digital ICT. Modern computing platforms possess vastly more memory (cloud-based computing threatens to reduce processing and storage bottlenecks), and richer networking options (wireless, intranets, wireless telephony hotspots). Modern computers have become much smaller, more robust, and more flexible as programmable components of a digital system.

The very earliest implementations of digital ICTs were narrow use machines, often employed in pure research or military activities (for example, code breaking efforts, modeling systems of non-linear equations for ballistic applications or atomic weaponry). Today, there are very few sectors that have not been impacted in a significant way by digital ICT. Digital ICTs have revolutionized telephony, banking, media, travel services, manufacturing, and a range of other sectors. They are central to modern developments in retail, health, and education. Computer-aided design (CAD), computer-aided manufacturing (CAM), and computer numerically controlled (CNC) precision machine tools along with a host of other technological innovations have dramatically revolutionized an array of manufacturing and distribution processes.

Digital ICTs may be creating the largest set of technological complementarities of any GPT to date. They have led to many important changes in the structure of the law, government policy, and business practices. It is already clear that the ultimate impact of digital ICTs on the political, social, and familial spheres will be profound and potentially transformational. Within labor markets, digital ICTs have been identified as a source of rising inequality in both labor incomes (Goldin and Katz, 2008) and access to information (the 'digital divide'). Digital ICTs have advanced as a key contributor to a range of important dynamics including profound changes in retail market structure and firm organization, the increased globalization of trade, and perhaps even an alteration in the nature of the state.

Digital ICT clearly fits the criteria of a GPT, as it displays scope for improvement, a growing range and variety of uses, and many technological complementarities. Regardless of whether the digital economy is in fact a 'new' economy, it is evident that digital ICT is a transformative GPT of historical significance.

IS DIGITAL ICT TRANSFORMING THE CREATIVE SECTOR?

We define 'transformational' changes as non-marginal, qualitatively unprecedented changes in the production and consumption of goods and services. They contrast with incremental evolutionary changes, which are refinements, improvements, and/or extensions of past techniques. Transformational effects often exist side by side with incremental effects within the same industry. The advent of digital photography has rendered the production of photographs cheaper, easier, and more flexible, and consumers have access to a range of technologies unthinkable a few decades ago. But the evolution of photography for decades has been towards making photography quicker, cheaper, and more portable (for example, Instamatics, point-and-shoots, the Polaroid one-step instant cameras). In the literary market, digital ICTs are having a dramatic impact on publishing models as distribution costs asymptotically approach zero (for example, e-books distributed to mobile e-readers). However, the act of artistic creation is much the same regardless of whether one works with a pen and paper, a typewriter, or a wireless keyboard. Further, the effects of ICTs often change over their lifetime. Early systems of writing, including pictographic forms, tended to have only modest impacts, as they extended agents' ability to engage in activities that were already important to the economy (early pictographs evolved in a continuous fashion from clay bullae used to record economic transactions in Mesopotamia). However, as pictographs evolved into a system of alphabetic writing, this ultimately had clear transformational effects (for example, see Dudley, 1991).

The long and complex history of ICTs and their impact on creative endeavors suggest that the range of effects resulting from contemporary digital ICTs will be similarly heterogeneous. To what extent can we infer anything about the distribution of these effects? In analyzing the impact of digital ICT on the broader economy, Kling and Lamb (2000: 297–9) argue that their effects can be grouped according to their effects on production, distribution, and consumption channels. 'Intensively digital' industries would simply not exist without digital ICTs (for instance, software engineering). All three channels in these industries have been transformed; they are also producing a range of technological innovations that feed back into the development of digital ICT. 'Highly digital' industries have had all three channels transformed by digital ICT (such as the online distribution of MP3s to smart devices). 'Mixed digital' industries exist as a mix of digital and non-digital channels (for example, the online delivery of flowers). The primary effects of digital ICT on mixed digital industries, according to Kling and Lamb, concern their distribution.

We find a similar distribution of effects in the creative sector. Consistent with Baumol and Bowen's (1966) findings on the distribution of 'cost disease' within the creative sector, and the effects of past ICTs, digital ICTs appear to be least transformative in those creative activities where artisanal production processes continue to dominate (for example, the production of oil paintings is largely untouched by digital ICTs). For core creative industries $\{c_0, c_1, \ldots\}$ digital ICTs' effects are small relative to more commercial creative industries and tend to impact distribution and consumption channels more than production channels (such as sales of paintings via the Internet). For more commercial industries $\{\ldots, c_{n-1}, c_n\}$ the effects are larger and tend to impact almost all aspects of the industry (for instance, Hollywood blockbusters), and in many instances the industries themselves are viable only as the result of digital ICTs (for example, videogames).

In sum, the literature suggests that: (1) artisanal production techniques in creative industries are relatively resistant to new technologies; and (2) there is a positive correlation between cultural content and artisanal production methods. Taken together these observations suggest a hypothesis: We should expect digital ICTs to have a broader range of transformational effects in the relatively commercial cultural industries relative to the core cultural industries. The reasoning is straightforward: The core cultural industries, those with relatively low commercial content, tend to have more artisanal production techniques. Consistent with the dynamic identified by Baumol – and the subsequent historical impacts of innovation (Heilbrun, 2011: 67–75 provides a discussion of this pattern) – artisanal production techniques tend to be relatively more resistant to automation and/or digitization. Consequently, production in these industries is impacted less dramatically. However, changes to the distribution and consumption channels may be very large and important. Our hypothesis suggests the following pattern of impacts of digitization on the creative sector: (1) core creative industries should find their distribution and consumption channels transformed far more than their production methods; (2) commercial creative industries are more likely to find their production, distribution, and consumption activities transformed, often in unprecedented ways, and such industries will often be vehicles for further innovation in digital ICT; and (3) industries of intermediate cultural content should display mixed effects.

CASE STUDIES

We now analyze three case studies: (1) a core creative industry (pictorial art); (2) a mixed case industry (architecture); and (3) a highly commercial industry (electronic entertainment).

Pictorial Art

Our main focus here, within the broader category of pictorial art, is traditional painting. Digital ICT has impacted the conceptual stage of painting (such as sketching possible compositions, exploring sources of inspiration such as past works) to a greater extent than the actual painting process. Artists have long used photographs and printed images as aids in the painting process. Today they have access to countless works by both recent artists and old masters, utilizing digital ICT regularly to better understand historical traditions and research contemporary developments. Through digital ICT, artists can bring up almost any image they desire (landscapes, objects, past works of art, the human form) in extremely high resolution. Digital ICT has dramatically lowered the marginal costs of accessing pictorial and intellectual/historical information. Past ICTs, such as the printing press and color photography, had similar effects (for example, through the production of art books and prints of museum collections). The contrast between past ICTs and digital ICT is with regard to the sheer magnitude of digital ICT's effects. For example, the marginal cost of accessing online databases of unprecedented size and quality is rapidly approaching zero (for example, see the extremely high-resolution images in ARTstor and the Google Art Project). Artists are even using off the shelf computer programs available at modest cost to catalog images, mix and match those images, and form digital collages to serve as references for the traditional painted canvas, sometimes using printed images in combination with more traditional painting methods to produce new hybrid mixed media creations.

Digital ICT has had a minimal impact on the production of traditional paintings (as distinguished from the conception of those same paintings). The painting process remains intensely artisanal. A painting is still created by hand with contact between a medium and a canvas, and this connection between the hand, medium, and surface has changed little. While differences arise in cost and quality of inputs (for example, oils are available in any desired color and require relatively less skill to prepare and mix), the method for producing an oil painting is, at its core, much as it was 500 years ago. One way in which digital ICT is impacting the production of paintings is as an emerging medium unto itself. Putting aside whether it is ultimately correct to call an artifact that is wholly or partially digital a 'painting', many artists from all levels and backgrounds are embracing the medium (for instance, the English artist David Hockney, active in the field since the late 1950s, has a number of articles and exhibitions illustrating this approach; see 'David Hockney's Fresh Flowers'). However, even digital art relies on artisanal production methods with little scope to economize on creative inputs. The same skills, training, and experience are applicable to 'digital' or oil painting. The difference between digital and oil as a medium is greater than that between acrylic and oil, but, after taking a class in oil, one can utilize other media and apply many aspects of the training in oil to that medium (while Hockney embraces digital technologies, he is still very active in traditional media). So, while digital ICT is an important new and innovative medium, this sort of development is clearly not without precedent in the field. Digital paintings are still subject to traditional issues such as composition, structure, brush stroke, and color choice. Like other media, the digital medium has special qualities of light, layer, technique, and detail. The particular characteristics of a medium make it well suited for particular uses. For example, Venetian painters embraced oil because it allowed them to capture the shimmering and reflective qualities of light so prominent in Venice.

Artists can use digital ICT to promote their work through online galleries and personal websites. Artists can process sales digitally, streamlining the transaction process with the possibility of bypassing the gallery (not unlike the online sale of flowers and travel services). But there are limits; an in-person viewing of the painting in a physical gallery or artist studio is often crucial. One interesting topic with regard to digital painting is how artists 'sell' their work. Unlike a physical painting where there is a clear original, there is no original, unique, digital painting. Any digital painting can be perfectly duplicated. This impacts sales and purchase prices, and is one reason why artists who embrace digital media are still active in traditional media. A digital painting is born as code: it is not the result of digitization; nothing is lost in the transmission of the code from the producer to the consumer. Furthermore, the marginal cost of transmission of digital paintings is effectively zero, and they suffer no degradation as a result of 'shipping'. The transmission of a painting in a traditional medium is a physical transfer of a unique object. There can only be one original painting. Any copy is not a perfect reproduction of the original. In any digital representation, aspects of the painting, tangible and intangible, are lost. An oil painting is not created from code. It is a physical object that can be digitized, with the digital image represented by code. Conversely, a digital painting begins as a code, and for it to be a tangible object it is translated into a printed image. In this transfer, original aspects of the painting are also lost: the nature of digital color, light, and resolution. It appears that there is an emerging synergy between the digital transmission of paintings and the digital medium. Digital paintings are increasingly designed for digital transmission – Hockney's digital flowers started as a way to send 'fresh flowers' via email.

Finally, digital ICT has dramatically expanded access to historical and contemporary artistic masterpieces to millions worldwide. While these developments are distinctive in scale and cost (for example, the cost of a book versus Google image search), photography in conjunction with printing and copying technologies (such as books, prints in magazines and newspapers, postcards, and slide projections) similarly expanded the possibilities for second hand consumption. Digital ICT has improved accessibility and the quality of the images (for instance, detail down to each individual brush stroke) and allowed for their transmission at a marginal cost approaching zero.

In sum, the effects of digital ICT on painting look mostly evolutionary in production. The effects of digital ICT on consumption and distribution, on the other hand, are much more likely to be transformative.

Architecture

Digital ICT has transformed how architects design their blueprints, the way in which materials for their designs are produced and assembled, and the actual construction of the installation (taking 'installation' as shorthand for any realization of an architectural blueprint including but not limited to office buildings, bridges, public art, houses, academic structures, etc.). In each of these areas, digital ICT has had a transformative impact. Architects used computers very early, but mostly in the production of blueprints and less in the construction of installations. In this early period, even when used in design, computers tended to be employed not as 'creative' tools but as an extension of traditional practices (a common occurrence for newly adopted technologies). Related industries (for example, aerospace and shipbuilding) utilized computers in design much earlier and to a much greater extent than did architecture. Aerospace engineers were quick to adopt wholly digital workflows that rarely employed traditional plans and drawings, creating three-dimensional models and designs (on CAD and CAM) and implementing CNC for prototyping (Kolarevic, 2003: 11).

Frank Gehry's *Fish Sculpture* in Barcelona (created in 1992) is an early example of digital ICT's transformative impact on architecture. Gehry's team needed to 'make the complex geometry of the project not only describable, but also producible, using digital means in order to ensure a high degree of precision in fabrication and assembly' (Kolarevic, 2003: 31). The team employed a program developed for the French aerospace industry 20 years earlier, CATIA (computer-aided three-dimensional interactive application). CATIA was also adopted early on by other industries such as shipbuilding. With CATIA, 'three-dimensional models were used in the design development, for structural analysis, and as a source of construction information, in a radical departure from the normative practices of the profession' (Kolarevic, 2003: 31).

For Gehry, digital technologies were initially used 'as a medium of translation in a process that takes as its input the geometry of the physical model and produces as its output the digitally-encoded control information which is used to drive various fabrication machines' (Kolarevic, 2003: 31). The process of 'reverse engineering', translating the physical into the digital, is done by scanning a three-dimensional model into a 'point cloud' that is then used to produce a digital representation of the model (Kolarevic, 2003: 31–2). The digital representation is converted into NURBS (non-uniform rational B-spline) curves and surfaces. Historically, the

> tradition of Euclidian geometry in building brought about drafting instruments, such as the straight edge and the compass, needed to draw straight lines and circles on paper, and the corresponding extrusion and rolling machinery to produce straight lines and circles in material. The consequence was . . . that architects drew what they could build, and built what they could draw. (Kolarevic, 2003: 32)

Today, architects know the scope of production capabilities of computer-controlled fabrication equipment, and their designs reflect the machines' capabilities. Architects are increasingly 'involved in the fabrication processes, as they create the information that is translated by fabricators directly into the control data the drives the digital fabrication equipment' (Kolarevic, 2003: 33). NURBS surfaces and curves precisely define complex, non-Euclidean geometric forms and allow for those forms to be computationally and physically possible through CNC production processes (CNC cutting, CNC milling and subtractive, additive, and formative fabrication). Ultimately, 'constructability in building design becomes a direct function of computability' (Kolarevic, 2003: 33).

Digital ICT has also transformed the assembly of buildings: 'After the components are digitally fabricated, their assembly on site can be augmented with digital technology. Digital three-dimensional models can be used to precisely determine the location of each component, move each component to its location and, finally, fix each component in its proper place' (Kolarevic, 2003: 38). An example of this is Gehry's Guggenheim Museum in Bilbao (established 1997). All components of the project were manufactured with bar codes encoding their placement data in the CATIA model. Next, 'Laser surveying equipment linked to CATIA enabled each piece to be precisely placed in its position as defined by the computer model' (Kolarevic, 2003: 38). The Japanese have taken this further with SMART (Shimizu Manufacturing system by Advanced Robotics Technology), a fully digitally enabled construction process (employed on a full-scale building project, the Juroku Bank Building in Nagoya).

Buildings, like paintings and sculpture, are first and foremost experience goods. That is, the principal form of consumption is through the first hand observation and interaction with and use of the good (walking through/around the building). Digital ICT does not transform or impact the way a viewer physically observes and interacts with a building or a painting. Advances in digital photography and imaging have allowed for high-resolution images of buildings to be distributed through the Internet at zero marginal cost. Digital ICT has dramatically expanded access to architectural masterpieces, buildings, and design to users worldwide. However, these important advances in ease of access to information impact the second hand viewing of buildings, and aspects about the buildings are undoubtedly lost in the translation process (tangible aspects, such as size of building, monumentality, materials, and walking through the space, and intangibles, such as emotional response). Furthermore, this is not an unprecedented impact. While it might be different in scope (say, the cost of a book versus the cost of a Google image search), photography in conjunction with printing and copying technologies similarly expanded the possibilities for second hand consumption.

In sum, the effects of digital ICT on architecture, especially in recent decades, are mostly transformational. These transformational effects have largely impacted production, while the impacts on distribution have been mixed, and the impacts on consumption hard to measure.

Electronic Entertainment

The electronic entertainment industry (that is, videogames) is perhaps the first purely digital cultural industry. It is certainly the first fully interactive medium in history.

Electronic entertainment is an example of a wholly new industry emerging with the introduction of digital ICT. Starting in the 1970s, the videogame industry has grown rapidly (in 2011, videogame revenues of $25 billion exceeded Hollywood box office receipts). Videogame companies have become more and more central to modern industrial policy as governments increasingly seek to attract critical clusters of game producers. They provide well-paying jobs to a range of high human capital individuals, providing a multitude of creative and lifestyle spillover effects that increasingly define vibrant 'livable cities'. In the US, important clusters of videogame producers are located in Austin, San Francisco, Seattle, and Boston.

Increasingly videogames are being recognized as cultural artifacts, with modern games tackling a range of emotional, political, and philosophical issues. Throsby (2010: 26) notes that there are often complementarities between core cultural industries and highly commercial industries like videogames: 'the plot of a novel or play may suggest ideas for a video or computer game, or a painter's work may inspire a fashion creation'. This is in fact an important dynamic in the contemporary industry, as a range of games have been made based on movies, books, and comics. And, increasingly, movies, books, and comics are being based on videogames. Videogames are digital, and it is clear that all channels in the industry (production, distribution, and consumption) have been technologically enabled by digital ICT. Here we touch briefly on some modern developments in videogames as cultural artifacts, and how, in a relatively unique way, the industry has driven many key innovations in digital ICT.

Paradoxically, despite the fact that its final product is born and consumed as code, the videogame industry still relies on distributing the majority of its product in physical form. Games were originally distributed on chips containing burned ROMs (read only memory) and floppy disks. Because of the large amount of data involved in producing a videogame, developers adopted optical technologies very early on. In 1991, game developer Cyan released *Myst*, a graphically intensive puzzle game that required a CD-ROM to play. *Myst* sold millions of copies and is widely considered a driving force in accelerating the adoption of digital optical drives in the North American PC market. The combination of widely available high-speed Internet and cheap digital storage has led to a sharp increase in the rate of digital delivery of electronic entertainment of all types. Again, videogames have led the way in the development and diffusion of technologies that are already impacting the delivery of film (Netflix), TV (Hulu), and books (e-readers).

The demand for higher-quality videogame graphics is one of the single largest factors driving the evolution of graphics technologies. The development of such graphics technologies has in turn had spillover effects for digital photography, digital video, and other graphics intensive activities. A range of software engineering challenges in videogame development has directly driven a range of software developments (for example, synchronizing data streams across multiple servers to tens of thousands of users, developments in artificial intelligence). And such developments are not strictly the domain of traditional videogaming. The presence of the Disney Corporation in Florida has led to the development of high-tech firms specializing in gaming-related simulation technology. This cluster of firms is central to a public–private research effort to leverage the developments in simulation technology to help solve a range of engineering problems in other industries.

Videogames are a clear example of how fundamental GPTs (ICT and otherwise) give rise to wholly new products, processes, and industries. As a medium, videogames did not exist five decades ago, but they have grown to be a crucially important industry in many modern industrial economies.

CONCLUSIONS

Given the nature of ICTs – that they concern the production, distribution, and consumption of information – it should come as little surprise that modern digital ICTs have such profound effects on the production, distribution, and consumption of creative goods and services. Inferring from our long experience with less transformative innovations, we suggest that the impacts of digital ICT tend to be largest on commercial creative industries and smallest on the core creative industries. Take film, for example:

> For most of their history . . . movies were only made from photographic film strips (originally celluloid) that mechanically ran through a camera, were chemically processed and made into film prints that were projected in theaters in front of audiences solely at the discretion of the distributors (and exhibitors). With cameras and projectors the flexible filmstrip was one foundation of modern cinema: it is part of what turned photograph images into moving photographic images. Over the past decade digital technologies have changed how movies are produced, distributed and consumed; the end of film stock is just one part of a much larger transformation. (Scott and Dargis, 2012: 41)

On the other hand, thousands of people today consume productions of *Swan Lake* in much the same way audiences did a century ago (even if the stage was designed by computer software and the lights are controlled from a digital panel). However, digital ICT is impacting the consumption and distribution of the performing arts, for example digitally streaming live performances from one theater to another (for example, the Metropolitan Opera Live performances are viewed simultaneously in hundreds of cities around the world, and the UK's famous National Theatre performances are also viewed live around the world). We are clearly just starting to glimpse the range of effects digitization will ultimately have on the creative arts as commercial enterprise.

REFERENCES

Baumol, W. and W.P. Bowen (1966), *Performing Arts: The Economic Dilemma*, New York: Twentieth Century Fund.

Caves, R. (2000), *Creative Industries: Contracts between Art and Commerce*, Cambridge, MA: Harvard University Press.

Dudley, L. (1991), *The Word and the Sword: How Techniques of Information and Violence Have Shaped Our World*, Cambridge, MA: Basil Blackwell.

Goldin, C. and L. Katz (2008), *The Race between Education and Technology*, Cambridge, MA: Belknap Press.

Greenwood, J. (1997), *The Third Industrial Revolution: Technology, Productivity, and Income Inequality*, Washington, DC: AEI Press.

Heilbrun, J. (2011), 'Baumol's cost disease', in R. Towse (ed.), *A Handbook of Cultural Economics*, 2nd edn, Cheltenham, UK and Northampton, MA, USA: Edward Elgar Publishing, pp. 67–75; EconLit with Full Text, EBSCOhost (accessed 4 October 2012).

Hockney, David, 'David Hockney's fresh flowers', Royal Ontario Museum, available at: http://www.rom.on.ca/hockney/ (accessed 4 October 2012).

Kling, R. and R. Lamb (2000), 'IT and organizational change in digital economies: a sociotechnical approach', in E. Brynjolfsson and B. Kahin (eds), *Understanding the Digital Economy*, Cambridge, MA: MIT Press, pp. 295–324.
Kolarevic, B. (2003), *Architecture in the Digital Age: Design and Manufacturing*, New York: Spon Press.
Lipsey, R., C. Bekar and K. Carlaw (1998), 'What requires explanation?', in E. Helpman (ed.), *General Purpose Technologies and Economic Growth*, Cambridge, MA: MIT Press, pp. 145–66.
Lipsey, R., K. Carlaw and C. Bekar (2005), *Economic Transformations: General Purpose Technologies and Long Term Economic Growth*, Oxford: Oxford University Press.
Lovejoy, M. (2004), *Digital Currents: Art in the Electronic Age*, New York: Routledge.
Mansell, R., C. Avgerou, D. Quah and R. Silverstone (eds) (2007), *The Oxford Handbook of Information and Communication Technologies*, Oxford: Oxford University Press.
Parry, R. (2010), *Museums in a Digital Age*, Oxford: Taylor & Francis.
Paul, C. (2008), *Digital Art*, New York: Thames & Hudson.
Scott, A. and M. Dargis (2012), 'Film is dead? Long live movies', *New York Times*, 9 September.
Steinmueller, E. (2007), 'The economics of ICTs: building blocks and implications', in R. Mansell, C. Avgerou, D. Quah and R. Silverstone (eds), *The Oxford Handbook of Information and Communication Technologies*, Oxford: Oxford University Press, pp. 196–222.
Throsby, D. (2001), *Economics and Culture*, Cambridge: Cambridge University Press.
Throsby, D. (2010), *The Economics of Cultural Policy*, Cambridge: Cambridge University Press.
UNCTAD (2008), *Creative Economy Report 2008*, Geneva: UNCTAD, available at: http://unctad.org/en/docs/ditc20082cer_en.pdf.
Wands, B. (2006), *The Art of the Digital Age*, New York: Thames & Hudson.

FURTHER READING

Jeremy Greenwood's *Third Industrial Revolution* (1997) is a good introduction to the digital economy. For an overview of developments in 'digital arts' see Lovejoy (2004), Wands (2006), Paul (2008), and Parry (2010).

2. Reining in those unstoppably rising costs

William J. Baumol

Most readers already will be familiar with my 'cost disease' theory, which seeks to explain why the costs of labor-intensive services experience dramatic increases. (Indeed, I have been introduced as the only economist whose name has been given to a disease.) In brief, my analysis asserts that some services, notably the live performing arts, health care and education, are vulnerable to decline in quality if there is a reduction in the time and effort devoted to them. Meanwhile, in other sectors of the economy, notably manufacturing, there have been impressive and continual savings in the labor time entailed in their production processes with no corresponding reductions in the quality of their outputs. As a result, the relative cost of producing labor-intensive services (that is, 'handicraft' or 'personal' services) must increase relentlessly and substantially, eliciting deep concern from the public and elected leaders, who predictably call for budget cuts and other punitive reforms.

Yet, as it is generally told, the cost disease story is more than a bit misleading for at least two reasons. First, we live in an economy with near universal productivity growth, though it is slower in some arenas than others. But, as Joan Robinson once pointed out to me, in such a situation all goods and services must be growing less costly in terms of the time and effort required for their production. Thus, we will need to work ever fewer hours to purchase a computer, a surgical operation or even a frankfurter. Of course, some items will become less expensive, in this sense, at a faster rate than others, but all outputs will require less labor and presumably a smaller amount of wages to acquire them. Indeed, as just one example, according to a 1997 report by the Federal Reserve Bank of Dallas, the work time required to buy a dozen eggs in 1919 – 80 minutes – had fallen to just five minutes by 1997 (Cox and Alm, 1997: 5). Moreover, in 1910, 345 hours of work time bought a kitchen range, and 553 hours bought a clothes washer. By 1997, those numbers had dropped to 22 and 26 hours, respectively (Cox and Alm, 1997: 8). But the most sensational decrease of all has been in the cost of computers. Computer capability is standardized in terms of the number of MIPS (millions of instructions per second) a computer can handle. In 1997, one MIPS of computer capacity cost about 27 minutes of labor at the average wage, while in 1944 the cost was a barely believable 733 000 lifetimes of labor (Cox and Alm, 1997: 19). The data cited here are more than a decade old, and I have not been able to find any studies of the subject that are more recent. Still, there is every reason to believe that the prices of these goods, by any measure, have continued to plummet.

With this explosion of purchasing power at our disposal, we can expect to afford even the sharply rising costs of services such as health care and education without cutbacks in quality or quantity. The only thing that will change, in terms of the cost to us, is how we will have to divide our money among these items. Because manufactures and agricultural products are growing steadily cheaper in real dollars while health care and education are growing more expensive, we will have to increase the share of money we devote to the latter services. Thus, the cost disease affects only the way in which we divide up the money we spend. It does not force us to decrease how much we buy or reduce our general standard of living.

The second modification that the cost disease story requires is recognition that increasing productivity in labor-intensive industries is not always impossible to achieve, nor is it necessarily detrimental to the quality of the resulting product or service. However, since productivity growth in these industries will be below the average for the economy, their costs, in terms of money, must rise. It follows that, in an economy beset by some degree of inflation, the monetary prices of all, or most, products will be rising. And, because of their slower productivity growth, the outputs affected by the cost disease will experience the greatest cost increases. This is arguably so in all three arenas – health care, education and the performing arts – that are cited most frequently as being vulnerable to the cost disease, as I will explain in detail next.

TOWARD LABOR-SAVING PRODUCTIVITY GROWTH IN EDUCATION

Education, the field in which I provide my own labor, is perhaps the easiest example of this. In education, enlargement of classes is the most obvious way to reduce labor input per student, though both parents and instructors argue that small classes are required for effective learning. There is some evidence (Finn and Achilles, 1999; Schanzenbach, 2007; Konstantopoulos, 2009) that smaller class sizes are helpful to the learning process. However, it was always my impression when I lectured to university classes of 250 to 400 students that their comprehension was not severely impeded by sharing a classroom with so many other students.

There are other promising ways to achieve similar or even greater labor savings in education, which may even improve student comprehension and learning retention. The use of recorded lectures is one possible idea. It should not be difficult to identify a small group of instructors whose lectures are unusually effective and engaging. Surely their expositions will be more comprehensible and memorable than those of the average, albeit 'live', faculty member. Moreover, the recordings need not be shown in a large assembly; a personal computer can carry out this task and do much more. For example, a student could choose to replay particular passages multiple times in order to clarify the lecturer's meaning. This is only one of many innovative, new ideas that have been proposed in education.

The bottom line here consists of two inexcusable facts. First, it is we, the instructors, who convince ourselves that our services are indispensable. Second, on the subject of effective education, the truth is that we often do not know (or know very little) whereof we speak. There is now some experimental and statistical evidence that begins to shed some light on the situation, but far more exploration is needed. We still lack definitive research results that tell us whether modification of current instruction methods in order to reduce the labor time expended per student – for instance, by expanding class sizes in particular subject areas – is damaging, neutral or even beneficial, and how best to implement such changes.

TOWARD LABOR-SAVING PRODUCTIVITY GROWTH IN HEALTH CARE

In health care the path to increasing productivity without damaging quality is more complicated. Still there are opportunities for saving labor in the provision of medical care. For example, computers, with their enormous capacity for information storage and analysis, conceivably can be very helpful in diagnosing illnesses from a set of symptoms entered by a doctor. This may lead to earlier diagnosis, reduced treatment costs and better outcomes for patients.

Indeed hospitals are already using technology and business services to improve their care of patients and simultaneously increase productivity by, for instance, improving the efficiency of nurses, intrahospital communication and the allocation of equipment (Wu, 2012, chap. 10). More specifically, the use of bar codes to dispense medications to hospital patients also is helping to reduce medication errors and improve productivity in hospitals. The bar codes affixed to each patient's wrist band and each medication automatically alert nurses if they have selected the wrong medication or the wrong dose for a particular patient (Wu, 2012, chap. 10). The result is a clear reduction in nurses' labor time, as well as a substantial reduction in the frequency of medical errors that are the bane of hospitals and the terror of patients. As Dr. Lilian Gomory Wu reports in our recent book, these methods contribute significant labor and cost savings in a health-care setting:

> Clearly, there is a critical place in the health-care industry for such business services that reduce the labor of personnel and the dramatic productivity losses and worse that result from medical mistakes. All of this can be improved by the business services that hospitals and other health-care institutions employ – though this benefit is usually insufficient to offset the steady rise in costs entailed in the provision of health care. (Wu, 2012: 142)

TOWARD LABOR-SAVING PRODUCTIVITY GROWTH IN THE LIVE PERFORMING ARTS

Finally, I come to the arts, the arena most pertinent to this volume, beginning with the live performing arts. Here, too, consumers are wary of the damage to the quality of a performance that productivity enhancements threaten.

First it is worth noting that in this arena the technology that offers economies in the use of labor did not exist, even in primitive form, before the twentieth century. Today, however, we are overwhelmed with the abundance of such technology, which has made the basic phonograph ancient history and which now endangers the future survival of live classical music performances. In the United States, for instance, concert audiences are usually characterized by a preponderance of hair that is grey, white or almost altogether absent. Indeed, a very reputable, recent study confirms the aging of classical music audiences in both the United States and Europe (Flanagan, 2012: 42, 156–7).

In the stream of ever more sophisticated technologies that have followed the phonograph, there are many near magical devices that allow people to watch or listen to musical or theatrical performances. Indeed, today's smartphones and tablet computers hardly could have been imagined even a few decades ago. If we think of productivity in terms of the labor expended to provide a performance to each audience member, the magnitude of the labor saving entailed in these new devices is difficult to comprehend.

Some people argue that no device can replicate the quality of live theatrical or musical performances. But others seem to be content with broadcasts and recordings of such performances. Indeed, there are grounds for the suspicion that the preference for live performance can be ascribed to habit, rather than superior taste. For instance, the noted modernist composer Milton Babbitt has likened live musical performance to reading a novel in a crowded, overheated room with the words of the book projected on a dimly lit screen. Babbitt once told me that he attended concerts only under duress; instead he preferred to listen to the music at home using his excellent stereo equipment. In short, there are ways to counteract the rising costs of live performance, as well as those of education and health care.

TOWARD LABOR-SAVING PRODUCTIVITY GROWTH IN THE VISUAL ARTS

I conclude with a discussion related to the subject closest to my own personal interests, the visual arts – to which I have devoted a substantial portion of my life as a painter, sculptor and lithographer.[1] Here, the invention of various powerful tools, beginning perhaps with electric drills and sanders, has made life much easier for sculptors. Meanwhile painters can now magically create works on the computer using specialized software, as I have recently done, rather than painting with actual oil-based paint. There are three immediate advantages to this method of painting. First, and evidently the most important, after completing a work it can be distributed throughout the world immediately via either the Internet or email, thereby eliciting comments and requests for reproduction rights from many sources. Second, this new medium offers novel techniques and opportunities for creative modification of composition and design. For instance, it is possible to test different colors or brush strokes side by side, with the mere click of the mouse, and then determine which is better. And third, not quite as trivial as it may sound, computer painting frees the painter from the unpleasant task of cleaning brushes.

These benefits also may contribute material cost savings to the visual arts. The painter need not purchase expensive canvas and costly paints. Moreover, distribution via email and the Internet ensures that the expense of transporting a painting to, say, China or even South America is zero, in sharp contrast to the costly process once required to transport an oil painting in safety. In short, digital innovations have contributed significant cost saving opportunities to the visual arts. In turn, the resulting works of art also are apt to grow ever more affordable in terms of the hours of labor needed to produce them. However, I remain skeptical that the resulting cost savings will slow the overall rise in the prices of these works of art, which remain heavily in demand, to a rate significantly beneath the economy's overall rate of inflation.

CONCLUSION: CAN THE COST DISEASE BE CURED?

The short answer is certainly not! Unless productivity ceases to grow altogether, or begins to grow at the same rate in every sector of the economy, the disease will remain with us – immune to any universal cure.

Why must this be so? In any economy the productivity growth rate of some activities, such as manufacturing, will be faster than average. In other labor-intensive activities, such as police protection or the performing arts, the reverse is true usually because of the difficulty of reducing the time and effort devoted to creating them without damaging their quality, as already noted. The economy's rate of inflation, by definition, is the average of all resulting costs. This means that the cost of manufacturing a computer, for instance, can be expected *not* to rise as fast as the average – that is, the cost of a computer must increase at a rate *slower* than the economy's rate of inflation. For the same reason, the opposite must be true of, say, the cost of performing a Mozart string quartet, because lagging productivity growth condemns the cost of that performance to rise more quickly than the average – that is, *faster* than the economy's rate of inflation.

In practice, however, the rising costs of the latter services are all too often ascribed to some form of misbehavior – usually the greed or incompetence of their suppliers. This misunderstanding underlies many ill-conceived government policies and deprives the economy of many valuable services whose prices are rising, though rising overall productivity ensures that such services actually are growing ever less expensive in terms of the amount of labor consumers must perform in order to pay for them. Surely such short-sighted policies will damage the general welfare, even as they fail to stem the rising costs of these often invaluable services.

Does all of this mean that the cost disease has been, or is about to be, cured? On the contrary, it means that the disease is likely to remain with us indefinitely. However, it can be harmless if it is not met by irresponsible responses, such as ill-considered budget cuts. Indeed, since the cost disease is the consequence of ubiquitous productivity growth – albeit uneven growth – it is *part* of a promising future in which the typical household should enjoy an abundance of goods and services and poverty should continue to decline. However, if governmental responses are poorly considered, citizens may suffer from great deterioration in public services, and impoverished members of society likely will bear the brunt of the suffering resulting from such private affluence and public squalor.

NOTE

1. Indeed, I can claim to be the only member of the Economics Department at Princeton University to have taught both economics (for some 40 years) and wood sculpture (for well over a decade), with its very pleasant, but heavy, physical labor.

REFERENCES

Baumol, W. with D. de Ferranti, M. Malach, A. Pablos-Méndez, H. Tabish and L.G. Wu (2012), *The Cost Disease: Why Computers Get Cheaper and Health Care Doesn't*, New Haven, CT: Yale University Press.

Cox, W. and R. Alm (1997), 'Time well spent: the declining real cost of living in America', in *Time Well Spent: 1997 Annual Report*, Dallas: Federal Reserve Bank of Dallas, pp. 5–17.
Finn, J. and C. Achilles (1999), 'Tennessee's class size study: findings, implications, misconceptions', *Educational Evaluation and Policy Analysis*, **21** (2), 97–109.
Flanagan, R.J. (2012), *The Perilous Life of Symphony Orchestras*, New Haven, CT: Yale University Press.
Konstantopoulos, S. (2009), 'Effects of teachers on minority and disadvantaged students' achievement in the early grades', *Elementary School Journal*, **110** (1), 92–113.
Nordhaus, W. (2008), 'Baumol's diseases: a macroeconomic perspective', *B.E. Journal of Macroeconomics*, **8**, article 9.
Schanzenbach, D.W. (2007), 'What have we learned from Project STAR?', *Brookings Papers on Education Policy*, 205–28.
Towse, R. (ed.) (1997), *Baumol's Cost Disease: The Arts and Other Victims*, Cheltenham, UK and Lyme, NH, USA: Edward Elgar Publishing.
Wu, L. (2012), 'Business services in health care', in W. Baumol with D. de Ferranti, M. Malach, A. Pablos-Méndez, H. Tabish and L.G. Wu, *The Cost Disease: Why Computers Get Cheaper and Health Care Doesn't*, New Haven, CT: Yale University Press, pp. 141–53.

FURTHER READING

Robert Flanagan's *The Perilous Life of Symphony Orchestras* (2012) analyzes recent cost data and concludes that the relentlessly rising costs facing symphony orchestras are rooted in limited opportunities for increasing productivity in the arts. Similarly, in an excellent 2008 article William Nordhaus examines a complete set of industry accounts data for 1948 through 2001 and concludes that the cost disease theory is confirmed by the data. Although it is more than a decade old, Ruth Towse's *Baumol's Cost Disease: The Arts and Other Victims* (1997) remains one of the most comprehensive and illuminating studies of the cost disease in the performing arts. Finally, see my own recent book *The Cost Disease: Why Computers Get Cheaper and Health Care Doesn't* (2012) for a good general explanation of the cost disease.

3. Evolutionary perspectives

Jason Potts

The standard approach to industrial economics starts with the industry's basic conditions, then runs through the structure–conduct–performance paradigm of industrial organization, and finally considers government regulation and policy. Most creative industries segments have been studied in this way, for example in Albarran (2002) and Caves (2000). These approaches use standard economic analysis to explain the particular properties and characteristics of a specific industrial sector.

The overview presented here is different again. It focuses on the creative industries and examines their economic effect, specifically their contribution to economic evolution. This is an *evolutionary systems* approach to industrial analysis, where we seek to understand how a sector fits into a broader system of production, consumption, technology, trade and institutions. The evolutionary approach focuses on innovation, economic growth and endogenous transformation. So, rather than using economics to explain static or industrial-organization features of the creative industries, we are using an open systems view of the creative industries to explain dynamic 'Schumpeterian' features of the broader economy. The creative industries are drivers of economic transformation through their role in the origination of new ideas, in consumer adoption, and in facilitating the institutional embedding of new ideas into the economic order. This is not a novel idea, as economists have long understood that particular activities are drivers of economic growth and development, for example research and development, and also that particular sectors are instrumental to this process, for example high-technology sectors. What is new is the argument that cultural and creative sectors are also a key part of this process of economic evolution. We will review the case for that claim, and outline purported mechanisms. We will also consider why policy settings in the creative industries should be more in line with innovation and growth policy than with industry policy.

FROM CULTURAL ECONOMICS TO ECONOMICS OF CREATIVE INDUSTRIES

It is useful to distinguish between the economics of arts and culture, or *cultural economics*, and the *economics of creative industries*. This is not a hard-and-fast distinction, and there is much overlap, but in broad terms this difference turns on whether the economic argument is built in the Paretian–Marshallian–Pigovian tradition (cultural economics) or in the Schumpeterian–Hayekian tradition (economics of creative industries). Much of cultural economics is an applied branch of neoclassical microeconomic theory built around the implications of 'Paretian' welfare economics and specifically the prospect of market failure in the production and consumption of arts and culture (Baumol and Bowen, 1966; Throsby, 1994; Blaug, 2003). In this chapter, the *economics of creative industries* refers to a branch of Schumpeterian evolutionary economics that is centred on the contribution of the creative industries to the national innovation system (Potts, 2011). The difference is not just analytic focus – market failure versus economic growth – but also a different set of industries. Traditionally, the economics of arts and culture tends to focus on the 'high-culture' visual and performing arts end of the spectrum, along with artistic and cultural heritage (museums, opera and so on). The creative industries are a wider and larger set that extends into the more popular, commercial and digital-media realm, including fashion, design, advertising, architecture, publishing and video games.

With a broad brush, cultural economics focuses on the social welfare problems that accrue to the undersupply of worthy but often financially non-viable cultural goods and services. It addresses policy solutions to these market failures, which invariably means treating them as public goods. In contrast, the economics of creative industries is focused on the opportunities of new technologies and business models and on the role of the creative industries sector in driving economic growth, entrepreneurship, job creation and regional development. The economics of creative industries is essentially part of the study of endogenous economic dynamics or economic evolution that is driven by entrepreneurship and innovation. It is an analysis of economic change, institutions and dynamic efficiency rather than economic welfare, public goods and static efficiency.

However, there are several overlapping analytic variations. Richard Caves (2000) develops an economics of the creative industries built on transactions costs analysis. He examines how firms in the creative industries face peculiar challenges in contracting and organization owing to the particular information and uncertainty characteristics of production and demand. To the extent that Caves emphasizes the special nature of the creative industries, this follows in the cultural economics line. But, in examining how firms actually address these challenges, Caves situates his analysis in line with market process theories. By taking the information and uncertainty problems associated with cultural and creative production seriously, Caves seeks to emphasize the value of entrepreneurial and market mechanisms in dealing with these information, knowledge and coordination problems. The upshot is a focus on *market process* solutions rather than market failure problems, a point also signalled by the likes of Alan Peacock (1993), William Grampp (1989) and Tyler Cowen (1998, 2002). Nevertheless, an evolutionary economics approach to creative industries thus directs the economic analysis of arts and culture away from market failure and welfare arguments for cultural protectionism and towards more open-market arguments based on consumer and producer uptake of new ideas, innovation dynamics and industrial evolution. This combines a microeconomic approach to the production of creative goods and services that is based on market process theory (entrepreneurship, market discovery) and a macroeconomic approach that is based on Schumpeterian growth theory (innovation, creative destruction).

The evolutionary economics approach to creative industries emphasizes different things to the cultural economics approach: private entrepreneurship not public intervention; market processes not market failure; innovation not conservation; technological opportunities not technological threats; coordination problems not allocation problems; a greater focus on income dynamics and creative destruction; globalization as an opportunity not a threat; and so on. Interestingly, the connection between creative industries and evolutionary economics was originally formulated to support a new line of policy arguing against treating the arts and cultural sector as a net economic drain producing positive cultural externalities, but instead seeing it directly as a source of economic growth and development (Hartley and Cunningham, 2001; Cunningham et al., 2004). This was advanced by a motley group of political economists and urban geographers who emphasized the role of cultural investment in creating the conditions under which innovation and growth can thrive (Cunningham, 2002, 2006; Hartley, 2005; Scott, 2006; see also Garnham, 2005). This focused on a wider set of industries than the core arts and cultural industries (heritage, music, film, visual and performing arts and so on), extending to the more commercial domains of fashion, design, new media and video games. An evolutionary framework for the economic analysis of the creative industries emerged through a focus on the growth and development benefits of creative industries at multiple levels, including amateur engagement, entrepreneurship, urban regeneration, regional development and national economic growth.

WHAT IS ECONOMIC EVOLUTION?

Economic evolution is not the same as economic growth. Economic evolution is an endogenous change in the knowledge base of the economic system and the structure of economic activities, including commodity sets, jobs, preferences, technologies and institutions. Growth, in turn, is an expansion of what already exists attributable to increased *activity* (for example investment or spending). Economic evolution, however, is a change in what the economy is made of and how it is ordered. Economic development, which refers to change in an economy's political, legal and social institutions, invariably overlaps with economic evolution, referring to policy-driven or planned changes in political and market institutions along the way.

Economic growth refers to a percentage change in measured aggregate economic output (for example gross domestic product, or the money-valued output of a sector or region). The theory of economic growth seeks to explain such an increase in output by an increase in factor inputs. This can occur by increased capital or labour, or by an improvement in the technology (also known as multi-factor productivity) that converts inputs into outputs. Or economic growth may result from an increase in output price, reduced input cost or increased demand (and thus improved economies of production).

Economic growth, in these models, is due to increased *investment* in factors of production, including technology, or from market growth driving increased specialization and gains from trade. To explain the connection between creative industries and economic growth, it is necessary to argue for: (1) increased investment in factors of production that are inputs into the creative industries sector; (2) increased demand for the output of this sector, bidding up prices or increasing trade; or (3) technological or productivity gains in this sector. Most accounts emphasize the first mechanism, including the DCMS mapping documents (DCMS, 1998, 2001) that showed the growth of jobs (factor inputs), as did Richard Florida's estimates of the size and growth of the 'creative class' (Florida, 2002). Tyler Cowen's (1998, 2002) work on globalization of creative industries and John Howkins's (2001, 2009) work on the creative economy emphasize increased demand for creative industries output. There is still little evidence either way on whether creative industries drive multi-factor productivity.

Economic evolution, however, is a more complex process. The standard theoretical model of economic evolution derives from the work of Joseph Schumpeter (1942) and is analytically represented in the work of Richard Nelson and Sidney Winter (1982), Kurt Dopfer and Jason Potts (2008) and Geoffrey Hodgson and Thorbjorn Knudsen (2010). In evolutionary economics, economic growth is less about an accumulation of resources or factors of production (for example capital) or changes in prices or demand, or even institutions ('rules of the game'), but refers to a deeper *growth of knowledge* process in which new ideas, habits and routines, organizations and institutions displace old ideas in an entrepreneur-driven process of innovation that Schumpeter called 'creative destruction'. In the evolutionary view of economic change the prime agents are entrepreneurial innovators who develop new businesses, create new markets, exploit new technologies and sources of supply, and experiment with new business models. The result is new sources of profit, new jobs and new sectors, but also the displacement of old sectors, jobs and business models. Evolutionary economics differs from new growth theory by placing less emphasis on human capital, ideas and technology spillovers per se, and more emphasis on the role of entrepreneurship, innovation trajectories, and the evolutionary market process of Schumpeterian *creative destruction*.

WHY CREATIVE INDUSTRIES?

A striking feature of the creative industries is their recent rise, or relative sectoral growth. The growth of the sector over the past several decades in value added, employment, exports and other measures has been widely, although not always consistently, observed (see DCMS, 2001; Higgs et al., 2008; UNCTAD, 2008). Economic evolution offers several explanations for this rise.

The most straightforward is embodied in the standard DCMS definition of the creative industries – creativity as input and intellectual property as output. The growth of the creative industries is growth in production and consumption of this creative output (Andari et al., 2007). But that begs the question of why this growth has occurred. For economists, the answer is equally straightforward. Income has increased, and consumers have substituted toward a higher-quality consumption bundle. Sociologists, however, have emphasized instead the growing 'culturalization' or 'aestheticization' of economic life (Lash and Urry, 1994; Lloyd, 2006). This is a form of progress that manifests itself beyond improvements in technical efficiency, but by the growth and embedding of cultural content and meaning into an ever-broader range of goods and services (Andersson and Andersson, 2006). This growing culturalization changes the mix of where and how value is added, thereby changing the broader economic ecology. This structural and preference shift is one of several proximate causes of the rise of the creative industries.

A second line of argument is that evolution in the broader economy has led to the rise of the creative industries through a complex of changes in: *technology* and in particular digital technologies (Hartley, 2009); *institutions*, with the trend toward more market-based societies; and *globalization* (Cowen, 1998, 2002). Potts (2011, chap. 11) notes that the rise of the creative industries broadly coincides with the rise of global market economies and can be in part explained by institutional enabling mechanisms.

The growth of the creative industries can also be seen as operating through the same forces that explain the rise of the service sector, namely increased specialization, lowered barriers to entrepreneurship, and significant productivity gains in primary and secondary sectors (the 'Baumol effect'). This is also related to increasing urbanization and post-industrial urban renewal. There is a tendency to couch this perspective in terms of *creative clusters* or *creative quarters*, emphasizing the urban geography and endogenous growth dimension that connects clusters to innovation (Pratt, 2004; Roodhouse, 2006; Currid, 2007).

However, from the perspective of evolutionary economics, and particularly from within the Schumpeterian framework of innovation systems, a fourth hypothesis on the cause of creative industries growth corresponds to what Potts and Cunningham (2008) called the evolutionary model of the creative industries, namely that the creative industries are best understood less as a sector per se, but rather as part of the innovation system. As John Hartley (2009: 50) explains:

> It can even be argued that 'creative industries' are the [manifest] form taken by innovation in advanced knowledge based economies, in which case their importance, like that of the media, exceeds their scale as a sector of the economy. It extends their role as a general enabling social technology. This would place creative innovation with other enabling social technologies like law, science and markets.

This approach is closely affiliated with the UK's National Endowment for Science, Technology and the Arts (NESTA), and the ARC Centre for Creative Industries and Innovation (CCI) in Australia (Hartley and Cunningham, 2001; Hartley, 2005). Amongst other quangos and research institutes, they have sought to emphasize the value of the creative industries explicitly in terms of their role in the innovation process. The creative industries provide services that are inputs into innovation by providing services that furnish the creative capital or by supplying the creative workers who are inputs into the innovation process. Bakhshi et al. (2008), Bakhshi and McVittie (2009) and Müller et al. (2009) have examined the innovation contribution of firms in creative industries to innovation in the wider economy, finding substantial contribution. Potts et al. (2008) and Hartley (2009) propose *social network markets* as a model of creative industries involved in the process by which consumers adopt novel goods and services. This approach can be seen in the work of Paul Stoneman (2011) on *soft innovation*, and Ian Miles and Lawrence Green (2008) on *hidden innovation* in the creative industries. The role of creative industries contributing to amateur innovation and open innovation through new media is discussed in Banks and Potts (2010), Benkler (2006), Shirky (2008) and Leadbeater (2008).

To see the creative industries as innovation drivers is not necessarily philistine instrumentalism. It does not deny that artistic, cultural and creative output is a socio-politically critical activity or an aesthetic or entertaining end-in-itself. It simply recognizes that the arts, cultural and creative industries have always played a key role in the origination, adoption and retention of new ideas into the socio-cultural and economic system. The economic value of the creative industries can of course be measured by short-run aggregates of consumer spending and jobs over economic activities associated with say heritage and entertainment or creative goods. But this innovation/transformation value may actually have greater long-run *economic* value than the short-run economic value of arts, cultural and creative sectors. The creative industries may thus be part of the investment and resources for change and therefore the growth in welfare that comes with that.

From the evolutionary perspective the creative industries are akin to science and technology sectors, which are also key elements of the innovation system and drivers of economic evolution. The difference is that, whereas science and technology deal mostly with the manipulation and development of new material forms and the economic opportunities this creates (Arthur, 2009), the arts, cultural and creative sectors deal with the human interface, with the new ways of being and thinking and interacting (Quiggin, 2006). They operate on the demand side of economic evolution, whereas science and technology operate mostly on the supply side. All innovation processes and trajectories as evolutionary processes of 'creative destruction' involve people originating and adopting new ideas, learning to do new things and experimenting with variations, and seeking to embed these new ideas into new habits, routines and even identities. New ideas can create new value only if existing structures and ways of doing things can change. The creative industries drive, facilitate and embed that process, resulting in externalities that may not be fully captured in market incentives.

The creative industries are part of the service sector. Yet they are unlike routine services that are based on known technologies and extant institutional structures (such as health, transport or insurance), which are often specialized outgrowths of primary and secondary sector operations. Instead, the creative industries are by definition involved in the process of new value creation because their business opportunities and value added derive from the very existence of novelty and innovation in other sectors. They provide services that are about that innovation process (Bakhshi and McVittie, 2009), and many of these services are business-to-business, rather than to consumer markets (Bakhshi et al., 2008). The creative industries are deeply engaged in the experimental use of new technologies (Müller et al., 2009), in developing new content and applications, and in creating new business models. They are broadly engaged in the coordination of new technologies to new lifestyles, new meanings and new ways of being, which in turn is the basis of new business opportunities (Hartley et al., 2012).

Perhaps this process-focused 'innovation services' view seems odd. But running through the list of creative industries – architecture, advertising, fashion, design, interface software, publishing and so on – the rationale for these services is much reduced in a closed, static or steady-state economy that is not continually transforming from within. Rather, most value creation possibilities of the creative industries arise because they solve problems in a market-economy context created by technological or socio-cultural change, further driving such endogenous change. The creative industries: monitor and analyse change and communicate that to people (publishing); create new products and facilitate their adoption (design); facilitate institutional adjustment to technical and cultural change (technical and community media); create new spaces for activities (architecture); develop new uses for new technologies (video games, film); facilitate embedding of new ideas into socio-cultural practices (television and film); and so on.

These dynamic or evolutionary functions are only manifest (indeed, are only necessary) in a world of endemic change. Along with their obvious aesthetic and entertainment value, part of the *raison d'être* of creative industries derives from processing innovation in the social and cultural context. In doing so they are of course heavy users of new communication technology, increasing its demand. This *evolutionary service* or *creative innovation service* has value proportional to the broader rate of economic evolution. Whatever their cultural value or static economic value, the creative industries are from the evolutionary economic perspective also a much underappreciated part of the explanation of long-run economic growth and development.

THE INNOVATION TRAJECTORY

We may connect the creative industries to different phases of an innovation trajectory, and to the institutions that facilitate this process, by considering them as part of the *innovation system* and as providing innovation services over an *innovation trajectory* (Handke, 2006; Bakhshi et al., 2008; Potts, 2009). An innovation trajectory has three broad phases: origination, adoption and retention (Dopfer and Potts, 2008). The creative industries are involved in all three phases.

First, the origination phase is the realm of the creative industries in providing the service of creativity and novelty generation. The obvious process here is the literal and poetic meaning of creativity, usually advanced as the major added value of this sector. Yet, from the evolutionary perspective, it is not obvious that this is indeed the major contribution. First, other sectors also produce creativity, and, second, the dynamic value comes not from the production of novelty, but from its adoption and retention.

Origination value in creative industries may be more indirect, relating instead to the development of innovation technologies (Dodgson et al., 2005). These are, for example, the platforms of gaming co-opted for commercial use, or the new social communication technologies adapted for commercial value (Burgess and Green, 2009). Where science and technology are unambiguously of value in the origination phase, the creative industries also add value in developing resources, methods and ways of doing things (in effect, technologies) of creativity generation. This can be observed in the way creative production methods, models and ways of working have been developed in the creative sector and then transferred to other sectors, for example creative production teams.

Creative industries also facilitate innovation in preparing the ground for new ideas (even when they don't originate them). The role of creative work in provocation and critique, or in rooting out and identifying problems with existing ways and creating dissatisfaction so as to make people open to new ideas is of course a part of the innovation process (Chai et al., 2007).

The adoption phase of innovation is perhaps the most important domain of creative industries contribution. For any new idea to succeed it must be adopted (chosen and adapted into their lifestyle) by many people. The determinants of adoption do not always come down to qualities of engineering. Usually attention and persuasion matter, as do the symbolic and expressive content and subjective value of an idea (Lanham, 2006). The creative industries provide knowledge and mechanisms to facilitate the adoption process in myriad ways. Significant gaps can arise between what is optimal and what actually gets adopted as a result of 'social network market' effects (Potts et al., 2008; Bentley, 2009; Ormerod, 2012), which arise when agents deal with novelty and uncertainty by copying other consumers, and 'novelty bundling market' effects (Potts, 2012), which arise when agents begin to specialize in dealing with particular novelties, and packing them for others. Creative industries add value to the innovation process by overcoming and amplifying these effects. The creative industries supply services that facilitate adoption and adaptation of the ways and means by which new markets and applications of new ideas are developed. This occurs in the context of 'choice under novelty' (Potts, 2010), in experimentation with the possibilities of new technologies, and with the emergence of new institutions.

The retention phase of innovation is when an idea becomes embedded for ongoing use, a process also known as habituation, normalization or institutionalization. This is a further service provided by creative industries, and again in multiple ways: for example, in respect of the construction and normalization of new identities associated with the particular innovation (Herrmann-Pillath, 2010). Note that mainstream economic theory widely assumes this process to be costless. Yet it is a significant investment for individuals, organizations and networks, the differential consequence of which shapes the knowledge base and institutions of societies.

The creative industries are properly part of the innovation system. This is not by any shoehorning into a science–technology matrix by measuring their innovative contribution in terms of, say, patents. Rather, it is because of their crucial role in the socio-cultural process of adoption and retention of new ideas (Earl and Potts, 2004). Obviously, creative industries produce art, culture and entertainment. Less obvious, however, is that they also produce the dynamic service of re-coordination of the socio-cultural and economic order to the ongoing growth of knowledge process. It is this latter aspect – this input into the innovation process – that properly connects creative industries to the arguments of innovation systems and policy.

POLICY IMPLICATIONS

The policy framework for cultural economics is lifted from the standard welfare economics playbook for dealing with positive externalities, namely subsidy, although sometimes couched in terms of 'cultural Keynesianism' or industry policy (Grampp, 1989). The theoretical and empirical work connecting creative industries to economic evolution suggests a very different policy framework that is based on a mix of competition policy and innovation policy.

From the perspective of evolutionary economics (Pelikan and Wegner, 2003), appropriate policy settings for the creative industries are not intended to correct a market failure, but to facilitate a market discovery process. Such a process will be entrepreneur driven and will involve demand uncertainty and uncertainty regarding what production methods work, what business models are most appropriate and so on. Firm experimentation and thus firm entry and exit, along with changes in jobs, technologies and other such aspects of what Joseph Schumpeter called 'creative destruction', are often an important part of this process.

In practice, this means limiting trade protection such as tariffs or quotas on cultural imports, which are sources of competition. It also means minimizing industry policy such as attempts to 'pick winners', grant monopolies (particularly in media) or otherwise provide public support to incumbents in response to lobbying or political considerations. It means letting markets work, both in allowing entry and exit to occur, and for price signals to be undistorted. However, there is a potential role for public support in facilitating the process of entrepreneur-driven market discovery (Bakhshi et al., 2011).

What of innovation policy in the creative industries? The results of innovation policy in the creative industries are somewhat mixed in the various countries that have endeavoured to develop policies to integrate the creative industries, and creative industries policies, into the national innovation system (for example, the UK, Australia, Singapore, New Zealand, Canada). This is in part due to the difficulty in making the transition from a subsidy model of industrial support to preserve existing cultural production based on the market failure model to one driven by very different objectives. Some of this difficulty arises because of an entrepreneurial and market-focused conception of policy support that is often at odds with dominant folkways and cultural norms in arts and cultural policy. So there has been no small amount of culture shock and resistance from lobby groups and academics representing the cultural and creative industries (Garnham, 2005). But the process of engagement is also hard because it is new and still partially experimental. A number of research and policy institutes (such as NESTA in the UK) have sought to add structure to this policy experiment by funding and analysing experiments in innovation policy in creative industries. Whether innovation policy continues to develop so as to integrate creative industries policy will thus depend on the outcomes of policy experiments, the results of policy learning and subsequent diffusion. If so, creative industries policy may become embedded as a core component of national innovation systems policy frameworks.

REFERENCES

Albarran, A. (2002), *Media Economics*, 2nd edn, New York: Wiley Blackwell.

Andari, R., H. Bakhshi, W. Hutton, A. O'Keeffe and P. Schneider (2007), *Staying Ahead: The Economic Performance of the UK's Creative Industries*, London: Work Foundation.

Andersson, D. and A. Andersson (2006), *The Economics of Experiences, the Arts and Entertainment*, Cheltenham, UK and Northampton, MA, USA: Edward Elgar Publishing.

Arthur, W.B. (2009), *The Nature of Technology*, New York: Free Press.

Bakshki, H. and E. McVittie (2009), 'Creative supply-chain linkages and innovation: do the creative industries stimulate business innovation in the wider economy?', *Innovation: Management, Practice and Policy*, **11** (2), 169–89.

Bakhshi, H., E. McVittie and J. Simmie (2008), *Creating Innovation: Do the Creative Industries Support Innovation in the Wider Economy?*, London: NESTA.

Bakhshi, H., A. Freeman and J. Potts (2011), *State of Uncertainty*, NESTA Provocation 14, London: NESTA.

Banks, J. and J. Potts (2010), 'Consumer co-creation in online games', *New Media and Society*, **12** (2), 253–70.

Baumol, W. and W. Bowen (1966), *Performing Arts: The Economic Dilemma*, New York: Twentieth Century Fund.

Benkler, Y. (2006), *The Wealth of Networks*, New Haven, CT: Yale University Press.

Bentley, R.A. (2009), 'Fashion versus reason in the creative industries', in M.J. O'Brien and S.J. Shennan (eds), *Innovation in Cultural Systems: Contributions from Evolutionary Anthropology*, Cambridge, MA: MIT Press, pp. 121–6.

Blaug, M. (2003), 'Welfare economics', in R. Towse (ed.), *A Handbook of Cultural Economics*, 2nd edn, Cheltenham, UK and Northampton, MA, USA: Edward Elgar Publishing, pp. 425–30.

Burgess, J. and J. Green (2009), *YouTube: Online Video and the Politics of Participatory Culture*, London: Polity Press.

Caves, R. (2000), *Creative Industries: Contracts between Art and Commerce*, Cambridge, MA: Harvard University Press.

Chai, A., E. Earl and J. Potts (2007), 'Fashion, growth and welfare: an evolutionary approach', in M. Bianchi

(ed.), *The Evolution of Consumption: Theories and Practices*, Advances in Austrian Economics, vol. 10, Bingley: Emerald, pp. 187–207.
Cowen, T. (1998), *In Praise of Commercial Culture*, Cambridge, MA: Harvard University Press.
Cowen, T. (2002), *Creative Destruction: How Globalization Is Changing the World's Cultures*, Princeton, NJ: Princeton University Press.
Cunningham, S. (2002), 'From cultural to creative industries: theory, industry, and policy implications', *Media Information Australia Incorporating Culture and Policy*, **102**, 54–65.
Cunningham, S. (2006), *What Price a Creative Economy?*, Platform Papers No. 9, Sydney: Currency House.
Cunningham, S., T. Cutler, G. Hearn, M. Ryan and M. Keane (2004), 'An innovation agenda for the creative industries: where is the R&D?', *Media International Australia Incorporating Culture and Policy*, **112**, 174–85.
Currid, E. (2007), *The Warhol Economy: How Fashion, Art, and Music Drive New York City*, Princeton, NJ: Princeton University Press.
DCMS (Department for Culture, Media and Sport) (1998), *Creative Industries Mapping Document*, London: DCMS.
DCMS (Department for Culture, Media and Sport) (2001), *Creative Industries Mapping Document*, London: DCMS.
Dodgson, M., M. Gann and A. Salter (2005), *Think, Play, Do*, Cambridge: Cambridge University Press.
Dopfer, K. and J. Potts (2008), *The General Theory of Economic Evolution*, London: Routledge.
Earl, E. and J. Potts (2004), 'The market for preferences', *Cambridge Journal of Economics*, **28**, 619–33.
Florida, R. (2002), *The Rise of the Creative Class*, New York: Basic Books.
Garnham, N. (2005), 'From cultural to creative industries: an analysis of the implications of the "creative industries" approach to arts and media policy making in the United Kingdom', *International Journal of Cultural Policy*, **11**, 15–29.
Grampp, W. (1989), *Pricing the Priceless: Art, Artists and Economics*, New York: Basic Books.
Handke, C. (2006), *Surveying Innovation in the Creative Industries*, Berlin: Humboldt University and Rotterdam: Erasmus University.
Handke, C. (2007), 'Promises and challenges in innovation surveys of the media industries', in H. van Kranenburg and C. Dal Zotto (eds), *Management and Innovation in the Media Industry*, Cheltenham, UK and Northampton, MA, USA: Edward Elgar Publishing.
Hartley, J. (ed.) (2005), *Creative Industries*, Carlton, VIC: Blackwell.
Hartley, J. (2009), *Uses of Digital Literacy*, Brisbane: University of Queensland Press.
Hartley, J. and S. Cunningham (2001), 'Creative industries: from blue poles to fat pipes', in M. Gilles (ed.), *The National Humanities and Social Sciences Summit*, Canberra: DEST, pp. 1–10.
Hartley, J., J. Potts and T. MacDonald (2012), 'Creative City Index', *Cultural Science*, **5** (1) (e-journal).
Hartley, J., J. Potts, S. Cunningham, T. Flew, M. Keane and J. Banks (2013), *Key Concepts in the Creative Industries*, London: Sage.
Herrmann-Pillath, C. (2010), *The Economics of Identity and Creativity: A Cultural Science Approach*, Brisbane: University of Queensland Press.
Higgs, P., S. Cunningham and H. Bakhshi (2008), *Beyond the Creative Industries*, technical report, London: NESTA.
Hodgson, G. and T. Knudsen (2010), *Darwin's Conjecture: The Search for General Principles of Social and Economic Evolution*, Chicago: University of Chicago Press.
Howkins, J. (2001), *The Creative Economy*, London: Penguin.
Howkins, J. (2009), *Creative Ecologies*, Brisbane: University of Queensland Press.
Lanham, R. (2006), *The Economics of Attention*, Chicago: University of Chicago Press.
Lash, S. and J. Urry (1994), *Economies of Signs and Space*, London: Sage.
Leadbeater, C. (2008), *We Think: Mass Innovation, Not Mass Production*, London: Profile Books.
Lloyd, R. (2006), *Neo-Bohemia: Art and Commerce in the Postindustrial City*, New York: Routledge.
Miles, I. and L. Green (2008), *Hidden Innovation in the Creative Industries*, NESTA research report, London: NESTA.
Müller, K., C. Rammer and J. Truby (2009), 'The role of creative industries in industrial innovation', *Innovation: Management, Practice and Policy*, **11** (2), 148–68.
Nelson, R. and S. Winter (1982), *An Evolutionary Theory of Economic Change*, Cambridge, MA: Harvard University Press.
Ormerod, P. (2012), *Positive Linking: How Networks Can Revolutionise the World*, London: Faber.
Peacock, A. (1993), *Paying the Piper: Culture, Music and Money*, Edinburgh: Edinburgh University Press.
Pelikan, P. and G. Wegner (eds) (2003), *The Evolutionary Analysis of Economic Policy*, Cheltenham, UK and Northampton, MA, USA: Edward Elgar Publishing.
Potts, J. (2009), 'Why the creative industries matter to economic evolution', *Economics of Innovation and New Technology*, **18** (7–8), 663–74.

Potts, J. (2010), 'Can behavioural biases in choice under novelty explain innovation failures?', *Prometheus*, **28** (2), 133–48.
Potts, J. (2011), *Creative Industries and Economic Evolution*, Cheltenham, UK and Northampton, MA, USA: Edward Elgar Publishing.
Potts, J. (2012), 'Novelty bundling markets', in D.E. Andersson (ed.), *The Spatial Market Process*, Advances in Austrian Economics, vol. 16, Bingley: Emerald, pp. 291–312.
Potts, J. and S. Cunningham (2008), 'Four models of the creative industries', *International Journal of Cultural Policy*, **14** (3), 233–48.
Potts, J., J. Hartley, J. Banks, J. Burgess, R. Cobcroft, S. Cunningham and L. Montgomery (2008), 'Consumer co-creation and situated creativity', *Industry and Innovation*, **15** (5), 459–74.
Pratt, A. (2004), 'Creative clusters: towards the governance of the creative industries production system?', *Media International Australia*, **112**, 50–66.
Quiggin, J. (2006), 'Blogs, wikis and creative innovation', *International Journal of Cultural Studies*, **9** (4), 481–96.
Roodhouse, S. (2006), *Cultural Quarters: Principles and Practice*, Bristol: Intellect Books.
Schumpeter, J. (1942), *Capitalism, Socialism and Democracy*, London: George Allen & Unwin.
Scott, A. (2006), 'Entrepreneurship, innovation and industrial development: geography and the creative field revisited', *Small Business Economics*, **26**, 1–24.
Shirky, C. (2008), *Here Comes Everybody*, New York: Allen Lane.
Stoneman, P. (2011), *Soft Innovation: Economics, Product Aesthetics and Creative Industries*, Oxford: Oxford University Press.
Throsby, D. (1994), 'The production and consumption of the arts', *Journal of Economic Literature*, **32**, 1–29.
UNCTAD (2008), *Creative Economy Report*, Geneva: UNCTAD, available at: unctad.org/en/docs/ditc20082cer_en.pdf.

FURTHER READING

The connection between creative industries and economic evolution is relatively recent and has been most extensively made in the work of Jason Potts. His collected papers in *Creative Industries and Economic Evolution* (2011) provide an overview of the scope of this concern. More broadly, the work of the creative industries and innovation research institute (the CCI) at Queensland University of Technology ranges across this space of applications of evolutionary and innovation economics approaches to creative industries studies and policy. Other evolutionary and innovation economists who have developed a body of work on creative industries economics are Christian Handke (2006, 2007), who focuses on intellectual property in the music industry, and Hasan Bakhshi (see e.g. Bakhshi et al., 2008), who focuses on creative industries policy frameworks and empiricism. A special issue of the journal *Innovation: Management, Practice and Policy* (volume 11, issue 2) on 'Innovation policy in the creative industries' provides a useful overview of studies of innovation in the creative industries. The entries on 'evolution', 'entrepreneur', 'innovation' and 'market' in Hartley et al. (2013), *Key Concepts in the Creative Industries*, are also a useful starting place.

4. Space and place

Andy C. Pratt

The aim of this chapter is to elaborate the role of space in relation to the digital creative economy. It will argue that the partial conceptualization of the digital creative economy, mainly focusing on consumption and the immaterial, has settled on a distorted and unhelpful understanding of the nature of current transformations. The net effect has been the simplification, or erasure, of the role that space plays in the digital creative economy. However, I point out that recent research has re-balanced this discussion by stressing the value of examining the (social, economic and cultural) embedded nature of the digital creative economy and so gives a more adequate account of the role of space.

Critics have been too quick to declare space irrelevant to digital economies. This was the result of seizing on just one 'revolutionary' aspect of change. Although notions such as the 'death of distance' do chime with our transformative experience of digital communications they are partial and we should be wary of such characterizations. First, there is the purist line of argument where a key distinction is stressed between the analogue and the digital, or atoms and bits (Negroponte, 1995). Such a binary has the unfortunate side effect of elevating in import and isolating the 'virtual' as an autonomous sphere, and marginalizing the social relations of the digital. I will argue that this formulation obscures the more subtle aspects of the digital creative economy: the hybridization of both the virtual and the real, and the technological and the social-cultural. Second, there is the commonly repeated use of the term 'digital economy' as if it were a (new) thing. Other contributions to this book discuss this definitional issue; however, what I want to point out is the differentiation *within* the creative economy: digitization is one part of a complex transformation. At least two dimensions impinge on our understanding and explanation: first, the different art/cultural/business form (for example, computer games are different from theatre); second, the different aspects of digital creative production (for example, creation is different from distribution, which is also different from consumption).

The chapter is divided into three main sections. It begins with, 'The death of distance', an account of the de-prioritization of space in accounts of the digital transformation. The second section, 'Place rediscovered', contrasts these expectations with a more mundane reality. This is not to suggest nothing changed. I outline two brief case studies of clustering in digital creative economy (new media and computer games) in order to illustrate that space does continue to matter, but in complex ways, which is discussed in the final section 'The embedded cultural economy'.

THE DEATH OF DISTANCE?

Alvin Toffler (1980), when writing about the social and economic transformations likely with the introduction of the 'third wave' technologies of computing, suggested that there would be a dramatic transformation of how we worked and where we worked. As it turned out Toffler only scratched the surface of the transformations that followed. He drew some insightful and pertinent conclusions, but the section on space and place was not one of them. For many people, he argued, the city could be abandoned, and the small town and the 'electronic cottage' would be the new future of work (Toffler, 1980: 120). He based this on the ability of communication technologies to distribute work without the need for co-presence. Toffler did acknowledge that, for a small number, face-to-face communication would still be important, but this would be a small elite.

This idea reappeared with some force in the late 1990s. Writing about the economic transformations likely to be wrought by the new communications technologies, Frances Cairncross (1998) declared the 'death of distance'. Others such as Diane Coyle (1998) offered an economic stylization as 'the weightless economy'. This upsurge of interest coincided with the popular fascination with virtual reality (Gibson, 1984). The essential idea is simple: when the costs of movement of goods fall to zero (in the hypothetical economy of bits) there will be no (monopoly) market power derived from unique location. By this I mean that, whilst two firms can set the same price, they cannot occupy the same 'perfect' location; hence, advantage is conferred to the one that does occupy that location. If there were no 'spatial monopolies' the world would operate as abstract economic theory has long argued that it should, oblivious of space. Such was the attraction of the neo-classical economic tradition that the new digital world seemed as if it might conform to a perfect economic model (devoid of transport costs, spatial 'distortions' and time lags).

Whilst there has undoubtedly been a transition from analogue to digital production, the implications in organizational and spatial terms should be carefully considered. What has been discussed above is primarily a technological change; moreover it has been assumed that certain consequences would necessarily flow in its wake. There are a number of counter-arguments. Underlying these is a conceptual issue that concerns space; it relates to what geographers characterize as the difference between absolute and relative notions of space (Gregory and Urry, 1985). When we use the formulation 'location in space' we are implicitly stressing the separation of intrinsic qualities of the object thus placed, and the space into which it is put. By contrast the relational view might be characterized as a socio-spatial relationship, whereby the qualities of object and space are constituted *in situ*.

Organizational changes have been written about in the field of the digital economy, and especially the digital creative economy (Castells, 1996). However, the question left unresolved here concerns causality in relation to social (organizational) and economic transactions and space. The material shift from larger corporations to smaller firms, to network organizations and to project-based firms is commonly documented (Grabher, 2004). All of these issues are without doubt important; however, to suggest that it was digitization that caused the change is more controversial. The counter-argument has more empirical support, namely that regulatory change, governance and market strategy play significant roles. In other words, digitization concerns an opportunity, not a necessary outcome.

Furthermore, there is the question of the 'distributional effects' of technology or, put simply, who gains from the application of new technologies. There has been much discussion of the 'digital divide' in terms of a one-off barrier to entry to this new world, the implication being that investment in technology alone can resolve the imbalance. Research points instead to the fact that what is at stake is the entry to an already asymmetrical marketplace, where rewards, and the goods on offer, are structured against the poor (Mossberger et al., 2003; Perrons, 2004). Hence, the digital divide will not be resolved with a quick (technical) fix, but requires instead a more extensive restructuring of economic and social relations.

What was missing from the simple accounts of digitization and economic transformation is that new technologies, like the old ones, are embedded in social structures. Moreover, there is not a 'pure' digitization, nor one shorn of its own production or consumption. This insight counters the notion that digitization has a determinate impact with the view that it is always mediated. Such a conclusion makes explanation a little more complicated, but it puts (relational) space firmly back on the agenda.

PLACE REDISCOVERED

The relevance of space was empirically underlined when the United Nations announced that 2010 would be the year in which for the first time in history more than 50 per cent of the world's population would live in cities. Moreover this trend is accelerating, not declining: we are an urban world. Co-location, and urban dwelling, is clearly important; cities and space are still important, in fact more so than in the past. The new media industry is the 'poster child' of the digital creative economy. These are the most intensely 'digital' of all industries. So, if the theory is right, we would expect the first signs of the death of distance to be evident here. Do we find workers with laptops in electronic cottages, or on the beach? No. In fact, we find them concentrated in some of the most expensive real estate in the world: in the fashionable districts of New York, London or San Francisco.

The question is: what is going on? Clustering, or co-location, alerts us to the nuances of space and the economy. Traditional debates of location highlighted common cost structures (such as transport costs); more recent debates have discussed a more inclusive notion of transactions costs which incorporate transport but in addition the economic and social benefits of close interaction (Scott, 2000). As we will note later, economic geographers, in particular, have argued that clustering is not the same in all industries. The a priori assumed cost-free and instantaneous exchange of information in the digital economy should have made co-location unnecessary. The fact that co-location was, and continues to be, necessary leads us to examine what is transacted in these clusters, and how. This prompts us to look at issues of the organization of the digital creative industries: two case studies are used to illustrate these points.

NEW MEDIA

San Francisco's 'Multimedia Gulch' (South of Market) and New York's 'Silicon Alley' (Lower Manhattan) – real places – were early examples for digital creative economy clusters: small firms co-located within the space of just a few blocks of the city concerned, many sharing the same buildings. Studies of those clusters highlighted a number of distinct issues associated with how the industry developed, and the organizational forms that adapted to accommodate it.

Proximity is important for firms in seeking to remain within relevant networks to access fast-changing and up-to-the-minute information. The quality of this information is fuzzy and uncodified, and usually requires face-to-face interaction to correctly interpret it. On one hand, project workers gain experience through working for other projects, and transferred knowledge and techniques. On the other hand, a labour pool can be maintained if workers can efficiently move from one job to another and build continuous employment. In these new media places both face-to-face networking and online networking are vital to gaining the next job. It is not simply skills matching but expertise and excellence that are required, mixed with an ingredient that is elusive, the ability to work with others.

The task of managing a project team is to choose the most talented people for a project, who do not antagonize others and who can deliver on deadlines; genius alone is insufficient. Project team management skill relies upon word-of-mouth and keeping close contact with changing reputations. Some of these things are possible via mediated communications; but face-to-face contacts are the most trusted. Likewise, project teams/firms locate in the same buildings, with competitors close by. A common work mode was for the whole team to be in one room – a fairly extreme form of co-location, which was found useful for immediate problem solving.

The companies formed are primarily project-based enterprises that are groups of experts brought together to work on a limited-life project. Thus the labour market, or labour pool of experts, is vitally important. Once that pool is developed, projects were attracted to it, as developers and designers would not have been able to easily or practically locate and relocate anywhere in the world for single projects. The diverse skill set found in such locations is critical to emergent new media companies, in part reliant on technical coding skills, in part on marketing and strategic management expertise, and in part on artistic skills of narrative and visual rendering. Cities with an already well-developed labour market in both the cultural economy and programming are obviously well placed to benefit from such conditions. To this had to be added a final element: money. It is not chance that the emergent hubs of new media are also places where venture capital is available and seeking investment.

COMPUTER GAMES

The computer (including video and console) games industry has been the star of the creative economy. US games sales now equal US cinema box office receipts. Beginning as an activity in which it was possible for a single person to write all the software for a game, the games industry grew in terms of market size and technical and financial requirements, which in turn led to bigger and more complex games that were produced by teams of more than a hundred persons. The games industry went through a number of organizational and market iterations in the 1980s with the advent of the home computer that could be programmed and used to play pre-programmed games. At first, the industry organization was similar to music, where a developer got signed up to a major company, which published and distributed the game. There was the added complexity that multiple publishers (for example Eidos or Electronic Arts) and the platforms (Sony, Microsoft, Nintendo) were in competition. Later, publishers and platforms merged, and games began to be released across all platforms. Games companies were restructured into large corporations and became, on the back of some phenomenally successful games franchises, significant players in the media landscape. Computer games display a contrasting locational pattern to new media.

Game developing has been a large project operation that requires the assembly of a team, and considerable specialized technology in-house. However, the commercial sensitivity of game releases has led to an internal-facing focus, if not complete secrecy. The business model for games is like that of recorded music, where the aim is to maximize sales within a short period of a market window of opportunity immediately after the release date of the game. The legacy is that games developers prefer to locate well away from one another and, in the UK, chose rural locations away from their competitors. So co-location in the same building is important, but not co-location with competitors.

More recently, the games industry has been through another organizational iteration, where massive capitalization has created new super-scale companies. These are games factories linked to single-game-dominated franchises. The particular physical location of these factories is determined by labour availability and cost, commonly modified by a range of state incentives in terms of tax and employment benefits. Many companies have, in recent years, moved to Canada, where the government has pursued a strategy to attract such companies and offered significant subsidies. Governments in France and Britain – places where the games industry initially grew strongly – were arguably slow to react and thus lost many key companies. The point is that, within reason, it does not matter where in the world the factory is located, as long as the skill base is there; with incentives, that skill base can also be substantially attracted to a location. Most new games factories are in fact in urban centres now. Moreover, the premises of games companies have the type of high-security systems found in the pharmaceuticals industry. The critical element is the organization of labour, secrecy, and strict project control.

THE EMBEDDED CULTURAL ECONOMY

Embedding is where 'it all comes together'. I have pointed out that technologies and organizations, not to mention people, have a material form, and thus they are in space, which confers certain accessibility and possibility (and the opposite). But space is not absolute. The relational nature of possibility is dynamized by interaction, which is itself a hybrid of the online and the offline. This is the lens through which the complex dynamics of industrial clusters have to be viewed. An emergent concern has been the differentiation of clustering processes by industry; the creative industries have been a particular focus of this concern.

Attention has been paid to the organization and operational characteristics of the creative industries. It can be argued that the creative economy represents an extreme case of a general variance across all industries. These characteristics have been discussed elsewhere, so I will only provide an elaborated list (Caves, 2000; Pratt and Jeffcutt, 2009). First, the industries of the cultural economy work in very high-risk environments; the rewards for the winners are massive, but so is the rate of failure. As the differentiator is quality, something decided *in situ*, it is very difficult to predict. Moreover, the entry bar is set very high. Second, in part as a consequence, the turnover rate of firms is great. Moreover, the product cycle tends to be very short. This is the 'fashion' element of the cultural economy, always changing. Third, while the majority of the content creators are very small, a few distributors on the whole tend to be very large. The presence of serial temporary project-based enterprises characterizes organization better than firms. Fourth, different cultural industries have different markets, and organizationally and strategically respond in a quite different manner, although some commonality exists as well. Finally, the cultural economy draws a significant part of its strength from the public sector, and from the not-for-profit and informal sectors. Often protagonists are working across all these divisions at once.

It would take a whole chapter, at least, to unpick these characteristics, but they converge on the role of space. The point that can be forcefully made here is that particular technologies may be deployed to quite different effect at various stages in the production process and in particular industries. Moreover this alerts us to another dimension: the creative economy is like any other in that it has a cycle of creation, manufacture, distribution, exchange and archiving. Thus, when examining clustering, attention has to be paid to the interaction of the creative ecosystem and the balance of emphasis to (for example) production or consumption, as these imply different types of embedding.

Drawing together the threads, three strands appear dominant. First, the organization and interaction afforded by space and technology create unique conditions of opportunity confined to particular places and times. There is a competitive and cultural advantage to be gained here: clustering and intense interaction favour activities that benefit from iterative and heuristic processes. Research indicates that creative industries can benefit from serendipity more than routinization and programmed connection, prioritizing the gains to the 'creative' aspects of digital creative production.

The second point is related, the creation of large and diverse cultural labour pools that generate economies of scope. These are especially relevant where intense innovation is required in an uncertain, or rapidly changing, product field. Labour mobility, often associated with freelance labour on very short contracts, embodies valuable knowledge that enables the circulation of tacit evaluations of performance and subtle differentiation of products. Moreover, proximity and connection keep labour and firms abreast of the rapid turnover of ideas that is characteristic of the fashion and design communities.

Third, there are economies of scale achieved in clustering that would otherwise not be available to individual micro-enterprises. In the case of the creative economy, scale economies enable the development of intermediaries that make the cluster function better collectively and individually. Further, a rather particular economy of scale can be created for investors who wish to minimize risk by creating portfolio investments. Such activities require a 'deal flow', that is, volume of activity, as well as specialized knowledge to balance the risk of investments: these are precisely the conditions that clusters afford.

CONCLUSION

The discussion of the spatialities of the digital cultural economy should alert us to a more complex picture than might otherwise have been assumed. The literature on the creative digital economy has stressed embodied interaction and spatialities, and the organizational steps taken to mediate risk. It is perhaps ironic that what is being stressed is the very materiality of the digital (whereas 'digital' is often equated with the immaterial). What can muddle our understanding is popular abstractions of the consequences of digitization: immediate and free communication. These are theoretical qualities; in practice they generate (in the context of local institutions and conditions) the particular qualities of knowledge. More often than not it is these qualities of knowledge, rather than simple quantities (the amount of knowledge, or its presence or lack), that are critically evident in embedding: the right knowledge at the appropriate time. The unique properties – arising from a unique composite of these elements – are present in particular places, industries and times.

The two case studies presented here illustrated the diversity of 'digital spaces', which were in part where particular technologies had been deployed in the production system. This application of digitization was not determinate, but admitted by, or blocked by, social processes, as we can see in the case of newspapers, and we might have seen in the music and film industries. Technologies were balanced by the unique character of the production process in each cultural industry, and the particular form of its audience or market.

The chapter has concluded that a spatial-technological dualism is the wrong way to frame the debate about the digital creative economy. In contrast, what I have argued in this chapter is that this relationship is better conceived of as a complex amalgam of the online and offline, of the digital and analogue, which are materialized in various ways in particular spaces, which in turn constitutes them as 'places'.

So, in order to understand the production of such spaces, we need to apprehend the embedded and embodied processes, in this case those of the creative economy. I have stressed that the cultural industries that constitute the creative economy are various but share some characteristics of organization, which are at the more extreme end of the 'normal' spectrum of forms aimed at managing high risk, a rapidity of turnover, and a range of changing and previously undefined qualities.

REFERENCES

Bathelt, H., A. Malmberg and P. Maskell (2004), 'Clusters and knowledge: local buzz, global pipelines and the process of knowledge creation', *Progress in Human Geography*, **28** (1), 31–56.

Braunerhjelm, P. and M.P. Feldman (eds) (2007), *Cluster Genesis: Technology-Based Industrial Development*, Oxford: Oxford University Press.

Cairncross, F. (1998), *The Death of Distance: How the Communications Revolution Will Change Our Lives*, Boston, MA: Harvard Business School Press.

Castells, M. (1996), *The Rise of the Network Society*, Cambridge, MA: Blackwell.

Caves, R. (2000), *Creative Industries: Contracts between Art and Commerce*, Cambridge, MA: Harvard University Press.

Cooke, P. and L. Lazzeretti (eds) (2008), *Creative Cities, Cultural Clusters and Local Economic Development*, Cheltenham, UK and Northampton, MA, USA: Edward Elgar Publishing.

Coyle, D. (1998), *The Weightless Economy*, London: Capstone.

Gibson, W. (1984), *Neuromancer*, London: Gollancz.

Grabher, G. (2004), 'Learning in projects, remembering in networks? Communality, sociality, and connectivity in project ecologies', *European Urban and Regional Studies*, **11** (2), 103–23.

Gregory, D. and J. Urry (1985), *Social Relations and Spatial Structures*, Basingstoke: Macmillan.

Hutton, T. (2008), *The New Economy of the Inner City: Restructuring, Regeneration and Dislocation in the Twenty-First Century Metropolis*, Abingdon: Routledge.

Indergaard, M. (2004), *Silicon Alley: The Rise and Fall of a New Media District*, New York: Routledge.
Johns, J. (2006), 'Video games production networks: value capture, power relations and embeddedness', *Journal of Economic Geography*, **6** (2), 151–80.
Mossberger, K., C.J. Tolbert and M. Stansbury (2003), *Virtual Inequality: Beyond the Digital Divide*, Washington, DC: Georgetown University Press.
Negroponte, N. (1995), *Being Digital*, London: Hodder & Stoughton.
Perrons, D. (2004), 'Understanding social and spatial divisions in the new economy: new media clusters and the digital divide', *Economic Geography*, **80** (1), 45–61.
Picard, R.G. and C. Karlsson (eds) (2011), *Media Clusters*, Cheltenham, UK and Northampton, MA, USA: Edward Elgar Publishing.
Pratt, A.C. (2000), 'New media, the new economy and new spaces', *Geoforum*, **31** (4), 425–36.
Pratt, A.C. (2002), 'Hot jobs in cool places: the material cultures of new media product spaces: the case of the South of Market, San Francisco', *Information, Communication and Society*, **5** (1), 27–50.
Pratt, A.C. (2009), 'Situating the production of new media: the case of San Francisco (1995–2000)', in A. McKinlay and C. Smith (eds), *Creative Labour: Working in Creative Industries*, London: Palgrave, pp. 195–209.
Pratt, A.C. (2011), 'Microclustering of the media industries in London', in C. Karlsson and R.G. Picard (eds), *Media Clusters*, Cheltenham, UK and Northampton, MA, USA: Edward Elgar Publishing, pp. 120–35.
Pratt, A.C. and P. Jeffcutt (2009), 'Creativity, innovation and the cultural economy: snake oil for the 21st century?', in A.C. Pratt and P. Jeffcutt (eds), *Creativity, Innovation and the Cultural Economy*, London: Routledge, pp. 1–20.
Pratt, A.C., R.C. Gill and V. Spelthann (2007), 'Work and the city in the e-society: a critical investigation of the socio-spatially situated character of economic production in the digital content industries, UK', *Information, Communication and Society*, **10** (6), 921–41.
Scott, A.J. (2000), *The Cultural Economy of Cities: Essays on the Geography of Image-Producing Industries*, London: Sage.
Toffler, A. (1980), *The Third Wave*, New York: William Collins.

FURTHER READING

Reviews of debates about industrial clusters can be found in Braunerhjelm and Feldman's (2007) edited book. The particularity of clustering in the creative industries is discussed in Scott's (2000) pioneering book, and in edited collections by Cooke and Lazzeretti (2008) and Picard and Karlsson (2011). Bathelt et al. (2004) is a pioneering work on clustering and relational space, using an example of a media cluster.

Specifics of debates about space and the computer games industry can be found in Johns (2006) and Pratt et al. (2007), and about the new media industry in Pratt (2000, 2002, 2009, 2011), Hutton (2008) and Indergaard (2004).

5. Business models

Nicola Searle and Gregor White

The emergence of business models as a research topic has coincided with the growth of online markets and changes to commercial practices in the digital era. As digital distribution, production and payment mechanisms have resulted in decreased transaction costs, business models previously economically unviable have become profitable. These decreased transactions costs have enabled innovation and experimentation in new business models, especially in the creative industries, where the ability to produce, distribute and sell digital media via the Internet has led to significant changes in business models and market structure.

This chapter presents an overview of business model theory and reports on research that examines changing business models in the digital media industry. As business model methodology matures, the literature moves towards a consensus on the definition and interpretation of business model theory and practice. The weaknesses of business models as both a method and an applied practice are discussed. The research focuses on firms based in the United Kingdom (UK), but the global nature of the Internet translates research themes internationally. The study examines three sectors of the creative industries: music, television and computer games. Case studies in each sector highlight the role of business models, innovation in business models and the overall change in the marketplace.

THEORY

Interest in business models has increased significantly in the last decade. As it is an emerging research topic, much of the debate over theory has centered on the fundamentals of business models, including definitions, taxonomy and the role of business models in business strategy. This section presents a review of competing business model definitions and theories.[1]

Definitions

A key point in discussing business models, as a relatively new research topic, is defining the term. The existing literature struggles to both define business models and distinguish the concept from similar terms such as process models. An early definition can be found in Chesbrough and Rosenbloom (2002: 532), who use a definition from the consulting firm KMLab, Inc.:

> a Business model is a description of how your company intends to create value in the marketplace. It includes that unique combination of products, services, image, and distribution that your company carries forward. It also includes the underlying organization of people, and the operational infrastructure that they use to accomplish their work.

This definition emphasizes value but is lacking in terms of parsimony. The literature has since debated alternate definitions, as noted in Jansen et al. (2007). They argue that competing definitions can be placed in two categories: the revenue model and the integrated model. The revenue model identifies the cash flows and finances on which a firm rests. The integrated model, in contrast, looks at the business more as a whole, and includes the strategy and structure of the firm's organization.

Zott et al. (2010) also note the absence of an agreed-upon definition of business models. They categorize ten business model definitions as fitting into three categories: e-business, strategic issues and innovation and technology management. Yet, as the authors note, these categories are not mutually exclusive and definitions overlap. This chimes with Linder and Cantrell (2000: 1), who argue that business models are distinct from change models and should be defined as 'the organization's core logic for creating value'. These authors also note that references to business models are often references to specific components of a business model rather than the whole business model.

The notion that business models are confused with, but are by definition distinct from, other types of models is shared by Gordijn et al. (2000). The authors argue that the process of mapping out business models (which they refer to as e-business modeling) is inappropriately conflated with business process modeling. Instead, the business model focuses on the *who* and *what*, whereas the business process deals with the *how*. Business models also focus on value instead of how processes are carried out. Edirisuriya and Johannesson (2009) agree that business and process models are separate and argue that a process model can be developed from a business model. Andersson et al. (2006) add goal models to the list of business and process models, where goal models focus on the *why*.

A popular definition of a business model is that of Osterwalder and Pigneur (2010: 14), where 'a business model describes the rationale and infrastructure of how an organization creates, delivers and captures value'. This is the preferred definition for this chapter. It focuses on value, and presents a parsimonious version of the integrated model definitions identified by Jansen et al (2007). However, it is likely that the Osterwalder and Pigneur definition will be further refined as the literature matures.

Osterwalder and Pigneur (2010) focus on the value created within the model (see also Gordijn et al., 2000; Linder and Cantrell, 2000; Amit and Zott, 2001; Chesbrough, 2007; Teece, 2010). Osterwalder and Pigneur conflict with Gordijn et al.'s (2000) definition of process models, but are more in line with later work on process models which examines the relationship between process and business models. The Osterwalder and Pigneur definition is a more succinctly expressed version than Chesbrough and Rosenbloom (2002) and conforms to Jansen et al.'s (2007) definition of integrated models. On balance, the Osterwalder and Pigneur definition acts as a working definition that is in line with the use of business models as a research topic.

Business Model as Research Method

Business models as a research method offer two tools as a unit of analysis: business model representation and decomposition. In business model representation, the business model is mapped into a table or graphic representing the model. According to

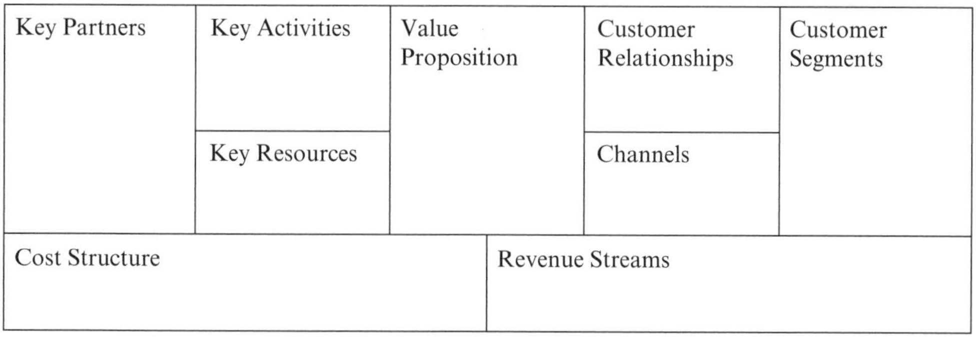

<table>
<tr><td rowspan="2">Key Partners</td><td>Key Activities</td><td rowspan="2" colspan="2">Value Proposition</td><td>Customer Relationships</td><td rowspan="2">Customer Segments</td></tr>
<tr><td>Key Resources</td><td>Channels</td></tr>
<tr><td colspan="3">Cost Structure</td><td colspan="3">Revenue Streams</td></tr>
</table>

Source: Osterwalder and Pigneur (2010).

Figure 5.1 The business model canvas

Andersson et al. (2006), business models are 'created in order to make clear who the actors are in a business case and explain their relations, which are formulated in terms of values exchanged between the actors'. Johannesson (2007) argues that business models serve as sketches for communication and as blueprints, which can then be executed. This facilitates organization and analysis of the business model. Itami and Nishino (2010) represent business models as equations. A heavily cited model is that of Osterwalder and Pigneur (2010). This model represents the business model as the 'business model canvas', as shown in Figure 5.1.

Casadesus-Masanell and Ricart (2010) argue that a business model can be represented as a whole or decomposed into different parts. Gordijn et al. (2000) argue that the decomposition of business models should focus on value-adding activities. The Osterwalder and Pigneur model allows for the business model to be broken down into nine components. These components are: key partners, key activities, key resources, value proposition, customer relationship, channels, customer segments, revenue streams and cost structure. They also note that a single firm may be a bundle of business models. This is in line with Hagel and Singer (1999), who argue that business models can be broken down into three core business types: product innovation, customer relationship management and infrastructure management. Business model representation aids in the identification and analysis of these business types.

The decomposition of the business model into components also offers a useful analytical tool. According to Casadesus-Masanell and Ricart (2010: 5), 'Some business models are decomposable, in the sense that different groups of choices and consequences do not interact with one another and thus can be analyzed in isolation. Depending on the question to be addressed, representing just a few parts of an organization's business model may be appropriate.' Baden-Fuller and Morgan (2010: 163) also suggest that business models can serve as a unit of analysis for research. They argue that 'exemplar case business models (such as McDonalds) are to management what the model organisms are to biology: real-life examples to study'. Baden-Fuller and Morgan also argue that business models can serve as blueprints and demonstrations of particular configurations. As the case studies presented in this chapter focus on change, this analysis of the business model in components is used.

A challenge for researchers is to delineate between bundled business models and identify individual models. Casadesus-Masanell and Ricart's (2010) approach allows the research question to inform this delineation by allowing the researcher to focus on the relevant, decomposed parts. However, the literature struggles with the theoretical delineation between models. Klang et al. (2010: 13) find: 'the literature review revealed no systematic and methodological grounded approach to cluster or group business model components'. They suggest that business models be delineated by the boundaries of the firm. Likewise, Lambert and Davidson (2012) note that this delineation is often applied in an industry-specific manner and continues to evolve as business models become a more developed research tool.

Osterwalder and Pigneur (2010) build on earlier scholarly work (for example Osterwalder and Pigneur, 2002; Osterwalder et al., 2005) to craft business models as a pragmatic tool for practitioners. Osterwalder and Pigneur (2002: 2) refer to the business model framework as a 'generic model with which companies can express the business logic of their firm or even the one of their competitors'. Like Osterwalder et al. (2005), this chapter uses the business model framework as a unit of analysis.

Strategy and Innovation

Business models, and innovation in business models, allow firms to innovate without investing in further resources and can lead to a competitive advantage (Amit and Zott, 2010). The literature has also examined the role of business models in promoting innovation. Particular attention has been paid to the failure of business models to adapt. Porter (2008) cites the example of retail video rental outlets, which struggle to adapt in the face of new business models based on online sales or distribution. The success of the Apple iTunes Store is cited as an example of a failure of the music business to adapt its model (Porter, 2008), in contrast to the triumph of Apple through its business model innovation (Amit and Zott, 2010).

A popular case study of the success of business model innovation is that of Xerox. The Xerox machine was initially considered too costly to achieve sufficient market penetration. Instead, as Chesbrough and Rosenbloom (2002) note, Xerox innovated its business model and chose to lease the machine. Rather than purchasing the machine, customers paid per copy, and the machines achieved significant market penetration. Eventually, however, this model became unsustainable, and Xerox switched its business focus to document and information management, as detailed in McGrath (2010).

The case studies in this chapter will show that experimentation is a key factor in innovation in business models. As McGrath (2010) notes, 'typically new models emerge when a constraint is lifted, and old ones often come under pressure when one [constraint] emerges'. She also argues that the absence of internal incentives reduces experimentation by industry incumbents. Chesbrough (2007) emphasizes the role of experimentation and cites case studies of Xerox company spin-offs and the self-distributed Radiohead album as examples of business model experimentation. He also suggests that industry incumbents may systematically ignore information that would encourage experimentation with business models. Instead, new models should develop and 'the organization's culture must find ways to embrace the new model, while maintaining the effectiveness of the current business model until the new one is ready to take over completely'.

Criticisms

Business models as a theory and strategy tool are not without their critics. Klang et al. (2010) contrast the relative enthusiasm of practitioners for business models with the criticisms by strategy researchers about business models' lack of clarity. In particular, Porter (2008) criticizes the approach as inappropriately isolating internal business structures from markets. As he argues, 'no business model can be evaluated independently of industry structure'. Focusing on the business model can distract from the larger picture that should consider competitive advantages and strategy. As noted in Teece (2010) and Leem et al. (2004), the dotcom bust of the early 2000s demonstrated the weakness of poorly executed business models, which failed to incorporate the extra-organizational market conditions. Practitioners have also noted these failings. Michael Dell, CEO of Dell Computers, as quoted in Meyer et al. (2011), noted, 'The biggest waste of money has been all the investment in companies with so-called new-economy business models. Business fundamentals haven't changed, and a lot of investors lost sight of that – and are paying for it.'

As noted in the section on definitions, further criticism of business models lies in their overlap with existing, more established bodies of theory. The value focus of business models struggles to distinguish itself from the resource focus of economic models and process focus of operational models. Klang et al. (2010) find that business models are often conflated with strategy but argue that the business model focuses on value creation whereas strategy focuses on the firm in relation to its competitors. Casadesus-Masanell and Ricart (2010) also note that, at the level of a single business model, the business model may be one and the same as the strategy. They argue, 'In the simple situations, where strategy is fully observable, strategy "coincides" with the organization's business model and little is gained from separating the two notions. An organization's business model is the reflection of its realized strategy.'

These criticisms of business models form the dialectical dialogue of point–counterpoint, which facilitates the maturation process of the business model methodology. As the literature moves towards a consensus on the definition of business models, we can anticipate that business model research will become better integrated into wider strategy and policy theory.

CHANGING BUSINESS MODELS IN THE CREATIVE INDUSTRIES

In the following, we draw on a study conducted on behalf of the UK Intellectual Property Office, which examines changing business models in the creative industries. The study is unusual in that it is a first look at the role of intellectual property (IP) in the digital media sector using business models as a research framework. It provides an insight into the use of business models in practice in the creative industries and highlights their transitive nature. This section briefly describes the methods used in the study.

Three sectors (music, television and computer games) were selected as ranking the highest by technology and digital output in the Technology Strategy Board's (2009) definitions. The firms chosen as case studies form a sample with ample variation to cover the diversity of business models. Each firm is identified via non-probability, quota sampling of firms with offices in Scotland. To capture some of the variety of the three sectors, two case studies per sector are analyzed. Data collection is conducted using a triangulation approach and includes semi-structured interviews with key personnel, participant observation and literature sources. The semi-structured interview questions were drafted using the Osterwalder and Pigneur (2010) business model framework. Following Casadesus-Masanell and Ricart's (2010) delineation approach, the business models of each firm are separated by product and distribution platform as the study focuses on digital content. The interviews illustrate the structure of the subject's business model and examine tensions around the role of digital technology, user-generated content and copyright. The interviews provided the majority of the qualitative data. Further details on the study's methods can be found in Searle (2011). A challenge to the use of business models as a research tool in qualitative studies is that these studies are highly resource-intensive and thus limited in scope; this naturally restricts the generalization of the research findings.

Illustrative Case Studies

Innovation in digital technologies has brought about rapid change for the creative media industries in particular. Where media were once bound to physical production and distribution formats, digital technology allows for inexpensive reproduction and distribution of media. This represents a process of Schumpeter's creative destruction as new opportunities emerge and older business models suffer. The creative media industries are struggling to adapt to the new regime. The classic illustration of this is that of music. As one interviewee put it, 'music is what brought people online'. The music industry has seen steep decline in sales of physical product that has not yet been matched by increased digital sales. Traditional music recording labels have lost market share to new entrants. Other sectors, such as television, which is less reliant on direct sales to consumers, have also undergone a revolution in the distribution of content and consumer expectations. Computer games, a sector borne by digital technology, are also changing from product-based business to focus on service. Collectively, these three sectors illustrate some of the challenges and transformations of the digital creative industries.

Music

That the business models of the music sector are the first in digital media impacted by online distribution should come as no surprise. The music industry has relatively small file sizes and unmet demand for singles, and consumer demand for digital music files was initially met through unlicensed services such as Napster. The music sector traditionally relied on music sales as physical sales of bundled goods in an album format. The transition to single-track sales and other models in the digital world has been problematic as the established industry struggles to compete with unlicensed, free services. As Porter (2008) notes, this change has allowed Apple, an industry outsider, to grow from launching an online music store in 2000 to claim 80 percent (Smith, 2005) of the digital market in Britain and 27 percent (Schramm, 2010) of all music sales in the US in 2010. The industry is currently experimenting with different, yet-unproven business models and attempting to expand areas such as merchandising and ticket sales.

Clash Music is a publishing company that operates in both the music and the publishing sector. As a case study, it offers an example as to how music can be leveraged into other media to generate revenues. Clash Music bundles three business models: music journalism, brand marketing and media partnerships. The firm expanded from its core music journalism business model into Clash Media, which develops and produces promotional events and content for its brand clients. The firm also curates events and is a media partner under the Clash Events umbrella. The key theme throughout these three business models is a strong brand image with magazine readers based on expertise in music and associated sectors. The firm leverages music into other formats to generate revenue and build its brand. This model, which is profitable and continues to evolve, may represent a future model for the music sector.

Heist Records is primarily the story of its owner, Ged Grimes. Grimes began his music career in what he described as a typical career path for recording artists in the 1980s when he signed with a major label that developed, managed, marketed and distributed his recorded music. He now focuses more on revenue from live performances and licenses (syncing). Grimes, who since the time of the study has begun touring with the band Simple Minds, has a second business model developing new musicians. This is a relatively new model for Heist Records and, as Grimes notes, 'Where and when to charge for music remains a question. The Internet gives fans a direct experience with artists.' In a third business model, Heist Records produces music for computer games, which forms another revenue stream and promotes the Heist record label. The evolution of Heist Records' business models mirrors that of the music industry as a whole. As direct sales of recorded music decline, the industry seeks to diversify its revenue streams through merchandising, ticket sales and syncing. However, a steady state has not yet been reached as the industry experiments with these new models.

Television

The television sector has been marked by an increase in the modes of distribution and increased competition for advertising funds. The digital age has widened the choice of screens for consumers, as content can now be consumed on TVs, computers and mobile telephones in streamed or downloaded formats. In the UK, television is funded primarily through advertising revenue and government license fees.

Tern TV is an independent television production company based in Glasgow. It focuses on two business models: television production for broadcasters and digital content production. Tern's core business model lies in the production of lifestyle and factual content for traditional broadcasters such as BBC and Channel 4. As dictated by the Communications Act 2003, Tern retains the residual IP rights of its commissioned work. This change has shifted the distribution of funding and revenues in the commissioning process. As Tern now owns these rights, up-front production payments from commissioners have decreased, as the firm is expected to profit from revenues from international distribution or merchandising. Tern's second business model takes its core competency in storytelling and applies it to the digital realm via online companion websites for television programs and other digital content. This business model is moving further away from television and into original digital content as, after the research study, the firm rebranded in 2012 as The Story Mechanics.

The second case study in television is the commissioning and online content activities of the BBC. Unique in the UK television sector, the BBC is a fully publicly funded broadcasting company, whose mission is to 'inform, educate and entertain'. Instead of being profit-driven, the BBC business model focuses on its public-service nature and seeks to be the 'most creative organization in the world'. BBC bundles many business models, which include broadcast television and online content. As the BBC is free at the point of consumption and funded by license fees, its pricing mechanism and business model are unusual in the television sector.

Changes in business models in the television sector are dominated by the ability to broadcast content in new formats and changes in consumer viewing patterns. The shift to digital consumption of television opens up new possibilities for both consumers and producers of this content. Tern TV and the BBC represent two types of firms in the sector, independent producers and broadcasters, and continue to adapt their business models to incorporate the opportunities offered by digital technologies.

Computer Games

Computer games, as the name of the sector indicates, is a digitally native sector (it exists only in the digital era). This relatively young sector exhibits extensive business model innovation in order to achieve profitability. Until recently, a common business model for computer games was based on developers working with publishers to produce boxed product games that were sold to consumers in retail shops. However, as the Internet has evolved, computer games are switching to a more service-based model in which games are distributed online, with and without a publisher, and pricing mechanisms have moved away from single purchase and towards subscription, tiered or micropayment (smaller, one-off payments) structures.

Dynamo games, founded in 2004, encompasses a number of business models found in the computer games industry. Its original business model is that of developer for larger publishers in which Dynamo is commissioned to develop games by a publisher that sells them to retailers or directly to consumers. Dynamo's other two business models involve self-publishing in which the publisher/commissioner role does not exist and the firm sells to consumers through retailers. In self-publishing for mobile games, Dynamo develops products for mobile devices that are sold under the freemium pricing mechanism where a limited version of the game is free and consumers can pay for more expanded versions. Dynamo's final model, self-publishing for online games, is a service-based structure in which games are available on Facebook under the freemium model. These business models are not self-publishing in a strict sense, as online and mobile retailers exert partial editorial control and dictate some terms of sale.

In contrast, YoYo Games has two business models: game development software developer and games publisher. The firm develops and sells the GameMaker software, which is a software aimed at amateur games programmers. Consumers then upload their games to a website where the GameMaker community interacts and evaluates uploaded games. The firm earns money from the sales of the GameMaker software. Its second business model is that of publisher of games developed by GameMaker users. Using the community to identify the best games, YoYo Games licenses these games from the GameMaker user, develops them further and then sells them via mobile device online stores such as iTunes, PlayStation Portable and Android stores. The YoYo business model is unusual in that it utilizes user-generated content from one business model to create a second business model.

Like the music industry, the computer games industry has seen big changes in the sales of boxed product in retail stores. As specialized stores lose market share to supermarkets and online marketplaces, the business models of the sector change. The relative youth of the computer games sector may allow it to avoid the sunk cost fallacy of other sectors and continue to innovate its business model; equally there may be intrinsic differences in the computer games market that afford it this flexibility.

Discussion

The analysis of the case studies of business models in the creative media industries presents three main themes: first, business models are still changing; second, differences between sectors exist; and, third, the role of intermediaries appears to be changing.[2] It should be stressed that the sectors in question as well as the business models themselves are far from stable. All of the case study firms have business units less than five years old. Since the time of data collection, they have continued to experiment with their business models. As new technologies emerge and consumer demand changes, so too will the business models of the creative industries. Researchers, practitioners and policy makers should expect data to date quickly.

The study also suggests that, while some parallels can be drawn between creative sectors, business models suitable for one sector of creative media may not be suitable for others, as technology, consumer expectations and production processes differ. At the same time, the lines between sectors are blurring, as evidenced by the expansion of Tern TV and Clash Music into sectors beyond their origins. Consumption of music, television and computer games offers different marginal utility for the consumer; where one track of music may be enjoyed by a consumer repeatedly, the same may not be true for an episode of a television series. This difference in consumer expectations and the limitations of technology platforms (for example broadcast television being relatively inflexible compared to online platforms) influence the ability of firms to adjust the pricing mechanisms of content. The case studies also suggest that some sectors, particularly computer games, are more readily able to adapt their content to new technology platforms. Thus, the pace and development path of business models in each sector may occur in different manners in response to the method of consumption and behavior of their audience.

Finally, a key theme is that the role of intermediaries in the digital media market is also changing. As noted earlier, music is an obvious example of how an industry outsider, Apple's iTunes, has now become a dominant player. At the same time, these new platforms, which include Facebook, Amazon and Google Android, take 30 percent (Searle, 2011) of revenues from sales of content through their platform. As new firms displace old in the process of innovation, the business models of the creative media industries must adapt.

CONCLUDING REMARKS

This chapter has presented an overview of business models as a research topic illustrated with application of the theory via case studies of digital media firms. In both the discussion of the literature and the case studies, the challenges associated with the on-going development of business models and their application have been stressed. As a practice, business models provide a useful analytical tool in sectors like the digital media industries. Characterized by disruption and innovation, business models in the media sector allow firms to adapt to changes in technologies, markets and consumer behavior, and researchers to study these changes.

Business models are an emerging research topic in the creative industries as the advent of the digital era disrupts traditional business models. However, as both the theory of business models and their application evolve, research in the area remains a challenge. The case studies presented in this chapter provide a snapshot of the business models of these firms. However, as we have discussed, even these business models have changed since the time of data collection. The business model literature review and study presented here suggest that further work can be done to advance both the understanding of business models and their practice in the creative industries.

NOTES

1. For comprehensive literature reviews on business models, see Zott et al. (2010) and the special issue on business models of the journal *Long Range Planning* (2010).
2. Note that the original study on which this chapter is based also considered IP; here we focus more on the business model.

REFERENCES

Amit, R. and C. Zott (2001), 'Value creation in e-business', *Strategic Management Journal*, **22**, 493–520.

Amit, R. and C. Zott (2010), 'Business model innovation: creating value in times of change', IESE Business School, University of Navarra Working Paper WP-870, July.

Andersson, B., M. Bergholtz, A. Edirisuriya, T. Ilayperuma, P. Johannesson, J. Gordijn, B. Greìgoire, M. Schmitt, E. Dubois, S. Abels, A. Hahn, B. Wangler and H. Weigand (2006), 'Towards a reference ontology for business models', in D.W. Embley, A. Olivei and S. Ram (eds), *Conceptual Modeling – ER 2006*, Lecture Notes in Computer Science, vol. 4215, Berlin: Springer, pp. 482–96.

Baden-Fuller, C. and M.S. Morgan (2010), 'Business models as models', *Long Range Planning*, **43** (2–3), 156–71.

Casadesus-Masanell, R. and J. Ricart (2010), 'From strategy to business models and onto tactics', *Long Range Planning*, **43** (2–3), 195–215.

Castronova, E. (2005), *Synthetic Worlds*, Chicago: University of Chicago Press.

Chesbrough, H. (2007), 'Business model innovation: it's not just about technology anymore', *Strategy and Leadership*, **35** (6), 12–17.

Chesbrough, H. and R. Rosenbloom (2002), 'The role of the business model in capturing value from innovation: evidence from Xerox Corporation's technology spin-off companies', *Industrial and Corporate Change*, **11** (3), 529–55.

De Vany, A. (2004), *Hollywood Economics: How Extreme Uncertainty Shapes the Film Industry*, New York: Routledge.

Edirisuriya, J. and P. Johannesson (2009), 'On the alignment of business models and process models', in D. Ardagna, M. Mecella and J. Yang (eds), *Business Process Management Workshops: BPM 2008 International Workshops*, Lecture Notes in Business Information Processing, vol. 17, Berlin: Springer, pp. 68–79.

Gordijn, J., H. Akkermans and H. van Vliet (2000), 'Business modeling is not process modeling', in S.W. Liddle, H.C. Mayr and B. Thalheim (eds), *Conceptual Modeling for E-Business and the Web*, Lecture Notes in Computer Science, vol. 1921, Berlin: Springer, pp. 40–51.

Hagel, J. and M. Singer (1999), 'Unbundling the corporation', *Harvard Business Review*, **2000** (3), 148–61.

Itami, H. and K. Nishino (2010), 'Killing two birds with one stone: profit for now and learning for the future', *Long Range Planning*, **43** (2–3), 364–9.

Jansen, W., W. Steenbakkers and H. Jagers (2007), *New Business Models for the Knowledge Economy*, Farnham: Gower Publishing.

Johannesson, P. (2007), 'The role of business models in enterprise modeling', in J. Krogstie, A.L. Opdahl and S. Brinkkemper (eds), *Conceptual Modelling in Information Systems Engineering*, Berlin: Springer, pp. 123–40.

Klang, D., M. Wallnofer and F. Hacklin (2010), 'The anatomy of the business model: a syntactical review and research agenda', DRUID Summer Conference 2010, Imperial College London Business School, available at: http://www2.druid.dk/conferences/viewpaper.php?id=501874&cf=43.

Lambert, S. and R. Davidson (2012), 'Applications of the business model in studies of enterprise success, innovation and classification: an analysis of empirical research from 1996 to 2010', *European Management Journal*, available at: http://dx.doi.org/10.1016/j.emj.2012.07.007.

Leem, C.S., H.S. Suh and D.S. Kim (2004), 'A classification of mobile business models and its applications', *Industrial Management and Data Systems*, **104** (1), 78–87.

Linder, J. and S. Cantrell (2000), 'Changing business models: surveying the landscape', Accenture, available at: course.shufe.edu.cn/jpkc/zhanlue/upfiles/edit/. . ./20100224120954.pdf.

McGrath, R. (2010), 'Business models: a discovery driven approach', *Long Range Planning*, **43** (2–3), 247–61.

Meyer, M., N. de Crescenzo and B. Russell (2011), 'In search of a viable business model', in M. Meyer and F. Crane, *Entrepreneurship: An Innovator's Guide to Startups and Corporate Ventures*, Thousand Oaks, CA: Sage, pp. 379–93.

Osterwalder, A. and Y. Pigneur (2002), 'An e-business model ontology for modeling e-business', 15th Bled Electronic Commerce Conference e-Reality: Constructing the e-Economy.

Osterwalder, A. and Y. Pigneur (2010), *Business Model Generation: A Handbook for Visionaries, Game Changers, and Challengers*, Hoboken, NJ: John Wiley & Sons.

Osterwalder, A., Y. Pigneur and C. Tucci (2005), 'Clarifying business models: origins, present and future of the concept, submitted to CAIS', Communications of the Association for Information Systems.

Passman, D. (2001), *All You Need to Know about the Music Business*, London: Penguin Books.

Porter, M. (2008), 'On competition', *Harvard Business Review*, **86** (1), 78–93.

Schramm, M. (2010), 'iTunes share of the US music market swells to 26.7%', *Tuaw*, 24 May, available at: http://www.tuaw.com/2010/05/24/itunes-share-of-the-us-music-market-swells-to-26–7/.

Searle, N.C. (2011), 'Changing business models in the creative industries', United Kingdom Intellectual Property Office, available at: http://www.ipo.gov.uk/ipresearch-creativeind-full-201110.pdf.

Sickels, R. (ed.) (2008), *The Business of Entertainment*, vol. 1: *The Movies*, Westport, CT: Praeger.

Smith, T. (2005), 'Apple touts iTunes' UK 80% market share', *The Register*, 7 September, available at: http://www.theregister.co.uk/2005/09/07/apple_responds_to_rivals/.

Technology Strategy Board (2009), 'Creative industries, technology strategy 2009–2012', available at: http://www.innovateuk.org/_assets/pdf/creative%20industries%20strategy.pdf.

Teece, D. (2010), 'Business models, business strategy and innovation', *Long Range Planning*, **43**, 172–94.

Zott, C., R. Amit and M. Massa (2010), 'The business model: recent developments and future research', available at: http://ssrn.com/abstract=1674384.

FURTHER READING

For further reading on business models, Osterwalder and Pigneur's (2010) book on *Business Model Generation* is a good starting point. Key academic papers are Zott et al. (2010) and the special issue on business models in the academic journal *Long Range Planning*, published in 2010. We have noted that, since the time of the study in 2011, interest in business models as a research topic continues to increase, and readers will likely find more recent literature reviews.

For sector-specific reading, Passman (2001 and later editions) provides a comprehensive look at the contractual infrastructure of the music industry and examines revenue streams, licensing practices and contracts. The literature on the economics of computer games is less developed than that of other sectors. The academic literature in particular is relatively sparse, but readers may find the work of Edward Castronova useful (for example Castronova's *Synthetic Worlds*, 2005). For television and film, Sickels's (2008) three-volume *Business of Entertainment* has interesting insights. De Vany's (2004) *Hollywood Economics* contains an excellent economic analysis of the film industry. While these authors do not explicitly deal with business models, they discuss business model components and strategy that can be read with a business model framework in mind.

6. Dynamic competition and ambidexterity

Hans van Kranenburg and Gerrit Willem Ziggers

Many firms operating in the creative industries are confronted with dynamic competition. The dynamic competition in many creative industries can be explained by worldwide deregulation and privatization trends that have contributed to a decrease in natural monopoly structures (Daidj, 2011). However, the key driver of dynamic competition over time is technological innovation (Evans and Schmalensee, 2002). Owing to the technological development, the structure of those creative industries has changed fundamentally (see, for example, Greenstein and Khanna, 1997; Kranenburg et al., 2001; Pennings and Puranam, 2001; Stieglitz, 2003; Christensen, 2011). The convergence of digital and interactive (information and communication) technologies accelerates the erosion of existing industry boundaries. This development raises the question of how competition and structural changes in the creative industries are accompanied by changes in industry boundaries and how firms can cope with these changes and forces of competition. In this chapter we start with a discussion on dynamic competition and industry convergence and how they affect the business landscape. Next we will discuss how firms can build a competitive position in a dynamic and convergent environment by addressing the strategy of ambidexterity and critical capabilities for firms to operate in the changing business landscape.

DYNAMIC COMPETITION AND CONVERGENCE IN THE CREATIVE INDUSTRIES

The creative industries refer to a range of economic activities which are concerned with the exploitation, exploration, and generation of knowledge and information (Hesmondhalgh, 2002). According to Howkins (2001) the creative industries comprise advertising, architecture, art, crafts, design, fashion, film, music, performing arts, publishing, R&D, software, toys and games, TV and radio, and video games. The education industry, including public and private services, can also be considered as part of the creative industries. Many of these industries are undergoing rapid technological change in which their boundaries are blurring. The development of new media has even speeded up the blurring of the boundaries and the convergence of different creative industries into one. Most technologies described as 'new media' are digital, often having characteristics of being manipulated, networkable, dense, compressible, and interactive. Some examples may be the Internet, websites, computer multimedia, video games, CD-ROMS, and DVDs. The great technological progress including new media upsets established market structures. For instance, Palmberg and Martikainen (2006) showed the ongoing integration of the Internet technology (IT) industry and the telecommunications manufacturing industry in Finland. Kranenburg and Hagedoorn (2008) documented similar patterns in the European telecommunications providers industry. Kranenburg et al. (2001)

investigated the integration of new media in the traditional publishing industry. They found that the publishing industry integrated new media technologies that were needed for Internet, e-business, and other electronic products and services.

In these creative industries, firms engage in dynamic competition for the market usually through research and development (R&D), competition to develop the 'killer' product, service, or feature that will confer market leadership and thus diminish or eliminate actual or potential rivals. Schumpeter (1950: 84) described dynamic competition centred on drastic innovation as the 'perennial gale of creative destruction' that changed the old market structure and order. Moreover, dynamic competition stimulates firms to find ways to invent new or better products and/or to identify cost savings through better processes or technologies. It lies at the heart of technological and economic progress and has more long-run effects than static competition, which mainly focuses on prices (Microeconomix, 2010). The key ingredient of competition over time is technological innovation. Firms attempt to gain a competitive advantage by either reducing costs or introducing new products. A new product which can do more things – or do the same things faster, more cheaply, or more safely – will replace inferior earlier products. The innovating firm will dominate the market at least until the next innovation comes along. In these dynamic competitive markets, temporary market power is inevitable – and indeed provides the incentive for firms to innovate. They compete to be the first to become or stay the market leader.

Within dynamic competitive industries, market leaders are generally contestable as a result of the constant threat of drastic innovations by firms in the industry or firms from other industries (Baumol, 1982). Even creative industries in which there is a limited number of competitors may behave competitively because of the discipline provided by potential competition. Hence, to assess the force of competition in creative industries, it is important to look not only at the number of existing firms but also at potential entrants or actual and potential innovative threats to leading firms, and in most instances there are many of these potential entrants or innovative threats (Stiglitz, 1987). Innovative threats generally involve technological and design approaches that differ radically from those by the incumbent (Evans and Schmalensee, 2002).

Nowadays, many creative industries show Schumpeterian features. Incumbent firms fear the threat that another incumbent firm or known or unknown new entrants will come up with a drastic innovation that causes demand for the incumbent's product to collapse. The potential threat forces firms in the creative industry with dynamic competition to bring out continuously new versions of old products or to bring out entirely different products that eliminate the demand for the old product. According to Evans and Schmalensee (2002) these threats also generally constrain the prices charged by incumbents. Furthermore, dynamic competition in the creative industries also increases with greater opportunities for combining features and services that were previously available separately to create products that are differentiated from existing offerings. The tying, bundling, and integration of services and products benefits consumers substantially, although it may destroy markets for previously separate products. For instance, product integration has been a major force in the computer, consumer electronics, information and communications, and telecommunications manufacturing industries in the last 15 years (Chan-Olmsted and Jamison, 2001). A current example is the integration of services and products in smartphones. A smartphone is built on a mobile operating system, with more advanced computing capability and connectivity than a feature mobile phone. Nowadays, smartphone manufacturers have integrated the functions of a personal digital assistant (PDA) within the mobile phone, the functionality of portable media players, low-end compact digital cameras, pocket video cameras, and GPS navigation units to form one multi-use device. The rapid development of mobile applications (so-called apps) markets and of mobile commerce has speeded up the smartphone adoption.

Hence, the market for standalone products has now almost entirely disappeared in the creative industries. Among the creative industries a convergence is still ongoing. Eventually only one or a few creative industries will remain. The studies on convergence emphasized different types of convergence. For instance, Greenstein and Khanna (1997) distinguish two types of convergence, substitution-based and complementary-based convergence. Convergence in substitutes occurs when different products share similar features and provide the same functions to end users, substituting each other. Convergence in complements occurs when previously unrelated products can be used together to create higher utility to customers or bundled into new products providing added value to end users. Pennings and Puranam (2001) expanded this typology by introducing demand-side and supply-side convergence. Demand-side convergence is related to the needs of customers. It happens when growing similarity of needs across different customer groups or markets occurs. In other words, the demand of customers becomes more homogeneous across different groups or markets. This type of demand-side convergence is part of substitution-based convergence. However, when customers at a one-stop shop try to obtain a product satisfying different needs, in other words product offering is preferred, then this type of convergence is part of complementarities. The supply-side convergence is also divided into two sub-categories. The first category is called technological substitution. It occurs when a technology overlaps and offers the same benefits as an existing technology. The other category is called technology bundling and occurs when different technologies are bundled to produce or improve new or existing products.

Although each type of convergence may occur autonomously, they are often mutually interlinked. For instance, the convergence in the information and multimedia industries is reflected in activities linked with one of the following three processes: (1) content production-related services (for example publishing, movies, gaming, and broadcasting); (2) content delivery-related services (for example glass fibre and cable); and (3) data processing services (for example software and programming). The changing technological realities and control of markets are illustrated by interactive multimedia production and delivery of content available via networks instead of the single media frameworks of the past (Daidj, 2011). Convergence itself does not cause any revolutionary change in content, but creates new economies of scope that permit the existing communication and distribution of content to be faster, more flexible, and more responsive to customer demand. Economies of scope depend also on the specificity of resources and capabilities and their transferability across industry boundaries (Picard, 2000; Borés et al., 2003). The result is a reconfiguration of the entire industry value chain with new and incumbent players (Wirtz, 2001). However, many incumbents have difficulty coping with these changes and the dynamic competition in their industry. An interesting example is the introduction of the smartphone by the computer company Apple, which changed the boundaries of the telecommunications industry. With the introduction of its smartphone, Apple caught up and passed the traditional manufacturers like Nokia and HTC in the telecommunications industry. These firms lost their leading and competitive positions within a few years and are now struggling to survive. They need to transform or reinvent themselves according to the new rules of the game or even set new rules.

THE STRATEGY OF AMBIDEXTERITY

The need for firms to redefine their core business and to acquire new capabilities means that firms must actively work to disrupt their own advantages and the advantages of competitors by continuously challenging existing capabilities. It implies a tension of how both to exploit existing business and markets and to explore new ones. This distinction between exploitation and exploration has long been addressed in terms of conflicting structures designed for efficiency and those designed for innovations (D'Aveni and Gunther, 1994; He and Wong, 2004). The former involves a continuous search for improvement along a fixed product offering, while the latter requires discontinuous shifts from one product offering to another that is more profitable. Both capture a number of fundamental differences in firm behaviour and strategy that have significant consequences for performance (March, 1991). On the one hand, adaptation to existing market demands may foster structural slowness and reduce a firm's capacity to adapt to future environmental changes and new opportunities (Hannan and Freeman, 1984). It implies a firm which is only incrementally adapting its existing business model, emphasizing process efficiency and effectiveness. On the other hand, experimenting with new alternatives reduces the speed at which existing capabilities are improved and refined (March, 1991). It may disrupt successful routines in a firm's existing markets to the point that it cannot compensate for the loss in the existing business (Mitchell and Singh, 1992). The challenge for firms is to develop and to incorporate new business models to fulfil new requirements and demands. This approach enables the firm to be really innovative and to develop new capabilities and resources to sustain its competitive position. For instance, Kranenburg and Ziggers (2013) emphasized that media firms have to rethink their traditional manner of revenue generation, the structure of the organization, the core competencies of the organization, and the way of creating value through new opportunities. They should redesign their business models or should develop an innovation-centred business model for the exploration of new opportunities and businesses. It is a transformation of the firm into an ambidextrous structure that facilitates firms to develop a capability both to compete in mature markets (emphasizing costs, efficiency, and incremental innovations) and to develop new products or services for emerging markets (emphasizing experimentation, speed, and flexibility) (Benner and Tushman, 2003). Ambidexterity in its most general definition refers to a company's ability to simultaneously or sequentially engage in two seemingly contradictory activities (Gupta et al., 2006). This can be achieved by creating separate structures for different types of activities (O'Reilly and Tushman, 2004) or by calling upon individuals to make choices in their day-to-day work (contextual ambidexterity). Han (2005) pointed out that the building of an ambidextrous organizational structure provides the platform for the organizational architecture that facilitates operations and increases efficiencies and innovation. The potential of ambidexterity is supported by the empirical findings of

He and Wong (2004), although their findings also suggest that there are limits to ambidexterity because both exploration and exploitation compete for scarce resources and require different strategies, and structure trade-offs have to be made (Smith et al., 2010). Inherently this implies tensions, and ambidexterity may become unmanageable when pushed to the extreme.

MARKET SENSING DYNAMIC CAPABILITY

To become an ambidextrous organization, to shape value chains, and to obtain sustainable competitive advantage in a dynamic competitive environment, dynamic capabilities are required that create, adjust, and keep a relevant stock of capabilities. The dynamic capability is the company's ability to purposefully create, extend, or modify its resource base (Helfat, 1997; Teece et al., 1997; Eisenhardt and Martin, 2000). A dynamic capability is a deeply embedded set of skills and knowledge exercised through a process enabling a company to stay synchronized with market changes and to stay ahead of competitors. It entails the capabilities that enable organizational fitness and that help to shape the environment advantageously (Day, 2011). Its main functions are: (1) sensing environmental changes that could be threats or opportunities, by scanning, searching, and exploring across markets and technologies; and (2) responding to the changes by combining and transforming available resources through partnerships, mergers, or acquisitions. The first type of dynamic capabilities are known as market sensing capabilities and the latter type are known as relational capabilities. These dynamic capabilities will help the firms to select the value chain model configuration for delivering value and capturing revenues. Dynamic capabilities in market sensing comprise the function of marketing intelligence and coordination of internal business processes to act quickly and effectively in response to information gathered from customers and other external stakeholders (Kohli and Jaworski, 1990; Narver and Slater, 1990). It entails a behavioural shift from a reactive to a sense-and-respond approach amplified with emerging technologies for seeking patterns and sharing insights quickly. But how can firms in dynamic competitive creative industries learn to make sense out of an increasingly volatile and unpredictable market? Drawing on complexity literature and organizational mindfulness literature may provide an answer. Complexity theory demonstrates that all successful adapting systems transform apparent noise into meaning faster than the apparent noise comes to them (Haeckel, 1999). This means that particularly adaptive firms have mastered a vigilant market sensing capability. Literature on organizational mindfulness (Fiol and O'Connor, 2003; Levinthal and Rerup, 2006) has addressed the capability of early sensing. Mindfulness is a heightened state of awareness, characterized by curiosity, alertness, and a willingness to act on partial information. Firms operating in creative industries with dynamic competition should develop a robust market sensing capability. They should develop an orientation towards the market, and, by means of multiple inquiry methods, ambiguous signals should be thoroughly analysed to learn more about promising patterns and new opportunities. They should use scenario thinking to consider multiple possible futures and should develop a high tolerance for ambiguity along with the ability to pose the right questions to identify what they do not know. Also these companies should develop a transparent organization avoiding organizational filters, and allowing them to notice threats or opportunities detected by someone deep in the organization or the network of partners. To develop a market sensing capability in a dynamic competitive environment requires learning, which entails a willingness to focus on: (1) past, present, and prospective customers and how they process information and respond to social networking and social media space; (2) an open-minded approach to new developments, needs, and technologies; and (3) an ability to sense and act on weak signals from the dynamic competitive business environment. It is the difference between testing copy versions with controlled experiments and continuously scanning the market for ideas, concepts, and formulations that are working or failing (Day, 2011). This learning from markets is not fully effectuated until the findings are accurately interpreted and adequately shared throughout the firm as well as among business partners. Similar to ambidexterity, developing a market sensing capability is problematic. Managers may misinterpret what they see in favour of what they want to see or ignore the results that challenge prevailing knowledge, and/or may not share information with their business partners. As a consequence, new market initiatives may fail.

RELATIONAL DYNAMIC CAPABILITY

The second type of dynamic capabilities for firms operating in the creative industries with dynamic competition are relational capabilities. Responding to market turbulence requires these firms to collaborate with network partners. This puts a premium on relational capabilities that extend the firm's resource base beyond firm boundaries and enable access to partners' resources (Dyer and Singh, 1998) and allow it to compete more effectively in the marketplace (Day, 1995; Ziggers and Henseler, 2009). A company that possesses relational capabilities places a high priority on present and prospective inter-organizational relationships and has advanced several skills. First, it possesses the skills to monitor and identify partnering opportunities. Those skills represent a deliberate and conscious investment to monitor the business environment that allows firms to have a more precise view on the kind of partners or resources from which to generate revenues and to form partner relationships with them (Dyer and Singh, 1998; Gulati, 1999). Second, it is able to systematically integrate strategies, to synchronize activities, and to disseminate knowledge across its network of partners. To leverage the uniqueness of its network, to combine the respective resources available, and to generate new capabilities that may be required, coordination skills become critical (Kandemir et al., 2006). If it is done properly the firm can exploit its competitive advantages more completely. Third, the ability to handle unforeseen contingencies in partner interactions requires partnering learning skills. This includes the internalization of direct experiences, successes, and failures when collaborating with other firms (Lyles, 1988) and the diffusion of learning effects across a network of partners (Kandemir et al., 2006). A means to this end are inter-organizational team meetings to exchange information about the progress and problems with regard to the collaboration. Finally, a firm has to emphasize the development of capable alliance managers who are capable of managing the key activities at every stage of the partnering life-cycle (for example Day, 1995; Kale and Singh, 2007). This requires a deliberate and conscious investment in training and educating alliance managers. If done properly, managers will be able to broker partner relationships so that partners develop and transfer knowledge that facilitates the pursuit of commercial opportunities. Nevertheless, the effective management of inter-organizational relationships is a complex endeavour. The main reason is the huge control and coordination problems that have to be dealt with when managing a network of businesses, including monitoring, accountability, and conflicts of interests. Relational capabilities enable to manage such networks effectively. As they are difficult to learn and far more difficult to copy, they can contribute to gaining advantage over competitors. Therefore firms operating in the dynamic competitive creative environment which have the right conditions and commitment will seize the best partners and succeed. Hence, to survive in the creative industries with dynamic competition, firms need to develop these two types of dynamic capabilities. For instance, media firms must control four main core resources and capabilities: creation of content; exclusive access to content; experience with marketing; and access to distribution channels (Daidj, 2011). Content access control is a key issue (Peltier, 2004). Content represents a scarce resource and a source of value for traditional media (books, newspapers, TV channels) as well as for new media (Internet, video games) or upstream vertical integration operations. Access to distribution networks for content also constitutes a resource, but the Internet favours media content owners rather than media content distributors: media companies can sell directly to consumers over the web, bypassing the cable, telecom, or satellite intermediary, which is therefore less critical (Daidj, 2011). However, it has to be noticed that particularly firms operating in creative industries with dynamic competition the capabilities and resources needed to compete are highly proprietary: Amazon has no chance of entering the market for local access provision or competing with Vodafone. eBay cannot compete with Cisco, and Nokia cannot compete with Amazon (Christensen, 2011). This means that resources or capabilities are non-imitable, mostly tacit, and bound within the firm. Therefore strategic moves have to be preceded by a thorough analysis of the internal resources and capabilities and the needed capabilities in the market. Moreover, an obvious strategy and development to enter adjacent markets are alliances and network building. Good examples are Apple, Google, and Microsoft, which signed agreements with different partners belonging to the ICT sector, but also with partners from the automotive and banking industry. These relationships concern research, production, marketing, and knowledge sharing, and in the process the firms are more and more involved in a variety of business networks (Daidj, 2011). As firms are embedded in these networks, their performance and survival are affected by all the member firms. This puts high demands on the critical market sensing and relational capabilities, especially when firms work cooperatively and competitively to support new products, satisfy customer needs, and eventually incorporate the next round of innovations (Moore, 1993).

CONCLUSION

Many firms in the creative industries are confronted with dynamic competition in their markets. These industries are showing Schumpeterian features. New techniques, new approaches, new technologies, and new competitors are changing the rules of the game. Consequently, all markets are subject to radical innovations, new requirements and demands, new competitors, and increasing complexity. That requires incumbent firms to respond and to adapt their organization, which depends on the competitive situation of the firms, the commitment and leadership within the company, and the ability to develop essential capabilities. Companies should embrace ambidexterity.

Many firms operating in the creative industries with dynamic competition have difficulty coping with the dynamic competition. Today's business reality is that firms have to be innovative and have to disrupt their own existing advantages by continuously challenging existing resources and capabilities. Using and entering business networks within and outside their industry allow firms to retain and sustain their competitive advantage, particularly because in creative industries with dynamic competition the capabilities and resources needed to compete are highly proprietary. Business networks connected through platforms allow firms to come up with tailored programs, mass customization, and adaptive and innovative economic models. However, these business networks will be highly dynamic, as they involve competition and cooperation, which induce new rounds of innovation and reconfiguration. Ambidexterity and market sensing and relational capabilities will increase firms' capacity to adapt to future environmental changes and new opportunities in the creative industries with dynamic competition.

REFERENCES

Baumol, W. (1982), 'Contestable markets: an uprising in the theory of industry structure', *American Economic Review*, **72** (1), 1–15.

Benner, M. and M. Tushman (2003), 'Exploitation, exploration, and process management: the productivity dilemma revisited', *Academy of Management Review*, **28** (2), 238–56.

Borés, C., C. Saurina and R. Torres (2003), 'Technological convergence: a strategic perspective', *Technovation*, **23** (1), 1–13.

Burgelman, R.A. and A.S. Grove (1996), 'Strategic dissonance', *California Management Review*, **38** (2), 8–28.

Chan-Olmsted, S. and M. Jamison (2001), 'Rivalry through alliances: competitive strategy in the global telecommunications market', *European Management Journal*, **19** (3), 317–31.

Christensen, J.F. (2011), 'Industrial evolution through complementary convergence', *Industrial and Corporate Change*, **20** (1), 57–89.

D'Aveni, R.A. and R. Gunther (1994), *Hypercompetition: Managing the Dynamics of Strategic Maneuvering*, New York: Free Press.

Daidj, N. (2011), 'Media convergence and business ecosystems', *Global Media Journal*, **11** (19), 1–13.

Day, G.S. (1995), 'Advantageous alliances', *Journal of the Academy of Marketing Science*, **23** (4), 297–300.

Day, G.S. (2011), 'Closing the marketing capabilities gap', *Journal of Marketing*, **75** (4), 183–95.

Dyer, J.H. and H. Singh (1998), 'The relational view: cooperative strategy and sources of interorganizational competitive advantage', *Academy of Management Review*, **23** (4), 660–79.

Eisenhardt, K.M. and J.A. Martin (2000), 'Dynamic capabilities: what are they?', *Strategic Management Journal*, **21** (10–11), 1105–21.

Evans, D.S. and R. Schmalensee (2002), 'Some economic aspects of antitrust analysis in dynamically competitive industries', *Innovation Policy and the Economy*, **2**, 1–49.

Fiol, C.M. and E.J. O'Connor (2003), 'Waking up! Mindfulness in the face of bandwagons', *Academy of Management Review*, **28** (1), 54–70.

Greenstein, S. and T. Khanna (1997), 'What does industry convergence mean?', in D. Yoffie (ed.), *Competing in the Age of Digital Convergence*, Boston, MA: Harvard Business School Press, pp. 201–26.

Grove, A.S. (2009), *Only the Paranoid Survive: How to Exploit the Crisis Points That Challenge Every Company*, New York: Currency Doubleday.

Gulati, R. (1999), 'Network location and learning: the influence of network resources and firm capabilities on alliance formation', *Strategic Management Journal*, **20** (5), 397–420.

Gupta, A.K., K.G. Smith and C.E. Shalley (2006), 'The interplay between exploration and exploitation', *Academy of Management Journal*, **49** (4), 693–706.

Haeckel, S.H. (1999), 'Winning in smart markets', *Sloan Management Review*, **41** (1), 7–8.

Han, M. (2005), 'Towards strategic ambidexterity: the nexus of pro-profit and pro-growth strategies for the

sustainable international corporation', paper presented at the JIBS Frontiers Conference, Rotterdam, The Netherlands.
Hannan, M.T. and J. Freeman (1984), 'Structural inertia and organizational change', *American Sociological Review*, **49** (2), 149–64.
He, Z.L. and K. Wong (2004), 'Exploration vs. exploitation: an empirical test of the ambidexterity hypothesis', *Organization Science*, **15** (4), 481–94.
Helfat, C.E. (1997), 'Know-how and asset complementarity and dynamic capability accumulation: the case of R&D', *Strategic Management Journal*, **18** (5), 339–60.
Hesmondhalgh, D. (2002), *The Cultural Industries*, London: Sage.
Howkins, J. (2001), *The Creative Economy: How People Make Money from Ideas*, New York: Penguin Press.
Kale, P. and H. Singh (2007), 'Building firm capabilities through learning: the role of the alliance learning process in alliance capability and firm-level alliance success', *Strategic Management Journal*, **28** (10), 981–1000.
Kandemir, D., A. Yaprak and S.T. Cavusgil (2006), 'Alliance orientation: conceptualization, measurement, and impact on market performance', *Academy of Marketing Science*, **34** (3), 324–40.
Kohli, A.K. and B.J. Jaworski (1990), 'Market orientation: the construct, research propositions, and managerial implications', *Journal of Marketing*, **54** (2), 1–18.
Kranenburg, H.L. van and J. Hagedoorn (2008), 'Strategic focus of incumbents in the European telecommunications industry: the cases of BT, Deutsche Post and KPN', *Telecommunications Policy*, **32** (2), 116–30.
Kranenburg, H.L. van and G. Ziggers (2013), 'How media companies should create value: innovation centered business models and dynamic capabilities', in M. Friedrichsen and W. Mühl-Benninghaus (eds), *The Handbook of Social Media Management: Value Chain and Business Models in Changing Media Markets*, Heidelberg: Springer, pp. 239–51.
Kranenburg, H.L. van, M. Cloodt and J. Hagedoorn (2001), 'An exploratory study of recent trends in the diversification of Dutch publishing companies in the multimedia and information industries', *International Studies of Management and Organization*, **31** (1), 64–86.
Levinthal, D. and C. Rerup (2006), 'Crossing an apparent chasm: Bridging mindful and less-mindful perspectives on organizational learning', *Organization Science*, **17** (4), 502–13.
Lyles, M.A. (1988), 'Learning among joint venture sophisticated firms', *Management International Review*, **23** (special issue), 85–97.
March, J.G. (1991), 'Exploration and exploitation in organizational learning', *Organization Science*, **2** (1), 71–8.
Microeconomix (2010), 'Horizontal mergers and dynamic competition', *Economic Focus*, December, 1–4.
Mitchell, W. and K. Singh (1992), 'Incumbents' use of pre-entry alliances before expansion into new technical subfields of an industry', *Journal of Economic Behavior and Organization*, **18** (3), 347–72.
Moore, J.F. (1993), 'Predators and prey: a new ecology of competition', *Harvard Business Review*, **71** (3), 75–86.
Narver, J.C. and S.F. Slater (1990), 'The effect of a market orientation on business profitability', *Journal of Marketing*, **54** (4), 20–35.
O'Reilly, C.A. and M.L. Tushman (2004), 'The ambidextrous organization', *Harvard Business Review*, **82** (4), 74–81.
Palmberg, C. and O. Martikainen (2006), 'Diversification in response to ICT convergence – indigenous capabilities versus R&D alliances in the Finnish telecom', *Journal of Policy, Regulation and Strategy for Telecommunications*, **8** (4), 67–84.
Peltier, S. (2004), 'Mergers and acquisitions in the media industries: were failures really unforeseeable?', *Journal of Media Economics*, **17** (4), 261–78.
Pennings, J. and P. Puranam (2001), 'Market convergence and firm strategy: new directions for theory and research', ECIS Conference, The Future of Innovation Studies, Eindhoven, The Netherlands.
Picard, R. (2000), 'Changing business models of online content services: their implications for multimedia and other content producers', *International Journal on Media Management*, **2** (2), 60–68.
Schumpeter, J.A. (1950), *Capitalism, Socialism, and Democracy*, 3rd edn, New York: Harper.
Smith, W.K., A. Binns and M.L. Tushman (2010), 'Complex business models: managing strategic paradoxes simultaneously', *Long Range Planning*, **43** (2), 448–61.
Stieglitz, N. (2003), 'Digital dynamics and types of industry convergence – the evolution of the handheld computer market in the 1990s and beyond', in J.F. Christensen and P. Maskell (eds), *The Industrial Dynamics of the New Digital Economy*, Cheltenham, UK and Northampton, MA, USA: Edward Elgar Publishing.
Stiglitz, J.A. (1987), 'Technological change, sunk costs and competition', *Brookings Papers on Economic Activity*, **18** (3), 883–947.
Teece, D.J., G. Pisano and A. Shuen (1997), 'Dynamic capabilities and strategic management', *Strategic Management Journal*, **18** (7), 509–33.

Wirtz, B.W. (2001), 'Reconfiguration of value chains in converging media and communications markets', *Long Range Planning*, **34** (4), 489–506.
Ziggers, G.W. and J. Henseler (2009), 'Inter-firm network capability: how it affects buyer–supplier performance', *British Food Journal*, **111** (8), 794–810.

FURTHER READING

Understanding further how firms should deal with dynamic competition can be achieved through the concept of the strategic inflection point. This relates to a point in time where fundamentals change so rapidly that the old way of doing business loses a great deal of its relevance. It signals a clear point when a decision must be made to change and remain relevant or ignore the signs and fade away. Burgelman and Grove (1996) is a good introduction to the concept of strategic inflection points. Grove (2009), founder and former CEO of Intel, clearly lays out his viewpoint on the importance of understanding dynamic competition and the importance of identifying the strategic inflection point by the top management team of a firm. Firms operating in creative industries with dynamic competition are confronted with high-speed follow-up of new strategic inflection points in their markets. New techniques, new approaches, new technologies, and new competitors are changing the rules of the game. Incumbent firms should respond to these developments and should adapt their organization to survive. The reaction of a firm depends on the competitive situation of the firm and the ability to develop essential capabilities. These firms should become and build an organization that embraces ambidexterity. Also crucial is a compelling vision, relentlessly communicated by senior management to build ambidextrous organizations and business models. The success of a firm to embrace ambidexterity and deal with dynamic competition depends on the commitment and leadership. Day (2011) and O'Reilly and Tushman (2004) provide a good overview of the primary qualities for ambidextrous leadership.

7. From prosumption to produsage

Axel Bruns

'WEB 2.0' AND THE RISE OF USER-LED CONTENT CREATION

Almost a decade after the 2004 O'Reilly Media conference which popularized the term 'Web 2.0', the impact of this concept on users and developers of the current generation of Web technology – and by extension on the digital economy overall – is undeniable. At the time, 'Web 2.0' promised an interactive, engaging online space in which users would be able to do more than surf from static, fixed Website to static, fixed Website. Although the implicit suggestion that the version change to 'Web 2.0' represented a clean break with this inflexible past must be read as mere marketing hype, the core principles which the concept outlined nonetheless form the operational basis for most mainstream Websites of the present day.

Such principles (see O'Reilly, 2005) included the customization of the user experience, by embracing on-the-fly Web page creation through AJAX and other database-driven Web technologies which provided an opportunity for users to actively query and select the available information on any given Website. By extension, such technologies offered an additional opportunity for users to become active as content creators and contributors, which fundamentally altered the vertical interaction between Website providers and Website users. In turn, this potential for user participation and content creation enabled the emergence of genuine horizontal collaboration between users, and for the formation of user communities, especially where sufficient functionality was available to make users aware of each other's activities and to help them coordinate their collaborations. This interplay around a shared content base in turn necessitated a reconceptualization of the content thus created or modified. As users work with and rework one another's contributions, the content development processes found in communities themselves no longer resemble those of organized industrial production, which also means that new models of content authorship and ownership, beyond standard copyright, must necessarily be found to facilitate them.

Benkler (2006: 2) summarizes the great sense of promise attributed to the emerging practices of collaborative, user-led 'Web 2.0' content creation[1] at the time:

> They hint at the emergence of a new information environment, one in which individuals are free to take a more active role than was possible in the industrial information economy of the twentieth century. This new freedom holds great practical promise: as a dimension of individual freedom; as a platform for better democratic participation; as a medium to foster a more critical and self-reflective culture; and, in an increasingly information-dependent global economy, as a mechanism to achieve improvements in human development everywhere.

But, at the same time, the novelty of this 'new' information environment should not be overstated. Collaboration in interaction and content creation has a long prehistory which predates the Internet age itself, and can be traced back to communal, commons-based activities that are at least as old as the Greek *agora*; commons-based approaches to managing and growing shared resources are likely to be as old as the history of farming itself, and persist to this day in various pockets of the world (see Ostrom, 1990). From this very long-term perspective, it is the industrial age and industrial economy, dominated by production companies each carefully guarding their trade secrets and positioning users merely as passive consumers, which presents a historical aberration. The emergence of Benkler's '"new" information environment', then, merely re-balances industry and user contributions in a way which is more suited to a digital creative economy.

PROSUMPTION?

Attempts to engage and harness user (or consumer) input in the industrial process are themselves far from novel, however; they did not emerge along with the 'Web 2.0' concept, but predate it by some margin. As early as 1970, Alvin Toffler coined his portmanteau of the 'prosumer' to describe industry attempts to involve especially knowledgeable, quasi-professional consumers in production processes – initiatives which he describes in *The Third Wave* (1980: 275) as 'the willing seduction of the consumer into production'. In the first place, this seduction is accomplished by offering to consumers the opportunity to customize and personalize the products they wish to purchase. This results in an on-demand, 'customer-activated manufacturing system' in which 'the consumer, not merely providing the specs but punching the button that sets this entire process in action, will become as much a part of the production process as the denim-clad assembly-line worker was in the world now dying' (1980: 274). This is similar to the limited options for content customization and personalization which are now available as a matter of course from many basic 'Web 2.0' Websites.

As such customization and personalization processes are established on a more comprehensive scale, however, they also generate an increasing amount of valuable information on user demands and ideas for further product development. In a second stage of the move towards prosumption, then, production companies are increasingly able to exploit this customer-contributed knowledge for commercial advantage:

> Producer and consumer, divorced by the industrial revolution, are reunited in the cycle of wealth creation, with the customer contributing not just the money but market and design information vital for the production process. Buyer and supplier share data, information, and knowledge. Someday, customers may also push buttons that activate remote production processes. Consumer and producer fuse into a 'prosumer'. (Toffler, 1990: 239)

Defined in this way, prosumption represents a highly uneven, off-centre and imbalanced 'cycle of wealth creation', however, which continues to substantially privilege industrial producers over their information-generating *and* money-spending users and prosumers. Their relationship remains a highly one-sided one, foreshadowing criticisms of 'Web 2.0' practices as merely exploiting the free labor of user-led content creation to the benefit of the corporations which operate the Websites (see Terranova, 2000 for an early example of such critiques).

Undoubtedly, such criticism is justified at least for a subset of such sites. Benkler

(2006: 76), for example, points out that corporations such as 'Google and Amazon . . . that have done immensely well at acquiring and retaining users have harnessed peer production to enable users to find things they want quickly and efficiently'. Both companies, and others like them, harness the activities of their users on a large scale in order to improve their products, for example to provide more accurate search results or more useful purchase recommendations. In essence, the Google index and Amazon recommendations are co-created between the companies which design the algorithms and the vast user bases which, more or less unwittingly, provide the interaction data from which the algorithms learn. Historically, both companies have thus been at the very forefront of current trends towards 'big data' analytics, which seek to generate new insights from the large-scale examination of patterns in online user data (see boyd and Crawford, 2012). Whether, given the free and fundamental search service which Google provides, such *harnessing* of user activities should necessarily be seen as *exploitation* in the negative sense of the term remains debatable, however.

Toffler's concept of prosumption does not provide a particularly useful model for many other practices of content creation in 'Web 2.0' environments. Most centrally, it cannot be applied successfully to wholly or predominantly user-*led* initiatives which proceed without oversight or coordination by commercial entities, or even to initiatives which build on platforms provided by distinct commercial or non-profit institutions, but whose content creation processes are managed predominantly by the user communities themselves. Ultimately, the prosumption model assumes that consumers contribute individually to prosumption practices, but that the task of coordinating and evaluating their contributions falls necessarily to the corporation, as the only entity which has the comprehensive overview required to carry out this task. This assumption holds for Google and Amazon, for example (no one external to these organizations is able to evaluate the totality of Google searches or Amazon purchase patterns), but it does not apply to the many 'Web 2.0' platforms which do enable their communities to publicly coordinate and evaluate communal activities amongst themselves. As Bauwens describes it, 'Whereas participants in hierarchical systems are subject to the panoptism of the select few who control the vast majority, in P2P systems, participants have access to holoptism, the ability for any participant to see the whole' (2005: 1).

It is unfortunate in this context that Toffler's prosumption model was seen for some time as a blueprint that described 'Web 2.0' practices in general. Tapscott's (1996) and Tapscott and Williams's (2006) updates on the concept of prosumption, though increasingly referencing 'Web 2.0' platforms such as Wikipedia, Flickr and YouTube (whose names even appeared on the cover graphic of their 2006 book *Wikinomics*), continued to position user-created content merely as intelligence to be harnessed and exploited by corporations; to them, this 'prosumptive approach to building a business offers advantages that tightly controlled business models can't replicate' (Tapscott and Williams, 2006: 127), but the central aim remains exactly this: to build a business on the basis of prosumer labor. Such teleological application of the industrial-age prosumption concept to information-age collaboration practices must necessarily fail to describe the full range of content creation activities associated with the 'Web 2.0' phenomenon. An exploration of alternative models is necessary.

TOWARDS PRODUSAGE

Adequate models must strive to describe the practices of 'Web 2.0' users from the inside, rather than from the perspective of commercial operators seeking to exploit user-led content creation for their own ends. From this embedded perspective, then, it is appropriate to describe such practices with reference to the major success stories of user-led content creation, regardless of whether they operate in commercial or non-profit configurations. Perhaps the most obvious object of investigation for such an analysis is Wikipedia, whose knowledge creation and management practices in turn build substantially on collaborative development models in open source software; however, the approaches described in the following discussion also apply to a vast range of other content creation projects across the digital creative economy.

Open Participation, Communal Evaluation

In the first place, of course, user-led content creation in 'Web 2.0' spaces such as Wikipedia relies on the participation of a sufficient number of users in these processes. Wikipedia and similar platforms therefore face the challenge of having to lower their barriers to participation as far as possible, while simultaneously mitigating the threat of disruptions from well-meaning but poor-quality contributions or deliberate interference, spam and defacement. New users coming to a site are inevitably reluctant to participate if they are required to make substantial investments of time and effort in learning site functionality or generating acceptable content, or if they feel that they are unwanted intruders into an already established community of participants on the site. Such concerns may be addressed by providing users with a clear and manageable pathway of stepping-stones towards full participation. Wikipedia users, for example, may move from mere usage towards more productive engagement by simply clicking the 'edit' button on any page and making minor adjustments to language and spelling in an anonymous capacity, but can build up from there towards full editorship by registering on the site, establishing a user profile, contributing more substantial edits to entries, engaging with other contributors through the discussion pages attached to each encyclopedia entry, and eventually assuming local or global administrative functions. Arguably, it is this granularity of participation which has contributed substantially to the site's impressive track record. As Benkler notes:

> The number of people who can, in principle, participate in a project is . . . inversely related to the size of the smallest-scale contribution necessary . . . The granularity . . . therefore sets the smallest possible individual investment necessary to participate in a project . . . If the finest-grained contributions are relatively large and would require a large investment of time and effort, the universe of potential contributors decreases. (2006: 101)

But such granularity also opens the door to abuse, and generates a stream of edits and updates which would be impossible to manage for any small team of dedicated moderators or reviewers. Wikipedia and similar content creation projects therefore extend their invitation to user participation beyond content contribution and towards moderation. Indeed, Wikipedia hardly distinguishes between the editing and reviewing process, and simply enables its contributors to revert previous changes, make further revisions, and initiate critical discussions about the merit of specific changes or change proposals. It is in this combination of *open participation* and *communal evaluation* that Wikipedia most obviously breaks with traditional content editing models – as do the collaborative software development models of open source, or many other projects inspired by such user-led approaches.

Such broad-based community models for the creation and evaluation of content necessarily also depend on the diversity of the community: a uniform group of participants with similar levels of knowledge, similar areas of interest, and similar beliefs and values will be unable to effectively create content through a process of iterative revision and evaluation, as its members lack the difference of skills and opinions required to investigate a problem from all sides. This builds on the open source maxim, formulated by Raymond (2000), that, 'given enough eyeballs, all bugs are shallow'. As bugs and other problems are encountered in software development, the number of participants concerned with beta-testing increases the speed and efficiency with which such bugs are addressed and eradicated. 'Adding more beta-testers may not reduce the complexity of the current "deepest" bug from the *developer's* point of view, but it increases the probability that someone's toolkit will be matched to the problem in such a way that the bug is shallow *to that person*' (Raymond, 2000, n.p.). Open source software development, in other words, pursues a probabilistic approach to problem-solving.

Unfinished Artifacts, Continuing Process

In effect, then, much like open source software, the user-created content of 'Web 2.0' sites such as Wikipedia exists in a state of 'permanent beta'. It remains constantly under development, and evolves along probabilistic lines which are determined by its users' diverse interests rather than coordinated by an overarching corporate strategy. Even more than open source software, however, whose maintenance and extension require specific technical skills in addition to intellectual knowledge and capacity, text-based content creation sites such as Wikipedia extend an invitation to a very wide potential contributor base: here, the specific technical skills are comparatively limited and relatively easy to learn, while the range of contributions possible is substantially wider. It is for this reason, then, that Wikipedia has been able to cover the traditional domains of encyclopedias as well as topics from obscure pop-cultural trivia to the latest breaking news, and has been able to do so with considerable success in more than 200 languages.

Given such probabilistic, user-driven models of content creation, our conceptualization of the material they generate must also be rethought. Conventional, industrial models of content creation generally result in distinct products, not least also in order to make them marketable. Importantly, especially in the emerging digital economy, such products represent a legacy of the industrial age, where distribution of intellectual content in physical form made it necessary to suspend development at specific points in order to package, distribute and sell products. While an encyclopedia volume, a software package, a newspaper issue or any other non-trivial product could always be improved and updated further, their development was thus segmented into product cycles which lasted, depending on the product, from less than one day (for newspapers) to several years (for less time-critical goods).

Distribution in non-physical form – especially through the Internet – removes any pressing necessity for such product cycles. The rise of subscription-based content update and software-as-a-service models points to the fact that services which feature incremental, continuous updates and upgrades are able to replace distinct, packaged products. At their simplest, such services may only seek to replace the delivery mechanisms for otherwise conventionally produced content. (Thereby they essentially speed up the product cycle to a point where regular product versions such as the annual encyclopedia edition or the daily newspaper become revisions which are provided on a much more frequent basis, or ad hoc whenever updates are necessary.) But the collaborative, distributed, probabilistic nature of user-led content creation further requires a shift from preparing such updates behind closed doors and releasing them when they are ready and approved towards developing and testing them in full view of the public. Only such public visibility of the work under development enables the project to enlist users encountering such new revisions in the process of testing, evaluating and further developing this 'public beta' content.

User-created content generated under such conditions must be thought of as consisting of *unfinished artifacts*, engaged in a *continuing process* of revision and development. Again, this can be seen as moving away from the industrial-age dominance of distinct, commercial, marketed products and back towards an older model of ongoing use and improvement of shared resources. This is a passing-along of materials from user to user, with modifications along the way – but now, in intangible, electronic form, and therefore directly available to a much larger range of participants in digital culture and the digital economy. This continuing process of development and – under the right circumstances – improvement also resembles the medieval palimpsest: a repeatedly overwritten text, which contains traces of each of its former iterations. In Wikipedia, but also in many other collaborative platforms which retain previous content revisions, this palimpsestic process of revising and overwriting can be retraced and made visible by accessing the 'page history' functions for any given entry.

Fluid Heterarchy, Ad Hoc Meritocracy

Such retracing of the contribution histories for individual artifacts or entire collections of user-created content also highlights the fact that, in spite of the equal *potential* for all users to participate in collaborative 'Web 2.0' platforms of this form or, as Bauwens describes it, 'the free participation of equipotent partners, engaged in the production of common resources' (2005: 1), *actual* participation by users is far from evenly distributed in most cases, owing to differing levels of interest, commitment and ability. While open participation in collaborative efforts and the probabilistic development activities that result from it do provide an equal opportunity for users to engage, these other factors nonetheless result in a stratification of the participant community into more or less important and influential members. The equipotentiality which Bauwens invokes, in other words, should not be misunderstood as claiming complete equality for all contributors and contributions, or as denying the existence of power differentials between different individuals and groups; rather, it simply 'means that there is no prior formal filtering for participation, but . . . that it is the immediate practice of cooperation which determines the expertise and level of participation. It does not deny "authority", but only fixed forced hierarchy, and therefore accepts authority based on expertise, initiation of the project, etc.' (2005: 1).

Further, the distributed, granular basis of collaborative content creation projects tends to mean that there are multiple centers of power and authority, rather than a clear hierarchical structure of command and control. In Wikipedia, for example, stratified communities of contributors can be found at various levels of complexity for every entry, every group of entries and every domain of knowledge in the encyclopedia. Such projects and platforms, in other words, constitute a community of communities, and in each case a user's positioning within the community is determined in the main by the user's contribution to the project, in the form of useful content or other interventions deemed valuable (including for example coordinating and documenting activities, guarding against disruptions and defacement, or welcoming new users into the fold). Users may therefore simultaneously be central members of a community to which they have contributed crucial content or services, but marginal participants of other communities on the same site to which they have made more minor contributions.

What emerges from this complex interplay of contributors and contributions, this ongoing evaluation, re-evaluation and repositioning of users on the basis of their latest contributions, is a highly changeable network of power relations which is best described as a *fluid heterarchy* and an *ad hoc meritocracy*. Though identified at times with highly visible individuals such as Wikipedia founder Jimmy Wales or Linux inventor Linus Torvalds, the communities which built these and other projects rely on a much broader range of leading users, distributed across the communities and subcommunities which exist around each granular element of their activities and potentially unaware of each other. Their leading positions are in turn determined on an ad hoc, meritocratic basis; should they drift away from the project or should the quality and relevance of their contributions decline over time, other currently more important members of the community will come to replace them as influential community leaders.

Communal Property, Individual Rewards

The meritocratic nature of such collaborative content creation communities also provides a further incentive for users to participate in the process. The opportunity to gain greater influence over the further direction of content creation processes, and to work in a leading role with like-minded users, serves as a reward for participation beyond the intrinsic satisfaction of making a useful contribution to the shared effort. By contrast, personal ownership in specific elements of the shared project cannot be a reward offered by sites which facilitate user-led content creation, as such ownership in intellectual property – if exercised to the exclusion of other users – would severely undermine the continuing development process. Sites such as Wikipedia therefore require the use of intellectual property licensing schemes which are modeled on the long-standing experience of the open source software development community with commons-based licenses. Such licenses affirm the rights of contributors to be acknowledged as (co-)authors of the shared, common resource, and may include stipulations against the unauthorized commercial exploitation of the resource, but explicitly give permission to other contributors to continue use and revision of their predecessors' work without further approval from these earlier contributors.

These licensing schemes, then, introduce a distinction between ownership and authorship: in essence, the work contributed by individual users becomes *communal property*, but these users retain the ability to extract *individual rewards* from their contributions at least in the form of personal standing within the community, and potentially of acclaim as notable contributors well beyond it. While the latter remains uncommon for Wikipedia contributors, whose personal contributions to any one entry are more difficult for outsiders to trace, several leading contributors to the development of open source software packages have used such acclaim to boost their professional careers as developers, consultants or authors, for example. Growing out of interest-driven collaborative content creation communities, such professionals begin as what Leadbeater and Miller describe as 'a new breed of amateur . . . the Pro-Am, amateurs who work to professional standards . . . The Pro-Ams are knowledgeable, educated, committed and networked, by new technology' (2004: 12). Over time, as opportunities become available, some Pro-Ams make the transition to professional work in a more conventional sense – but even those who continue simply as Pro-Ams maintain a quasi-professional commitment to their work.

Conventional, industrial content production draws only on a narrow range of fully professional participation, then (as well as – under the prosumption model – on a much more low-grade form of product input from consumers). By contrast, user-led content creation is open to a much wider continuum of user involvement which ranges from quasi-professional Pro-Am activities by lead users through diverse forms of less intensive participation at various levels of granularity all the way to a mere usage of the artifacts of collaborative content creation. 'Web 2.0' sites which build on the four key principles outlined here – (1) open participation, communal evaluation; (2) unfinished artifacts, continuing process; (3) fluid heterarchy, ad hoc meritocracy; and (4) communal property, individual rewards – appeal to their users at every step to be more than just users. They encourage them instead to make productive contributions to the ongoing content creation project, at whatever level of granularity and engagement is possible for them. This positions participants in a hybrid position as potentially both users and producers of content at the same time – that is, as *produsers* (Bruns, 2008).

The practices which these produsers engage in – the collaborative and continuous building and extending of existing content in pursuit of further improvement, or *produsage* – do not resemble conventional content production. The outcomes of their work, as temporary artifacts of an ongoing process, do not resemble conventional products, but neither do they represent a form of prosumption as it has been defined above. Instead, produsage processes can operate with significant success independently of commercial entities, as Wikipedia and many other community-driven projects have demonstrated. Whatever terms we use to describe them, the principles and processes of produsage must be understood on their own terms, rather than through the lens of industrial-age producer–consumer relationships.

BEYOND PRODUSAGE

Produsage concepts and communities are inherently connected to the sociotechnical environments which enable and support them. While some produsage projects and communities, such as Wikipedia, date back at least to the turn of the millennium and have been afforded an opportunity to evolve independently, others have been tied more closely to changing corporate, industrial and intellectual environments. The software and especially the computer games industries, for example, have embraced produsage practices with considerable enthusiasm, especially in the field of game mods and add-ons, where prodused additions to commercial games have contributed substantially to the commercial success of these games (see Banks, 2009). Indeed, the contribution of commercial production in some such collaborative settings has focused centrally on developing the underlying games engine, while the creation of content and storylines to be played using such engines has strongly drawn on the work of produsage communities.

A recent trend towards industry and community co-funding of games development using crowdfunding services such as Kickstarter or Pozible ostensibly formalizes such collaborative relationships by creating partnerships between production houses and produsage communities. At present, such models may be most prevalent in computer games and related software projects, but they also outline possibilities which may be applicable well beyond this sector of the creative industries. Especially where considerable fan enthusiasm may be mobilized to support and contribute to the maintenance and development of a project, approaches which combine formal production and communal produsage teams may well be successful. Even in more resource-intensive industries, which generate physical rather than digital artifacts, such collaboration may be possible, especially if the line between produsers and producers is drawn at the digital–physical interface. In the open innovation communities of kitesurfers and circuit board designers, which von Hippel (2005) discusses, for example, design is a largely communal activity, while the physical production of artifacts is carried out by specialized companies. Unlike the case for Toffler's prosumption, however, the community retains the rights to its designs; giving up such rights is no longer the default price of admission.

In some industries, relationships between industry and community have been considerably more problematic. Interactions between professional news organizations and their produser counterparts (including news bloggers and citizen journalists) have long been characterized by animosity rather than collaboration, for instance (see, for example, Highfield and Bruns, 2012). Such mutual disdain is gradually being replaced both by in-house projects which seek to invite produser collaboration and by engagement between professional journalists and citizen journalists in neutral, third-party spaces. On the one hand, projects such as the *Guardian*'s MPs' Expenses site (which invited readers to collaboratively hunt through a vast collection of UK MPs' expenses claims in order to detect any signs of impropriety) acknowledge that many data journalism activities cannot be conducted by a small staff of professional journalists alone, but instead require wide community participation (Bruns, 2012). On the other, the growing role of social media spaces such as Facebook and – especially – Twitter for the rapid dissemination, discussion and evaluation of journalistic content particularly in the context of breaking news means that journalists, newsmakers, experts, news enthusiasts and the general news public are increasingly frequenting the same, public social media spaces, and that existing boundaries between professionals, Pro-Ams, produsers, users and consumers are further eroded.

In such spaces, then, all parties with an interest in a specific story are able to come together, to exchange ideas and information, to discuss, argue and attempt to make sense of the news. On an ad hoc basis, they collaborate more or less effectively to 'work the story' (Bruns and Highfield, 2012). Such activities retain many of the principles of produsage as we have encountered them above: when discussing the news in a Twitter hashtag, for example, participation is open to all, and contributions are communally evaluated by @replying and retweeting (which confers visibility and, to some extent, endorsement). From this, a heterarchy of especially visible and influential participants emerges, based on the merit of their contributions. The discussion of the news is ongoing and unfinished, and the arrival of new information can fundamentally shift the dynamics of the process. Even without formal licensing agreements beyond Twitter's terms of service, an assumption persists that any message may be passed along, retweeted and commented upon by others.

In essence, therefore, such social media spaces are now positioned to become universal third-party platforms to support produsage processes involving a diverse set of contributors, potentially replacing extant dedicated sites for produsage (or at least duplicating and augmenting some of their functionality). Rather than users conducting discussions about a specific issue or problem in the shared produsage project through the project's own platform (for example, Wikipedia's discussion pages), social media may increasingly serve as universal backchannels for collaborative content creation communities. This embedding of social media platforms into produsage practices would be nothing more than a logical extension of the fundamental principles of produsage. The 140-character limitation of a single tweet positions it as one of the most granular contributions possible, and the open-access, globally visible nature of tweets means that the potential for open participation is guaranteed.[2]

At the same time, however, any shift towards Twitter and other social media as supporting technologies for produsage processes also creates a series of potential pitfalls. First, while such social media technologies are readily available, their use alongside a dedicated produsage platform such as Wikipedia may increase the complexity of participation and thereby raise rather than lower barriers to participation. Users would now need to maintain multiple accounts (Wikipedia, Twitter, Facebook and so on) and would need to coordinate their activities across these accounts in order to participate fully. Second, any splintering of discussion across built-in contributor fora and third-party social media spaces may undermine community coordination of activities. It could no longer be assumed that discussions and decision-making take place only within the fora provided by Wikipedia; parallel, possibly conflicting debates may also exist in other spaces, between a different subset of contributors. Third, a longer-term side-by-side operation of different discussion and coordination technologies may lead to the formation of opposing factions within the community, centered around each technology, which pursue competing and irreconcilable agendas. And finally, a reliance on services such as Twitter and Facebook to facilitate community discussion and coordination means a reliance on commercial operators and proprietary technology for potentially crucial activities within the community. This could jeopardize the long-term documentation, archiving and preservation of community activities. Operating on its own platform since 2001, Wikipedia has preserved a full archive of all contributions and discussions since its first encyclopedia entries were published to the Web; this constitutes a rich historical resource. Relying on a third-party provider for part of its functionality, the longevity of that archive could not have been guaranteed – and information of significant value, about the history of Wikipedia, of its community, of contemporary knowledge processes, and of produsage and 'Web 2.0' as such, might have been lost as a result.

Such concerns illustrate the challenges ahead for produsage and its communities. As 'Web 2.0' practices have matured, and as produsage models have become established as legitimate, successful processes for collaborative content creation, the number of non-profit and commercial providers of third-party platforms to support produsage activities has also increased. Commercial approaches to working with produsage communities – sometimes in genuine partnerships, sometimes in settings which pursue prosumption by stealth – have multiplied (Bruns, 2012). In this complex, diversified environment, individual contributors and produsage communities overall will need to continue to guard their integrity and ensure that the fundamental principles of produsage are maintained. This means: (1) continued openness to new contributors, but also the maintenance of community standards through the constant evaluation of activities; (2) organizational and technological frameworks which support the constant, gradual revision and extension of existing content; (3) flexibility in community structures which enables proven contributors to assume positions of leadership; and (4) ownership and authorship rules which recognize individual contributors but deny them the power to control the further use and development of their work. Such principles are well enshrined in a number of the leading produsage projects which are currently under way; other projects would do well to study these principles in detail and to translate them to their own contexts.

NOTES

1. 'User-created content' and 'user-generated content' are generally used with no clear distinction in the literature. For consistency, I use the former throughout this chapter; however, part of the point here is also to note the lack of consistent language in describing this phenomenon, which is why I suggest 'produsage' as an alternative term.
2. By contrast, the focus on strong ties and the lack of universal visibility for Facebook messages serve as somewhat greater barriers to widespread community participation in produsage activities.

REFERENCES

Banks, J. (2009), 'Co-creative expertise: Auran Games and Fury – a case study', *Media International Australia*, **130**, 77–89.

Bauwens, M. (2005), 'Peer to peer and human evolution', *Integral Visioning*, 15 June, available at: http://integralvisioning.org/article.php?story=p2ptheory1 (accessed 1 March 2007).

Benkler, Y. (2006), *The Wealth of Networks: How Social Production Transforms Markets and Freedom*, New Haven, CT: Yale University Press.

boyd, d. and K. Crawford (2012), 'Critical questions for big data', *Information, Communication and Society*, **15** (5), 662–79.

Bruns, A. (2008), *Blogs, Wikipedia, Second Life and Beyond: From Production to Produsage*, New York: Peter Lang.

Bruns, A. (2012), 'Reconciling community and commerce? Collaboration between produsage communities and commercial operators', *Information, Communication and Society*, **15** (6), 815–35.

Bruns, A. and T. Highfield (2012), 'Blogs, Twitter, and breaking news: the produsage of citizen journalism', in R.A. Lind (ed.), *Producing Theory in a Digital World: The Intersection of Audiences and Production*, New York: Peter Lang, pp. 15–32.

Highfield, T. and A. Bruns (2012), 'Confrontation and cooptation: a brief history of Australian political blogs', *Media International Australia*, **143**, 89–98.

Jenkins, H. (2006), *Convergence Culture: Where Old and New Media Collide*, New York: NYU Press.
Leadbeater, C. and P. Miller (2004), 'The Pro-Am revolution: how enthusiasts are changing our economy and society', *Demos*, available at: http://www.demos.co.uk/publications/proameconomy/ (accessed 20 December 2012).
O'Reilly, T. (2005), 'What is Web 2.0? Design patterns and business models for the next generation of software', *O'Reilly*, available at: http://oreilly.com/web2/archive/what-is-web-20.html (accessed 20 December 2012).
Ostrom, E. (1990), *Governing the Commons: The Evolution of Institutions for Collective Action*, Cambridge: Cambridge University Press.
Raymond, E.S. (2000), 'The cathedral and the bazaar', available at: http://www.catb.org/~esr/writings/cathedral-bazaar/cathedral-bazaar/index.html (accessed 20 December 2012).
Tapscott, D. (1996), *The Digital Economy: Promise and Peril in the Age of Networked Intelligence*, New York: McGraw-Hill.
Tapscott, D. and A.D. Williams (2006), *Wikinomics: How Mass Collaboration Changes Everything*, New York: Penguin.
Terranova, T. (2000), 'Free labor: producing culture for the digital economy', *Social Text*, **18** (2), 33–57.
Toffler, A. (1980), *The Third Wave*, New York: Bantam.
Toffler, A. (1990), *Powershift: Knowledge, Wealth, and Violence at the Edge of the 21st Century*, New York: Bantam.
Von Hippel, Eric (2005), *Democratizing Innovation*, Cambridge, MA: MIT Press.

FURTHER READING

The concept of produsage is explored in much greater detail in *Blogs, Wikipedia, Second Life and Beyond: From Production to Produsage* (Bruns, 2008). Several significant alternative conceptions of user-led content creation also exist: Yochai Benkler's *The Wealth of Networks* (2006) describes what he calls 'peer production' from a perspective that is more strongly influenced by law and economics; Henry Jenkins outlines the broader impact of the move to user-led content creation for the media and entertainment industries in *Convergence Culture* (2006).

8. Consumption patterns

Koen van Eijck and Max Majorana

This chapter addresses changing consumption patterns. Starting with a brief presentation of classical sociological insights in the generation and manifestation of cultural consumption patterns, we will present the emergence of the cultural omnivore and the voracious consumer and how this inspired a re-thinking of the structuring and meaning of consumption. Subsequently, shifts in the mode of cultural production will be addressed. The consequences of such shifts will be illustrated in a discussion of digital services for music distribution.

CONSUMPTION AND SOCIAL CLASS

Cultural consumption has solid social roots. Ever since the early work of economist and sociologist Thorstein Veblen ([1899] 1953), it has been recognized that social class is a crucial determinant of cultural taste and behaviour. The reasons for the existence of this strong relationship are subject to a continuous stream of studies. The theoretical accounts can be subsumed under two broad points of view. On the one hand, it is argued that it takes resources, or capital (economic, cultural, social), to develop and implement a certain (cultural) consumption pattern. Unequal command of these resources leads to consumption patterns that differ in terms of both quantity (how many resources are available) and quality (the type of resources invested in consumption). Lack of money, knowledge, or social contacts can be barriers for realizing certain consumer lifestyles.

This implies that consumption reveals which social connections, skills, knowledge, financial means, or tastes people have, which points at the second explanation. Through consumption, people show who they are or where they come from, and thereby allow others to assign honour or prestige. Such prestige may be a mere by-product of a consumer lifestyle, but it may also be a reward that is consciously sought after by engaging in certain esteemed activities. Status motivations have been argued to occasion specific cultural behaviours among people who are eager to (demonstrate that they) belong to a certain social group. Thus, consumption patterns are always more than just an innocent or neutral reflection of skills or tastes that people happen to have acquired. They make visible status distinctions between people and function as sources of social in- and exclusion. Therefore, much research on cultural consumption focuses on the affirmation of social status for its explanation.

Although 'affirmation of social status' might suggest otherwise, the close link between status and cultural consumption does not imply that drawing social boundaries is a conscious act of snobbish elitism. Rather, the resources members of a certain class have available strongly but unconsciously shape their habitus, which Pierre Bourdieu (1984: 101) describes as a 'practice-unifying and practice-generating principle . . . [or] the internalized form of class condition and of the conditionings it entails'. Thus, the habitus is argued to be always class-specific, as it depends on the resources available to members of a certain class and, by thus conveying class membership, carries social status connotations. Since the habitus develops from early socialization onwards, its implications may remain largely unknown to those whose opportunities and worldviews it so thoroughly structures. For Bourdieu, a habitus is typically reproduced from generation to generation, making sure that the power relations between classes are also reproduced and often taken for granted as natural differences.

Cultural capital, the embodied form of which is part of the habitus, has typically been found to be the most powerful concept for explaining cultural consumption (Bennett et al., 2009; Kraaykamp and van Eijck, 2010). For Bourdieu (2002: 282), embodied cultural capital can be conceived of as 'long-lasting dispositions of the mind and body' or embodied schemes of perception, appreciation, and action that determine people's consumer preferences. From this supposedly strong link between the class-based habitus and cultural consumption, it can be inferred that consumption patterns will be rather homogeneous or predictable. This follows from both theoretical viewpoints mentioned above, as Lizardo (2010: 310) makes clear. On the one hand, he argues that embodied cultural capital can indeed be defined as 'an aptitude or a generalized, transposable (across contexts) skill acquired in the combined realms of the upper-middle class family and the school system'. In addition, he distinguishes what he calls the boundaries point of view, where cultural capital is defined as the institutionalized repertoire of high status signals useful for purposes of marking and drawing symbolic boundaries in a given social context. In the boundary perspective, the link with specific cognitive requirements as to the content of cultural capital is absent, since 'whatever counts as cultural capital are those symbolic resources that are actively mobilized by members of groups or class fractions to establish their difference from other groups and thus to devalue the cultural resources and symbolic practices of outsiders' (Lizardo, 2010: 310). Bourdieu (1984) similarly talked about a 'cultural arbitrary', indicating that it does not matter much which exact cultural taste the upper classes display; it is in large part the fact that *they* have this taste that makes it a form of cultural, or symbolic, capital. Although, at a more general level, consumer lifestyles reflecting the ability to spend money or to deal with complexity will be more prestigious than lifestyles requiring no such investments, the symbolic value of cultural consumption patterns is defined as a function of the social prestige of its possessors as much as in terms of properties of specific cultural items or activities such as complexity or rarity.

THE CULTURAL OMNIVORE

The above implies that the symbolic value of consumer goods, including cultural activities, is not inherent in the objects of consumption. For example, traditional distinctions between highbrow and lowbrow culture are therefore prone to change or erosion. Today, the intergenerational reproduction of cultural lifestyles is less obvious than a few decades ago owing to interrelated social processes such as the education expansion and the concomitant rise in social mobility rates, growing cultural relativity, greater availability of cultural products through the media, commercialization, and the legitimization of popular culture since the baby-boom generation. As a result, cultural boundaries have become blurrier, and the consumption patterns of the members of the higher classes are no longer characterized by an exclusive preference for highbrow culture but rather by an inclusive set of preferences for a broad range of cultural items. In the 1990s, cultural sociologists discovered the cultural omnivore as a new type of consumer who did not care much about the distinction between highbrow and popular culture. Peterson and Simkus (1992: 169) found that 'there is mounting evidence that high-status groups not only participate more than do others in high-status activities but also tend to participate more often in most kinds of leisure activities'. Through appreciating popular culture alongside highbrow activities, cultural omnivores are thought to demonstrate their openness to a wide range of cultural products. The importance of such openness is not limited to the personal sphere, but has also become a more important attribute in light of the expansion of the creative industries (Brooks, 2000; Florida, 2002). This means that it is increasingly difficult to link a single cultural product to a social position, and also to gauge a person's social position from any single consumer item or preference a person displays. Consumption patterns have become more of a mixture of elements that were previously deemed incompatible. According to Lahire (2008), today such 'cultural dissonance' is in fact the rule rather than the exception.

Omnivorous consumption patterns have been found to be prevalent among the upper-middle classes all over the Western world – for an overview, see Peterson (2005). The phenomenon of the cultural omnivore does not, however, mean that cultural stratification is disappearing. In fact, cultural omnivorousness may very well be a new sign of symbolic status, as it is clearly linked to the highly educated segments of the population. Although highly educated omnivores are not snobbishly shunning popular culture like the traditional highbrow consumers, they differ substantially from those in lower classes by their interest in highbrow or legitimate culture alongside popular items. In short, they distinguish themselves from others through their:

> ability to *decode* and *incorporate* all manners of aesthetically produced materials and performances . . . cutting across previously institutionalized boundaries separating the largely non-profit organizational network of production of the 'fine arts' from the for-profit 'cultural-industry system' . . . in charge of disseminating the popular arts through market mechanisms. (Lizardo, 2008: 20)

It can be argued that increased education levels have enhanced people's ability to aesthetically appropriate objects by considering form rather than (just) substance, but did little to further the traditional highbrow taste of the higher educated. This makes sense if we realize that this aesthetic disposition defines how people relate to objects rather than being an intrinsic property of the object itself. Following Bourdieu (1984) and Park (1993), it can be argued that the aesthetic disposition resembles a keenness to engage in an aesthetic relation with virtually any cultural product or object. For Park, the concept of the aesthetic has to do with the quality of experiences, not with the cultural objects themselves. This quality of experiences is determined by one's ability and willingness to engage with objects in an aesthetic manner. And, as there is no reason to reserve this aesthetic approach for (highbrow) culture, this disposition encourages people to constantly appropriate new objects to enter into an aesthetic relation with, extending well into the world of consumer goods. Lury (1996: 77) claims that a process of stylization takes place regarding the 'design, making and use of goods . . . as if they are works of art, images or signs and as part of the self-conscious creation of lifestyle'. Using cultural and material goods as means of representation turns much of everyday life into a symbolic display of taste and social affiliation, or a site of ongoing cultural discrimination. The application of aesthetic criteria has become both more prevalent and more nuanced. It has become more prevalent because popular culture and consumer goods have become important sites for displaying aesthetic dispositions. It has become more nuanced because more subtle distinctions are to be made owing to the booming of cultural, designer, and fashion products. Instead of a few clear-cut boundaries between, for example, highbrow and lowbrow, we now have a fine maze of numerous boundaries even within the highbrow and the popular, and even within genres.

THE VORACIOUS CULTURAL CONSUMER

Omnivorousness implies having a broad cultural consumption pattern across any boundaries between highbrow and popular culture. Omnivores also tend to be very active. They do not just have a broad taste, but also act upon it by engaging in many leisure activities. Sullivan and Katz-Gerro (2007) therefore argue that contemporary cultural consumption patterns of members of the upper-middle classes may be related to notions such as hedonism, with its high turnover of goods and experiences, and with people's increased ability or eagerness to switch from one activity to the next while avoiding profound investments in any single pursuit (see also Bell, 1976; Bauman, 1996). An explanation for the increasing volume of cultural activities among cultural omnivores may lie in the high pace of contemporary life, with its increasing time pressure (Linder, 1970; Gershuny, 2000). While drawing different cultural boundaries may lead to omnivorousness in terms of composition, arguably the increased pace of leisure activities is another crucial element of present-day cultural consumption patterns.

The growing availability and accessibility of culture due to the increase and geographical spread of cultural venues, cultural subsidies by governments, online access to (information on) cultural products, and increased geographical mobility may all have contributed to the increased pace of consumption. These developments, too, make consumers more likely to try out many different cultural experiences rather than specialize in a single discipline or a few genres. Sullivan and Katz-Gerro (2007) coined the term 'cultural voraciousness' to describe these cultural practices where many activities are undertaken with a high frequency.

The rise of the voracious consumer can be understood from an economic perspective by resorting to Becker's (1965) classical theory on the allocation of time. Conceiving each household, or person, as a producer of its own leisure by investing time and money, it follows that growing prosperity encourages people to increase the productivity of their leisure time by investing more money and less time per activity. Fewer hours of leisure time similarly shift the balance of time–money investments, which means that more activities will be undertaken in less time, either by shortening them or by doing them simultaneously. Since high income and time scarcity tend to go hand in hand among highly qualified workers, the culturally active segments of the population especially have a high pace of leisure.

A number of studies have shown that average leisure time has indeed diminished and that women who combine paid work with taking care of children and the household especially have become busier during the last few decades (Gershuny, 2009). Others have found no clear overall trends in leisure time but do report increasing leisure inequality, where lower educated adults generally have seen their leisure time increase since the 1970s while the higher educated have seen this less, if at all (Gimenez-Nadal and Sevilla, 2012).

As people's more complex and diverse role obligations (van den Broek et al., 2002) are in part facilitated by new communication technologies such as the Internet or mobile telephones, these media have also enabled people to make more complicated or last-minute leisure arrangements. The insatiable consumer identified by Campbell (1987) and Baudrillard (2002) is looking for new experiences. He or she is also better able than ever to arrange these experiences in a tight schedule and, if necessary, on the spot. And since the people who have the busiest, most rewarding jobs simultaneously attempt to maximize the productivity, or 'time yield', of their leisure consumption, it turns out that less leisure time often leads to more dense leisure activity in the remaining time available. As a consequence, the relation between the amount of available leisure time and the number of (cultural) activities engaged in is found to be either absent or positive rather than negative (van Gils, 2007; Sullivan, 2008).

According to Sullivan, voraciousness is not the only consequence of increasing business among what she calls the 'income rich, time poor'. Another temporal strategy she observes for this social segment is inconspicuous consumption, where people buy expensive consumer goods without the time to use them or the intention of displaying them. Sullivan (2008: 20) argues that 'inconspicuous consumption is defined by purchases of expensive leisure goods which are then stored away due to lack of time, with an intention to use them at some imagined future time when *there will be time*'. Thus, voracious consumers are busy working, busy optimizing the utility of their leisure time and busy buying stuff the use of which must be deferred as a result of their overall busyness. The higher educated are at the forefront of these developments as they have the longest working hours and are the most affluent. Partly as a consequence of the increased confluence of cultural and economic capital, high levels of material and cultural consumption go together more often (van Eijck and van Oosterhout, 2005). Overall, this makes consumption an increasingly cultural or creative activity, as those with money to spend are increasingly rich in cultural capital too.

FADING GENRE BOUNDARIES IN CULTURAL PRODUCTION AND DISTRIBUTION

The above-mentioned studies on changing cultural tastes and consumption patterns all seem to agree that traditional cultural boundaries are either shifting or eroding. They also have in common that explanations for this development are sought in social changes and their impact on the characteristics of consumers. Below, we will take the reverse angle and focus on the impact of developments in the modes of cultural production and distribution as drivers of change in consumption patterns.

In his classic article on classification in art, DiMaggio (1987: 441) seeks to systematically explain how and why such classifications change:

> I consider processes by which genre distinctions are created, ritualized, and eroded, and processes by which tastes are produced as part of the sense-making and boundary-defining activities of social groups. The two sets of processes come together in what I call *artistic classification systems* (ACSs): the way that the work of artists is divided up both in the heads and habits of consumers *and* by the institutions that bound the production and distribution of separate genres.

According to DiMaggio, cultural systems vary along several analytically distinct dimensions rather than being classifiable into discrete types. He argues that artistic classification systems differ in terms of differentiation (number of genres distinguished), hierarchy (degree of vertical ordering of genres according to prestige), universality (general versus group-specific salience of genres), and ritual strength (defence of genre boundaries by both producers and consumers). Each dimension contains a cognitive *and* an organizational component in order to guarantee that production and consumption patterns will not be studied in isolation. Focusing solely on consumption patterns implies that the menu from which consumers make their selections remains unexplained. Looking only at cultural production, on the other hand, suggests that 'art can be explained simply as an imposition upon consumers' (DiMaggio, 1987: 442). We therefore need to look at production and distribution as well if we wish to understand changes in cultural consumption patterns.

When looking at the production of culture, Taylor (2009) argues that highbrow culture has tried to position or present itself as more like popular culture since the late 1980s, which he demonstrates with examples from best-selling classical album lists, which are dominated by celebrity musicians such as Yo-Yo Ma, Lang Lang (who has an Adidas sneaker named after him: the Lang Lang Gazelle), André Rieu, and even Sting. Of course, we can also think of examples from the contemporary visual arts, where renowned artists such as Damien Hirst, Jeff Koons, Takashi Murakami, and Maurizio Cattelan are using the language of popular (media) culture to generate moments of global attention. The booming amount of literary and art prizes is similarly indicative of the increasing longing for mass publicity in the art world. Highbrow culture is seeking the companionship of popular culture and using its publicity strategies and imagery in order to get a foothold in the struggle for (media) attention. Taylor (2009) demonstrates that the way classical music is produced today reduced the ritual strength with which cultural boundaries are defended and that the production system plays a significant role in this. DiMaggio (1987: 450) speaks of commercial classification to point to the same phenomenon: producers seem less inclined to target specific markets for their highbrow products as they seek larger markets and economies of scale, 'often at the risk of reducing the ritual value of the products they sell'. The same process increases differentiation, which is another way of expanding the market by introducing new or cross-over genres. The competition among artists that is generated on such a commercial market also contributes to cultural differentiation. DiMaggio (1987) refers to this as professional classification, which is the result from struggles among artists for status and material success. As competition encourages artists to develop new styles in order to distinguish themselves both from their peers and from their predecessors, this leads to artistic classification systems that are highly differentiated yet poorly institutionalized. This lowers ritual strength and hierarchy, weakening the significance of genre boundaries for consumers.

TECHNOLOGY AND DISTRIBUTION: THE CASE OF MUSIC

New ways of marketing and producing cultural goods have thus contributed as well to weaker cultural classifications and less homogeneous consumption patterns. A final important development held responsible for changing cultural consumption patterns to be dealt with in this chapter is technological innovation and its implications for the production and, especially, distribution of culture goods. Technological change, particularly digitization, has not only greatly enhanced accessibility but also affected the meaning of genre as a label to organize cultural supply. According to Rimmer (2012: 303), 'In essence, music's contemporary status as a commodity of which a hugely increased amount (and information and opinion about) can now be relatively easily accessed by, and exchanged between, listeners, has set into motion more exploratory modes of music listening and consumption for many.' For one thing, research on the topic 'strongly suggest[s] that greater internet use increases the range of musical tastes' (Peterson and Ryan, 2004: 234). Moreover,

> It might well be the case that the development of MP3 technology and its associated listening apparatus (such as computers and mobile phones) facilitates not only a broadening of taste regimes but also shifts in the nature of music 'choosing'. Effectively, then, by enabling listeners to acquire and exchange an abundance of music quickly and easily, and subsequently allowing them to experience it in novel ways (e.g. random shuffle, personalized playlists), the digitization of music has built into it a tendency to weaken the experiential relationship binding listening practices to works from single genres. (Rimmer, 2012: 303)

In order to illustrate the consequences of such developments, we will consider Internet streaming services such as Spotify, Grooveshark or YouTube that provide audiences with a seemingly limitless back catalogue of music. The instant availability of a huge share of the history of recorded music is likely not only to change the perception of genre differentiation, as mentioned above, but also to change the way in which people listen to – or use – music.

The name Spotify arose from a contraction of the terms spot and identify, referring to the detection and recognition of ostensibly long-lost music. The site thus seems to cater for nostalgia, or 'a complex and sentimental longing to the past . . . characterized by a mixture of positive and negative feelings' (Davis, 1979). In the seventeenth century nostalgia was associated with neurological disease, followed in the twentieth century by melancholy and depression. The evolution of the notion is remarkable; recent psychological studies have demonstrated that nostalgia has a number of specific functions that have a positive effect on one's emotional state, enhancing self-esteem and a sense of belonging (Barrett et al., 2010). The fact that certain music, especially the works encountered during one's teenage years, is likely to be autobiographically salient and thus elicit lively memories makes it a particularly strong trigger of nostalgia.

But it can also be argued that Spotify and similar services mark the beginning of the end of nostalgic listening, owing to the fact that they make music less scarce and less bound to time. In addition, nostalgia is linked mostly to shared experiences, while Spotify marks a shift in experiencing music from a collective to an individualized practice. Furthermore, Spotify's distribution of an immense archive will result in a future generation of music-lovers who will grow up with not merely contemporary music.

In *Retromania* (2011), American music journalist Simon Reynolds claims that contemporary pop culture reaches back on its history in multiple ways. Reynolds points out the many recent band reunions, genre revivals, re-releases of old albums, and popularity of mash-ups and remixes. According to Reynolds (2011: 14), this pull towards nostalgia in or through music can be easily explained by music's strong attachment to a certain Zeitgeist or period in one's life, but given its current shape and size he argues it is becoming a worrying addiction in our digital culture: 'Are we heading towards a sort of cultural-ecological catastrophe, where the archival resources of rock history have been exhausted? What happens when we run out of past?'

To support his argument that contemporary pop culture is addicted to its own past, Reynolds brings out many examples. The most captivating is probably Chris Anderson (2006) and his theory of the so-called long tail, which Anderson first tried to use back in 2004 in order to analyse online retailers' business models. According to this model, products with a relatively low demand can still take a great segment of a business. According to Anderson (2004), 'Hit-driven economics is a creation of an age without enough room to carry everything for everybody . . . This is the world of scarcity. Now, with online distribution and retail, we are entering a world of abundance.' The shelf space of web shops is enormous, allowing customers to go deeper into the catalogue browsing a long list of available titles that would not be there if the shop were to rely on a local market. In line with the arguments put forward by Peterson and Simkus (1992), Lahire (2008) and Rimmer (2012), Anderson (2004) argues that people's tastes are not as mainstream or limited as we might think and that, once distribution and consumption go online, there is room for many niches, ready to be embraced by worldwide audiences and to be catered for by the industry.

The concept of the long tail not only provides insight into how money can be made in today's music industry (that is to say, how online web shops have a great advantage over traditional retailers, as they can attend to thinly spread audiences), but also exposes the consequences of a market that is being controlled by its back catalogue. Contemporary releases cannot outweigh pop music's nostalgic legacy. Not only is the musical past growing day by day; the imbalance between old and new will increase accordingly, as the past does not need to make room for the present in a digital culture characterized by the absence of spatial limits. This can be noted, for instance, in the recommendations a customer receives on Amazon based on his or her purchase. When one buys a recently released album, similar records are being offered on an algorithmic basis. Needless to say, the majority of these suggestions will turn out to be older than the new album one has just purchased. The decrease in the proportion of new releases sold relative to the back catalogue, which has been going on for years, leaves no room for misunderstanding: the past is indeed 'outselling the present', as stated by Reynolds (2011). Recent statistics by *OC Weekly* even point out that, over the first half of 2012, the American record industry sold more old than new albums for the first time in history (Kornelis, 2012).[1] Nielsen analyst David Bakula argues that another reason for this, apart from their availability, is that record labels and retailers keep lowering the prices of older albums, which attracts new customers with a specific interest in the older releases. Therefore the scenario in which contemporary music will play an increasingly marginalized role not only has economic implications, but also affects consumers' buying and listening behaviour.

The gradual ageing of pop culture is a side effect of a broader tendency, which is manifest not only in retail industry, but in pop culture as a whole – be it mainstream or alternative. For the time being, Spotify and its like seem to be the culmination of a development, initiated around the turn of the century, when digitization resulted in the emergence of a rhizomatic network culture. The notion of the rhizomatic network, originally from biology and referring to the underground expanding roots of plants such as strawberries, was introduced by Deleuze and Guattari (1988) to articulate a new way of critical thinking. It refers to an anarchistic, horizontal root structure containing a large amount of connections and new shoots, lacking a distinct beginning or ending and having no central authority. Applying this conception to pop culture, one could interpret the diminishing influence of record shops and hit charts, which were once vital for the connection between music fans and the industry, as an example of the decay of a more centralized structure. The rapid emergence of peer-to-peer networking, YouTube and Spotify may well signify the establishment of a knowledge rhizome where everything is immediately accessible while hierarchy, universality, and ritual strength are virtually absent.

Reynolds (2011) speaks of the insatiable hunger of music consumers, reminiscent of the voracious consumer described by Sullivan and Katz-Gerro (2007). He argues: 'As with the Internet as a whole, our sense of temporality grows even more brittle and inconstant: restlessly snacking on data bytes, we flit fitfully in search of the next instant sugar rush' (Reynolds, 2011: 61). This individualized mode of voracious consumption is also acknowledged by author Douglas Coupland:

> I'm starting to wonder if pop culture is in its dying days, because everyone is able to customize their own lives with the images they want to see and the words they want to read and the music they listen to. You don't have the broader trends like you used to. (Solomon, 2010: 12)

The decreasing impact of central hit lists that function as the soundtrack to people's lives makes shared cultural experiences – at least through recorded music – less likely to occur. At the same time, it has become easier to share music with others through the seamless integration of Spotify with Facebook, or the massive sharing of personal listening statistics with your online contacts through interactive websites such as Last.fm. Personally, music remains very important, but, given the multiple channels through which everything is made accessible to everybody, it does seem to lose quite a bit of its social urgency or impact. The 'imagined communities' of music fans become ever more fragmented and differentiated.

CONCLUSION

We have identified a number of major shifts in cultural consumption patterns that took place during the past few decades. First, the link between social stratification and cultural consumption has blurred. The rise of the cultural omnivore has demonstrated that both highbrow and popular culture are consumed most by those with the highest levels of education and occupational status. The distinction between cultural brow-levels has become less clear, which is also reflected in the fact that popular culture is increasingly receiving serious critical attention while highbrow culture is seeking a public beyond its traditional elitist audience. In addition, inequality in the volume of cultural consumption has risen as a result of a polarization of time budgets between households. Those inclined to be omnivores are also the most voracious cultural consumers, and studies indicate that, instead of being geared towards either highbrow or popular fare, cultural consumption patterns are primarily structured by differences in breadth and intensity. Instead of effacing cultural differentiation, digitalization may have increased differences between socio-cultural groups, as those without much interest in culture are hardly encouraged by its mere accessibility, while those rich in cultural capital experience few limits to their voraciousness. In the meantime, digital distribution channels have made cultural consumption patterns so differentiated that their role as sources of social cohesion or distinction is becoming increasingly hard to grasp.

NOTE

1. 'Old' was defined as released at least 18 months ago.

REFERENCES

Anderson, C. (2004), 'The long tail', *Wired*, 12.10, available at: http://www.wired.com/wired/archive/12.10/tail.html.

Anderson, C. (2006), *The Long Tail: Why the Future of Business Is Selling Less of More*, New York: Hyperion Books.

Barrett, F.S., K.J. Grimm, R.W. Robins, T. Wildschut, C. Sedikides and P. Janata (2010), 'Music-evoked nostalgia: affect, memory and personality', *Emotion*, **10** (3), 390–403.

Baudrillard, J. (2002), 'The ideological genesis of needs', in J.B. Schor and D.B. Holt (eds), *The Consumer Society Reader*, New York: New Press, pp. 57–80.

Bauman, Z. (1996), 'From pilgrim to tourist, or a short history of identity', in S. Hall and P. du Gay (eds), *Questions of Cultural Identity*, London: Sage, pp. 18–36.

Becker, G.S. (1965), 'A theory of the allocation of time', *Economic Journal*, **75** (299), 493–517.

Bell, D. (1976), *The Cultural Contradictions of Capitalism*, New York: Basic Books.

Bennett, T., M. Savage, E. Silva, A. Warde, M. Gayo-Cal and D. Wright (2009), *Culture, Class, Distinction*, London: Routledge.

Bourdieu, P. (1984), *Distinction: A Social Critique of the Judgement of Taste*, London: Routledge & Kegan Paul.

Bourdieu, P. (2002), 'The forms of capital', in N.W. Biggart (ed.), *Economic Sociology*, Oxford: Blackwell, pp. 280–91.

Brooks, D. (2000), *Bobos in Paradise: The New Upper Class and How They Got There*, New York: Simon & Schuster.

Campbell, C. (1987), *The Romantic Ethic and the Spirit of Modern Consumerism*, Oxford: Blackwell.

Davis, F. (1979), *Yearning for Yesterday: A Sociology of Nostalgia*, New York: Free Press.

Deleuze, G. and F. Guattari (1988). *A Thousand Plateaus: Capitalism and Schizophrenia*, London: Athlone Press.

DiMaggio, P. (1987), 'Classification in art', *American Sociological Review*, **52** (4), 440–55.

Florida, R. (2002), *The Rise of the Creative Class*, New York: Basic Books.

Gershuny, J. (2000), *Changing Times: Work and Leisure in Postindustrial Society*, Oxford: Oxford University Press.

Gershuny, J. (2009), 'Veblen in reverse: evidence from the multi-national time-use archive', *Social Indicators Research*, **93** (1), 37–45.

Gimenez-Nadal, J.I. and A. Sevilla (2012), 'Trends in time allocation: a cross-country analysis', *European Economic Review*, **56** (6), 1338–59.

Kornelis, Chris (2012), 'Old records are outselling new ones for the first time', available at: http://blogs.ocweekly.com/heardmentality/2012/07/old_records_are_outselling_new.php (accessed 29 March 2013).

Kraaykamp, G. and K. van Eijck (2010), 'The intergenerational reproduction of cultural capital: a threefold perspective', *Social Forces*, **89** (1), 209–31.
Lahire, B. (2008), 'The individual and the mixing of genres: cultural dissonance and self-distinction', *Poetics*, **36** (2–3), 166–88.
Linder, S.B. (1970), *The Harried Leisure Class*, New York: Columbia University Press.
Lizardo, O. (2008), 'The question of culture consumption and stratification revisited', *Sociologica*, **2** (September–October), 1–32.
Lizardo, O. (2010), 'Culture and stratification', in J.R. Hall, L. Grindstaff and M. Lo (eds), *Sociology of Culture: A Handbook*, London: Routledge, pp. 305–15.
Lury, C. (1996), *Consumer Culture*, London: Routledge.
Park, S.B. (1993), *An Aesthetics of the Popular Arts: An Approach to the Popular Arts from the Aesthetic Point of View*, Uppsala: Almqvist & Wiksell International.
Peterson, R.A. (2005), 'Problems in comparative research: the example of omnivorousness', *Poetics*, **33** (5–6), 257–82.
Peterson, R.A. and J. Ryan (2004), 'The disembodied muse: music in the Internet age', in P. Howard and S. Jones (eds), *Society Online: The Internet in Context*, London: Sage, pp. 223–36.
Peterson, R.A. and A. Simkus (1992), 'How musical tastes mark occupational status groups', in M. Lamont and M. Fournier (eds), *Cultivating Differences: Symbolic Boundaries and the Making of Inequality*, Chicago: University of Chicago Press, pp. 152–86.
Reynolds, S. (2011), *Retromania: Pop Culture's Addiction to Its Own Past*, London: Faber & Faber.
Rimmer, M. (2012), 'Beyond omnivores and univores: the promise of a concept of musical habitus', *Cultural Sociology*, **6** (3), 299–318.
Solomon, D. (2010), 'Dreaming of a white Olympics: questions for Douglas Coupland', *New York Times Sunday Magazine*, 7 February, 12.
Sullivan, O. (2008), 'Busyness, status distinction and consumption strategies of the income rich, time poor', *Time and Society*, **17** (1), 5–26.
Sullivan, O. and T. Katz-Gerro (2007), 'The omnivore thesis revisited: voracious cultural consumers', *European Sociological Review*, **23** (2), 123–37.
Taylor, T.D. (2009), 'Advertising and the conquest of culture', *Social Semiotics*, **19** (4), 405–25.
Van den Broek, A., K. Breedveld and W. Knulst (2002), 'Roles, rhythms and routines: towards a new script of daily life in the Netherlands?', in G. Crow and S. Heath (eds), *Social Conceptions of Time*, Basingstoke: Palgrave Macmillan, pp. 195–214.
Van Eijck, K. and R. van Oosterhout (2005), 'Combining material and cultural consumption: fading boundaries or increasing antagonism?', *Poetics*, **33** (5–6), 283–98.
Van Gils, W. (2007), *Full-Time Working Couples in the Netherlands: Causes and Consequences*, Wageningen: Ponsen & Looijen.
Veblen, T. ([1899] 1953), *The Theory of the Leisure Class: An Economic Study of Institutions*, New York: Viking Press.

FURTHER READING

Pierre Bourdieu's *Distinction: A Social Critique of the Judgement of Taste* (1984) is the most influential sociological publication on cultural consumption. Using all sorts of empirical material to illustrate his points, Bourdieu lays out his theory of inequality and reproduction of taste, using his central theoretical notions of habitus, (cultural) capital, and field. *Culture, Class, Distinction*, published in 2009 by Tony Bennett and five co-authors from the UK, is equally eclectic in its methods. Organizing their material both around artistic disciplines and around different classes, they sketch more contemporary cultural lifestyles and complement Bourdieu's work with more recent theoretical accounts of differentiation in cultural consumption.

9. Digital divide

*Pippa Norris and Ronald Inglehart**

The discussion of a digital divide in Internet access and use is the most recent manifestation of the broader phenomenon of information poverty (Hayward, 1995; Wresch, 1996; Arunachalam, 1999). Across and within societies, enduring and widespread disparities exist in access to all forms of information and communication technologies, including the audiovisual sector (radio, television, and films), telecommunications, and the print sector, as well as dial-up and broadband connections to the Internet. Moreover, today it makes little sense to focus only on the digital divide in computing technology, because industries that once used to be regarded as distinct sectors are increasingly converging into multimedia enterprises; smart phones access DVD movies, the *New York Times* video can be watched on TiVo-enabled televisions, while cell phone snaps of breaking events appear on network TV. Indeed even the conventional notions of 'news' and 'the news media', or 'access to the Internet', once clear cut, have fuzzier boundaries today.

To address these issues and clarify trends, we will start by comparing national indicators based on official statistics, standardized by international agencies such as UNESCO, the International Telecommunication Union (ITU), the OECD, Eurostat, and the World Bank, based on time-series data for all countries where data is available. These sources are limited in certain important respects, however, including problems of systematic bias in missing data, whether measurements relate to availability or use, and deciding upon the most appropriate unit of analysis for comparison.

Missing data is most prevalent in the least developed economies; for example, a third of these societies have no recent official statistics monitoring the proportion of households that have a radio or TV set (Partnership on Measuring ICT for Development, 2008). Missing data can arise from the absence of a recent official household or market research survey monitoring ownership of consumer durables. National statistical offices in the least developed economies often lack the capacity and resources to conduct reliable and comprehensive household surveys and censuses, and surveys are also problematic in fragile post-conflict states that contain a highly mobile population of displaced persons and refugees. As a result, estimates of the worldwide distribution of ICTs contain a systematic bias in missing data, potentially inflating these estimates.[1] Rapid changes in the use of ICTs also make it important to have up-to-date annual estimates, and decennial population censuses are insufficient for this purpose. Some major ICT indicators are now more than five years out of date, or suffer from poor reliability and incomplete time-series. UN agencies, including UNESCO, are currently taking steps to improve the international comparability and accuracy of statistical indicators for communication and information technologies, but these initiatives will take time to bear fruit (Puddenphatt, 2007). There have also been attempts to generate comprehensive multidimensional indicators. For example, since 2001 the World Economic Forum (2007) has published a composite Networked Readiness Index for 127 countries, providing benchmarks to monitor the economic environment, technological infrastructure, and patterns of usage. Given serious limits of missing data in many developing countries, however, and systematic biases that can arise from their exclusion, more reliable and comprehensive coverage is provided by analyzing separate indices (Barzilai-Nahon, 2006).

There is also greater emphasis on gathering ICT statistics measuring the availability of technological products than on collecting more complex data on their use and impact. While counting the physical number of radio or television sets or personal computers in households is relatively straightforward, it is more difficult to assess levels of use, for example whether the location for Internet access is counted at work, at home, or elsewhere, such as brief use in public libraries, cafes, or community tele-centers. Some people have 24/7 WiFi and fast broadband connections; others may access a particular website once a week or once a month. Some may use the Internet intensively every day for research and education, email and file sharing, financial transactions, online gaming, and entertainment, while others send an occasional text message. Are all these equally 'Internet users'? The convergence of ICTs, and the rapid development of new functions for ICTs, complicates the measurement and comparisons over time. For example, YouTube was invented in a Menlo Park garage in February 2005; just a few years later, in a Fall 2007 survey, almost half of all American Internet users (48 percent) reported that they had used YouTube or other video sharing websites. This figure rose to almost three-quarters (70 percent) of the younger generation (under 30-year-olds) (Pew Internet, 2008a). The available surveys monitoring access to and use of ICTs have improved in reliability over time, but we are still a long way from standardizing these measures.

The unit of analysis also needs consideration; for example, official records document the number of SIM cards that are issued by the telecommunications industry for mobile cell phone accounts, or the number of cell phones sold, but this does not tell us the number of individual users (Sutherland, 2008).[2] Many people may have multiple SIM cards, exchanging them to gain coverage in different areas or countries, or separate mobiles for use at home and at work. Convergence across sectors has also complicated the estimates of trends; for example, official statistical offices maintain records of newspaper sales and circulation figures, but this does not take account of the millions of online readers. The number of television sets per household is also a poor proxy for the number of viewers, not simply because of time-displacement technologies such as VCRs and TiVo, but also because of cross-national variations in social patterns of TV viewing and in the average size of households (Bongaarts, 2001).

If these important qualifications are borne in mind, the available international data provide the best available estimate of trends over time and comparisons across countries and global regions.

TRENDS IN ACCESS TO THE INTERNET

Has access to the Internet gradually widened to include most of the population in developing societies, or do deep-rooted inequalities persist? During the 1990s, many interpretations of the digital divide envisaged the necessity of distributing personal computers and wired broadband connections (DSL, cable) to access the Internet and email – and thus the need for reliable electricity sources, keyboard skills and computer literacy, landline telephone infrastructure, and the like. But recent years have witnessed important technological innovations in this field that have reduced some of the technological hurdles to information access in poorer societies, bypassing some of the obstacles. This includes the availability of wind-up radios, solar-power batteries, wireless connectivity (WiFi, WiMax), $100 rugged laptops, Internet cafes, community telephone and Internet centers, and cell phones with data services, email, and text messaging.[3] All these developments may help to close the global digital divide. At the same time, some observers suggest that the core inequalities in information poverty have persisted and may even have deepened (James, 2008). Post-industrial societies and emerging economies that invested heavily in advanced digital technologies have reaped substantial gains in productivity. This may encourage them to build on their success and expand this sector of the economy still further. Moreover, it still remains the case that, beyond isolated pockets of innovation, many of the poorest societies in the world continue to lack the basic infrastructure and resources to connect their rural populations to global communication networks and markets.

Despite a substantial surge in access to mobile cell phones with data services, the available evidence on the distribution of Internet users suggests that the relative size of the gap between rich and poor societies worldwide has widened in recent years, rather than narrowed (World Bank, 2008). From 1996, Internet use accelerated steadily and rapidly in high-income nations until 2007, when the majority of the population living in these countries was online. Nevertheless, substantial variations remain even among high-income nations; for example, within the European Union only one-fifth of all households in Bulgaria and Romania were connected to the Internet in 2007, compared with more than three-quarters of all households in such countries as the Netherlands, Sweden, and Denmark (Eurostat, 2007[4]). The diffusion of the Internet among middle-income economies accelerated later than in richer nations, and most of these societies continue to lag far behind in Internet connectivity. By 2007, just over one-quarter of the population (29 percent) was online in middle-income societies, compared with almost twice as large a share of the population (58 percent) in high-income societies. In rapidly growing emerging markets that are large consumers and producers of ICT goods and services, such as Brazil, China, Russia, South Africa, and India, Internet access and email have spread most widely among the professional urban middle classes, often by data services via smart mobile cell phones rather than by traditional computers (OECD, 2007).

By contrast, despite these trends, the populations living in the least developed societies around the world, such as Mali, continue to lack Internet access, including connectivity via computers with fast broadband connections as well as data service cell phones. The starkest contrasts are in Africa and Asia. Thus, the ITU (2008a) estimates that fewer than five out of every 100 Africans used the Internet in 2006, compared with one out of every two people living in the G8 nations. Internet use has expanded at a modest rate in low-income societies, increasing from 0.06 percent of the population in 1997 to 6 percent a decade later. This is a large proportional increase, but the absolute level remains low. The size of the global gap can be calculated to summarize the difference in the proportion of Internet users in the low- and high-income societies; although the relative size of the gap diminished, the absolute size of the gap quintupled from almost 10 percentage points in 1997 to 53 points in 2007. Access is gradually growing in most poor nations; places such as Mali are not untouched by these developments, but they lag far behind the rapid rate of diffusion of PCs, laptops, and smart phones, with fast wireless and LAN broadband connections common in most affluent post-industrial nations, along with the related digital technologies through iPods, TiVos, PDAs, and similar devices.[5]

ACCESS TO LANDLINE AND CELLULAR TELEPHONES

Access to the Internet used to be limited by the availability of personal computers and the need for dial-up connections through landline telephones. This represented a major bottleneck; in particular, state controlled telecommunication monopolies in Central and Eastern Europe, Africa, and Asia often failed to provide universal services to many remote rural areas, and the demand for landlines lagged far behind supply (Barnett and Choi, 1995; Barnett et al., 1996, 1999; Barnett, 2001). Today WiMax and WiFi connections have reduced the need for landline connections for Internet access, and the surge in the use of mobile cell phones, many of which provide data services, has been dramatic. Nevertheless, although mobile cell phones have reduced the barriers to Internet use, and although they represent a cheaper technology, they have not totally eliminated global inequalities in Internet access or in telephony.

Between 1975 and 1999, there was an increasing disparity between high- and low-income economies in access to telephony (World Bank, 2008). Since then, however, these disparities may be changing, as the number of fixed telephone lines in OECD countries largely stabilized from 2000 to 2006, while the number of landlines grew in developing societies (ITU, 2008b). Moreover the spectacular growth in mobile cellular phones has transformed the market.

The 2008 World Summit on the Information Society reports that there were 3.3 billion mobile subscribers at the end of 2007 – an increase of more than a billion users during the two years since 2005 (ITU, 2008b). India, Brazil, Russia, and China each added many millions of users. Mobile cell phones are important for inter-personal communication and for extended social networks, especially in developing societies that lack the infrastructure for reliable and affordable landline telephones. Cell phones with data services facilitate email, SMS texting, gaming, music, photos and videos, and mobile access to websites, bypassing the need for keyboard skills and expensive computing equipment. With small screens and limited speeds, however, they are most useful for brief information updates using the Internet, such as for checking weather, bank accounts, flight departures, social networks, email, or map directions, and they are still not as flexible, effective, or fast as personal computers and laptops for surfing the Internet for an extended period of time. Nielsen Mobile, tracking usage in 16 countries, found that in mid-2008 nearly 40 million Americans (16 percent of all US mobile users) used their handset to browse the web, twice as many as in 2006, as higher speed connections and unlimited data packages spread (Nielsen Mobile, 2008). The UK, Italy, and Russia were the next highest in usage. But in Indonesia, Taiwan, India, and New Zealand, for example, less than 2 percent of mobile subscribers used their phone handset to surf the web. Moreover Nielsen Mobile found that, while PC users surfed about 100 individual websites per month, mobile users visited about six domains. The most popular websites were Yahoo Mail, Google Search, and the Weather Channel. Data connection speeds will be faster with third generation (3G) connections, once network footprints for this service expand. It is also difficult to estimate the exact number of individual customers who have mobile phones, since many people have duplicate SIM cards (Sutherland, 2008). The ITU estimates that mobile penetration is now over 100 percent in Europe. Nevertheless, even with the aggressive surge in cell phones, global disparities in access persist. In the poorest countries, such as Burundi, Ethiopia, Sierra Leone, and Niger, for example, in 2007 less than 3 percent of the population subscribed to a mobile phone (ITU, 2008a). On average, in 2007 there were 152 telephone subscriptions per 100 people in high-income societies (including fixed and mobile accounts at home and work), compared with 31 in low-income societies. Again, poorer developing countries have indeed expanded their telephone connectivity, but they started from a very low base, and the absolute gap between rich and poor nations has grown over the last three decades.

ACCESS TO RADIO, TELEVISION, AND NEWSPAPERS

How do these trends compare with the diffusion of radio and television sets (whether connected via terrestrial, cable, or satellite signals), as well as the circulation and sales of newspapers? Television being an older technology, it might be expected that the diffusion rate should be flatter over time than for new ICTs. For television absolute and relative inequalities in access persist; in low-income societies, there were only six television sets per 100 people. From 1975 to 1999 a modest rise in the availability of radios occurred in low-income societies. According to the latest available figures, on a per capita basis, almost five times as many radios are available in rich as in poor societies (World Bank, 2008).

UNESCO gathers statistics on the number of outlets and the distribution, circulation, and sales of printed daily newspapers and periodicals. On a per capita basis, this is the only medium where there has been growing convergence between high- and low-income societies (World Bank, 2008). But this is not due to a significant expansion of readership or circulation figures in the low- or medium-income countries; it reflects an erosion of newspaper circulation figures in the high-income countries ever since the mid-1980s (predating the rise of online newspapers on the Internet). Even with this closure, the gap in newspaper circulation between rich and poor nations remains substantial. Multiple readers may compensate somewhat for the contrasts in sales and circulation, but, on a per capita basis, one newspaper per 100 people is sold in low-income countries, compared with 26 per 100 in rich countries. Rising levels of literacy in developing societies have not substantially increased sales of daily papers. Evidence comparing the distribution of published books and periodicals shows similar disparities.

INDIVIDUAL-LEVEL INEQUALITIES IN ACCESS TO INFORMATION

In addition to these cross-national disparities, a related and even more complex form of inequality exists; even within affluent post-industrial economies, such as the United States, some citizens have far greater access and use of information sources than others. Even among those with access, people differ significantly in their regular exposure and attention to mass communications and news sources, and this pattern usually reflects social and demographic groups within each country. The 'social information gap' refers to individual-level disparities that are closely associated with socioeconomic resources (measured by levels of household income, household savings, work status, and occupational social class), cognitive skills (measured by education and language), demographic characteristics (age and gender), and motivational factors (such as interest and trust in the media). During the early 1970s, the 'knowledge gap' was first noticed by scholars, in which a group of better-educated people, who already have more information, also acquire more information from the media. By contrast, those with low education, who already know less, acquire less information (Tichenor et al., 1970; Holbrook, 2002). Educational and socioeconomic differences are thereby thought to be reinforced over time, not mitigated, by attention to the news media.

Broader patterns of selective exposure to news are not simply linked to education and social class, however, as there are also generational and age-related patterns of news consumption. For example, cell phones with data services and wireless handheld PDA devices are particularly common among the younger generation in the US. These devices facilitate on-the-go texting and email communications, sharing digital photos, and viewing YouTube videos, Accuweather forecasts, or GPS maps, in coffee shops, airports, and traffic jams. But this does not necessarily mean that the younger generation have become heavy consumers of online news from traditional media sources, such as CNN and the *New York Times* (Althaus and Tewksbury, 2000; Horrigan, 2008). Readership of newspapers is usually found to be skewed towards the older population, as is viewership of television, owing to the sedentary nature of this activity. A comparative study of newspaper readership in European societies since the early 1970s found that social background continued to be an important predictor of who does, and does not, regularly read the press, although educational and gender differences diminished gradually over time (Norris, 2000). The same study reported that the social profile of who watches the news in Europe also widened over the same period, as the audience expanded in size. Race-related patterns of news consumption have also been widely documented in the United States, while there are continuing debates about the role of race and ethnicity in determining the contemporary use of the Internet (NTIA, 2000).

Most of the recent evidence concerning these issues has been derived from studies of Internet access in post-industrial societies, especially the United States and Europe. During the 1990s, for example, the US Department of Commerce generated a series of reports, *Falling through the Net*, drawing attention to patterns of exclusion among the African-American and rural communities (NTIA, 1999; 2000). It was claimed that some of these social gaps had closed over time, although this interpretation was subsequently challenged by other studies (Martin, 2003). Similar concerns about the digital divide became common in other affluent post-industrial societies; for example, the OECD and the European Union reported parallel social disparities in computer and Internet access, reflecting differences in income, social class, employment status, education, generation, ethnicity/race, gender, and urbanization (Loader, 1998; OECD, 2000). Debate during the 1990s focused on whether the social gaps in the online population would persist as a permanent feature of information societies, with social and employment status determined by levels of technological connectivity and skills, or whether these differences in the online population would gradually fade away over time, or 'normalize', as costs fell and the technology became more user-friendly, for example once text messaging and web browsers became more easily available and cheaper through smart cell phones (Birdsell et al., 1998; Resnick, 1998; Margolis and Resnick, 2000). Tracking surveys conducted by the Pew Internet and American Life Project since 2000 have documented a gradual closure of the gender gap in Internet use in the US, but the persistence of a 10-point gap between black and white populations, and a massive 55-point generational gap between the under-30s and the over-65s (Ono and Zavodny, 2003; Pew Internet, 2008b). Scattered research has explored similar disparities in other countries, but it remains to be seen whether they reflect a general pattern (van Dijk, 2005; Ono and Zavodny, 2007).

We therefore need to explore not just patterns of Internet access but also reported use of the broader range of news media. Identifying the most effective public policies designed to overcome the global and social information gaps, and predicting the future direction of trends, requires a careful examination of the underlying drivers behind this phenomenon. What are the most plausible explanations of these disparities? During the last decade, research in the scholarly literature and in the policy community has focused attention on understanding the reasons for lack of connectivity to new ICTs, including the economic, political, and sociological drivers (Guillen and Suarez, 2005; Yu, 2006). In understanding the drivers, as an analytical strategy it helps to distinguish between explanations of the digital divide in access to new ICTs (typically focused on indicators such as the distribution of Internet hosts and users, the availability of hardware such as laptops, PCs, and smart phones, and the price and availability of broadband or wireless connectivity) and explanations that seek to understand broader patterns of information inequality across a wide range of media sources (Yu, 2006). If global inequalities are confined to access to new ICTs, then the most appropriate explanation (and the most effective policy interventions) might rest on factors closely related to this medium. Thus, one might focus on: the cost of computer hardware, software, and Internet service connection charges; the telecommunication infrastructure supporting dial-up, broadband (such as DSL, cable, fiber), and 3G wireless (WiFi/WiMax) connections; the regulatory environment in telecommunications and bottlenecks arising from government policies and lack of market competition; the need for computing and literacy skills so that people can use these technologies effectively, including creating their own community websites, networks, and blogs; and the predominance of English-language contents, which characterized much of the World Wide Web during the early years, limiting non-English speakers.[6]

But if lack of connectivity to ICTs closely reflects disparities that persist across and within societies in access to other more traditional mass communication channels – including access to radio and television sets, and the sales and circulation of newspapers, magazines, and books, as well as use of telephones and the audience for movies – then this suggests that the causes are not specific to the medium itself (Hayward, 1995; Wresch, 1996). For one doesn't need keyboard or literacy skills to switch on a radio or TV. Instead, information inequalities in poorer societies may be attributed to more deep-rooted problems of economic development – such as endemic poverty and a low level of educational and cognitive skills, not to speak of the lack of a reliable electricity supply, which would also hinder access to traditional mass media such as television (Norris, 2001). Major utilities supplying power grid and transmission lines often fail to serve remote areas; in Ethiopia, for example, less than 1 percent of the rural population, and only 13 percent of all households, has access to electricity (Central Statistical

Agency, 2000). The growth in population is outstripping the growth in connections, increasing the proportion of Ethiopians without electricity. Half the population cannot read. Innovative projects, such as distributing $100 laptops, free wind-up radios, and solar-powered flashlights, will not solve these basic developmental problems. Unless broader social and economic inequalities are addressed, it will remain difficult for the flow of information and news about the world to penetrate beyond elites in poorer societies and to affect public opinion more generally. It may well be the case that journalists, politicians, diplomats, and business leaders in Dubai, Kuala Lumpur, and Addis Ababa are densely networked via laptops and smart phones, and informed about world affairs from headline stories in CNN International, Google News, and BBC World. But this does not mean that poorer people in these countries will have affordable access to these resources.

How do we explain these disparities? The fifth wave of the World Values Survey (2013) monitored regular use of a variety of news and information sources in a wide range of 51 societies.[7] Norris and Inglehart (2009) studied the relationship between media use and a number of demographic and socioeconomic variables, cognitive skills, and motivational factors. Education is one of the strongest predictors of news media use, being significant across all societies, with the strongest impact in low-income societies. This relationship is to be expected, since education is closely related to literacy, essential for newspaper reading and use of the Internet, as well as providing the cognitive skills and background knowledge which help in processing complex information about current events and public affairs in the news. After education, household income proved strong and significant across all types of society; as already observed, resources are necessary to buy TVs and radio sets and to access cable, satellite, and pay TV services, as well as to subscribe to the Internet and to buy newspapers on a regular basis. More affluent households are more likely to be able to afford these services. Among the demographic predictors, the age effect proved significant and negative across all types of economies, indicating that the younger generations were more likely to use the news media than are older people. The gender gap reversed by type of economy, with women predominating in the news media audience in the rich countries, but men predominating in poorer societies. Social class, however, demonstrated a more complex pattern; it played a significant role in predicting news use in medium- and low-income nations, but not in more affluent societies. This pattern suggests that the class disparities in use of the news media are greatest in countries where access is relatively limited, but that, once access diffuses more widely throughout the population, then class differences in use of the news media may fade. Finally, both political interest and trust in the news media have a significant impact in all societies, and in the expected direction, with high levels of interest and trust being linked with high media use – with one exception: in low-income nations, the relationship between trust and media use was reversed. Overall, social and demographic characteristics (especially education, income, and occupational class) proved the strongest predictors of news use in low-income societies, with all these factors playing less important roles in high-income societies.

How does this pattern vary by type of media? The importance of education and cognitive skills associated with it has a significant and strong impact on all types of media use, especially newspapers and the Internet. The age profile shows the familiar pattern which has been widely observed for affluent societies: while newspapers and TV and radio news draw on an older readership, Internet use is skewed towards the younger generation. But by contrast in low-income societies the younger generation is more likely to access all forms of news media information than the older age groups, including newspapers and TV and radio news as well as the Internet. This suggests that the rapidly rising levels of primary and secondary schooling, literacy, and access to higher education which accompany human development in poorer societies have probably strengthened the cognitive skills, and thus media use, of the younger generation more than their parents and grandparents. Familiarity with the English language was also linked with use of the Internet and newspapers, but not audiovisual media, which are more likely to be available in local languages. The material resources of household income and savings were also related to use of all types of media, while social class was also significant across most categories, with higher class being particularly closely linked with greater use of newspapers and the Internet. Men were more likely than women to use newspapers and the Internet, while there was no gender gap in using TV and radio. Lastly political interest was also significant, and trust in newspapers and TV was linked with use of these media. Norris and Inglehart's (2009) analysis further suggests that these social characteristics are better at predicting Internet and newspaper users, while radio/TV attracted a broader and more socially diverse audience. The general pattern is that the social profile of newspaper readers and Internet users is relatively similar in its resources, motivation, and skills, with the most important difference in affluent nations linked with age. The strong similarity across use of these two media suggests that similar factors are driving usage in both cases, and that we do not need to focus on factors that are unique to the nature of the Internet, such as the cost or availability of broadband ISP services, keyboard skills, or the availability of computers.

CONCLUSIONS

The evidence considered in this chapter leads to three main conclusions. First, in *absolute* numbers, recent years have seen a remarkable surge in access to communication and information technologies, especially in post-industrial societies. The worldwide expansion of the online population has obviously been most striking; the first working prototype for the World Wide Web was built in 1990; by 2007, an estimated 1.4 billion people were using the Internet (ITU, 2008b). There has also been rapid growth in adoption of mobile cell telephones; the ITU (2008b) estimates that there are about 3.3 billion mobile cell phone subscribers, in a world of about 6.7 billion people. Cell phones expand social networks and communications, and those with data services facilitate flexible access to email, text messages, and the Internet. Major emerging markets, such as Malaysia, Russia, China, Brazil, and India, have registered some of the most substantial gains in connectivity, but growth has also been relatively rapid across Africa. In affluent post-industrial societies and in many emerging middle-income economies such as South Africa, growth has occurred in the availability of many other related ICTs, including landline telephones, household radios and TVs, facsimile and DVD machines, personal and laptop computers, and dial-up, broadband, and wireless Internet connections.

Secondly, despite these important developments, far from catching up and gradually closing over time, the *absolute* gap in levels of access to mass communications and information technologies, which divides the richest and poorest societies, has expanded in recent years. Advanced industrialized economies have pulled further ahead of the rest of the world in information access, as they started the technology race with greater financial resources available for investment in high-tech industries, a deeper and wider pool of people with technical and scientific skills, and an established telecommunications infrastructure. The public has relatively little access to information and communication technologies in the least developed and poorest countries in Africa and Asia. Nevertheless, many of the new digital information and communication technologies have become widely available only during the last decade and, in the longer term, patterns of technological diffusion, societal development, and generational turnover may help to close the global and social information divides. But there is little evidence from the available international statistics that this process is already under way, even with access to the old media such as radio, TV, and newspapers. Up to 2007, the poorest nations had not been catching up; they had been falling further behind.

Lastly, social inequalities in access to ICTs have also persisted in recent years, and most show no signs of closure, even within the richest nations, such as the United States, Sweden, and Australia (Norris, 2001; Warschauer, 2004; van Dijk, 2005). Many recent technological accounts imply that information gaps are confined to the use of personal computers and the Internet, but the evidence indicates that these inequalities apply to other kinds of media exposure and reflect more deep-seated social inequalities based on cognitive skills, socioeconomic resources, and motivational attitudes. Norris and Inglehart (2009) identify some of the most important variables, including education, income, and age, that are associated with media use and Internet use in particular. Further research seems desirable regarding the impact of media use on social and political attitudes and values.

NOTES

* This chapter is an edited excerpt of Norris and Inglehart (2009).

1. For the problem arising from missing, non-random data, see McKnight et al. (2007).
2. Subscriber identity module (SIM) cards are issued by cellular service providers to charge calls to users' accounts.
3. WiMax is important for development because it is a new 802.16 IEEE standard designed for point-to-point and point-to-multipoint wireless broadband access that is cheaper, smaller, simpler, and easier to use than any existing broadband option (such as DSL, cable, fiber, 3G wireless), and it also bypasses the existing wired infrastructure and legacy service providers (i.e. the telephone and cable companies).
4. For the EU policy background see EurActiv, http://www.euractiv.com/en/infosociety/bridging-digital-divide-eu-policies/article-132315.
5. Data of the ITU (2008b) also indicates a pronounced, positive correlation between a population's affluence and Internet use.
6. See for example Hargittai (1999), Afullo (2000), Norris (2001), Warschauer (2004), and Guillen and Suarez (2005).
7. In the analysis reported on here, the data was used to construct an overall summary media use scale, generated by summing self-reported weekly use of newspapers, radio/TV news, the Internet, books, and magazines as information sources 'to learn what is going on in your country and the world', with the mean 5-point score standardized to 100 points for ease of comparison.

REFERENCES

Afullo, T. (2000), 'Global information and Africa: the telecommunications infrastructure for cyberspace', *Library Management*, **21** (4), 205–13.
Althaus, Scott L. and David Tewksbury (2000), 'Patterns of Internet and traditional news media use in a networked community', *Political Communication*, **17** (1), 21–46.
Arunachalam, S. (1999), 'Information and knowledge in the age of electronic communication: a developing country perspective', *Journal of Information Science*, **25** (6), 465–76.
Barnett, G.A. (2001), 'A longitudinal analysis of the international telecommunication network, 1978–1996', *American Behavioral Scientist*, **44**, 1638–55.
Barnett, G.A. and Y. Choi (1995), 'Physical distance and language as determinants of the international telecommunications network', *International Political Science Review*, **16**, 249–65.
Barnett, G.A., T. Jacobson, Y. Choi and S. Sun-Miller (1996), 'An examination of the international telecommunications network', *Journal of International Communication*, **32**, 19–43.
Barnett, G.A., J. Salisbury, C. Kim and A. Langhorne (1999), 'Globalization and international communication networks: an examination of monetary, telecommunications, and trade networks', *Journal of International Communication*, **62**, 7–49.
Barzilai-Nahon, K. (2006) 'Gaps and bits: conceptualizing measurements for digital divide/s', *Information Society*, **22** (5), 269–78.
Birdsell, D., D. Muzzio, D. Krane and A. Cottreau (1998), 'Web users are looking more like America', *Public Perspective*, **9** (3), 33.
Bongaarts, J. (2001), 'Household size and composition in the developing world in the 1990s', *Population Studies*, **55** (3), 263–79.
Central Statistical Agency (2000), *Ethiopia Demographic and Health Survey 2000*, Addis Abba: Central Statistical Agency, available at: http://www.measuredhs.com/pubs/pdf/FR118/00FrontMatter.pdf.
Dijk, J.A.G.M. van (2005), *The Deepening Divide: Inequality in the Information Society*, London: Sage.
Eurostat (2007), *Internet Usage in 2007: Households and Individuals*, available at: http://epp.eurostat.ec.europa.eu.
Guillen, M.F. and S.L. Suarez (2005), 'Explaining the global digital divide: economic, political and sociological drivers of cross-national Internet use', *Social Forces*, **88** (4), 681–708.
Hargittai, E. (1999), 'Weaving the Western Web: explaining differences in Internet connectivity among OECD countries', *Telecommunications Policy*, **23** (10–11), 701–18.
Hayward, T. (1995), *Info-Rich, Info-Poor: Access and Exchange in the Global Information Society*, London: K.G. Saur.
Holbrook, T. (2002), 'Presidential campaigns and the knowledge gap', *Political Communication*, **19**, 437–54.
Horrigan, J. (2008), *Mobile Access to Data and Information*, March, Washington, DC: Pew Internet and American Life Project, available at: http://www.pewinternet.org/pdfs/PIP_Mobile.Data.Access.pdf.
ITU (2008a), 'Facts and figures', available at: http://www.itu.int/ITU-D/connect/africa/2007/bgdmaterial/figures.html.
ITU (2008b), *World Summit on the Information Society Stocktaking 2008*, Geneva: ITU, available at: http://www.itu.int/wsis/stocktaking/index.html.
James, J. (2008), 'Digital divide complacency: misconceptions and dangers', *Information Society*, **24** (1), 54–61.
Loader, B.D. (ed.) (1998), *Cyberspace Divide: Equality, Agency and Policy in the Information Society*, London: Routledge.
McKnight, P.E., K.M. McKnight, S. Sidani and A.J. Figueredo (2007), *Missing Data: A Gentle Introduction*, New York: Guilford Press.
Margolis, M. and D. Resnick (2000), *Politics as Usual: The Cyberspace 'Revolution'*, Thousand Oaks, CA: Sage.
Martin, S.P. (2003), 'Is the digital divide really closing? A critique of inequality measurement in *A Nation Online*', *IT and Society*, **1** (4), 1–13.
Nielsen Mobile (2008), 'Critical mass: the worldwide state of the mobile web', July, available at: http://www.nielsenmobile.com/documents/CriticalMass.pdf.
Norris, P. (2000), *A Virtuous Circle*, New York: Cambridge University Press.
Norris, P. (2001), *Digital Divide*, New York: Cambridge University Press.
Norris, P. (2003), *Digital Divide: Civic Engagement, Information Poverty, and the Internet Worldwide*, Cambridge: Cambridge University Press.
Norris, P. and R. Inglehart (2009), *Cosmopolitan Communications*, New York: Cambridge University Press.
NTIA (1999), *Falling through the Net*, Washington, DC: US Department of Commerce, available at: www.ntia.doc.gov.ntiahome/fttn99.

NTIA (2000), *Falling through the Net*, Washington, DC: US Department of Commerce, available at: http://search.ntia.doc.gov/pdf/fttn00.pdf.
OECD (2000), *Information Technology Outlook*, Paris: OECD, pp. 85–8.
OECD (2007), 'Communications in the emerging BRICS economies', in *Communications Outlook 2007*, Paris: OECD, pp. 277–305.
Ono, H. and M. Zavodny (2003), 'Gender and the Internet', *Social Science Quarterly*, **84**, 111–21.
Ono, H. and M. Zavodny (2007), 'Digital inequality: a five country comparison using micro-data', *Social Science Research*, **36**, 1135–55.
Partnership on Measuring ICT for Development (2008), *The Global Information Society: A Statistical View*, Santiago, Chile: United Nations.
Pew Internet and American Life Project (2008a), 'Increased use of video-sharing sites', 1 January, available at: http://www.pewinternet.org/pdfs/Pew_Videosharing_memo_Jan08.pdf.
Pew Internet and American Life Project (2008b), 'Usage over time', June, available at: http://www.pewinternet.org/trends.asp.
Puddenphatt, A. (2007), *Defining Indicators of Media Development: Background Paper*, Paris: UNESCO.
Resnick, D. (1998), 'Politics on the Internet: the normalization of cyberspace', in C. Toulouse and T.W. Luke (eds), *The Politics of Cyberspace*, New York: Routledge, pp. 48–68.
Sutherland, E. (2008), 'Counting mobile phones, SIM cards and customers', LINK Centre, available from the ITU at: http://www.itu.int/ITU-D/ict/statistics/material/sutherland-mobile-numbers.pdf.
Tichenor, P.G., A. Donohue and C.N. Olien (1970), 'Mass media flow and differential growth of knowledge', *Public Opinion Quarterly*, **34**, 159–70.
Warschauer, M. (2004), *Technology and Social Inclusion: Rethinking the Digital Divide*, Cambridge, MA: MIT Press.
World Bank (2008), *Internet Users (per 100 People)*, available at: http://data.worldbank.org/indicator/IT.NET.USER.P2.
World Economic Forum (2007), *Global Information Technology Report 2007–8*, available at: http://www.insead.edu/v1/gitr/wef/main/home.cfm.
World Values Survey (2013), available at: http://www.worldvaluessurvey.org.
Wresch, W. (1996), *Disconnected: Haves and Have-Nots in the Information Age*, New Brunswick, NJ: Rutgers University Press.
Yu, L. (2006), 'Understanding information inequality: making sense of the literature of the information and digital divides', *Journal of Librarianship and Information Science*, **38** (4), 229–52.

FURTHER READING

Norris (2003) and Warschauer (2004) are widely acknowledged works on the digital divide.

PART II

DEVELOPMENT OF TECHNOLOGIES IN THE CREATIVE ECONOMY

10. Copying technologies

Cliff Bekar

Copying – the ability to duplicate discrete packets of information with varying levels of fidelity – is a function of all information communication technologies (ICTs). The importance of ICTs derives from their impact on the production, manipulation, and distribution of information – activities central to any economy. Producing an exhaustive narrative history of such technologies is not feasible here (a partial list includes lithography, etching, photography, telegraphy, telephony, the typewriter, motion pictures, magnetic tape recording, radio, photocopiers, digital scanners, and video cassette recorders). Instead, the framework of general purpose technologies (GPTs) – a set of technologies defined by their characteristics that tend to have transformative effects on the structure of economies – is employed to structure a historical analysis of a select few ICTs.

The focus is on text-based copying technologies and not those designed to copy visual and/or aural information. The chapter attempts to place aspects of digitization within the historical context of past copying technologies. It starts by defining GPTs and developing the concept of 'technological principles', turning next to analyze the channels through which ICTs typically impact the economy. A central argument of the chapter is that, by altering the cost structure of producing, manipulating, and distributing information, ICTs that are themselves GPTs (general purpose ICTs) are often socially and economically 'transformative'. The chapter then considers three important copying technologies – writing, printing, and digitization.

GENERAL PURPOSE TECHNOLOGIES AND PRINCIPLES

Technology here is defined as:

> the set of ideas specifying all activities that create economic value. It comprises: (1) knowledge about product technologies, the specifications of everything that is produced; (2) knowledge about process technologies, the specifications of all processes by which goods and services are produced; (3) knowledge about organizational technologies, the specification of how productive activity is organized in productive and administrative units for producing present and future goods and services. (Lipsey et al., 2005: 58)

In this view, technology is best understood as a set of blueprints (nominal or actual) that are productive in the economic sense only when embodied in the structure of the economy (for example, the technology of hydraulics contributes to increased economic output once embodied in a waterwheel). GPTs are a subset of technology, specifically 'a single generic technology or technology system, recognizable as such over its whole lifetime, that initially has much scope for improvement and eventually comes to be widely used, to have many uses, and to have many spillover effects' (Lipsey et al., 2005:

98). Important general purpose ICTs include writing, printing, telegraphy, and computers. GPTs are fundamental to the process of economic growth and typically produce transformational changes in the structure of the economy. Transformational changes are massive, qualitatively unprecedented changes in the production and consumption of goods and services that contrast with incremental refinements and/or improvements of existing techniques.

Central to the analysis of copying as one of the functions of ICTs is the distinction between a 'technology' and a 'technological principle'. Whereas technologies are embodied in a set of blueprints, technological principles are embodied in a set of theoretical propositions. Principles are not embodied directly in the structure of the economy. Technological principles reside either: (1) in technological blueprints which are in turn embodied in the structure of the economy; or (2) formally or informally in human capital. A technological principle with a large number and variety of uses – one that comes to be deployed in a vast range of technologies – is called a general purpose principle (GPP). A GPP

> shares many of the characteristics of a mature GPT, with the main exception that it is not the specification of a single generic product, process or organizational technology that is identifiable as such over its lifetime. It is a concept that is employed in many different technologies that are widely used across the economy for many purposes and has many spillover effects. (Lipsey et al., 2005: 99)

Consider the principle of mechanical advantage, a principle describing the relationship between energy inputs and outputs from a mechanical system (that is, the system's force multiplier). Vast numbers of technologies (levers, pulleys, systems of gearing, and so on) embody the principle of mechanical advantage. An understanding of mechanical advantage is central to the development of functioning of steam engines, electric dynamos, and a range of transportation technologies. Early humans were able to exploit mechanical advantage through trial and error with little formal understanding of the science behind it. By the time of the early Greek philosophers, the science behind mechanical advantage was well established, formalized, and taught.

As a class of technology, ICTs embody – to varying degrees – the technological principles behind producing, manipulating, and accessing information. These three broad principles in turn include many more specific sub-principles (for example, transcribing, editing, and copying information). All general purpose ICTs embody these principles in different degrees, and therefore all have implications for copying. For example, while telegraphy primarily lowered the cost of transmitting information, it also created copies of information in new locations. The relative size of the effects is key. Telegraphy dramatically lowered the cost of transmission relative to the cost of copying; it was therefore never employed primarily as a copying technology. Analyzing ICTs as the embodiment of specific technological principles reveals a continuity in their historical evolution that is not always evident when they are analyzed as physical artifacts. A central issue in developing early telegraphy systems concerned how to efficiently encode alphanumeric information in a form that could be transmitted over a telegraph line. Multiple systems were developed over a long period of trial and error. This experimentation phase had roots in earlier information technologies where agents attempted to transmit information over long distances via sets of lights, smoke, and loud sounds (Standage, 1998). Eventually an eight-bit encoding of alphanumeric information was developed and standardized (the American Standard Code for Information Interchange, ASCII). ASCII was first used on teleprinters and eventually on telegraphy systems. ASCII is still widely used to encode alphanumeric information in modern computers and is a key aspect of digital ICT systems. So, while modern networked computers bear little resemblance to the nineteenth-century telegraph networks, their functioning relies on shared theoretical principles, namely that information can be represented as eight-bit codes that in turn can be encoded as a series of dots and dashes or binary ones and zeros.

Defined loosely, copying is the reproduction, with varying levels of fidelity, of a piece of information. The effects of general purpose ICTs depend in large part on the fidelity of the copies they produce, along with their impact on the cost of producing, reproducing, and accessing information. While the range of effects is typically large, five are highlighted here: (1) the sectoral makeup of an economy; (2) the structure of industries within economies (including those in the creative sector); (3) the institutional and organizational makeup of economies or societies; (4) the dynamics of how knowledge is produced and accessed; and (5) the rate of innovation.

A BRIEF HISTORY OF COPYING TECHNOLOGIES

Three key general purpose ICTs that had historically important implications for copying – writing, printing, and digital ICT – are now considered.

Writing

Writing is the first great general purpose GPT in history and is relatively unique as a GPT in a couple of important ways. First, it quickly displaced its forerunners (pictographs, knots tied in different patterns, notching, brands, heralds, and so on). The competition between writing and the precursor systems it came in contact with was brief and one-sided. The usual pattern of existing technologies prolonging their lifespan by undergoing a burst of development in response to a new competitor is not observed (Gelb, 1963). The technological precursors to writing neither cooperated with nor continued alongside writing as an information system. Second, the core technological principles that lie behind writing as an ICT have changed relatively little over time:

> In the last twenty-five hundred years the conquests of the alphabet have encompassed the whole of civilization, but during all this period no reforms have taken place in the principles of writing. Hundreds of alphabets throughout the world, different as they may be in outer form, all use principles first and last established in Greek writing. (Gelb, 1963: 197–8)

Despite being in existence for over two millennia, and experiencing relatively little development, modern writing (that is, alphabetic writing) has never been displaced or even seriously challenged; it remains a crucial copying technology today.

The first instance of writing is difficult to date with precision. Writing was once thought to have developed independently in multiple geographically distinct regions.

Scholars now agree that most writing systems – Indus Valley (2200 BC), Egypt (3000 BC), Crete (2000 BC), and Anatolia (1500 BC) – derived from Sumerian cuneiform (developed around 3100 BC; fully modern alphabetic systems of writing emerge later in the Mediterranean). It is 'generally agreed that other scripts developed later, independently, in China and Mesoamerica. The origin of Chinese and Mesoamerican writing is still enigmatic' (Schmandt-Besserat, 1996: 1). Early scholars believed that Sumerian cuneiform evolved out of early proto-pictographic systems. Recent research has firmly established that writing actually evolved out of a token-based counting system in use since the Neolithic period (Schmandt-Besserat, 1996). These (mostly) clay tokens represented physical objects in early accounting systems. The use of tokens evolved over millennia:

> First, during the Middle and late Upper Paleolithic, ca. 30,000 – 12,000 B.C., tallies referred to one unit of an unspecified item. Second, in the early Neolithic, ca. 8000 B.C., the tokens indicated a precise unit of a particular good. With the invention of writing, which took place in the urban period, ca. 3100 B.C., it was possible to record and communicate the name of the sponsor/recipient of the merchandise, formerly indicated by seals. (Schmandt-Besserat, 1996: 99)

Clay tokens, and the facilities to produce and distribute them, have been prominent features in archeological sites across the Near East. By the time of Sumer, they were distributed in clay cylinders called bullae. Tokens were placed in a bulla bearing the seal of a specific person so 'it was possible to identify both the sender and the number of objects being sent' (Dudley, 1991: 24). Through time the number of tokens contained in the bulla came to be recorded on its exterior. Eventually, it was realized that the tokens were redundant, and logographic writing was born:

> There can be no doubt that later, when accountants had the idea to imprint the tokens themselves on the surface of the envelopes in order to make visible the type and number of tokens held inside, they were copying the practice of pressing seals on the envelopes . . . These negative impressions of tokens on the surface of envelopes were the first signs of writing. (Schmandt-Besserat, 2007: 4)

The final step 'in the development of writing, known as phonetization, assigns a symbol to a sound, rather than to an object or idea' (Dudley, 1991: 23).

Given that the core elements of the token accounting system were in place since the Neolithic period, why did writing take so long to fully develop? One common framework in the expansive literature on the emergence of writing is the challenge–response model. The development of settled agriculture produced higher population growth, larger political units, and denser trade networks. These developments, in turn, led to an increase in required administrative complexity. This complexity increased the payoffs to information systems that could expand record keeping and administrative functions: 'developments in farming and industry played a major role in the development of the token system. Cultivation of cereals was directly related to the invention of plain tokens, and the complex counters are linked to the beginning of industry' (Schmandt-Besserat, 1996: 103). Solving the emerging problems associated with the provision of public goods (especially large-scale irrigation systems) required a relative increase in the power of the central authorities. In Dudley's (1991: 22–3) view, these developments increased the rate of return on improving the token system: 'Around 4000 B.C., the communities of Sumer were small in terms of population and apparently decentralized in their power and structure. Information storage, based largely on human memory, was also dispersed. However, the potential gains to an innovation that permitted more efficient centralized information processing were great.' So it seems that there was a positive feedback effect where token-based accounting systems allowed for greater social and economic complexity, which in turn increased the return on refining the system: 'Plain tokens were linked to the rise of rank society, but it was the advent of the state which was responsible for the phenomenon of complex tokens' (Schmandt-Besserat, 1996: 107). As finer divisions of labor emerged within the economy, and the number of goods to be accounted for proliferated, the system of complex tokens fell under pressure and eventually gave way to writing, whose great innovation was to use combinations of phonetic symbols to represent a potentially infinite number of objects and/or ideas.

As with other historical ICTs, writing's primary channels of effect were through the institutional or organizational makeup of the economy: 'Each change of reckoning device – tallies, plain tokens, complex tokens – correspond to a new form of economy: hunting and gathering, agriculture, industry. Each change in reckoning device also corresponded to a new political system: egalitarian society, rank society, the state' (Schmandt-Besserat, 1996: 122). Writing has been linked to an increase in market activity, population density, and city size, along with the emergence of a 'temple bureaucracy' and ultimately the 'state' (Dudley, 1991). Writing expanded the network of those who contributed to knowledge and rendered those contributions more cumulative, impacting the rate of institutional and technological innovation over the very long run: 'Whenever writing appears it is accompanied by a remarkable development of government, arts, commerce, industry, metallurgy, extensive means of transportation, full agriculture and domestication of animals' (Gelb, 1963: 221).

Writing impacted the evolution of artistic expression:

> In the ancient Near East, art broke new ground with the advent of literacy. Following writing's lead, the major forms of art, namely pottery, seals, stone vases, steles, and wall paintings, changed from evocative to narrative. They went beyond the mere repetition or association of symbols to depict complex scenes involving multiple interrelated participants. (Schmandt-Besserat, 2007: 59)

The interaction between writing and art produced increasing abstraction, increasing information density, and ultimately a sort of 'visual language'. Schmandt-Besserat (2007: 105) argues that the effects of these developments were crucial: 'The development of a comprehensive written visual language caused civilization to grow more complex. Literacy makes for larger organizational and political units, allowing empires to push their domination farther afield.'

In sum, writing is one of the central innovations in human history. Writing increased the fidelity and permanence associated with copies, while dramatically lowering the costs of storing and accessing those copies. The impact of writing on these core technological principles is associated with: (1) an increasingly complex economy with a wider range of goods being produced and traded; (2) the emergence of a temple economy providing public goods; (3) an increase in the rate at which knowledge was produced and distributed; and (4) an increase in the rate of innovation.

Printing

Printing has existed in myriad forms since shortly after the introduction of writing (forms of xylography including block books, lithography, and the like). 'Printing' is often used as shorthand for the press itself. Here we use the term to refer to movable type. Whereas inventions prior to the modern period are typically not identified with a single individual, most people know Johannes Gutenberg as the 'inventor' of the printing press. While Gutenberg was responsible for key printing innovations – the development of new inks and the standardization of printing dies – the claim that he 'invented' the printing press is problematic. Eisenstein (1983: 13) notes that its development was based on a cluster of disparate innovations 'entailing the use of movable type, oil-based ink, wooden hand press, and so forth'. These innovations involved a large number of individuals over a period of centuries. The mechanical hand press used by early printers evolved out of those employed by contemporary winemakers. Paper was another crucial innovation in the development of printing: 'It would have been impossible to invent printing had it not been for the impetus given by paper, which had arrived in Europe from China via the Arabs two centuries earlier and came into general use by the late 14th century' (Febvre and Martin, 2000: 30). With earlier access to paper, printers in China (and parts of the Korean peninsula) experimented with movable type long before their European counterparts, but they relied on wooden blocks, which lacked the durability required for repeated use and the ability to hold particular inks (for a dissenting view see Chow, 2007). The need for more durable and precise dyes meant that the loci of innovations in the early European presses were around metalworkers and, in particular, goldsmiths.

Gutenberg was responsible for the 'the scientific synthesis of all these different trends and trials. He fulfilled the need of the age for more and cheaper reading matter by substituting machinery for handicraft' (Steinberg, 1996: 8). A goldsmith by trade, Gutenberg established the first commercial printing press in Mainz, Germany between 1445 and 1450. By the fifteenth century, commercial printing was in the air:

> Gutenberg was not the first to grasp the need for, and the potentialities of, large-scale production of literature. On the contrary, his invention was prompted largely by the fact that the multiplication of texts was not only a general want but had also, by the middle of the fifteenth century, become a recognized and lucrative trade. (Steinberg, 1996: 7)

The market for manuscripts had been growing for centuries before the introduction of the press and is generally broken into two periods, the Monastic Age (roughly 500–1200) and the Secular Age (1200–1500). In the Monastic Age, manuscripts were produced in scriptoriums, where groups of scribes would hand-copy texts onto vellum and/or parchment: 'In the 700 years between the Fall of Rome and the 12th century, it was the monasteries and other ecclesiastical establishments associated with them which enjoyed an almost complete monopoly of book production and so of book culture' (Febvre and Martin, 2000: 15). Scribal production displayed almost no scale economies: increasing manuscript production simply increased required scribal inputs (scribal copying hours) in a linear way. Further, given the relatively low levels of European literacy and education during the period, human capital was among the scarcest of productive inputs. Scribal copying techniques, employing relatively high levels of human capital and lacking scalability, meant manuscripts remained relatively expensive throughout the period. Personal and institutional libraries therefore remained modest, with the market for manuscripts confined to the wealthy. Despite the high costs and inelastic supply of copying services, the market for books, manuscripts, and other types of documents (for example bills of sale and lading, receipts, contracts, and so on) continued to grow.

Most historians now agree that a rapidly expanding market for manuscripts significantly pre-dated the introduction of movable type. Not only was the market for copying services growing, but its character was changing as well. It was becoming more secular and more closely associated with early universities, and it increasingly serviced a lay audience. From 1200 onward,

> The social and intellectual changes which are particularly clearly reflected in the founding of universities and the development of learning among the laity, and which occurred at the same time as the bourgeoisie emerges as a class, had profound repercussions on the ways in which books came to be written, copied, and distributed. (Febvre and Martin, 2000: 15)

This early expansion in the book market was facilitated by a key materials innovation, the introduction of paper. European paper production grew tremendously throughout the later fourteenth century. The decline in paper prices along with its increasingly elastic supply (relative to vellum and parchment) facilitated a dramatic expansion in the market for copying services (as well as the market for new manuscripts and documents) throughout the fifteenth century.

Many developments typically associated with the introduction of printing actually pre-date its introduction: 'from the mid-13th century, copyists were forced to improve their methods to meet the growing demand, and this in turn led in some workshops to something very like standardised mass production' (Febvre and Martin, 2000: 28). While such organizational innovations allowed for increased output, they did little to reduce unit costs (and therefore the price of manuscripts). Despite some important continuities between manuscript production in the latter Secular Age and manuscript production in the incunabula period, it is important to remember that the long term effects of printing were anything but continuous: 'while acknowledging the significance of changes affecting the twelfth-century book production, we should not equate them with the sort of "book revolution" that occurred in the fifteenth century. The latter, unlike the former, assumed a cumulative and irreversible form' (Eisenstein, 1983: 11).

The early effects of printing were manifold, and this chapter touches on only a couple of them. With the growing market for manuscripts, books, and documents of all types, the mechanization of printing allowed for a sustained rapid growth in these markets over the next few centuries. First, a division of labor grew, where printing was used for books and science while forms of xylography tended to be used for artistic endeavors. Printing was an important complement to writing: 'From indulgences to interleaved almanacs to bills of lading to bank checks to those great twentieth-century bestsellers . . . printing has become the great means of eliciting writing by hand' (Stallybrass, 2007: 340–41). Printing dramatically altered the fundamental stability of the text, altering notions of authority of knowledge and authorship. These developments ultimately led to an increase in 'textual trust' and helped facilitate the development and spread of the scientific revolution in the early modern period: 'A new confidence in the accuracy of mathematical constructions, figures, and numbers was predicated on a method of duplication that transcended older limits imposed by time and space and that presented identical data in identical form to men who were otherwise divided by cultural and geographical frontiers' (Eisenstein, 1983: 271).

Printing also impacted the development of the creative arts. Interestingly, Scherer notes that one of the earliest applications of average total cost was in the music printing industry (Scherer, 2012). The concept of average unit cost was employed to help choose between different technologies (engraving dominated for small print runs, printing for larger runs). The scalability of printing dramatically increased the size and nature of the market for sheet music, as well as the industrial organization of its firms, although, unlike the case for some markets, all the dominant forms of copying – hand copying, engraving, and printing – continued to compete in the market for sheet music for a long time. Chan (2002: 137) notes that, in seventeenth-century London, 'Engraving was now regarded not so much as a luxury but as the new way of printing, though copying continued to flourish both privately and as a way of publishing.'

Being complementary to the new energy and materials technology of the Industrial Revolution as well as the information processing capabilities of modern digital ICT, the printing press remains an important technology today. Printing has been central to every 'information revolution' since it was introduced. As a technology, printing dramatically altered the cost structure of information, increasing the fixed costs of copying while lowering its marginal cost. Further, printed copies of material enjoyed a higher degree of fidelity relative to their hand-copied alternatives. The technological principles embodied in printing had a range of effects, including: (1) dramatically larger markets for manuscripts, books, and documents of all types; (2) spillover effects on a range of sectors; (3) the development of an early state bureaucracy and linguistic communities associated with emerging nation-states; and (4) an accelerated rate of scientific development and ever widening spheres in which science was applied.

Digitization

Despite causing historically significant declines in copying costs (for textual, visual, and aural information), digitization is not even primarily a copying technology. An exhaustive exploration of digitization would thus take us quite far afield. Further, it is not yet time to write the 'history' of digitization. Instead the focus here is on putting some of those developments in the historical context of copying technologies.

Digitization is not a set of blueprints for a specific technology, or even a set of technologies; it can be understood only as a technological principle. The principle of digitization is embodied in the technology of 'digital ICT', a collection of electronic components embodying binary logic. The system is composed of networked PCs, tablets, phones, and other smart devices. The core technology of digital ICT is the electronic computer: 'Like all transforming GPTs, the computer started as a crude, specific-purpose technology and took decades to be improved and diffused through the whole economy' (Lipsey et al., 2005: 114). The essential components of any digital technology system are digital storage (memory), digital logic (CPU), digital arithmetic (GPU), and input/output channels (networking). These basic subsystems continue to be the building blocks of all modern computing systems and are known as the von Neumann architecture (Dyson, 2012: 78). Today's digital ICTs are usually characterized by high degrees of connectivity (networking) and are embodied in a range of relatively low cost components (PCs, tablets, phones, smart devices, and so on).

As a technological principle, digitization concerns encoding, analyzing, and transmitting information. The extent to which the principle of digitization is separable from the technology of digital ICT is seen in the vast gap between the start of the research trajectory into binary logic and its embodiment in digital computers. In 1623, Bacon established that just two symbols could encode any numerical information. In 1679, Leibniz postulated that zero and one were sufficient to encode the logic of arithmetic. By the late 1930s Alan Turing had formalized the relationship between systems of binary logic and computation. With the development of early computing systems during and shortly after the Second World War, the digital became mechanical and ultimately electronic. Until relatively late in the computer's development Turing was still focused on the principle of binary logic as opposed to the technology of the computer: 'Being digital should be of more interest than being electronic' (Turing, quoted in Dyson, 2012: 5).

There are interesting continuities between the development of the technological principles underlying the development of writing, printing, and digitization. Recall that the early writing (and therefore, ultimately, phonetic alphabets) developed out of counting systems that date back to the Neolithic period. Writing emerged out of these early counting systems as they evolved to represent information in an increasingly abstract manner. Central to this process was the manner in which information was encoded. Concrete counting relied on a one-to-one correspondence between a group of specific objects and the number of those objects, whereas 'Our numbers 1, 2, 3 express the concepts of oneness, twoness, threeness as abstract entities . . . Abstract counting thus marks the beginning of arithmetic and the point of departure of modern mathematics' (Schmandt-Besserat, 1992: 187). This move towards abstraction was important in the development of writing:

> It was not by chance that the invention of pictography and phonetic writing coincided with that of numerals; instead, both were the result of abstract counting. The abstraction of the concept of quantity (how many) from that of quality, which merged inextricably in the token prototypes, made possible the beginning of writing. (Schmandt-Besserat, 1992: 194)

The principle of digital logic is, in important ways, simply the extension of this move towards abstract counting. Like the development of writing, digitization evolved out of a counting system (binary) that had been evolving for centuries.

The long run evolution of binary logic, digitization, and ultimately electronic computers is consistent with the historical development of past GPPs and their embodiment in GPTs. To an economist studying the impact of digitization – in contrast to Turing – 'being electronic is at least as interesting as being digital'. That is, the economic impact of digitization is largely manifested insofar as it is embodied in digital ICT. The digitization of information is clearly one of the central features of the twentieth century's technological landscape. At the broadest level, there is little doubt that many of the impacts of the 'digital' or 'new economy' are fundamental and unprecedented. Examples of such developments, much discussed in the literature, include the death of distance, the rise of the knowledge economy, and the post-industrial society.

As a copying technology, digitization has forever altered the information landscape. It has dramatically lowered the fixed and marginal costs of copying (compare the cost of producing a 'Gutenberg Bible' today relative to the cost of the original). Digitization has so dramatically increased the fidelity of copies that there are now broad swaths of the modern economy being forced to rethink what it even means to be an 'original'. Copies of photos, operas, movies, texts, and videogames are now indistinguishable – in the literal sense – from their originals, causing whole industries to grapple with core issues like the legal concepts surrounding intellectual property rights. And, as with the impact of past ICT GPTs, digitization has been transforming the evolution of institutions and organizations:

> the new technologies of information processing and control came to be applied to a wide range of bureaucratic functions: purchasing records, inventory, overhead allocations, payroll analysis, shipping costs, sales projections, market forecasting. The larger the enterprise and the more diversified the throughputs it controlled, the more crucial the new technologies would become. (Beniger, 1986: 424)

CONCLUSION

It is clear that the introduction of copying technologies has transformed the makeup of economies throughout history. This chapter has focused on text-based copying technologies. Not including aural and visual copying technologies in this study means we have left aside a range of important historical effects related to those technologies.

Copying technologies have very specific channels of effect by which they tend to impact the economy. Unlike some other important technologies (materials and energy, for example), ICTs tend to primarily impact the organizational and institutional structure of economies. These institutional changes feed back to economic growth over the very long run, and thus the primary effects of general purpose ICTs tend to be indirect. An implication of this is that they tend to have implications far beyond their economic impacts. These include changes in governance, law, science, and the arts. Just as printing and writing both fundamentally altered the development of literature, music, and the visual arts, it appears that we are living through the earliest stages of the next great copying revolution: digitization.

REFERENCES

Beniger, J. (1986), *The Control Revolution: Technological and Economic Origins of the Information Society*, Cambridge, MA: Harvard University Press.

Chan, M. (2002), 'Music books', in J. Barnard and D.F. McKenzie (eds), *The Cambridge History of the Book in Britain*, vol. IV: *1557–1695*, Cambridge: Cambridge University Press, pp. 127–38.

Chow, K. (2007), 'Reinventing Gutenberg: woodblock and movable-type printing in Europe and China', in S. Alcorn, E. Lindquist and E. Shevlin (eds), *Agent of Change*, Boston: University of Massachusetts Press.

Dudley, L. (1991), *The Word and the Sword: How Techniques of Information and Violence Have Shaped Our World*, Cambridge, MA: Basil Blackwell.

Dyson, G. (2012), *Turing's Cathedral: The Origins of the Digital Universe*, New York: Pantheon Books.

Eisenstein, E. (1983), *The Printing Revolution in Early Modern Europe*, Cambridge: Cambridge University Press.

Febvre, L. and H.J. Martin (2000), *The Coming of the Book*, London: Verso.

Gelb, I.J. (1963), *A Study of Writing*, Chicago: Chicago University Press.

Johns, A. (1998), *The Nature of the Book: Print and Knowledge in the Making*, Chicago: University of Chicago Press.
Lipsey, R.G., K. Carlaw and C. Bekar (2005), *Economic Transformations: General Purpose Technologies and Long Term Economic Growth*, Oxford: Oxford University Press.
McMurtrie, D. (1948), *The Book: The Story of Printing and Bookmaking*, New York: Oxford University Press.
Parshall, P. and R. Schoch (2005), *Origins of European Printmaking: Fifteenth-Century Woodcuts and Their Public*, New Haven, CT: Yale University Press.
Scherer, F.M. (2012), *Quarter Notes and Bank Notes*, Princeton, NJ: Princeton University Press.
Schmandt-Besserat, D. (1992), *Before Writing*, vol. I: *From Counting to Cuneiform*, Austin: University of Texas Press.
Schmandt-Besserat, D. (1996), *How Writing Came About*, Austin: University of Texas Press.
Schmandt-Besserat, D. (2007), *When Writing Met Art: From Symbol to Story*, Austin: University of Texas Press.
Stallybrass, P. (2007), '"Little jobs": broadsides and the printing revolution', in S. Alcorn, E. Lindquist and E. Shevlin (eds), *Agent of Change*, Boston: University of Massachusetts Press, pp. 315–41.
Standage, T. (1998), *The Victorian Internet*, New York: Walker & Company.
Steinberg, S.H. (1996), *Five Hundred Years of Printing*, New Castle, DE: Oak Knoll Press.

FURTHER READING

See Johns (1998) for interesting material on the social construction of the book and what it meant for print culture in early modern Europe. McMurtrie (1948) is a classic work on printing, with much material relating to the early scholarly position on writing. See Parshall and Schoch (2005) for a wonderful study of early printmaking (lithography).

11. Technological change and cultural production

Peter Tschmuck

The term 'technology' comes from the Greek expression *téchnē*, meaning the study of 'skill', 'craft' and 'art'. Technology has always had a close relationship to the arts and culture. It had already been applied when the first known cave paintings of the upper Palaeolithic Age in El Castillo in northern Spain (*circa* 40 800 years old)[1] and in Chauvet in France (*circa* 37 000 years old) (Clottes, 2003) were produced.

Technology played a crucial role when cultural symbols were fixed on media such as stone, clay, skins, textiles, metal, papyrus and paper. The earliest known example of music notation – the cuneiform tablet from Nippur in the Sumerian realm – can be dated back to 2000 BC. The cuneiform fragments of the earliest Sumerian poems, which foreshadow the famous *Epic of Gilgamesh*, belong to the same period (Dalley, 2009). The use of technology enabled 'high' cultures around the world to fix symbols as a form of cultural expression, thus setting themselves apart from purely oral cultures. However, the reproduction of texts, scores and images was the domain of skilled craftsmen and a very small class of intellectuals from these ancient times until the Middle Ages.

Thus, the invention of movable type printing by Johannes Gutenberg in the mid-fifteenth century was a milestone in human development. It enabled the mass production of texts and subsequently also of notations and images at relatively low marginal costs. The technological improvement fuelled the social and cultural change from a medieval to an early modern European society.

As the industrial revolution intensified throughout the nineteenth century, technological achievements such as the mass production of sheet music, literature, photography and sound recording/reproduction once again revolutionized production, dissemination and consumption in the cultural sector. For the first time in human history, these inventions enabled the fixation and storage of visual and auditory impressions.

In the twentieth century the ability to reproduce and disseminate cultural products, such as film and broadcasting, dramatically increased with electrical and magnetic recording technologies. An economic aspect of these developments in the twentieth century was the formation of large media and entertainment conglomerates that exercise oligopolistic and even monopolistic control over the production, dissemination and consumption of cultural and artistic goods.

The Millennium marked a further paradigmatic technological shift in the cultural sector. Digitization, which can be dated back to the mid-1960s, unfolded its full potential, first in the music industry and later in other entertainment sectors (Henten and Tadayoni, 2011). Compared to prior structural breaks, the digital revolution has not only facilitated copying and exchanging cultural and artistic content but also enables private users of content to participate in its production and dissemination, resulting in a networked society that has transformed markets and market structures (for example Benkler, 2006).

Thus, technology and cultural development go hand in hand in human development.

In the following the co-evolution of technological innovation and cultural change is further highlighted and the concept of mediamorphosis is introduced.

TECHNOLOGICAL AND CULTURAL CHANGE AND THE CONCEPT OF MEDIAMORPHOSIS

Tschmuck (2012) takes the view that technology is only one driver of innovation processes –others are social actors, business and artistic practices – and conceptualizes creativity as the main force of paradigmatic shifts. He talks about 'cultural paradigms', in which, for example, the music industry develops along 'creative paths'. If 'system-alien' creativity breaks up the paradigmatic frame, new possibilities of interaction occur that cause a 'cultural paradigm shift'. Thus, technological change within the cultural sector should be interpreted as a social process that is not independent of historical conditions, and technological change within the cultural sector is interpreted as a social process that is embedded in the historical process of the production, dissemination and consumption of cultural goods.

The concept of mediamorphosis, which was introduced by Blaukopf (1989) and further developed by Smudits (2002), helps to explain how the cultural sector evolves over a long period of time in a long-term perspective. A mediamorphosis consists of a transformation of communication that occurs as a result of technological innovation in the media (Smudits, 2002: 44). It dramatically changes artistic or cultural production, which results in altered conditions of production, dissemination and consumption of cultural goods. Smudits (2002: 204) identifies four partially overlapping mediamorphoses since early modern times – graphic mediamorphosis (writing), mediamorphosis (printing), chemical-mechanical and electrical mediamorphosis, and digital mediamorphosis (see Figure 11.1). The next section uses the conceptual framework of

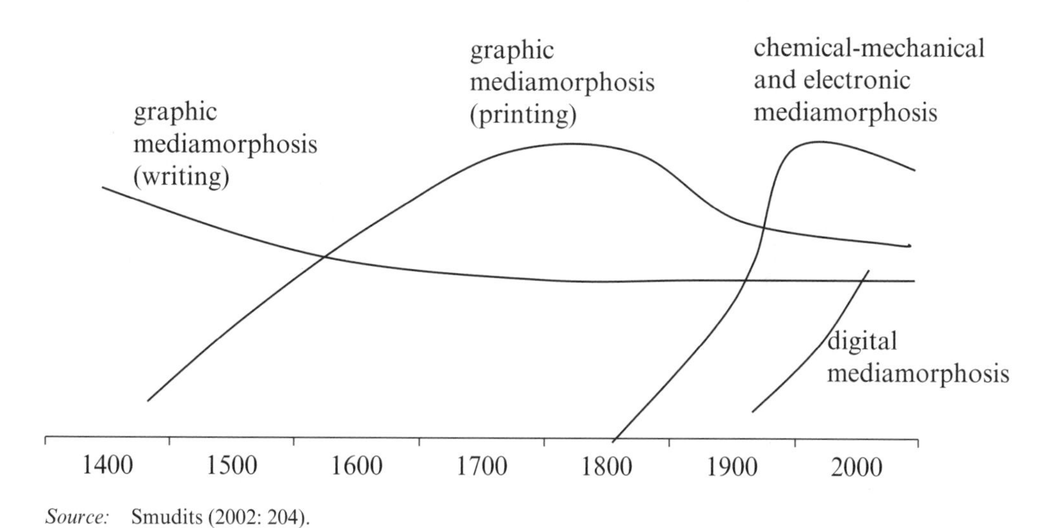

Source: Smudits (2002: 204).

Figure 11.1 The mediamorphoses of cultural production

mediamorphoses to discuss technological change in the cultural sector with a focus on the last two mediamorphoses.

TECHNOLOGICAL CHANGES IN THE CULTURAL SECTOR

Chemical-Mechanical and Electrical Mediamorphosis

Whereas the first and second graphic mediamorphoses, the invention of writing and printing, revolutionized the dissemination of ideas, the chemical, mechanical and electrical technologies made it possible to fix and store otherwise ephemeral processes. Although all inventions such as photography, sound recording and film initially were invented with non-artistic uses in mind, they indirectly and directly influenced the art sector. When early inventors such as Nicéphore Niépce, Louis Daguerre, William Henry Fox Talbot and John Herschel brought forward the chemical process of camera photography in the first half of the nineteenth century, naturalistic paintings became less popular. Instead, Impressionism tried to express sentiments and moods instead of depicting an objective reality. Expressionism in painting went a step further by presenting the world solely from a subjective perspective and extended into a wide range of the arts, including literature, theatre, dance, film, architecture and music. Apart from the indirect impact on the arts, photography became an art form by itself in the twentieth century, and attracted the attention of collectors, art galleries and museums (for example Becker, 1982). Sound recording and film were also invented for non-artistic purposes. When Thomas A. Edison invented the phonograph in 1877, he saw it as a dictaphone and a device to store telephone calls rather than a machine to store and reproduce music (Gelatt, 1955: 29). In the following years, the phonograph was transformed to a 'coin-in-the-slot machine' for cheap public entertainment (Garofalo, 1997: 19). Early movie performances were also spectacles for mass consumption in amusement parks. It is worth noting that some of the pioneers of sound recording (namely Thomas A. Edison and the Pathé brothers in France) also experimented with film technology. Like photography, film and sound recording became recognized art forms only during the twentieth century (see for example Bakker, 2008).

New technologies also revolutionized art consumption. Hand in hand with the player piano, the phonograph replaced the piano in private households, causing a steep decline of the piano manufacturing industry in the early twentieth century (Ehrlich, 1990: 176–93). Practising music became more and more the field for professionally educated performers, and musical amateurs were transformed into passive consumers. This development was exacerbated by the emergence of radio in the 1920s and television in the 1950s. These new media channels required the passive consumption of cultural goods and, coincidentally, established a new advertisement-based business model. The economic logic of mass media also enabled the emergence of state monopolies for broadcasting in Europe and entertainment industry conglomerates in the US, explained by Caves (2000) as the integration of several value-added processes into large business conglomerates as a means to reduce transaction costs that would otherwise be present with multiple suppliers of different stages in the chain of production.

However, electronic technology not only influenced the cultural and artistic (re)production in mass media; it also was integrated into musical composition by avant-gardists such as John Cage and Karlheinz Stockhausen (Ungeheuer, 2002) and into the visual and applied arts by, for example, Nam June Paik, who is considered to be the first video artist (Meigh-Andrews, 2006).

The mass reproduction of artistic works, however, was opposed and fundamentally criticized by cultural theorists. In the essay *The Work of Art in the Age of Mechanical Reproduction*,[2] Walter Benjamin ([1936] 2008: 214–18) argues that the 'sphere of authenticity is outside the technical' so that the original artwork – Benjamin speaks of the 'aura' of an artwork – is independent of the copy. In the act of reproduction the 'aura' is taken from the original by changing its context and therefore the reproduction is devaluing the artwork.

To sum up, the chemical-mechanical and electrical mediamorphosis enforced the strict separation of the production and consumption of artistic expressions and brought forward the professionalization of artists. It also laid the foundation of the mass consumption of cultural goods and the emergence of large entertainment conglomerates and, according to some, the debasement of culture.

Digital Mediamorphosis

The origins of digitization as process to convert information (image, sound, text and so on) into a binary code can be traced back to the invention of transistors in the late 1940s, which enabled the construction of more powerful digital computers. However, digitization could become widespread only with the diffusion of compact personal computers, which were marketed for the first time by Apple in 1977 (Linzmayer, 2004). Another major landmark was the transition from analogue to digital recorded music in 1982, when Sony and Philips presented the compact disc (CD) and the CD player to the public (Tschmuck, 2012: 166). Thus, the digital format supplanted analogue formats, such as vinyl records as well as music and video cassette tapes.

This format shift met the emergence of computer networks that emerged in 1969 with the science-based ARPANET, later called Internet. With the invention of a unique language for the Internet – the Transmission Control Protocol/Internet Protocol (TCP/IP) – Internet participation grew exponentially from the mid-1980s on. Early in the 1990s, MP3[3] technology appeared as a method of compressing audio signals to enable the easy transfer of audio and video data on the Internet (Haring, 2000). The next step was to program a software (called Napster) that allowed users to share digital music and video files stored on the computers' hard drives (Alderman, 2001; Menn, 2003).

Peer-to-peer file sharing[4] not only revolutionized the distribution of digital content but also emancipated consumers as well as artists to a certain degree from intermediaries such as record labels, film production firms, book publishers and media companies. With further developments such as the integration of mobile phones into the computer networks and the emergence of social media, user-generated content and cloud computing services, the barrier of active production and passive consumption blurred and gave way to a process of 'prosumption'. Alvin Toffler used the term 'prosumer' for the first time in his bestseller *The Third Wave* (1980) to describe a person who consumes what she/he produces with the help of technology. Berman (2004) applied the concept to media usage and argued that media users would be increasingly involved in the creative process.

Thus, artistic and cultural production is no longer shaped by the traditional division of labour and instead occurs in circular value networks in which the artistic output is at the same time artistic input.[5] Good examples for this phenomenon are the sampling and mash-up culture in the music scene, as well as the concept of collaborative writing and gaming on the Internet. New value is created in these participative and collaborative networks that results in a higher social welfare, as Benkler (2006) has already argued in *The Wealth of Networks*.

This development towards a digital networked society, however, also challenges the role of entertainment conglomerates as intermediaries between artists and consumers, as well as the role of 'professional' artists. Since it is possible to disseminate digital content directly over the Internet, the physical products (CDs, DVDs, books, newspapers/magazines and so on) lose their relevance in the value-added process. Therefore, market power shifts from former gatekeepers such as record labels, book publishers, film production firms and media houses to the providers of the technical infrastructure – mainly Internet service providers and telecommunication companies. But content aggregators and online or mobile platforms such as Apple/iTunes, Amazon.com and Google/YouTube, whose business models are not based on the purchase of digitized artistic and cultural content, also play an increasingly important role in the cultural industries (for the music industry see Tschmuck, 2012: 193).

To sum up, digital mediamorphosis enables cultural and art consumers to distribute, share and use digitized content. The formerly passive consumer, thus, enters the sphere of artistic production and emerges as a prosumer. This, however, challenges current copyright rules that were developed for strictly separated production and consumption spheres. When there was a need for intermediaries to connect production and consumption of cultural goods, artists had to assign copyright to an intermediary if they wanted their work to be copied and disseminated. With the boundaries between production and consumption blurring, the dominance of the intermediaries vanishes and gives way to a network of prosumers, in which cultural and artistic output is used immediately as input for further artistic and cultural production.

CONCLUSION

The concept of mediamorphosis helps to explain the impact of technology on the cultural sector in a long-term perspective. In this perspective, technology is not a simple driver of cultural and artistic change but a constitutive element of cultural and artistic production, which is embedded in a socio-historical context. In the second graphic mediamorphosis (printing), cultural and artistic expressions became copiable as texts such as books, sheet music, engravings and etchings. It was not necessary any more to directly participate in the process of cultural and artistic production. The new technologies separated the trained and skilled artists from the educated art connoisseurs. As a result the production and consumption of cultural and artistic expression became separated spheres.

In the chemical-mechanical and electrical mediamorphosis, ephemeral cultural and artistic expressions such as music and the performing arts could be stored and reproduced for mass consumption. Specialized intermediaries emerged that bridged the widening gap between professional cultural and artistic production and passive consumption. These intermediaries benefited from their gatekeeping power and grew into large entertainment conglomerates that control each aspect of the production and dissemination of cultural and artistic goods. As a result the production and consumption spheres were even more separated in the twentieth century than in the previous centuries.

However, the digital mediamorphosis enables consumers to distribute, use and alter cultural and artistic goods in their social contexts. This turns passive consumers into prosumers who control the means of cultural and artistic production. As a result the production and consumption spheres merge into a network of simultaneous prosumption of cultural and artistic goods.

NOTES

1. For a detailed account, see National Geographic (2012).
2. The German original, published in 1936, is entitled 'Das Kunstwerk im Zeitalter seiner technischen Reproduzierbarkeit'.
3. MP3 stands for 'Motion Picture Expert Group/Layer 3'.
4. Peer-to-peer file sharing allows users to download media files such as music, movies and games using a software client that searches for other connected computers.
5. The circularity of artistic production is not the outcome of the digital revolution, but digital technology enables a new quality of artistic expression by disseminating new (derivative) works immediately on the Internet.

REFERENCES

Alderman, John (2001), *Sonic Boom: Napster, MP3 and the New Pioneers of Music*, London: Fourth Estate.

Bakker, Gerben (2008), *Entertainment Industrialised: The Emergence of the International Film Industry, 1890–1940*, Cambridge: Cambridge University Press.

Becker, Harold S. (1982), *Art Worlds*, Berkeley: University of California Press.

Benjamin, Walter (1936), 'Das Kunstwerk im Zeitalter seiner technischen Reproduzierbarkeit', *Zeitschrift für Sozialforschung*, **5** (1), 40–66.

Benjamin, Walter ([1936] 2008), *The Work of Art in the Age of Mechanical Reproduction*, trans. J.A. Underwood, London: Penguin.

Benkler, Yochai (2006), *The Wealth of Networks: How Social Production Transforms Markets and Freedom*, New Haven, CT: Yale University Press.

Berman, Saul (2004), *Media and Entertainment 2010: Open on the Inside, Open on the Outside – The Open Media Company of the Future*, IBM Institute for Business Value Future Series Report, Somers, NY: IBM.

Blaukopf, Kurt (1989), *Beethovens Erben in der Mediamorphose: Kultur- und Medienpolitik für die elektronische Ära*, Heiden, Switzerland: Niggli.

Caves, Richard E. (2000), *Creative Industries: Contracts between Art and Commerce*, Cambridge, MA: Harvard University Press.

Clottes, Jean (2003), *Chauvet Cave: The Art of Earliest Times* [*La grotte Chauvet: L'art des origines*], trans. Paul G. Bahn, Salt Lake City: University of Utah Press.

Dalley, Stephanie (ed. and trans.) (2009), *Myths from Mesopotamia: Creation, the Flood, Gilgamesh, and Others*, rev. edn, Oxford: Oxford University Press.

Ehrlich, Cyril (1990), *The Piano: A History*, rev. edn, Oxford: Clarendon Press.

Garofalo, Reebee (1997), *Rockin' Out: Popular Music in the U.S.A.*, Upper Saddle River, NJ: Prentice Hall.

Gelatt, Roland (1955), *The Fabulous Phonograph: From Tin Foil to High Fidelity*, Philadelphia, PA: J.B. Lippincott.

Gustavson, Todd (2009), *Camera: A History of Photography from Daguerreotype to Digital*, New York: Sterling Press.

Haring, Bruce (2000), *Beyond the Charts: MP3 and the Digital Music Revolution*, Hollywood, CA: Off the Charts.

Henten, Anders and Reza Tadayoni (2011), 'Digitalization', in Ruth Towse (ed.), *A Handbook of Cultural Economics*, 2nd edn, Cheltenham, UK and Northampton, MA, USA: Edward Elgar Publishing, pp. 190–200.

Hesmondhalgh, David (2002), *The Cultural Industries: An Introduction*, London: Sage.
Linzmayer, Owen W. (2004), *Apple Confidential 2.0: The Definitive History of the World's Most Colorful Company – The Real Story of Apple Computer, Inc.*, San Francisco: No Starch Press.
McLuhan, Marshall (1962), *The Gutenberg Galaxy: The Making of Typographic Man*, Toronto: University of Toronto Press.
Meigh-Andrews, Chris (2006), *A History of Video Art: The Development of Form and Function*, Oxford: Berg.
Menn, Joseph (2003), *All the Rave: The Rise and Fall of Shawn Fanning's Napster*, New York: Crown Business.
National Geographic (2012). 'World's oldest cave art found – made by Neanderthals?', 14 June, available at: http://news.nationalgeographic.com/news/2012/06/120614-neanderthal-cave-paintings-spain-science-pike/ (accessed 15 October 2012).
Read, Oliver and Walter L. Welch (1976), *From Tin Foil to Stereo: Evolution of the Phonograph*, 2nd edn, Indianapolis, IN: Howard W. Sams & Co.
Smudits, Alfred (2002), *Mediamorphosen des Kulturschaffens: Kunst und Kommunikationstechnologie im Wandel*, Wien: Braumüller.
Toffler, Alvin (1980), *The Third Wave*, New York: William Collins.
Tschmuck, Peter (2012), *Creativity and Innovation in the Music Industry*, 2nd edn, Heidelberg: Springer.
Ungeheuer, Elena (2002), 'Elektroakustische Musik 1945–1975: Prisma musikalischer Originalität', in Hans-Werner Heister (ed.), *Geschichte der Musik im 20. Jahrhundert: 1945–1975*, Laaber, Germany: Laaber Verlag, pp. 96–101.
Vogel, Harold L. (2010), *Entertainment Industry Economics: A Guide for Financial Analysis*, 8th edn, Cambridge: Cambridge University Press.

FURTHER READING

The impact of technology in the fine and applied arts is further discussed in Becker (1982). The impact of printing technology on the cultural development is best described by McLuhan (1962). For a history of technological development in the music industry see Gelatt (1955), Read and Welch (1976) and Tschmuck (2012). The technological history of the early film and movie industry is outlined in Bakker (2008). The history of photography is highlighted in Gustavson (2009). Technological and economic implications in the entertainment industries are analysed in great detail by Caves (2000), Hesmondhalgh (2002) and Vogel (2010).

12. Media convergence

*Michael Latzer**

'Convergence' is an ambiguous term used by various disciplines to describe and analyse processes of change toward uniformity or union. Its application in the communications sector, often referred to as 'media convergence', also encompasses valuable approaches and insights to describe, characterize and understand the digital creative economy. A certain amount of fuzziness combined with the broad, multipurpose character of convergence leads to both a general and a wide range of very specific understandings of the convergent communications sector. This sector substantially overlaps with the digital creative economy, which is also characterized by a degree of vagueness. Common subsectors and subjects between communications and digital creative industries such as broadcasting, publishing, advertising, music, film and games are even growing because of convergence. Beyond that, the consequences of media convergence are also discussed for other parts of the creative industries, such as museums, libraries and design in particular. New digital media technology and services are considered as central drivers of creative industries. Altogether, this makes studies of media convergence, both its approaches and its results, highly relevant for the understanding of the digital creative economy.

VARIOUS DISCIPLINES AND SUBJECTS

For centuries, concepts of convergence have been used in various academic disciplines to describe and analyse manifold processes of change. Like other analytical concepts, the term 'convergence' was first used in the natural sciences and then introduced to the social sciences and humanities. In the social sciences, various disciplines use the concept of convergence to describe different phenomena. The term is applied, for example, in political science to the convergence of political systems, especially of the Western capitalist system and the Eastern socialist one. In technology research, the approximation and fusion of nano-, bio- and information technologies with the cognitive sciences is called NBIC convergence, or converging technologies. In communications research, the concept of convergence is employed to analyse different sorts of blurring boundaries. Research into the growing uniformity between the programming of public and commercial broadcasters in dual-order models, for example, is discussed as convergence, as are transformations in national media systems in general, focusing on whether they are becoming more similar (Kleinsteuber, 2008).

Further, convergence refers to the blurring of boundaries between media, more precisely the blurring of the traditional demarcation between telecommunications (point-to-point) and the mass media. This is identified in this chapter as the core piece and meaning of convergence.

In addition, in the telecommunications policy debate, the integration of wired and wireless communications is called convergence. The process of blurring boundaries between sub-sectors of communications is also central to the formation of a digital creative economy, and it has a crucial effect on various of its sub-sectors, thus making convergence concepts even more interesting to an understanding of this formation process.

Another common feature of media convergence is the interdisciplinarity of its research topics, which also holds true for the digital creative economy. The strength and at the same time the weakness of convergence is its fuzziness and its multipurpose character, which it shares with other successfully and widely used terms that bridge disciplinary discourses and research, for example the term 'governance' (Schuppert, 2005; Schneider, 2012).

DIFFERENT PERSPECTIVES AND FUNCTIONS

Convergence is used and discussed not only in academia but also by policy-makers and the industry, however with differing objectives, interests, definitions and accentuations. For the industry, convergence is predominantly a strategic objective and a business challenge. For policy-makers it is a policy goal and challenge. For academics it is mainly an analytical concept applied to understand and explain important aspects of media change in general and numerous detailed developments in particular.

Concepts of convergence fulfil different purposes and functions. They provide the analytical framework for various aspects of change, and bridge different disciplinary discourses of the subjects involved. They explore the big picture of change but also very detailed parts of it. By doing so, they integrate conflicting processes of convergence and divergence as two sides of the same phenomenon (Pool, 1983; Jenkins, 2006). In other words, concepts of convergence embrace both blurring traditional boundaries between old media and novel diversification and differentiation of new media. Convergence as a metaphor has the function of simplifying the complexity of media change. It fits nearly all aspects of digital media development, and it is also used as a 'rhetorical tool' to convince stakeholders of certain reforms (Fagerjord and Storsul, 2007).

The industry has been discussing the inevitability and desirability of convergence of telecommunications and broadcasting since the 1980s. In telecommunications circles, the pursuit of strategic objectives due to convergence has taken place more intensively than in media circles. Even three decades ago, the telecommunications industry had high hopes for integrated ISDN broadband networks and fibre-optic technology as central infrastructure for the convergent communications sector (Garnham and Mulgan, 1991), hopes which have been only in part fulfilled. Media representatives were more reserved in their interpretation of the convergence trend, equating it with deregulation or commercialization, and occasionally gave the impression that convergence exemplifies a hostile takeover by telecommunications (Latzer, 2009).

In the policy field convergence became a hot topic for international organizations such as the OECD, ITU and WIPO, for nation states and on the supranational level for the European Union as of the 1990s. Initially it was also discussed as a collision between the worlds of telecommunications and broadcasting, which had very different corporate and political cultures. Accordingly, in 1992 the OECD raised the significant question of whether this really was convergence or a collision between the two sectors (OECD, 1992). The EU officially took the issue up in 1997, with the Green Paper on the convergence of the telecommunications, media and information technology sectors, and the implications for regulation (European Commission, 1997). Harmonization and liberalization of the national European telecommunications sectors started in the mid-1980s and was largely accomplished within a decade. With convergence, the EU then embarked on another explosive reform topic, which was even more complex than the liberalization debate, and resulted in convergence-related institutional reforms at the supranational level. For example, political competencies for telecommunications and broadcasting were integrated in the Directorate General for Communications Networks, Content and Technology (Latzer, forthcoming a). On the national level, policy-makers likewise took up the convergence topic, focusing on regulatory consequences in particular.

Since the 1980s, communications research, too, has concerned itself with the characteristics and possible consequences of the convergence trend (Pool, 1983; Baldwin et al., 1996; Latzer, 1997, 2009; McQuail and Siune, 1998; Marsden and Verhulst, 1999; Bohlin et al., 2000; Storsul and Stuedahl, 2007; Drucker and Gumpert, 2010). The resulting literature covers a wide range of topics and approaches, from technological and economic aspects of convergence to political and socio-cultural features, which are outlined as levels of convergence below.

In the 1990s, industry, politics and research together made convergence one of the central buzzwords in the communications field and beyond, alongside and often combined with digitization, liberalization and globalization. With the rapid proliferation of Internet-based services, especially with Web 2.0, digital TV, social media and wireless communication, the convergence phenomenon has attracted even more attention since the beginning of the twenty-first century.

BLURRING BOUNDARIES BETWEEN TELECOMMUNICATIONS AND MEDIA

The beginnings of research on media convergence (Pool, 1983) and the subsequent large bulk of the convergence literature concentrate on the process of blurring lines between individual and mass communication. They focus on the convergence of modes of communication and the blurring of boundaries between traditional media and their sub-sectors in the communications sector. More precisely, convergence between telecommunications and the traditional mass media, in particular with broadcasting, is analysed.

From an analytical point of view it is helpful to conceptualize the blurring of boundaries between telecommunications and mass media narrowly as the core piece and meaning of media convergence. Furthermore, as convergence continues and is even increasingly used as a buzzword for talking about a very wide range of phenomena and changes, its time dimension should be considered. It is neither an endless nor a steady process, as is sometimes misleadingly implied, but a temporary one. It peaked at the end of the twentieth century, even though there are significant offshoots for communications and the digital creative economy well into the twenty-first century. For example, such implications of convergence include the proliferation and application of Internet-based services throughout the economy. These offshoots should not be mistaken for the core element of a narrowly defined convergence. The bursting of the Internet bubble around the turn of the millennium slowed the process down in the short term but did not halt it. Further, such a time-sensitive perspective on convergence, and the distinction between a core process and its offshoots, does not support the notion that every single consequence of media convergence is to be a transformation towards unity and uniformity.

In other words, it would overstretch the concept of media convergence to expect that every future implication associated with the blurring of boundaries between media will go in the direction of uniformity. Central convergence processes towards uniformity have already happened at the end of the twentieth century, and stakeholders are still struggling with their consequences, which have disrupted business and regulatory models, strategies, classifications and laws that have been used for decades in politics, the economy and research. Not surprisingly, the media convergence process is followed by divergence processes as well, by novel differentiations within the convergent communications sector. In any case, there is no way back to the old structures. Changes by convergence can primarily be considered as structural change, with wide-ranging second-level effects for content and creativity (Kolo, 2010; Potts, 2011). Pace, intensity and details of change vary between countries, depending on different starting positions and peculiarities of national communications systems and structures.

Seen historically, the twentieth-century communication sectors, which were nationally organized and essentially characterized by more or less universal distinctions between telecommunications and mass media, formed the starting point for media convergence. The commercial use of telegraphy and telephony began in the second half of the nineteenth century and became known as the telecommunications sector. The broadcasting sector established itself commercially a few decades later and was classified together with the press as part of the media sector. These two sub-sectors – telecommunications and the media – used different technologies and separate networks. They were run by different companies, there were distinct political competences and separate regulatory agencies and legal foundations, and they had different underlying regulatory models (Latzer, 2009).

By the end of the twentieth century this technology-oriented subdivision into media and telecommunications, into mass communication and individual communication, was crumbling. Traditional categorizations, analytical frameworks, separate regulatory bodies and regulatory models for telecommunications and the mass media were challenged, driven by a combination of digitization, mobile communications, the Internet and digital television.

As the difficulty of classifying the online communication sector shows, the result of the convergence of telecommunications and broadcasting is more than just the sum of its parts. The way the trend is formulated conceptually and terminologically varies depending on the research perspective. The result of convergence is variously described as multimedia, TIME (telecommunications, information technologies, media, entertainment) or cross-media, emphasizing its media-overlapping character.

From a structural perspective, convergence changes the techno-social, societal communication systems towards mediamatics (Latzer, 1997). It is the computer sector that connects the previously separate sub-sectors of telecommunications and the mass media. This process has gone through two main stages, starting at the end of the twentieth century (see Figure 12.1). In a first step, data communication and the digitization of telephony marked the arrival of computer technology (infor*matics*) into *tele*communications, which was coined as 'telematics' (Nora and Minc, 1978). This was followed by the convergence of the likewise digitalized mass *media* with tele*matics* toward an integrated

until 1970s since 1970s since 1980s

telecommunications

computers (informatics)

broadcasting (electr. mass media incl. CATV)

TELEcommunications + inforMATICS = TELEMATICS

electr. MEDIA + teleMATICS = MEDIAMATICS

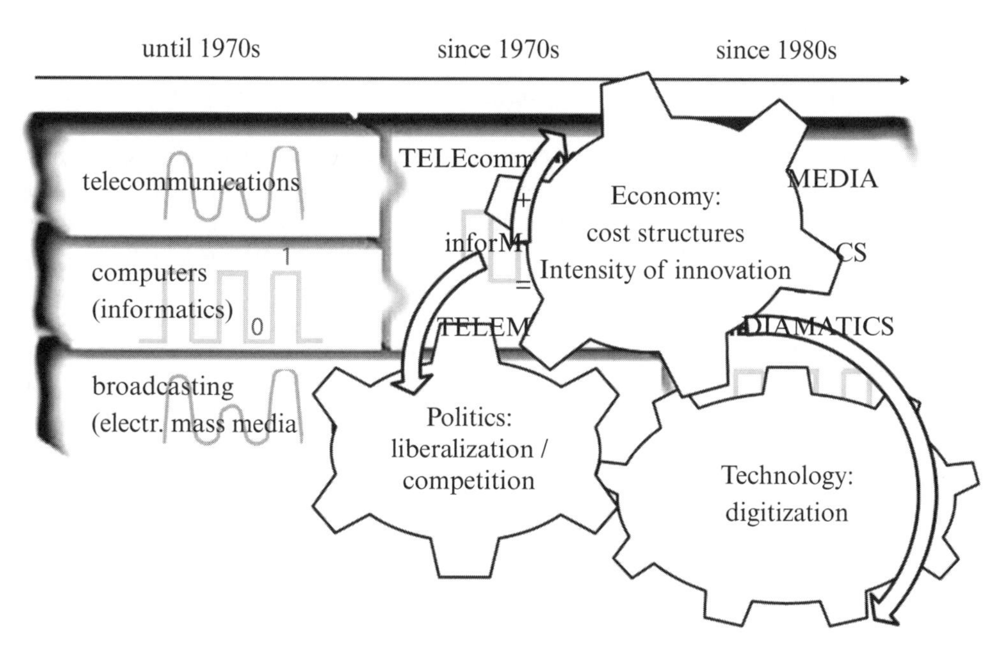

Note: The impact of convergence in communications is not limited to the electronic sub-sectors, but, for example, also affects the press sector.

Figure 12.1 Co-evolutionary convergence steps in electronic communications

societal communications system called 'mediamatics' (Latzer, 1997). The literature on media convergence is mainly concerned with this second stage. The convergence process was co-evolutionary; that is, its direction and pace were determined by the reciprocal interplay of technological innovations, corporate strategies, political-legal reforms and changes in media reception patterns as sketched in a simplified manner in Figure 12.1.

LEVELS AND IMPLICATIONS OF CONVERGENCE

It follows that convergence is taking place at different levels. The numerous terms, definitions and classifications used in the literature can be summarized in four categories: technological, economic, political and socio-cultural convergence. Because of overlaps with the communications sub-sectors and structural similarities, all of these are instructive for the understanding of the digital creative economy as well.

Technological convergence is playing a leading role. It stands for a universal digital code across telecommunications and electronic mass media, and for common protocols (IP), which are used for different technological (hybrid) platforms/networks (fixed-wire and mobile) and lead to service-integrating devices, such as TV-capable smart phones. These changes are also referred to as network convergence and terminal convergence (Storsul and Fagerjord, 2008). Digitization is one important part of the convergence phenomenon, one of its enabling factors, characteristics and driving forces. Despite its importance, however, it would be inappropriate and misleading to reduce convergence to technological convergence alone, as is often done. Above and beyond this, it should not be combined with naïve expectations of an all-embracing uniform medium, of future households with only one network or one terminal per person for all communications purposes. To the contrary, convergence creates better technical and economic conditions for a plurality of integrated networks, services and terminals. Technologically, it creates a digital modular construction system (Latzer, 1997), which offers great flexibility for innovatively assembled services, and economically it lowers the cost compared to analogue, electromechanical technology. Convergence leads to increased flexibility on the supply side, and hence to increased product variety as the previously rigid combination of technology and content is dissolved.

Combined with technological convergence there is *economic convergence* in the communications sector (Wirth, 2006). This includes market convergence on the meso- and macro-level and corporate convergence, characterized by new business models, and organizational change within companies at the micro-level. Market convergence raises important questions: How should relevant markets be defined, for instance for integrated broadband networks? Does convergence lead to increased competition because products converge in substitutes and compete with each other, or does it lead to reduced competition because products converge in complements, which implies more cooperation (Greenstein and Khanna, 1997)?

Further economic topics (Wirth, 2006) include: the transformation from traditionally vertical businesses such as television and telephony to horizontal segments such as content production, packaging and transmission; the impact of convergence on mergers and acquisition strategies (Chan-Olmsted, 1998); the implications for strategic management; and the consequences for demand. For example, in what is described as triple play, corporate convergence has led to the same companies now being active in telecommunications and broadcasting as well as on the Internet. If fixed and mobile telephony are included, this becomes quadruple play. Following core businesses such as search engines (for example Google) and electronic trading (for example eBay), new kinds of convergence enterprises are emerging. Traditional media and telecommunications companies, including public broadcasters, are changing to new business models, which is further combined with internal reorganizations. Press and television companies are good examples (Killebrew, 2005). Their current dilemma is basically that old business models no longer work, and ready-made, tested new ones are not yet available. An example of organizational convergence is the experimentation with integrated multimedia newsrooms, which, in a next-level effect, calls for changes regarding qualifications and skills. Ultimately, all of these structural changes have an impact on the product, and on the quality of the content produced, with ramifications for public communication in national and supranational communications systems in general.

Political convergence is mainly discussed as policy and as regulatory convergence. The traditional policy model, with its fundamental division into telecommunications and the mass media came under pressure. While the industry proceeded quickly into the convergence era, policy-makers and researchers remained largely stranded in the traditional separation of telecommunications and the media. Policy convergence discusses the transformation from traditionally separate telecommunications and media policies towards one national or supranational communications policy (Cuilenberg and Slaa, 1993; Latzer, 1998). This overlaps with regulatory convergence, which reflects integrated regulatory agencies and laws for the convergent communications sector. Alongside obsolete demarcations, convergence means that new regulatory responsibilities are emerging or growing in importance (Bohlin et al., 2000; Drucker and Gumpert, 2010), including the protection of intellectual property, freedom of speech and the regulation of domain-name systems. Further, the challenge of balancing socio-cultural and economic regulation increases with the blurring boundaries between media.

After a period of unrest caused by convergence, a dominant new design of governance for convergent communications markets is becoming apparent which constitutes major building blocks for worldwide reforms (Latzer, 2009). Constituent components include: integrated political strategies for telecommunications, the Internet and the media; integrated control structures (regulatory authorities) and laws for the convergent communications sector; a technology-neutral, functional taxonomy; a subdivision into transmission and content regulation; and a growing reliance on alternative models of regulation such as self- and co-regulation.

Socio-cultural convergence is also discussed as rhetorical, cultural, socio-functional, receptional and spatial convergence, and as convergence culture. All of these aspects are closely linked to the digital creative economy. The media can be conceived of as being constituted by technology and social/cultural practice. Research on socio-cultural convergence focuses on changes in social practice, phenomena such as transmedia storytelling, content and genres that are used across channels and platforms. Rhetorical convergence focuses on language and refers to the creation of new genres by remixing traits of genres of different media (Fagerjord and Storsul, 2007). Under the term 'convergence culture' (Jenkins, 2006), academics discuss the impact of convergence on popular culture with consequences on how we learn, connect and work, the change towards a stronger participatory culture, the transformation from audience to 'prosumers', and the co-production of media texts by integrating user-generated content and collective intelligence. The consequences of convergence are thus not only top down but also driven from the bottom up.

Cultural convergence, understood as the impact of convergence on media culture, is also of interest from a media-economic point of view (Wirth, 2006). Research focuses on: the repurposing of existing media content; cross-media formats; managerial challenges of convergence in newsrooms; changing working conditions through convergence in the newsroom; multi-skill requirements; the redesign of content; and the impact on creativity in changing workplaces (Killebrew, 2003).

It has also become apparent that, in the form of socio-functional convergence, telecommunication is now increasingly used in the private-entertainment sector and that broadcasting is used for business communication (for example internal corporate business TV). Demand-side analyses of convergence examine the way in which the media time-budgets, daily routines, leisure activities and job profiles are changing (Oehmichen and Schröter, 2000). There have been shifts, substitutions and combinations in the way services are used – which is also known as receptive convergence, as it concerns the change in reception patterns and a convergence of usage patterns (Höflich, 1999). Finally, there is spatial convergence, which refers to the globalizing effect of rapidly growing cross-border services and uniform technology (Latzer, 2009).

CO-EVOLUTIONARY PERSPECTIVE

Convergence is addressed from a variety of theoretical perspectives. A co-evolutionary approach is particularly useful in dealing with the interdependencies of different levels of convergence as described above (see Figure 12.1), to draw different conclusions regarding the implications of the convergence phenomenon, and to adequately deal with the underlying complexity and evolutionary character of media change and the convergence phenomenon. Thus it strengthens a scientific foundation which is more appropriate for dealing with the specific attributes of the research subject (Latzer, 2013, forthcoming b).

Media change in general and convergence in particular can be conceived of as innovation-driven, co-evolutionary processes in a complex environment. Innovations that are analysed as co-evolutionary, adaptive cycles of renewal are the nucleus of change. They are the central driving forces of dynamic developments in communications and the digital creative economy. From an evolutionary economic perspective, creative industries are not only the outcome of innovations. With an infrastructure role, creative industries also contribute to the origination, adoption and retention of new technologies in open complex innovation systems (Potts, 2011). Convergence is driven by different kinds of innovations, as reflected in the different levels of convergence. Reciprocities between these different levels are of particular importance. A co-evolutionary approach takes into account the interdependencies of technological, economic, political and socio-cultural convergence processes (innovations), and by doing so leads to additional insights and different implications for political and corporate strategies. While evolution can be characterized as design without a central designer, co-evolution means designing and being designed at the same time. Alternative terms are 'co-construction' and 'confluence' (Benkler, 2006). These concepts overcome the long and fierce debates about technological and social determinism in the literature on media change and convergence.

A co-evolutionary approach to the Internet, a system that is central to change both in communications and in the digital creative economy, provides a good example. This perspective presents the Internet as a complex, adaptive system, characterized by non-linear developments, emergence and decentralized structures. It explains the interplay – more precisely, the mutual selective pressure and adaptive behaviour – of technology, organization and business models that nurture each other (Beinhocker, 2006). Coincidences are included in these developments as another constitutional characteristic. Co-evolutionary processes can be found, for example, in the World Wide Web (WWW), where the simple web behaviour of individuals – who are not centrally controlled – leads to an emergent, unforeseeable, complex behaviour of the total, self-organized WWW social system. There is also a co-evolutionary relation between the search engines and the link structure of the web, which altogether results in an adaptive behaviour of the WWW social system (Mitchell, 2009).

The co-evolutionary approach highlights not only the content of the Internet but, in combination, its infrastructure. Thus the Internet is best described as a modular, open system with an end-to-end design that allows innovations at every node of the network, in other words by any user. Altogether, this offers great flexibility and scope for innovatively assembled services. In this way the previously rigid combination of technology and content is dissolved.

To sum up, the Internet is a modular construction system, essentially an innovation machine. This co-evolutionary perspective leads to various implications for political and corporate strategies. It provides a different conceptual framework. Compared to other approaches, the predictability and controllability of developments are much more limited, leading to different conclusions on the role of the state regarding policy guidelines and corporate strategies. For example, more adaptive policies are used – including feedback loops such as periodic review processes – and trial-and-error methods are increasingly applied (Latzer, forthcoming b).

Co-evolutionary models are in particular applied for the analysis of complex systems. One of their characteristics is emergence, that is, the unforeseen formation or appearance of new structures and characteristics in a system that are not directly derivable from existing, old characteristics. The convergent communications sector and the digital creative economy can both be considered as emergent phenomena. The result is more than the sum of its parts and cannot simply be understood in terms of those parts. A convergence analysis from a co-evolutionary perspective therefore promises additional interesting insights. The co-evolutionary approach not only contributes to the scientific foundation of an analytically sound convergence concept in the narrow sense but also provides the theoretical basis to better understand the various offshoots of convergence in a wide sense. While convergence in a narrow sense is well suited to analysing changes of already existing parts, co-evolution and complexity approaches are also helpful to explaining the 'new', the outcome and implications of convergence for various (other) parts of society.

CONCLUSIONS

The relevance of the concept of media convergence for the understanding of the digital creative economy stems from structural similarities and growing overlaps with convergent communications markets. New digital media are the outcomes of convergence, and they are central drivers of the digital creative economy. Various stakeholders use convergence concepts to convey different aspects of media change. A narrow definition of convergence concerns the blurring of boundaries between traditional sub-sectors of communications. Broader definitions of convergence, especially those that do not consider its time dimension, narrow its merit as an analytical concept, basically because of growing vagueness and less reference to the core piece and mechanisms of convergence. Convergence-induced changes in communications and the digital creative economy are driven by the interplay of technical, economic, political and socio-cultural factors. A co-evolutionary approach takes growing complexity and interplay into account. It compensates for a general weakness of the narrow convergence concept, which is strong in the analysis of the 'old' converging parts but weak in the explanation of the emerging 'new', for example on implications of the Internet throughout the economy. A combined co-evolutionary and complexity perspective sketches, among other things, the outcome of convergence, the formation of a transformed societal communications system. Convergence can be understood as an innovation-driven, co-evolutionary process in a complex environment. It is a process of structural change with a wide range of implications for content and creativity. Concepts of convergence provide the big picture but also allow for detailed analyses throughout the digital creative economy.

NOTE

* Parts of this chapter build on Latzer (2009, 2013, forthcoming b). I would like to thank Johannes M. Bauer and Natascha Just for their comments and Katharina Hollnbuchner for research assistance.

REFERENCES

Baldwin, T.F., S.D. McVoy and C. Steinfield (1996), *Convergence: Integrating Media, Information and Communication*, London: Sage.

Beinhocker, E. (2006), *The Origin of Wealth*, London: Random House.

Benkler, Y. (2006), *The Wealth of Networks*, New Haven, CT: Yale University Press.

Bohlin, E., K. Brodin, A. Lundgren and B. Thorngren (eds) (2000), *Convergence in Communications and Beyond*, Amsterdam: Elsevier.

Chan-Olmsted, S.M. (1998), 'Mergers, acquisitions, and convergence: the strategic alliances of broadcasting, cable television, and telephone services', *Journal of Media Economics*, **11**, 33–46.

Cuilenberg, J. van and P. Slaa (1993), 'From media policy towards a national communications policy', *European Journal of Communication*, **8** (2), 149–76.

Drucker, S.J. and G. Gumpert (eds) (2010), *Regulating Convergence*, New York: Peter Lang.

European Commission (1997), 'Towards an Information Society approach: Green Paper on the convergence of the telecommunications, media and information technology sectors, and the implications for regulation', COM(97)623 final, 3 December, available at: http://aei.pitt.edu/1160/01/telecom_convergence_gp_COM_97_623.pdf (accessed 27 August 2012).

Fagerjord, A. and T. Storsul (2007), 'Questioning convergence', in T. Storsul and D. Stuedahl (eds), *Ambivalence towards Convergence: Digitalization and Media Change*, Göteborg: Nordicom, pp. 19–31.

Garnham, N. and G. Mulgan (1991), 'Broadband and the barriers to convergence in the European Community', *Telecommunications Policy*, **15** (3), 182–94.

Greenstein, S. and T. Khanna (1997), 'What does industry convergence mean?', in D.B. Yoffie (ed.), *Competing in the Age of Digital Convergence*, Boston, MA: Harvard Business School Press, pp. 201–26.

Höflich, J.R. (1999), 'Der Mythos vom umfassenden Medium: Anmerkungen zur Konvergenz aus einer Nutzerperspektive', in M. Latzer, U. Maier-Rabler, G. Siegert and T. Steinmaurer (eds), *Die Zukunft der Kommunikation, Phänomene und Trends in der Informationsgesellschaft*, Innsbruck: Studien Verlag, pp. 43–60.

Jenkins, H. (2006), *Convergence Culture: Where Old and New Media Collide*, New York: New York University Press.

Killebrew, K.C. (2003), 'Culture, creativity and convergence: managing journalists in a changing information workplace', *International Journal on Media Management*, **5** (1), 39–46.

Killebrew, K.C. (2005), *Managing Media Convergence: Pathways to Journalistic Cooperation*, Oxford: Blackwell.
Kleinsteuber, H.J. (2008), 'Convergence of media systems', in W. Donsbach (ed.), *The International Encyclopedia of Communication*, Blackwell Reference Online, available at http://www.communicationency clopedia.com/subscriber/tocnode?id=g9781405131995_yr2012_chunk_g97814051319959_ss39-1 (accessed 20 July 2012).
Kolo, C. (2010), 'Online Medien und Wandel: Konvergenz, Diffusion, Substitution', in W. Schweiger and K. Beck (eds), *Handbuch Online Kommunikation*, Wiesbaden: VS Verlag, pp. 283–307.
Latzer, M. (1997), *Mediamatik: Die Konvergenz von Telekommunikation, Computer und Rundfunk*, Opladen: VS Verlag für Sozialwissenschaften.
Latzer, M. (1998), 'European mediamatics policies: coping with convergence and globalization', *Telecommunications Policy*, **22** (6), 457–66.
Latzer, M. (2009), 'Convergence revisited: toward a modified pattern of communications governance', *Convergence: International Journal of Research into New Media Technologies*, **15** (4), 411–26.
Latzer, M. (2013), 'Medienwandel durch Innovation, Ko-Evolution und Komplexität: ein Aufriss', *Medien & Kommunikationswissenschaft*, **2013** (2), 235–52.
Latzer, M. (forthcoming a), 'Convergence, co-evolution and complexity in European communications policy', in K. Donders, C. Pauwels and J. Loisen (eds), *Handbook on European Media Policy*, Houndmills: Palgrave Macmillan.
Latzer, M. (forthcoming b), 'Towards an innovation–co-evolution–complexity perspective on communications policy', in S. Pfaff and M. Löblich (eds), *Media and Communication Policy in the Era of the Internet and Digitization* (working title), Baden-Baden: Nomos.
McQuail, D. and K. Siune (eds) (1998), *Media Policy: Convergence, Concentration and Commerce*, London: Sage.
Marsden, C.T. and S.G. Verhulst (eds) (1999), *Convergence in European Digital TV Regulation*, London: Blackstone Press.
Mitchell, M. (2009), *Complexity: A Guided Tour*, Oxford: Oxford University Press.
Nora, S. and A. Minc (1978), *L'informatisation de la société*, Paris: La Documentation Française.
OECD (1992), *Telecommunications and Broadcasting: Convergence or Collision?*, Paris: OECD.
Oehmichen, E. and C. Schröter (2000), 'Fernsehen, Hörfunk, Internet: Konkurrenz, Konvergenz oder Komplement?', *Media Perspektiven*, **8**, 359–68.
Pool, I.d.S. (1983), *Technologies of Freedom*, Cambridge, MA: Belknap Press.
Potts, J. (2011), *Creative Industries and Economic Evolution*, Cheltenham, UK and Northampton, MA, USA: Edward Elgar Publishing.
Schneider, V. (2012), 'Governance and complexity', in D. Levi-Faur (ed.), *Oxford Handbook of Governance*, Oxford: Oxford University Press, pp. 129–42.
Schuppert, G.F. (ed.) (2005), *Governance-Forschung: Vergewisserung über Stand und Entwicklungslinien*, Baden-Baden: Nomos.
Storsul, T. and A. Fagerjord (2008), 'Digitization and media convergence', in W. Donsbach (ed.), *The International Encyclopedia of Communication*, Blackwell Reference Online, available at: http://www.com muni cationencyclopedia.com/subscriber/tocnode?id=g9781405131995_yr2012_chunk_g97814051319959_ss39-1 (accessed 20 July 2012).
Storsul, T. and D. Stuedahl (eds) (2007), *Ambivalence towards Convergence: Digitalization and Media Change*, Göteborg: Nordicom.
Wirth, M.O. (2006), 'Issues in media convergence', in A.B. Albarran, S. Chan-Olmsted and M.O. Wirth (eds), *Handbook of Media Management and Economics*, Mahwah, NJ: Lawrence Erlbaum, pp. 445–62.

FURTHER READING

See Pool (1983) for an early and basic convergence concept and its impact on political culture. Wirth (2006) provides a systematic literature review on various aspects of economic convergence. Latzer (2009) focuses on governance issues resulting from political-regulatory convergence. Bohlin et al. (2000) as well as Drucker and Gumpert (2010) provide insights in the multitude of policy and regulatory challenges. Storsul and Stuedahl (2007) and Jenkins (2006) discuss a wide range of socio-cultural convergence issues. The academic journal *Convergence: The International Journal of Research into New Media Technologies* keeps track of current debates regarding convergence.

13. Has digitization delivered? Fact and fiction in digital TV broadcasting

Reza Tadayoni and Anders Henten

Media and communication technologies have during the past couple of decades gone through one of the most radical changes in their history, namely the transformation from analogue to digital technologies. This change has massive influence on the whole chain from production to consumption of media and communication services. However, digitization is a gradual process and can have different development speeds in different part of this chain. We can, for instance, have a digitized backbone but an analogue access network, and we can have analogue production while deploying digital technologies in the distribution systems. The development patterns will, furthermore, vary in different parts of the world.

In this chapter, we focus on the digitization of broadcast TV and examine the implications in terms of variety, quality, interactivity, convergence, and mobility. In the late 1980s and early 1990s, when the discussions concerning the new digital paradigm for broadcasting started, there were a number of myths and promises. Presently, we are in a position to assess the results of the processes of digitization in a more realistic manner and forecast some of the major elements of future developments.

In discussions and the literature in the pre-digital era and the beginning of the digitization processes, a number of advantages of digitization were put forward. The major ones were the following:

- Variety: Digital technologies deployed in the distribution and delivery networks are spectrally more efficient. Digital technologies require less spectrum to transmit the same services. Owing to spectrum scarcity, the number of TV channels in the terrestrial networks has been limited. Enhanced spectral efficiency in the digital era results in an increasing capacity for delivering more TV channels in the same frequency bands and allows for the creation of multi-channel platforms, strengthening competition in broadcast markets.
- Quality: Digital technologies will enhance the quality of the signals and provide a higher video quality and will improve the user experience.
- Interactivity: Broadcast has hitherto been a one-way mass communication technology. Digitization will change this and enable interactivity, which will be a major element in the formation and development of new services and formats.
- Convergence: In the digital era, the broadcast and Internet worlds will converge and bring the innovation dynamics of the Internet and the IT industry to the broadcast sector.
- Mobility: Digitization will enable the mobility and development of mobile broadcast services. This has been seen as one of the important competitive advantages of terrestrial networks compared to satellite and cable networks.

In this chapter, we first discuss these parameters in the different phases of the development of TV (the TV paradigms) and, afterwards, we examine the parameters in a more focused manner with respect to different digital TV infrastructures and the degree to which these parameters have been drivers of actual developments or just fictions and wishes. In the overview of the TV paradigms, we include three other parameters, which provide important characteristics of the changes and transitions between the different paradigms. The additional parameters are:

- Terminal: This is concerned with the changes and the specificities of the receiver devices relating to the different paradigms.
- Infrastructure: This parameter deals with the underlying network infrastructure for bringing TV to the end users. In the terrestrial networks, there is a particularly close interdependence between the infrastructure and the available spectrum.
- Ownership: In the different development phases or paradigms, there are different types of ownership – in particular with regard to the infrastructure part, with the entry of new business actors.

The chapter includes two major sections. In the following section, the partly sequential and partly overlapping technological TV broadcast paradigms are presented: from analogue TV, through digital TV including mobile TV, to IPTV (Internet Protocol TV), HBBTV (Hybrid Broadcast Broadband TV), and OTT TV (Over-the-Top TV). In the subsequent section, the most important basic elements of broadcasting, from content production and contribution to distribution, are presented and the implications of digitization with respect to spectral efficiency and so on are discussed. The chapter ends by bringing out the conclusions to be drawn.

TV BROADCAST PARADIGMS

This section examines TV broadcast from analogue TV to OTT and its different development paradigms with respect to the parameters: terminal, infrastructure, ownership, spectrum, variety, quality, interactivity, convergence, and mobility. We identify the following paradigms: analogue TV, digital TV including mobile TV, IPTV, HBBTV, and OTT TV.

Analogue TV

Analogue TV has mainly been based on three standards: the PAL (Phase Alternating Line) standard, the SECAM (Séquentiel Couleur à Mémoire) standard, and the NTSC (National Television System Committee) standard. The standards are incompatible, and different countries or regions have each used their own standard. The terminal was a TV receiver connected to an in-house or roof-top antenna, a satellite dish, or a cable TV connection point. The TV receiver had a long life-cycle of about 15 years based on the electronic components used in the TV equipment. There were different providers of TV receivers, and competition was mainly on price and design.

At the beginning of TV broadcast, the terrestrial infrastructure was the sole platform, and it was only later that cable and satellite infrastructures came to the market – first as complementary and later as competitive infrastructures. At first, parts of the VHF (very high frequency) spectrum were allocated and used for TV and, later, portions of the UHF (ultra-high frequency) spectrum were added to enable more TV channels in the terrestrial networks. Three ownership models were used: (1) state-owned public service TV; (2) non-commercial public service, financed by licence fees; and (3) private/commercial advertisement-based TV. In the US, for instance, commercial TV was also subject to public interest regulation. This resulted in a monopoly market organization in many countries and duopoly or oligopoly in others. However, satellite and cable TV changed this situation, and many actors entered the market. With digital TV, more channels in the terrestrial networks were also made possible.

Owing to spectrum scarcity and the technology deployed, the capacity in terrestrial networks was very limited and, therefore, only a few TV channels were offered in these networks. The low capacity and lack of variety related only to the terrestrial infrastructure, while satellite and cable TV had much better opportunities for increasing the number of TV channels, delivering multi-channel packages to their customers. The quality of the received TV service followed the deployed standard and, apart from the change from black and white to colour TV, the quality was almost the same over a long period of time. Obviously, there were small differences between different brands in the way they developed the hardware, but there was not much room for major differentiations. TV was a non-interactive broadcast platform, and convergence and mobility were not important issues in the analogue era. However, with the introduction of text TV, broadcast TV started to enhance its services beyond traditional audio and visual services by integrating other information elements.

Digital TV

Digital TV, indeed, has changed many of the parameters discussed in this chapter in a radical manner. As with analogue TV, there are different standards in different countries and regions, the main standards being DVB (Digital Video Broadcasting), ATSC (Advanced Television System Committee), and ISDB (Integrated Services Digital Broadcasting). However, different digital TV standards use some of main components regarding coding and compression standards or technologies, such as the MPEG (Moving Picture Experts Group) family of standards. The terminal used for digital TV broadcast can be an analogue TV receiver connected to a digital set-top box converting from digital to analogue signals. In such a set-up with a digital set-top box and the use of analogue TV receivers, it is possible to view only qualities corresponding to standard-definition TV (SDTV), while an integrated digital TV receiver provides the possibility for viewing higher-quality signals with different variations of high-definition TV (HDTV). The digital TV receiver is thus very different from an analogue TV, as there is considerable room for improvement and differentiation within the different standards. Figure 13.1 shows that, in the years to come, there will most likely be three different versions of TV receivers on the market at any given point of time.

All the analogue TV infrastructures (terrestrial, satellite, and cable) can be digitized to deliver digital TV to the users. The spectrum allocations for broadcast TV are the same

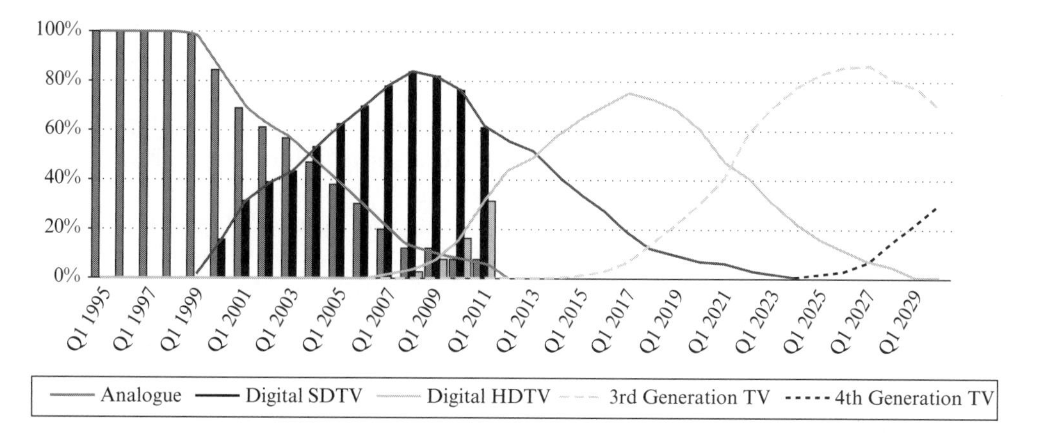

Source: McCann and Mattei (2012).

Figure 13.1 Trends in development of digital TV standards

as with analogue TV. Moreover, as with analogue TV, there are different categories of ownership; however, the number of commercial TV operators, in particular on terrestrial platforms, has increased dramatically. Furthermore, a new type of actor has entered the market for terrestrial digital TV, namely multiplex or platform operators and gatekeepers that organize the provision of TV channels to the users.

Owing to more efficient utilization of spectrum resources, part of the spectrum can be taken out of TV allocations and be redistributed to other service areas, which is why the term 'digital dividend' is being used. As a result of the more efficient utilization of spectrum and different coding and compression technologies, the number of TV channels in a given spectrum band can also be increased radically compared to analogue TV. The possibility of increasing the quality of the TV signals is, furthermore, one of the important potentials of digital TV. Whether this capability is used depends on several issues, but already in the present we are beginning to see more and more providers delivering HDTV and even 3D TV. Interactivity was envisioned as an important quality of digital TV before its launch. However, as seen later in this chapter, interactivity and convergence have not been important drivers in the current deployment. Mobility is possible in some of the digital TV standards, but they generally have been modified in order to be able to better handle the delivery of mobile TV services.

Mobile TV

Mobile TV is a part of digital TV but is dealt with here as a separate issue, as it has specific characteristics. All major terrestrial TV standards have adopted a mobile TV standard in order to be able to deliver TV to handheld devices. The infrastructure is mostly the terrestrial broadcast TV infrastructure (for mobile TV broadcast) and mobile broadband infrastructures (for streaming mobile TV). The terrestrial spectrum is the same as for digital TV in general, where different countries allocate part of the TV spectrum for mobile TV provision. With respect to mobile broadband infrastructures, the mobile operators allocate part of their spectrum for mobile TV services. Ownership of mobile broadcast TV has been a big issue and is identified as one of the barriers to the development of mobile TV in Europe (CEC, 2008). The issue is not so much the ownership of TV channels but, to a larger extent, the ownership of infrastructures. The issue of ownership regarding mobile streaming services is quite straightforward, as this is a part of the services provided by the mobile operator.

Concerning capacity and variety, mobile TV has the potential to deliver a huge number of TV channels, as the mobile devices have small displays and the capacity required to deliver TV to these devices is low. Interactivity is inherent in the mobile TV concept, as the majority of mobile devices are interactive devices with return channels through mobile broadband networks. However, it does not seem that interactivity has been an important aspect of the development of mobile TV. When it comes to the convergence issue, it could likewise be said that all conditions for reaping the benefits of the convergence processes have been in place with respect to mobile TV but that, in reality, the provisions consist of regular TV delivered to mobile devices. Convergence still entails the potential to provide mobile TV with its own 'language' and change from 'TV on mobile' to 'mobile TV', which really could be a new media form (Tadayoni et al., 2009).

IPTV

IPTV is another aspect of the development of digital TV. IPTV denotes the provision of TV through broadband networks using the Internet Protocol (IP). IPTV can be provided both as managed services and as a best effort service. However, the term 'IPTV' generally denotes the delivery of TV through managed broadband networks. In such a managed provision, the IPTV provider guarantees a specific degree of quality of service (QoS), while in the case of best effort it is offered through the general Internet with varying levels of QoS guarantees. There are a number of different IPTV technologies available on the market, and the International Telecommunication Union (ITU) has specific focus on developing standards for IPTV.

The terminal in IPTV is an IP-enabled TV receiver or a regular analogue or digital TV receiver connected to an IP set-top box. The infrastructure is the broadband infrastructure, and a certain minimum capacity is necessary to deliver IPTV through broadband. Part of the spectrum in the broadband infrastructures must be dedicated for IPTV, and this decision is made by the broadband operator. With regard to ownership, the platform is owned by the broadband provider, and the service provisions follow similar provisions to those of other multi-channel platforms like cable TV.

The number of TV channels in IPTV can be unlimited, as the selection of channels is not done locally but centrally at the server site or head-end. In reality, the number is limited because of business considerations. The quality discussion is the same as with digital TV. There are possibilities for delivering different qualities, and the choice regarding the quality to be provided is based on the available capacity and on business considerations. Broadband is an interactive medium; therefore IPTV has the greatest potential to deliver interactive TV. However, time will show the degree to which this potential will be utilized. Concerning convergence, there are potentials for all the gains from the convergence processes as in the case of mobile TV. IPTV can be offered via fixed as well as mobile broadband networks and, therefore, the potential for mobile IPTV is present. However, whether this will be a success depends on how much the industry learns from past experiences with mobile TV.

Recently, the TV sector has witnessed changes following two different tracks: (1) a wave of changes integrating broadcast with the Internet world under the heading of HBBTV; and (2) a pure Internet-based approach under the heading of OTT TV technologies and platforms.

HBBTV Platforms

The idea of HBBTV platforms is to connect smart TVs to broadcast as well as broadband networks, enabling the broadcasters to develop advanced services related to their content (Merkel, 2010). The HBBTV idea is to make it possible for the broadcasters to retain control of the platforms and the associated applications. The European Broadcasting Union (EBU) has been one of the main promoters of this development (EBU, 2010). It is possible in a transition phase to use regular TV equipment together with a set-top box. Depending on what the broadband infrastructure will be used for, there will be certain capacity requirements. The spectrum issue is a combination of what has been discussed above regarding digital TV and IPTV. The ownership issue is quite complicated with HBBTV, as content owners as well as distribution platforms and even TV receiver providers can act as gatekeepers for the platform.

Variety and convergence

The variety question follows the same combination of issues. Owing to the broadband connection, it is possible to deliver personalized services and on-demand types of services like video on demand (VoD). The quality in HBBTV is the same as with digital TV with the addition that extra information can be transmitted through the broadband network, which can enhance the experience of the TV received through the broadcast infrastructure. A certain minimum package can be broadcast to all consumers, and additional services can be delivered to specific users based on different conditions. With regard to interactivity, one can say that HBBTV establishes a real interactive medium in the broadcast world, which has been difficult to establish with digital TV. To what degree this possibility will be used is something to be seen in the coming years. With regard to convergence, there is the issue of combining the broadcast part with the broadband part. Even though the TV receiver has access to both broadcast and broadband networks, new developments are necessary to make the integration at the level of services. This integration is obviously included in the HBBTV standards, and huge developments are taking place to bring the new HBBTV services to the market. Mobile broadcast TV is an example of HBBTV, and mobile provision is potentially possible and relevant regarding the HBBTV platforms.

OTT

With OTT, the Internet is seen as the generic platform for the provision of broadcast content and the associated advanced services regarding, for instance, the integration with social networking and community services taking advantage of the proven dynamics and huge innovation potentials of the Internet. This track is US-centric, with companies like Google and Apple as the main actors.

With OTT, the terminal can be TV receivers or other kinds of devices such as PCs or laptops with connectivity to the broadband infrastructure. It is not necessary to have any connection to the broadcast infrastructure, and this may be seen as a huge change. The viewers will use their current TV devices together with a converter box delivered by the OTT providers like Apple and Google or they use their gaming devices, for example PlayStation or Xbox, to access the TV content. The infrastructure is the general Internet, and the spectrum issue is as in the case of broadband. Ownership of the OTT platforms is in the hands of companies from the IT sector like Google, Apple, and Microsoft, and the new actors in the OTT TV market like Netflix and Hulu. So far, the majority of provisions are on-demand services, and linear TV is not part of these provisions.

The number and variety of channels also follow the same discussions as with IPTV. The quality, as with any other digital platform, depends on the amount of capacity that can be assigned for the TV services, and OTT has a big challenge here, as there is no guarantee for QoS in the general Internet. An attempt is being made to solve this problem by establishing content delivery networks (CDNs) and by bringing the content as close to the users as possible. Interactivity is an inherent part of OTT, as it is based on the Internet.

The different paradigms and parameters are summarized in Table 13.1.

IMPLICATIONS OF DIGITIZATION

In this section, an analysis of the digitization processes and their advantages and disadvantages is provided focusing on four major delivery networks: terrestrial, satellite, cable TV, and broadband. An evaluation is made concerning the characteristics mentioned above (variety, quality, interactivity, convergence, and mobility) in order to conclude to what extent the proclaimed advantages of digitization have turned out to be fact or fiction.

First there is an overview of digital broadcasting, and then the listed characteristics relating to the different broadcast infrastructures are discussed in the following subsections.

Digital Broadcast Technology

Broadcast services include several different types of services in the value chain – from content creation to the final delivery of the services (TV programmes) to the end users. We can distinguish between two main parts:

- Contribution: The information or service exchange between broadcast stations, between broadcast stations and service providers, and internally between different departments of broadcast stations. Contribution denotes the business-to-business and inter-firm part of the chain.
- Distribution and delivery: The provision of the final broadcast services to the

Table 13.1 TV paradigms and assessment parameters

	Basic parameters			Characteristics				
	Terminal	Infrastructure	Ownership	Variety	Quality	Interactivity	Convergence	Mobility
Analogue TV	Analogue TV receiver	Terrestrial, satellite, and cable	State, public services, or commercial TV	Limited in the terrestrial infrastructure	PAL, SECAM, or NTSC	Not possible	Not an issue	Not applicable
Digital TV	Integrated digital TV receivers or analogue TV connected to set-top box	Terrestrial, satellite, and cable	The same as analogue but a new actor, a platform operator, is introduced	Increase in the number of TV channels in all infrastructures	Depending on the profile of the deployed standard	Possible but not used	Potentially important, but the effects still to be seen	Only relevant in the terrestrial networks
Mobile TV (broadcast)	Handheld devices	Terrestrial	A conflict between broadcasters and mobile operators on platform ownership	Many TV channels due to small displays	Different depending on the standard used	Possible but not used	Potentially important, but the effects still to be seen	Only relevant in the terrestrial networks
Mobile TV (streaming)	Handheld devices	Mobile broadband	Mobile operator owns the platform	The number is not an issue as the selection or change of channel is done centrally	Different depending on the capacity allocated	Possible and used for on demand	Potentially important, but the effects still to be seen	Applicable in all mobile broadband infrastructures

Table 13.1 (continued)

	Basic parameters			Characteristics				
	Terminal	Infrastructure	Ownership	Variety	Quality	Interactivity	Convergence	Mobility
IPTV	IPTV devices or regular TV with IP set-top box	Broadband	Broadband operator	The number is not an issue as the selection or change of channel is done centrally	Different depending on the capacity allocated and standard used	Possible but not used; on-demand services are important here	Potentially important, but the effects still to be seen	Relevant in mobile and wireless broadband infrastructures
HBBTV	Smart TV with both broadband and broadcast connectivity or regular TV with IP set-top box	Broadcast (terrestrial, satellite, or cable) and broadband	Complicated gatekeeper situation, where content providers, infrastructure providers, and device manufacturers can play a role	Combination of issues related to IPTV and digital TV	Quality enhance-ment to broadcast can be offered as add-on through broadband	Possible and time will show how much it will be used; on-demand services are important here	A good example of convergence	Relevant in mobile broadcast platforms
OTT TV	Smart TV with broadband connectivity or regular TV with OTT converter	Broadband	Dominated by actors from IT world	Like IPTV	Like IPTV, but as there is no QoS guarantee the quality can depend on network conditions	Possible and very much used when it comes to on-demand services	A good example of convergence	Applicable in the mobile broadband infrastructures

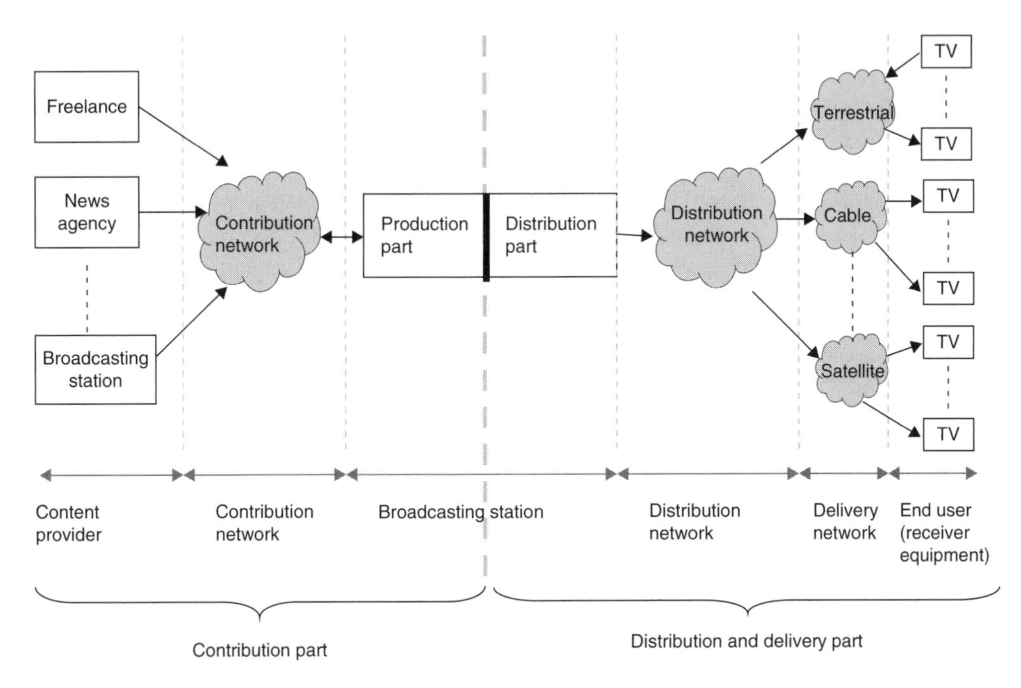

Figure 13.2 Contribution and distribution in analogue TV networks

end users. Distribution and delivery denote the business-to-consumer part of the chain.

Different requirements are applied to the contribution and distribution/delivery networks, and the content provided on these networks. In Figure 13.2, the value chain for analogue broadcasting is depicted.

In the contribution part, the majority of services are not finalized and are in an unedited (raw) format. However, the level of finalization of the services varies. In some cases, they are source materials, such as video materials from a news agency, and in other cases they are more finalized, such as a movie or a documentary bought from a content provider. Even in the latter case, some further processing such as putting on subtitles, removing parts, or adding other information, such as advertisements, within the programmes may be necessary. The rawness of material and the necessity for further processing require that the material must be of a high technical quality, as the editor must be able to work with the material almost frame by frame. Consequently, the requirements on the technical quality of the signal in the contribution network are very high.

Regarding the distribution and delivery part, the service is in a finalized state and is ready to be consumed, and the consumer does not need to process the information. Therefore the requirements on the technical quality of the signal in these networks are lower than in the contribution network.

Elements of the Value Chain

In the following, different parts of the value chain are briefly described.

- Content: The content providers are broadcast stations and different specialized and general bureaus. The broadcast station itself creates some of the content but it also acquires content from commercial firms such as news agencies, freelance journalists, and firms specialized in the creation of entertainment and educational programming. Furthermore, the broadcasters provide content to each other; for instance, members of the EBU provide news materials from their own countries to other countries.
- Contribution network: This is a network that carries the content to the broadcast station. It can be a high-speed telecommunication link, often optical fibres, as well as other high-speed links, like radio-wave and satellite links.
- Broadcast station: The broadcast station can be divided into two different parts: the production part, connected to the contribution network, with the task of editing and finalizing programmes, and the distribution part that handles the final distribution of the programmes to the end users.
- Distribution network: The network between the broadcast station and the delivery network is called a distribution network. At the beginning, dedicated high-speed telecommunication networks were used as distribution networks. However, in recent years, the telecommunication links are mainly based on non-dedicated optical fibres – and, when optical fibres are unfeasible, satellite and radio links are deployed. The distribution network is necessary only when there is a geographical distance between the delivery network and the broadcast station. A local TV station which operates only one transmitter and is physically located near the transmitter connects directly to the transmitter and, consequently, to the delivery network. Another example is a broadcast station that directly feeds the signal into cable TV networks.
- Delivery network: Delivery networks are the infrastructures that the end users are connected to. The major delivery networks are terrestrial, satellite, and cable TV networks. These networks, in the analogue world, are one-way networks transmitting to the end users.
- End user: The end user is the person who consumes the service. The reception of the service is done using receiver equipment consisting of a TV receiver and possibly an outdoor antenna.

Digitization of broadcasting has radical impacts on the whole value chain. Furthermore, new units such as service provision and multiplex functions are added to terrestrial digital broadcasting compared to analogue broadcasting. In Figure 13.3, the value chain of digital broadcasting is depicted, and, in the following, different parts of the value chain are analysed in the light of the digitization process.

Regarding content, the digitization of broadcasting has influences on the content itself as well as on the entry of new actors into the market. If exploring the potentials of interactivity, the programmes will evolve into involving the end users actively in the programmes they consume. Furthermore, new types of data services are created and services

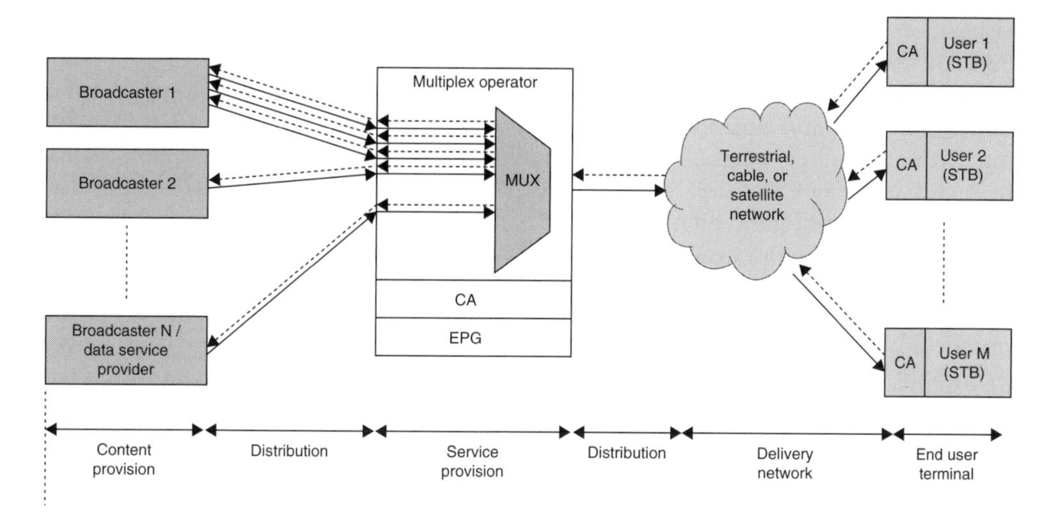

Notes: STB: set-top box; CA: conditional access; EPG: electronic programme guide; MUX: multiplex.

Figure 13.3 Value chain of digital broadcasting

known from the Internet are provided in the broadcast networks. The possibility of the transmission of programmes in different technical qualities further provides the content producers with new possibilities in their content creation and provision.

Distribution networks will carry the compression coding from content providers to service providers and from service providers to delivery networks. The distribution networks are used only if there is a distance between the content provider, service provider, and delivery networks. Distribution networks were one of the first parts of the value chain that were digitized using different compression standards, for instance MPEG standards.

Delivery Networks

In this sub-section, we discuss the parameters indicated in the introduction with respect to different delivery networks.

Terrestrial

Transmission of TV signals using ground-based transmitters in a terrestrial network is traditionally the most used and known delivery form. In markets where other multichannel platforms are not well developed, digital terrestrial broadcast platforms are important for the delivery of broadcast services. In markets where other infrastructures, including broadband, are well developed, it can be discussed to what degree radio spectrum should be used for the delivery of broadcast services via terrestrial networks.

The major advantage of using terrestrial distribution for digital TV is that the signal can be received using a simple roof-top or in-house antenna. The possibility of simple portable and mobile reception and the possibility of regionalization of the signals are also among the advantages. By regionalization is meant that targeting the signal to limited areas (regional and local TV) can be done in a cost-efficient form.

The disadvantage in comparing to satellite or cable distribution, for example, is the limited capacity immediately available for digital TV. This is a problem because the number of digital services is a parameter that has an influence on the interest in investing in digital equipment. Another disadvantage is that the radio spectrum used for terrestrial TV is optimal for mobile communication.

When it comes to the variety issue, it is a fact that, because of spectral efficiency, the number of TV channels in all digital infrastructures increases dramatically compared to analogue TV. The recent technological developments show that there is still room for improvement and we witness newer standards with even higher spectral efficiency. As an example in the DVB family of standards, we see a technical development where the spectral efficiency of the new DVB-T2 standard is almost 40 per cent better than DVB-T (Fischer, 2010). This makes it possible to pack even more TV channels or TV channels with better quality (for instance HDTV) in a given frequency band.

This has resulted in the creation of multi-channel platforms in the terrestrial networks in many countries. However, the precondition for multi-channel broadcast in terrestrial TV is that the political environment accepts using the resources to send more TV channels rather than better-quality TV channels. This has definitely been the case in European countries, which have created real multi-channel platforms in the terrestrial networks. In other countries such as the US, the multi-channel development has not been as important and there has been priority on using the resources for HDTV from the beginning.

Quality enhancement, that is, the enhancement of technical video quality, is a potential in the terrestrial networks but goes against multi-channel development. In the terrestrial infrastructure, there is limited spectrum available and it can be used to deliver higher-quality video, but this limits the possibilities for expanding the number of TV channels. In later deployments, a combination strategy is used where parts of the resources are used for delivering channels with higher quality and parts are used for providing more TV channels.

The interactivity in terrestrial networks is more a myth than a fact. To begin with, many experiments were done to develop interactive services. However, the results were not convincing. As an example, EPG (electronic programme guide), which is a 'local' interactive service and is delivered on all digital TV platforms, has had a very insignificant use, and the end users still apply zapping to select and change programmes. However, the new development of hybrid broadcast broadband (HBB) platforms can change the situation radically, and real interactivity can have a future here.

In addition, the effects of convergence are yet to be seen in the terrestrial networks. DTT (digital terrestrial television) networks are to date used only for the delivery of regular TV in digital form. But, as with the discussions on interactivity, the HBB development will have important impacts on this issue.

Mobility, on the other hand, has been a very important parameter in the discussions prior to the transition to digital terrestrial TV. From the beginning, it was obvious that mobility could only be achieved in the terrestrial networks, and this was seen as a competitive advantage of terrestrial TV compared to satellite and cable TV. Mobile reception was in itself a major argument in the discussions of legitimizing terrestrial networks to be used for broadcast also in the digital era. In the course of further developments, a change happened from mobile TV primarily envisioned to be offered in cars, buses, trains, and so on to mobile TV services offered to personal mobile devices. This resulted in the development of new mobile versions of broadcast standards such as DVB-H (Digital Video Broadcasting – Handheld) in the DVB family of standards (Penttinen et al., 2009). Mobile TV in a broadcast network has, however, not been a success, at least in the US and Europe (Evens and Prario, 2012), and can, to a large extent, be categorized as a myth.

Satellite (DTH)

Direct-to-home (DTH) satellite networks were the first delivery networks to be digitized. At first, digitization of satellite networks for broadcast purposes was driven purely by cost reductions related to the transition to digital technology. In recent years, broadcast services requiring large amounts of capacity (like HDTV and 3D TV) are driving the further development of these networks. A major advantage of using satellite distribution for digital TV is the coverage issue. Satellite infrastructures cover large geographical areas.

As with terrestrial networks, spectral efficiency and the possibility of increased capacity is a very important parameter for satellite networks. In reality, this was the most important parameter in the transition to digital technology. Spectrum and airtime are very expensive resources, and the ability of digital technology to allow for packing more channels in a given frequency band has been crucial. Also, the further development from the DVB-S standard to DVB-S2 has resulted in more spectral efficiency (Fischer, 2010).

The quality issue, that is, the possibility of offering higher audio and video quality, has been a more realistic option in the satellite networks as compared to the terrestrial networks. However, at first, this was not an important question with respect to satellite networks, and the new capacity enhancements were basically used to provide more broadcast services or reduce the cost of delivering the current ones. In recent years, we have seen that HDTV and 3D TV are finding their way to the satellite networks and the quality parameter is becoming increasingly important.

The interactivity and convergence issues are basically myths in much the same manner as for the terrestrial networks. However, as with the terrestrial infrastructure, the HBB platforms can change the situation radically, and real interactivity can be an important driver in the future.

When it comes to vehicle mobility, that is, the provision of mobile broadcast to cars, buses, and trains, where there is the possibility of the installation of external antennae, this was considered an important parameter at the beginning of the development of digital satellite broadcast. Later, a standard combining satellite and terrestrial networks for offering mobile services was developed in the DVB family of standards. This standard is called DVB-SH and, as in the case of mobile TV in the terrestrial networks, has not been successful in either the US or Europe.

Cable TV

Cable TV is one of the last delivery networks going through the digitization process. In many countries where terrestrial and satellite networks are already digitized, the cable

TV networks either remain analogue or have both analogue and digital provisions in a simulcast form. The simulcast of analogue provision is used because many customers have not changed their old analogue TV terminals or at least still have one analogue TV among their TV sets. Cable network providers compile their services by gathering TV services from satellite and terrestrial distribution.

As with the other infrastructures, the spectral efficiency achieved in digital cable TV enables the provision of more services. However, this has not been an important issue in cable TV networks. The modern analogue systems have plenty of capacity, and the need for transition to digital technology has not been an obvious requirement. However, cable TV networks have in recent years been important infrastructures for the provision of broadband services, and here spectral efficiency is an important driver for enabling higher broadband capacity.

As with DTH satellite systems in the first phases, enhanced quality was not an important driver in the transition to digital cable TV networks, and the expansion of resources as a result of spectral efficiency was used to provide even more TV channels and the development of cable TV networks as broadband infrastructures. Lately, however, we have seen that HDTV and 3D TV are also finding their way to the cable TV networks, and the quality parameter has been an important factor in the competition with terrestrial and broadband networks.

From the beginning, interactivity was seen as an important competitive advantage of cable TV networks owing to the fact that a return channel can be integrated within the cable TV networks, and there was no need for establishing a return channel through other networks. In this case also, however, interactivity has not shown itself to be an important issue in the development of cable services. With respect to HBB, cable TV networks are much better positioned than terrestrial and satellite networks owing to the fact that cable TV is a broadcast as well as a broadband platform.

As with terrestrial and satellite networks, the effects of convergence are yet to be seen in cable TV networks. And, in the discussions concerning interactivity, the HBB development will play an important role in the impact of convergence for the development of digital cable TV networks. When it comes to mobility, this has for obvious reasons never been an issue in the cable TV networks.

Broadband

Broadband networks are becoming a complementary but also competing infrastructure for the provision of broadcast services. Broadband is used for managed IPTV as well as best effort OTT services, and broadband delivers one of the infrastructures in the HBB platform, namely the broadband part.

The main advantage of broadband is that, when the available broadband capacity can handle the delivery of TV as well as Internet services, the end users need only one infrastructure for all services. This is a materialization of infrastructure convergence and can have far-reaching effects on the sector.

The major disadvantage is that broadband is not dimensioned for distributive services, and a waste of network resources is obvious in many provisions. However, there is a possibility of implementing multicast in both the managed and the best effort platforms.

Spectral efficiency and the possibility of delivering increased capacity has been a major driver for all broadband technologies in their development towards higher and higher

Table 13.2 Summary of facts and myths

Delivery networks	Parameters				
	Variety	Quality	Interactivity	Convergence	Mobility
Terrestrial	Fact	Myth*	Myth	Myth	Myth
Satellite	Fact	Myth*	Myth	Myth	Myth
Cable	Fact	Myth*	Myth	Myth	Never been an issue
Broadband	Fact	Myth*	Fact	Fact	Fact

Note: * The quality issue became important later in the development. It is also important to mention that this analysis is based on the introduction strategies in Europe. In the US, for example, HDTV was a major driver of digitization of broadcast TV from the outset.

capacity. The high capacity in the broadband networks is a major parameter for making them a valid infrastructure for broadcast delivery. In many economically developed countries, we see broadcast and on-demand TV delivered through broadband networks. High-capacity broadband is, in reality, a major threat to terrestrial broadcasting, as many countries have specified goals for reaching high-capacity broadband for their populations (for example European Commission, 2010).

High-speed broadband infrastructures are definitely positioned well for the delivery of high-quality audio and video services. However, the provisions we have seen until now have mostly been of low or standard qualities.

When it comes to interactivity, broadband is very well positioned. Broadband is an interactive infrastructure in its original design. However, many broadband technologies provide asymmetric downstream and upstream capacities, but the asymmetric nature is not a problem for interactive broadcast, as interactive broadcast demands much more downstream than upstream capacity.

Furthermore, convergence is a major driver for broadband. High-speed broadband is materialization of infrastructure convergence and, when it comes to service convergence, broadband has the best of conditions for achieving real convergence.

With regard to mobility, the mobile broadband service providers very early in their development delivered video and TV services as managed streaming services as well as through best effort Internet services. The new mobile broadband technologies like LTE (long-term evolution) will provide better possibilities for mobile broadcast. In Table 13.2, the parameters discussed are summarized.

CONCLUSIONS

The chapter first introduces the different TV paradigms from analogue TV, to digital TV including mobile TV, and further to IPTV, HBBTV, and OTT. The issue discussed here is the extent to which the digitization of broadcasting with respect to these different paradigms and concerning the different distribution networks (terrestrial, satellite, cable, and broadband) has achieved the goals that were originally set out: increased variety, quality, interactivity, convergence, and mobility. Before digitization of broadcasting, or at the beginning of the digitization process, these were the aims that supposedly were to be reached with digitization. How did it then go?

The results are mixed. The absolutely most important goal reached is an increased variety in the number of channels that users can get access to. The increased capacity has, hitherto, mostly been used for increasing the number of channels instead of enhancing the technical quality of transmissions. This applies to all distribution networks discussed. With respect to interactivity, convergence, and mobility, it is difficult boldly to state that the proclaimed goals have been reached – at least, up until now, with the exception of broadband.

The reason for broadband being an exception is that broadband, per se, is an interactive medium – in contrast to terrestrial, satellite, and cable networks, which from their origin were one-way networks. Cable was later developed to be used for Internet two-way communications, and interactivity can also take place via satellites if this facility is deployed in satellite communications. However, especially terrestrial, but also satellite, and to some extent cable networks bear the birthmark of their origin as being one-way networks, and interactivity based on convergence and mobility has developed only on a very small scale.

With HBBTV and OTT this could potentially change in the coming years. As of now, the interactivity regarding OTT is mostly to decide which programme to select, but the potential for interactivity and other developments based on convergence is considerable. This also applies to HBBTV, which to some extent can be considered the answer of broadcasters to Internet development, which can be seen as a threat to them. However, the combination of broadcast and broadband, which is the essence of HBBTV, may lead to the realization of some of the goals originally presented as the potential of digitization.

REFERENCES

CEC (2008), *Legal Framework for Mobile TV Networks and Services: Best Practice for Authorisation – The EU Model*, COM(2008)845 final, Brussels: Commission of the European Communities.

EBU (2010), *Requirements for the Standardization of Hybrid Broadcast/Broadband (HBB) Television Systems and Services*, EBU – TECH 3338, Geneva: EBU.

European Commission (2010), *Digital Agenda for Europe*, COM(2010) 245 final/2, Brussels: European Commission.

Evens, T. and B. Prario (2012), 'Mobile TV in Italy: the key to success, the cause for failure', *International Journal of Digital Television*, **3** (1), 53–68.

Fischer, W. (2010), *Digital Video and Audio Broadcasting Technology: A Practical Engineering Guide*, 3rd edn, Heidelberg: Springer.

McCann, K. and A. Mattei (2012), *Technical Evolution of the DTT Platform: An Independent Report by ZetaCast, Commissioned by Ofcom*, Bath: ZetaCast.

Merkel, K. (2010), *HbbTV: A Hybrid Broadcast–Broadband System for the Living Room*, EBU Technical Review, Geneva: European Broadcasting Union.

O'Leary, S. (2000), *Understanding Digital Terrestrial Broadcasting*, London: Artech House.

Penttinen, J., P. Jolma, E. Aaltonen and J. Väre (2009), *The DVB-H Handbook: The Functioning and Planning of Mobile TV*, Chichester: Wiley.

Simon, Jean Paul (2012), *The Dynamics of the Media and Content Industries: A Synthesis*, European Commission, Joint Research Centre, Institute for Prospective Technological Studies, Luxembourg: Publications Office of the European Union.

Tadayoni, R., A. Henten and I. Windekilde (2009), 'Mobile TV: an assessment of EU policies', Paper presented at EuroCPR Conference 2009, Seville, Spain.

FURTHER READING

Understanding Digital Terrestrial Broadcasting by Seamus O'Leary (2000) is still a good reference book for acquiring an understanding of different technical elements of digital broadcasting. For readers with more advanced technical skills, *Digital Video and Audio Broadcasting Technology* by Walter Fischer (2010) is an excellent resource to refer to. For readers with an interest in a broader insight into the development of broadcasting and media industries, the IPTS report by Jean Paul Simon (2012) titled *The Dynamics of the Media and Content Industries: A Synthesis* is highly relevant.

PART III

POLICY AND COPYRIGHT ISSUES IN THE DIGITAL CREATIVE ECONOMY

14. Cultural policy

Terry Flew and Adam Swift

The development of cultural policy during the twentieth century is underscored by three key developments. First, the formation of the Arts Council of Great Britain in 1946, first headed by the Cambridge economist Lord Keynes, saw the scaffolding developed for ongoing government support for the arts. In doing so, it established the principle of an "arm's length" relationship between the government of the day and individual artists, through the development of independent arts boards engaged in the peer review of creative works. Second, the formation of the Fifth Republic in France in 1958 saw the creation of a Ministry of Culture, headed by the writer André Malraux. Malraux and his successors have seen three major tasks for a national cultural policy: government support for the creation of new artistic and cultural works; the promotion and maintenance of cultural heritage; and enabling equitable access to creative works and creative opportunities through all segments of society. Finally, at a global level, agencies such as UNESCO have sought to promote national cultural policies as an element of national sovereignty, particularly in the developing world, and this has involved addressing sources of structural inequality in the distribution of global cultural and communications resources.

The appropriate scope and breadth of cultural policy is a subject of ongoing debate. In the Arts Council tradition associated with the English-speaking world, there has tended to be a relatively narrow definition, with creative, performing and visual arts policies being bracketed off from areas such as communications and media. By contrast, the European tradition has tended towards a broader definition, aligning arts and media policy along with areas such as language policy, policies towards citizenship and identity, multiculturalism, policies towards cities and regions, and cultural heritage and tourism.

The cultural economist David Throsby (2010) has identified cultural policy as a form of public policy applied to the production, distribution and consumption of, as well as international trade in, cultural goods and services. Throsby observed that these *explicit* cultural policies co-exist with a series of *implicit* cultural policies, or policies in other domains that have a cultural dimension or have cultural impacts. Such areas of policy include economic policy, trade policy, urban and regional development policies, education policy, and policies towards intellectual property. The realm of implicit cultural policies could of course be enlarged even further: attitudes towards organized religion, systems of media regulation, international legal and economic agreements, and overall levels of taxation and government spending all impact upon the levels and forms of government support for culture (Ahearne, 2009). Moreover, the distribution of responsibilities between city, regional and national governments influences cultural policy: the 'creative cities' movement of the 2000s saw local governments taking far more of a leadership role in cultural policy development (Isar, 2012).

Debates about technology policy have frequently occurred in isolation from those of cultural policy. This is reflective, at least in part, of a tendency to think about technology in terms of the devices we use, or the infrastructures that enable access to these technologies, as distinct from social practices involved with using the technologies, or the cultural content that is produced and distributed through these technologies. In technology policies associated with Internet development, for instance, there is often much talk about the transformative power of high-speed broadband, and much less discussion about the policy settings that would enable new forms of digital content to be developed.

If policy discourses surrounding technology often neglect the cultural dimension, it is also the case that cultural policy has often demonstrated ambivalence bordering on antagonism to the influence of media technologies. Murphie and Potts (2003: 7) have traced this to the Romantic conception of culture, emerging in the early nineteenth century in opposition to industrial society, which 'embraced culture as the positive dimension of civilized societies . . . [and] established a dichotomy between culture and technology'. In the twentieth century, the influential theory of the culture industry developed by the Frankfurt School of neo-Marxist philosophers promoted the idea that the 'industrialization of culture' through large-scale industries and technologies of mass distribution led to 'the degradation of the philosophical and existential role of culture as authentic experience' (Mattelart, 1994: 190). Moreover, the economics of cultural policy have been dominated by 'market failure' rationales (Towse, 2010: 169–71), where the economic case for government support for the arts and culture has been based around:

- the benefits of access to the arts and culture as public goods that are not adequately captured by relative prices;
- the question of whether arts and culture are merit goods, and the public needing to be persuaded to consume cultural goods and services; and
- the need to safeguard the arts and cultural heritage for future generations, which would not be adequately secured on the basis of current demand.

A related series of dichotomies have remained in cultural policy discourse, including divisions associated with:

- high culture and mass entertainment, with technology associated primarily with the latter, to the detriment of the former;
- aesthetic value and the commercial market, with large corporations using technology to distribute the 'debased' products of mass media culture; and
- cultural excellence and the democratization of culture, with the latter seen as perpetually threatening the former; the primary role of cultural policy in this context was seen as one of preventing cultural excellence from being overwhelmed by the mass market.

This conceptual divide between discourses of technology and culture has often played itself out as a policy divide between media and communications policy, on the one hand, and arts and cultural policy, on the other (van Cuilenburg and McQuail, 2003). This was particularly apparent from the 1990s onwards, where realization of the centrality of communications infrastructure to the knowledge economy (David and Foray, 2002) combined with an accelerated take-up of the Internet. This saw a much stronger focus upon economic priorities in the communications field, and a bringing together of broadcasting and telecommunications law and policy.

During the late 1990s and 2000s, however, the rise of creative industries policies around the world has given greater attention to the digital content industries as drivers of innovation and economic growth, challenging historical 'market failure' rationales underpinning government support for the arts and culture (Potts, 2011; Flew, 2012). Cultural policy statements also began to give greater attention to digital content, as well as to the contribution of the cultural sectors to economic development; the Australian government's 1994 *Creative Nation* cultural policy statement was a pivotal document in this regard, coming as it did on the cusp of the take-off of the Internet as a mass medium (Flew, 2012: 50–51).

The challenges of retooling cultural policy for digital environments can be seen with consideration of the games industry. The games industry was estimated to be worth at least $US65 billion worldwide in 2012, is expected to grow to $US82 billion by 2016 (DFC Intelligence, 2011), and has become one of the dominant global entertainment industries of the twenty-first century. Games development operates on the basis of global production networks, and its products are sold on global markets, making it difficult – although not impossible – to promote the games industry through nationally based industry and cultural policy paradigms (Kerr and Cawley, 2012). It is also an industry characterized by the rapid prototyping of new products, a high failure rate that is compensated for by the huge returns that can accrue to successful games, and a resulting experience of uncertain employment and the scope for exploitation among its workforce. Critics of the creative industries (for example Miller, 2008) have termed this 'precarity', arguing that the industries exploit the enthusiasm of user communities in order to attract skilled employees prepared to take risks, despite relatively low wages and labour conditions that offer little predictability or security, with resulting adverse material and psychological effects.

While the craft and creative skills associated with games development draw upon the digital arts and design, as well as information technology skills, the fast-paced, globalized and highly commercialized environments of games development is very different to that of more traditional arts and cultural sectors, or even film and broadcast media. The performing arts, film, television and recorded music industries have moved from arrangements where long-term commitment to studios by actors, directors and production teams was the norm, towards what Caves (2000: 98) defines as 'one-shot deals', where more short-term and flexible contracts are the norm. Nevertheless, these industries have been able to draw from established markets for their products in adapting to these new realities. On the other hand, the games industry has always faced uncertain and still-maturing markets, uncertain vesting schedules, and the production of 'perishable' goods in an over-abundant market, while operating under conditions where vertical specialization of companies in the value chain and greater financing of companies by venture capital are the norm.

The games industry is typically defined as one of the leading creative industries. It captures many of the signature elements of the creative industries, such as: the combination of art, design and technology; an appeal to those younger consumers termed 'digital natives'; and the association of work in the games industry with play and the exercise of individual creativity. But, as what Banks and Cunningham (this volume, Chapter 37) describe as the definitive 'born digital' creative industry, it is also characterized by more rapid innovation cycles, a truly global market, and the need to rapidly adapt to the emergence of new platforms – such as smart phones – that contrast it to other media and audiovisual content industries. Moreover, games have increasingly engaged their users as active co-creators of content through networked gaming environments. Larissa Hjorth (2010: 260) has observed that, 'as opposed to the older model of media packaged for consumption, twenty-first century networked media promotes audiences to be active co-producers of meaning and of content'.

In this respect, games are best viewed in terms of what Potts and Cunningham (2008) refer to as the 'innovation model' of the creative industries, where creative industries are test beds of innovation in terms of new business models, new applications of digital technology, and new engagements of the consumer. This is in contrast to the cultural policy 'heartland' sectors such as the creative and performing arts, where government support is seen primarily in terms of compensating for the failings of the market. The European Commission has discussed this in terms of the unique roles that creative industries sectors, such as games, can play in fuelling demand for ICT applications, consumer electronics and telecommunication devices, and innovation in other industries (European Commission, 2012).

This is not to say that government support for the arts should be abandoned as an element of cultural policy, and that cultural policy should become a *de facto* innovation policy. Haseman and Jaaniste (2008) identified a range of ways in which traditional arts practices can make a positive contribution to a national innovation system. These include: the role played by the arts in fostering a culture of innovation and experimentation; the continuing significance of peer-defined excellence; the promotion of interdisciplinary collaboration; the ability to commercialize products and processes arising out of creative practice; and the growth of the creative sectors themselves. In relation to the creative economy generally, Bakhshi et al. (2013) estimated that creative economy employment in the United Kingdom now accounted for 8.7 per cent of the total workforce, as compared to 8.4 per cent in 2004, and that the rate of creative economy employment growth over 2004–10 was five times that of the rest of the UK economy. Importantly, of the 2.5 million employed in the UK creative economy, almost half (1.2 million) worked outside of the creative industries, in a diverse array of creative occupations across manufacturing, retail, tourism, telecommunications and other sectors (Bakhshi et al. 2013: 1–2, 67–70).

Cultural policy in the twenty-first century differs from its original structure as a result of growing attention to the economic dimensions of culture. Throsby has observed that UNESCO's cultural policy statements of the 1970s 'contained few if any references to the economics of culture, beyond an occasional reference to the administrative means for obtaining and deploying cultural funds for cultural purposes' (Throsby, 2010: 2). Growing attention to the economic dimensions of cultural policy first emerged in the 1980s, with studies such as that undertaken by Augustin Girard for UNESCO (Girard, 1982), which observed that national cultural policies had promoted state-funded cultural activities with limited impact on the cultural consumption patterns of the wider population, while largely ignoring and often condemning the commercial sector. Such studies began to promote the concept of cultural industries, observing the policy gulf between treatments of state-funded and commercial popular culture, and arguing that 'far more is done to democratise and decentralise culture with the industrial products available on the market than with the "products" subsidised by the public authorities' (Girard, 1982: 25).

For better or worse, cultural policy is now seen as an aspect of economic policy, not simply in the sense of economic evaluation of the impacts of spending on the arts and culture, but in the sense that investments in culture are seen as direct contributors to economic growth. As David Throsby (2008: 229) has observed:

> Traditional concepts of cultural policy as arts policy have been changing . . . Now the arts can be seen as part of a wider and more dynamic sphere of economic activity, with links through to the information and knowledge economies, fostering creativity, embracing new technologies and feeding innovation. Cultural policy in these circumstances is rescued from its primordial past and catapulted to the forefront of the modern forward-looking policy agenda, an essential component in any respectable economic policy-maker's development strategy.

The other major set of forces transforming twenty-first-century cultural policy are those involving digital technologies. The Internet and digital media, and associated developments such as networking, convergence and media globalization, compel policy-makers to change priorities, by radically shifting the very forms through which we understand culture to be produced, distributed and consumed. Cultural policy-makers have been seeking to adapt to such changing media and communications landscapes. For example, the *Creative Australia* national cultural policy tasked Screen Australia with initiatives to promote digital storytelling, and the administration of an Australian Interactive Games Fund to 'support independent games studios to create innovative digital content in Australia and strengthen Screen Australia's program to build sustainable multiplatform content and distribution businesses' (Australian Government, 2013: 80), alongside its more traditional forms of support for film and television production.

Four elements of the rise of social production and the networked information economy particularly impact upon cultural policy. First, the *demand for participation* suggests that, as digital media enable multi-level communication and interaction, those engaged in digital cultural activities no longer expect simply to be audiences or consumers, but to be able to actively participate in the process of developing social meaning. Second, *user-created content* shows that, as the capacity to create, distribute and use content has been radically decentralized across communications networks, professional content creators (journalists, artists, photographers, designers, etc.) increasingly face competition from 'amateur' content creators. Third, *content co-creation* shows that both the demand for participation and the capacity of users to create and distribute their own content mean that cultural forms are increasingly multi-layered and 'open', with the original 'professional' content (a game, a film, a TV programme and so on) existing alongside a range of user-created content associated with that work. In *Convergence Culture: Where Old and New Media Collide*, Henry Jenkins (2006) referred to this as 'transmedia storytelling', which he saw as increasingly central to convergent media and cultural content. Finally, the shifting sources of *value* in a network can be noted, whereby cultural policy was historically framed by management of the tension between cultural value, on the one hand (the value ascribed to a cultural work by experts, tradition, history and interpretation), and economic value (the monetary value that any cultural product could attract in a market), on the other. In a digitally networked environment, the nature and sources of value are more diverse, and value chains are no longer linear. With content co-creation, for example, it is often the ability to participate among users themselves that is part of what is valued in the content.

What can be observed is a decentralization of cultural production and distribution, and the associated rise of what Wilson (2010) terms 'social creativity'. Wilson proposes that the concept of social creativity provides 'a means of understanding how interaction across boundaries (including those of the creative industries) enables, motivates and constrains the reproduction and/or transformation of social values, and the realisation of human beings' creative potential' (Wilson, 2010: 368). While discourses of creative industries and the creative economy have broadened the concerns of cultural policy beyond a focus on individual artists and cultural institutions, there is the risk that the creative industries themselves come to be seen as the exclusive sites of creativity, and cultural policy primarily focused upon what those representing these industries see as important – for example more public subsidy or stronger copyright laws. Wilson instead proposes that social creativity can reanimate cultural policy discourses and link them more closely to principles of equitable access, empowerment and participation, through a 'practical idealism' that has been better enabled by digital technologies. In this respect, Wilson (2010: 374) proposes that:

> The practical idealism of social creativity offers a replacement to our current conception of the creative economy, characterised by the uneasy relationship between creativity and the economy. This brings with it a vision (that is, the possibility) of a better and fairer society and economy for all to share. To the extent that cultural policy can enable this process, it will need to focus on supporting encounter, on learning from difference and on crossing boundaries. It will also need to extend its vision well beyond its habitual target market of those working in the creative industries, to include those voices that are otherwise left out and marginalised.

Cultural policy in its twentieth-century forms, associated with arts management, market failure, and the less technology-intensive ends of the creative industries, requires retooling for the twenty-first-century environment of technological convergence, social creativity and distributed cultural value. As authors such as Yochai Benkler (2006), John Holden (2008) and John Hartley (2012) have observed, much of the innovation associated with technological creativity has been coming either from commercial cultural organizations (YouTube, Facebook and so on) or from a new generation of 'pro-am' cultural producers taking advantage of distributed digital platforms to promote their own social creativity. In this respect, cultural policy can appear as a laggard, shoring up dubious and historically derived notions of 'excellence' and policing the divide between cultural experts and non-experts. At the same time, the development of innovative new forms of cultural policy not only requires an awareness of the value of creativity in a digital networked global creative economy. It also requires a commitment to forms of cultural policy that actively engage the participants in everyday culture, and not simply the representatives of established arts sector producer interests, in order to achieve a more democratic cultural remit.

REFERENCES

Ahearne, J. (2009), 'Cultural policy explicit and implicit: a distinction and some uses', *International Journal of Cultural Policy*, **15** (2), 141–53.

Australian Government (2013), *Creative Australia: National Cultural Policy*, available at: http://creativeaustralia.arts.gov.au/ (accessed 8 June 2012).
Bakhshi, H., A. Freeman and P. Higgs (2013), *A Dynamic Mapping of the UK's Creative Industries*, London: National Endowment for Science, Technology and the Arts.
Benkler, Y. (2006), *The Wealth of Networks*, New Haven, CT: Yale University Press.
Caves, R. (2000), *Creative Industries: Contracts between Art and Commerce*, Cambridge, MA: Harvard University Press.
Cuilenburg, J. van and D. McQuail (2003), 'Media policy paradigm shifts: towards a new communications policy paradigm', *European Journal of Communication*, **18** (2), 181–207.
David, P. and D. Foray (2002), 'An introduction to the economy of the knowledge society', *International Social Science Journal*, **171**, 9–23.
DFC Intelligence (2011), *Worldwide Market Forecasts for the Video Game and Interactive Entertainment Industry*, available at http://www.dfcint.com/wp/?p=312 (accessed 9 April 2013).
European Commission (2012), *Promoting Cultural and Creative Sectors for Growth and Jobs in the EU*, Communication from the Commission to the European Parliament, the Council, the European Economic and Social Committee and the Committee for the Regions, Brussels, 26 September, available at http://ec.europa.eu/culture/our-policy-development/documents/com537_en.pdf (accessed 3 April 2013).
Flew, T. (2012), *The Creative Industries, Culture and Policy*, London: Sage.
Girard, A. (1982), 'Cultural industries: a handicap or a new opportunity for cultural development?', in *Cultural Industries: A Challenge for the Future of Culture*, Paris: UNESCO, pp. 24–39.
Hartley, J. (2012), *Digital Futures for Media and Cultural Studies*, Malden, MA: Wiley-Blackwell.
Haseman, B. and L. Jaaniste (2008), *The Arts and Australia's National Innovation System 1994–2008: Arguments, Recommendations, Challenges*, CHASS Occasional Paper No. 7, November, available at http://www.chass.org.au/papers/PAP20081101BH.php (accessed 9 April 2013).
Hjorth, L. (2010), 'Computer, online and console gaming', in S. Cunningham and G. Turner (eds), *The Media and Communication in Australia*, Sydney: Allen & Unwin, pp. 259–70.
Holden, J. (2008), *Democratic Culture: Opening Up the Arts to Everyone*, London: Demos.
Isar, Y.R. (2012), 'Cultures and cities: some policy implications', in H.K. Anheier and Y.R. Isar (eds), *Cities, Cultural Policy and Governance*, Los Angeles: Sage, pp. 330–39.
Jenkins, H. (2006), *Convergence Culture: Where Old and New Media Collide*, New York: NYU Press.
Kerr, A. and Cawley, A. (2012), 'The spatialisation of the digital games industry: lessons from Ireland', *International Journal of Cultural Policy*, **18** (4), 398–418.
Mattelart, A. (1994), *Mapping World Communication: War, Progress, Culture*, Minneapolis: University of Minnesota Press.
Miller, Toby (2008), 'Anyone for games? Via the new international division of "cultural" labor', in H.K. Anheier and Y.R. Isar (eds), *The Cultural Economy*, Los Angeles: Sage, pp. 227–40.
Murphie, A. and J. Potts (2003), *Culture and Technology*, Basingstoke: Palgrave Macmillan.
Potts, J. (2011), *Creative Industries and Economic Evolution*, Cheltenham, UK and Northampton, MA, USA: Edward Elgar Publishing.
Potts, J. and S. Cunningham (2008), 'Four models of the creative industries', *International Journal of Cultural Policy*, **14** (3), 233–47.
Throsby, D. (2008), 'Modelling the cultural industries', *International Journal of Cultural Policy*, **14** (3), 217–32.
Throsby, D. (2010), *The Economics of Cultural Policy*, Cambridge: Cambridge University Press.
Towse, R. (2010), *A Textbook of Cultural Economics*, Cambridge: Cambridge University Press.
Wilson, N. (2010), 'Social creativity: re-qualifying the creative economy', *International Journal of Cultural Policy*, **16** (3), 367–81.

FURTHER READING

David Throsby's *The Economics of Cultural Policy* (2010) provides a good overview of the range of issues associated with cultural policy from an economic perspective, although his focus is more on the traditional arts than on the digital economy. Ruth Towse's *A Textbook of Cultural Economics* (2010, chap. 10) takes a more critical perspective on cultural policy, including a critique of conventional 'market failure' arguments. Nick Wilson's 'Social creativity: re-qualifying the creative economy' (2010) provides a refreshing rethink of debates about arts and cultural policy in the context of creative industries and the creative economy, drawing upon lessons from the UK experience.

15. Measuring the creative economy

*Peter Goodridge**

In the UK and elsewhere, the creative sector has been growing in both size and the attention it receives, leading policy-makers and researchers to frequently ask: 'What is the contribution of the creative sector to the UK economy?' As a result a new research area has emerged that is still in the early stages of development and has employed a variety of methods. Using the UK as a case study, this chapter reviews the more popular methods used, highlights their limitations, and proposes the use of an alternative approach that is fully grounded in the economics literature and neatly overcomes those same limitations.

In assessing the contribution of creative or innovative activity, two approaches are common. The first is to compile a set of time-series indicators, often weighted to form a composite, as in the European Innovation Scoreboard (2007). But: (1) interpretation of an index composed of subjectively chosen, correlated indicators is problematic; (2) the choice of weights is also subjective; and (3) a change in the weighting scheme can produce different results.

A second is to aggregate output across pre-selected creative industries, as in: the UK's Department for Culture, Media and Sport annual report (DCMS, 2011); a similar Office for National Statistics (ONS) study based on the Input–Output tables (Mahajan, 2006); and the World Intellectual Property Organization's framework for estimating the contribution of copyright industries (WIPO, 2003). For want of a better term, this will be referred to as the 'aggregation method'. Owing to its economic approach and use of official data, the method attracts perceived credibility, but actually has severe limitations and does not truly measure what it is intended to. In highlighting why, and because it is a regular analysis so relevant to the topic of this chapter, the annual DCMS study will be used for illustrative purposes. Please note however that the limitations apply in general, and are not particular to the DCMS results.

LIMITATIONS OF THE STANDARD APPROACH

Within the annual DCMS report, economic estimates of the creative sector are produced using data on gross value added (GVA) and employment. The former is the more widely cited measure and is used to illustrate the following discussion.

Subjectivity in Application: Industry Does Not Equate to Activity

The first task for this approach is to identify just which industries ought to be included. The DCMS defines creative industries as those 'which have their origin in individual creativity, skill and talent and which have a potential for wealth and job creation through the generation and exploitation of intellectual property'. This is ambiguous in terms of practical application. An illustration of this can be found in the latest DCMS report (2011):

> This release has had two key changes . . . SIC07 codes 62.02 and 62.01/2 have been removed . . . and the scaling factor that was previously applied to the GVA estimates has been dropped. The impact of these has caused a considerable reduction in the estimate of GVA, but these changes make the estimates in this release a more accurate representation of the Creative Industries.

'Computer consultancy activities' (62.02) and 'Business and domestic software development' (62.01/2) were excluded as they were considered 'more related to business software than to creative software' (DCMS, 2011). But it is not clear why business software does not meet the definition of creative industries given above. Even if some threshold were used to select industries, there is: (1) subjectivity in setting the threshold; and (2) implicit disregard of the cumulative activity in industries that fall below the threshold, which is often substantial.

Furthermore, industries as defined by the Standard Industrial Classification (SIC) do not neatly correspond to economic activity, so there is no clear boundary between creative and non-creative activity. The DCMS definition of the creative sector incorporates the 13 industry groups presented in Table 15.1. In the final column are the proportions applied to industry GVA, with the intent of removing non-creative activity.

From Table 15.1 we can see that the creative sector includes the provision of services in advertising, architecture, design and leisure software. This directly implies that the output of, say, a designer working in the design industry is 'creative', which is reasonable, but also that the output of a designer working in, say, a manufacturing firm is not. Why should this be the case? The same applies for market researchers, architects, software writers or any other 'creative worker' outside this narrow list of SICs.

The very value of creative output is derived from each unit being in some sense unique. To retain that value and appropriate revenue from it, firms often choose to produce it in-house rather than contract it out to the list of industries in Table 15.1. Continuing with the example of design, most is actually undertaken on their own account by firms outside the design industry itself (Galindo-Rueda et al., 2008). But this in-house activity is not considered in Table 15.1, with the notable exception of that in the fashion industry. It is not consistent to attempt to capture the design activity used in clothes manufacture but not in the production of other final goods.

A second direct implication of Table 15.1 is that the output of, say, an administrative worker outside this group of industries is not creative, but that of an administrative worker within this group of industries is creative. Why should the standard operational and administrative activities in these industries count as creative? Many might say they should not.

Both the DCMS (2011) and WIPO (2003) reports include recognition that: (1) industries normally assigned to the creative sector also engage in non-creative activity; and (2) excluded industries also produce creative output. In attempting to circumvent this problem, WIPO classifies industries as 'core', 'partial', 'interdependent' and 'non-dedicated support', depending on the prevalence of creative activity in business processes. But, to maintain the ability to fully appropriate revenues from unique outputs, there is often a considerable degree of vertical integration in firms and industries that

Table 15.1 DCMS mapping from SIC07 to creative sector

Industry	Sub-industry (SIC07 code)	Proportion applied (%)
1. Advertising	Advertising agencies (73.11)	100
	Media representation (73.12)	100
2. Architecture	Architectural activities (71.11)	100
	Specialized design activities (74.10)	4.50
3. Art and antiques	Retail sale in commercial art galleries (47.78/1)	100
	Retail sale of antiques including antique books, in stores (47.79/1)	100
4. Crafts	*Majority of businesses too small to be picked up in business surveys*	
5. Design	Specialized design activities (74.10)	95.40
6. Designer fashion	Clothing manufacture (14.11–14; 14.19–20; 14.31; 14.39; 15.12; 15.20)	0.50
7. Film, video and photography	Reproduction of video recording (18.20/1)	25
	Photographic activities (74.2)	25
	Motion picture and video production activities (59.11/1 and 59.11/2)	100
	Motion picture, video and TV post-production activities (59.12)	18.40
	Motion picture and video distribution activities (59.13/1 and 59.13/2)	100
	Motion picture projection activities (59.14)	100
8. and 12. (Interactive leisure) software/ electronic publishing; digital and entertainment media	Reproduction of computer media (18.20/3)	25
	Other software publishing (58.29)	100
	Publishing of computer games (58.21)	100
	Ready-made interactive leisure and entertainment software development (62.01/1)	100
9. and 10. Music and visual and performing arts	Sound recording and music publishing activities (59.20)	100
	Reproduction of sound recording (18.20/1)	25
	Performing arts (90.01)	100
	Support activities to performing arts (90.02)	100
	Artistic creation (90.03)	100
	Operation of arts facilities (90.04)	100
	Motion picture, television and other theatrical casting (78.10/1)	0.07
11. Publishing	Printing of newspapers (18.11)	100
	Pre-press and pre-media services (18.13)	100
	Book publishing (58.11)	100
	Publishing of newspapers (58.13)	100
	Publishing of journals and periodicals (58.14)	100
	Other publishing activities (58.19)	50
	News agency activities (63.91)	100
13. Radio and TV	Radio broadcasting (60.10)	100
	Television programming and broadcasting activities (60.20)	100
	TV programme production activities (59.11/3)	100
	Motion picture, video and TV post-production activities (59.12)	81.60
	TV programme distribution activities (59.13/3)	100

Source: DCMS (2011, annex A, table 6).

produce and use creative outputs, making it difficult to separate 'core' processes from other activities and introducing a further element of subjectivity into the methodology.

For the same reason, the DCMS also disaggregates estimates of creative sector employment into: (1) those with a creative job, working in the creative industries; (2) those with a non-creative job, working in the creative industries (support employees); (3) those with a creative job, not working in the creative industries. The DCMS also attempts to remove 'support activity' from the GVA estimate, using the factors in the final column of Table 15.1:

> In certain sectors the SIC codes do not map directly to the Creative Industries. This is generally due to either the SIC code capturing non-creative elements (e.g. designer fashion SIC codes includes the manufacture of the clothes) or where elements of other non-creative industries are captured by the code (e.g. photographic activities SIC codes include elements such as 'passport photos'). Proportions are applied to the SIC group so that only the creative elements are included. (DCMS, 2011: 12)

But despite such adjustments, by only counting creative output from the industries in Table 15.1, the majority is missed and a host of non-creative output is erroneously included. Furthermore, as will be shown below in the context of the music industry, application of this method actually results in the unintended inclusion of additional non-creative output. As industries defined by the SIC differ greatly in degrees of vertical integration, and therefore the extent to which they produce, distribute and use creative output, aggregation across industries results in all three activities being treated identically, which is not appropriate. For example, the publishing industry either fully or partly includes all of the following functions: (1) the creation of artistic, literary or musical works; (2) their distribution; and (3) their use.

Example: Estimating the GVA of 'Music', Using the SIC

Consider the music industry, but note the same argument can be made for other industry groups in Table 15.1. The SIC does not classify 'music' as a distinct industry; instead its components are dotted around the SIC in publishing, live entertainment, artistic creation and so on. Aggregation across these industries actually results in a severe miscounting of what most would consider 'music output'.

To highlight this, note from Table 15.1 that 'Music GVA' includes the value added generated in the 'Operation of arts facilities', that is, live venues. What does this value added equate to? On the production side,[1] value added is sales less intermediate expenditures. On the income side, it is the sum of incomes of employees and owners of capital that reside in the industry. Therefore, from the production side, GVA in live venues is gross revenues, largely ticket sales, less payments, including those made to the musicians, that reside in 'Performing arts' and 'Artistic creation'.

So the element that acts as a return to creative output (in this case live music) is actually subtracted from GVA in live venues. What remains is used to compensate industry labour and capital, including administrators, security guards, and the owners of venues and set equipment. But the income earned by, say, the owner of the venue is simply a return to tangible capital (the building), rather than a return to creativity. The returns to creative output appear in the incomes of artists and musicians, and also of record labels and publishers, which also own some share of the creative good. All of these agents reside outside the 'Operation of arts facilities'.

The DCMS estimate of music GVA also includes 25 per cent and 100 per cent of GVA in 'Reproduction of sound recording' and 'Sound recording and music publishing activities' respectively. But not all the value added in music publishing is a return to creativity, and in the case of reproduction what is being counted is the compensation to factory buildings, equipment and employees that manufacture CDs. The subtracted intermediates include the payments to labels, artists and publishers for the right to produce copies, which act as a return to the music itself.

This misidentification of creative output occurs throughout. Table 15.1 also includes both the production of film and its projection in cinemas. But value added in cinemas is sales (for example tickets and popcorn) less intermediates. Intermediate payments include those for the right to project, which flow to the owners of the film original, usually the studio and production company. What remains is the incomes earned by cinema employees and owners of cinema capital, including the popcorn machine. It is difficult to accept that the margins earned on film projection and popcorn sales represent genuine creative activity. Equally it does not seem appropriate to consider cinema ushers and ticket hall attendants as generating creative output.

Therefore, by aggregating value added in industries in which creation takes place with that in industries which use creative output, the DCMS method explicitly includes additional non-creative output, the very outcome it intended to avoid through the use of the proportions in the final column of Table 15.1.

Classification Issues

The DCMS method allocates firms to the creative sector according to the SIC. As well described in Hellebrandt and Davies (2008), the industrial classification of firms is based on their primary activity of engagement. The first point to note is that economic activity is surveyed on a reporting unit (RU) basis. Often an entire firm or enterprise is classified as a single RU even if it has several sites. Suppose an enterprise that is a single RU actually operates from two separate sites or local units (LUs). Site 1 is a factory say, whilst site 2 undertakes all design activity. If site 1 is larger, the entire enterprise and its output are classified in manufacturing, even though a proportion of output is actually design. This is the 'dominance rule'. In this scenario, none of the output produced on site 2 will feature in estimated creative sector GVA.

Now suppose the firm contracts out all design to a firm in the design industry. That same output will now fall within the DCMS definition of the creative sector, even though it is exactly the same in nature as that excluded in the previous scenario. Of course, design is a specific example; the function could be any form of creative activity.

Impact of Firm Size on Classification

For the purposes of data collection, larger enterprises are sometimes broken up into several RUs. If the two sites described above were treated as distinct RUs, RU1 and its output would be allocated to manufacturing, and RU2 to design. But this shows that the aggregation method is more likely to capture creative activity undertaken in large firms, making the result dependent on (1) the sizes of firms that produce creative output, (2) the structure of individual firms and (3) classification decisions taken by the statistical office, all of which can change over time. Small and medium-sized enterprises (SMEs) make up the majority of the UK market sector and are considered important in the context of creative activity. However, the creative output of SMEs is not well accounted for in this methodology.

Granularity and Bias in the SIC

Further classification issues arise from the greater granularity of the manufacturing breakdown compared to that for services. Let us extend the example so that a single RU operates out of three LUs. Two are plants dedicated to the manufacture of distinct products in the SIC, and the third designs both products. Employment on each site is shown in Table 15.2.

As classification is conducted at the two-digit SIC level, even though the majority of employees are engaged in manufacturing (80 compared to 55 in design), the entire RU would be classified in the design industry. A slight adjustment of the employment numbers would result in it being classified as SIC 24, even though the manufacture of SIC 25 and design form large parts of the firm's activity. Now suppose the manufacture of basic and fabricated metals was classified as the same industry in the SIC. Then the firm would be allocated to manufacturing, even though it undertakes a considerable amount of design. Classification is partly a product of the differential level of detail across the SIC, and measuring the creative sector using SICs is affected by this.

Table 15.2 Example of UK industrial classification procedure

Local unit or site	Industry (SIC)	SIC code	No. of employees
LU1	Manufacture of basic metals	SIC 24	50
LU2	Manufacture of fabricated metal products	SIC 25	30
LU3	Specialized design activities	SIC 74.1	55

Source: Author's example.

The Growing Servitization of the Manufacturing Sector

Some of the limitations of the aggregation method are due to the fact that the nature and composition of manufacturing output are changing, and can include design, consulting and other services, which may or may not be bundled with manufactured products. Rolls-Royce is a commonly cited example; see for example Neely (2008) and Hellebrandt and Davies (2008), with the latter noting the statement on the Rolls-Royce website that 53 per cent of global annual revenues derive from the sale of services. Other firms traditionally viewed as part of the production sector that offer a range of services include Dell, IBM, BP and Shell. Manufacturing (and other) firms also produce a range of 'creative outputs' for use in the production of final goods. This is particularly true of manufacturing sectors in advanced economies such as the UK.

Potential for Double-Counting

Although not inevitable, the aggregation method also introduces great potential for double-counting. Methods that seek to isolate specific activities without proper consideration of how they integrate into wider measures of output can be easily misinterpreted. As stated in the DCMS report (2011), 'There is considerable overlap between the Digital Industries and the Creative Industries. Therefore any estimates that attempt to measure the Digital Industries should not be compared to or aggregated with estimates of the Creative Industries.'

With regard to the digital economy, in *The Connected Kingdom Report*, Boston Consulting Group (BCG, 2010) sought to measure the contribution of the Internet to UK GDP using a variant of the aggregation method. GDP can be estimated using data from the production side, the income side (both described above) and the expenditure side. Ignoring international trade, GDP expenditure equates to final household consumption plus business investment plus government expenditure. Summing across data for final consumption mediated by e-commerce and private/public ICT investment due to the Internet, BCG concluded that: (1) £100 billion, 7.2 per cent of UK GDP, was due to the Internet; and (2) were the Internet an industry it would be the fifth largest in the UK.

There are many problems with this interpretation, and the result actually says very little about the contribution of the Internet to UK GDP. Were we to discover that consumers spent £100 billion after catching a bus to their local high street, then, by adding that to the money spent on bus fares and investments in buses themselves in the transport services industry, we could estimate the contribution of buses to GDP in a similar way. But most would recognize it would not be sensible to do so.

Furthermore, there is inherent double-counting in the BCG method. First, treatment of the Internet as a separate industry, with say Amazon implicitly part of both the retail and the Internet sector, is pure double-counting. Second, a large part of retail revenues flow back to original producers. Only the margins earned by online retailers ought to count as value added in what BCG term the Internet sector. Again, this is double-counting.

The real question is whether the Internet has increased the volume of consumption or efficacy of production. If the Internet has increased the quantity or quality (and therefore volume) of goods and services consumed, or reduced the cost of their production, then it has made a positive contribution to GDP. Although it is not the focus of this chapter, it will be clear that the framework proposed further below can be applied to estimate the contribution of the telecommunications capital used to deliver Internet services, in a way that is consistent with measurement of GDP and with no double-counting.

International Trade

A final limitation of the DCMS approach is the inadequate consideration of international trade. If the aim is to estimate the UK creative sector, appropriate treatment of international payments is crucial. Consider again the production and projection of film. The output of UK film production has two broad components: the first produces UK-(part-)owned[2] films which generate long-lived UK revenues via payment for the rights to project, distribute on DVD, broadcast on television, use on merchandise and so on over multiple years; the second produces non-UK-owned films, for a one-off fee for the 'rest of the world' sector.

The first is the production of a UK asset and the second is an export. By using the value added of film production in estimation, the DCMS method makes inadequate consideration of UK asset creation (investment), implicitly treating it as equivalent to production of an asset not owned in the UK. If we are interested in the magnitude of the UK creative sector, UK ownership of creative assets surely matters.

Furthermore, UK (and worldwide) projection uses both UK and non-UK films as inputs. As we know, payments for the right to project are subtracted from value added in projection. What matters from the perspective of *UK creative activity* is the value of payments to project, distribute, broadcast or use which flow to *UK (part-)owners (investors)* from both UK and non-UK sources. The method proposed below will properly account for this issue and all the others described above.

THE 'INTANGIBLES APPROACH'

It is clear that measurement of creative activity using the aggregation method is conceptually and practically problematic. The remainder of this chapter will present an alternative approach, grounded in the economics literature and national accounting framework, to facilitate improved understanding of the contribution of creative activity.

As already noted, and particularly for more advanced economies, much creative output is produced in-house for repeated use in final production, and therefore occurs outside the industries in Table 15.1. One such output is design, but consider also software, market research, and product or process development as a few examples. A superior approach is therefore to consider creative *activity* itself, regardless of the industry it occurs in.

Where such activities contribute to production over a period greater than one year, they are usually described as intangible (or knowledge) capital in the wider literature. Where their contribution is for less than one year, they ought to be viewed as intermediates that do not appear in final value added. By estimating investment in intangible (or 'creative') capital and treating it consistently with long-lived tangible capital (for example buildings, plant and machinery, vehicles), we can estimate both its share in GDP and its contribution to growth.

The symmetry between intangible capital and what is commonly described as the creative sector is clear. One element of intangible capital is the film, television, musical and literary originals[3] used repeatedly in the production of copies and other output. Some intangible capital (software, mineral exploration, artistic originals and soon R&D) is explicitly recognized in the measurement of GDP, but other intangible capital is not so must be integrated in a way consistent with official measurement.

The precise advantages of the intangibles approach are that: (1) the identification of investment provides a measure of output classified by asset or *activity*, rather than industry; (2) the estimated contributions of creative activity are consistent with wider measures of investment and output, and so are distinct from the contributions of other factors of production, thus avoiding the incorporation of the returns to non-creative activity

Table 15.3 Identified forms of intangible capital

Computerized information	Innovative property	Economic competencies
Computer software	Scientific R&D	Firm-specific training
Computerized databases	Non-scientific R&D	Reputation (advertising and market research)
	Design	Organizational capital
	Artistic originals	
	Mineral exploration	

Source: Corrado et al. (2005).

and the usual inability to compare across studies; (3) all estimation is conducted within an internally consistent accounting framework with no double-counting; and (4) rather than having to make subjective judgements on what (proportions of) industries are creative, the only valuation used is that assigned by the market, as present in the estimated returns from the commercialization of creative capital.

The first study to present a comprehensive categorization of the intangible capital used in the generation of final output was Corrado et al. (2005), summarized in Table 15.3. Clearly it is broader than that implicit in Table 15.1. For comparability purposes, the following analysis will consider two alternative measures of creative activity. The first most closely aligns with the DCMS application, and will consider artistic originals; design, advertising and market research as 'creative intangibles'. The second will incorporate all other forms of intangible capital in Table 15.3.

Conceptual Framework

In this proposed framework, creative activity is identified regardless of industry, so manufacturers that produce long-lived design blueprints are also considered in their role as producers of design, in addition to firms in the design industry itself. Therefore it is helpful to consider the economy in terms of sectors of activity rather than the SIC, thus allowing estimation of distinct activities that occur in the same firm or industry.

Consider an economy broken down into two sectors: first, the creative (or upstream) sector produces creative assets such as long-lived design, artistic originals, reputational capital and so on; second, the final output (or downstream) sector uses creative assets in the production of final goods.[4] The nominal value of creative (upstream) gross output, $P^N N$, can be expressed as equivalent to the factor and intermediate costs in the creative sector times any producer mark-up (μ), as in (1):

$$P^N N = \mu(P^L L^N + P^K K^N + P^M M^N) \tag{15.1}$$

where $P^L L^N$, $P^K K^N$ and $P^M M^N$ are payments to labour, capital and intermediates respectively and P their competitive prices. The factor μ accounts for the unique nature of creative outputs and the monopoly power of creators or owners in revenue appropriation. Note this market power is sometimes legally protected by intellectual property rights (IPRs) such as copyrights, trademarks, patents or design rights.

Consider next the downstream which uses the creative asset in production. Perhaps it is a manufacturer employing some blueprint either created in-house or purchased, or perhaps an artistic original is being used to produce copies for sale. If the downstream purchases the asset (rights) outright, then the investment is equivalent to upstream (gross) output, $P^N N$. Unlike most tangible assets, the rights to creative assets are somewhat divisible, so assets can also be sold in components. For example, the full rights to a film include cinema, DVD and television rights.

Although full (or partial) transactions for creative assets do occur, they are rare. Instead downstream users typically rent creative assets from the owners. For instance, film rights tend to be owned by the studio and production company. Payments by downstream users (for example cinemas and broadcasters) take the form of licence fees[5] that act as a return to the creative asset. Whilst payment of explicit rentals is common for artistic originals, in cases where the upstream is located in-house the rentals are unobserved and form an implicit part of firm revenues. Considering the asset value from the perspective of (implicit or explicit) rentals, upstream output can be expressed as:

$$P^N N = \frac{\sum P^R R}{(1+r)^t} \qquad (15.2)$$

which says that the asset value must equal the value of discounted rentals, where r is the real interest rate, R the stock of creative assets, P^R the rental price, and t the expected asset life-length. R can be modelled as in (3), where δ is the rate at which the asset depreciates or decays in terms of its ability to generate revenue:

$$R_t = N_t + (1 - \delta)R_{t-1} \qquad (15.3)$$

The downstream sector produces (gross) output, $P^Y Y$:

$$P^Y Y = P^L L^Y + P^K K^Y + P^R R^Y + P^M M^Y \qquad (15.4)$$

where L, K and M represent labour, capital and intermediates, and $P^R R^Y$ are the (implicit or explicit) rental payments for using creative capital. The downstream is assumed competitive so there is no mark-up. The mark-up earned by the creative sector (μ) is inherent in the rentals, $P^R R^Y$. This framework is summarized in Figure 15.1, which uses the example of the film industry.

For clarity, consider the measurement of literary originals as an alternative example. Say that, in the past year, a UK resident[6] produced one new title and, over its lifetime, the net present value (NPV) of the book's total earnings, from primary and secondary, domestic and international sources, is £10 million. Those earnings form part of current and future UK income. The value of investment made in producing that title (including any mark-up generated from the unique talent and input of the writer) is equivalent, so, on the expenditure side, investment in literary originals is also estimated at £10 million. Current and future UK output is also generated from the stock of originals produced in past years, of which the latest investment is a component. The contribution of that stock can be estimated by integrating the above model into standard national and growth accounting frameworks.[7]

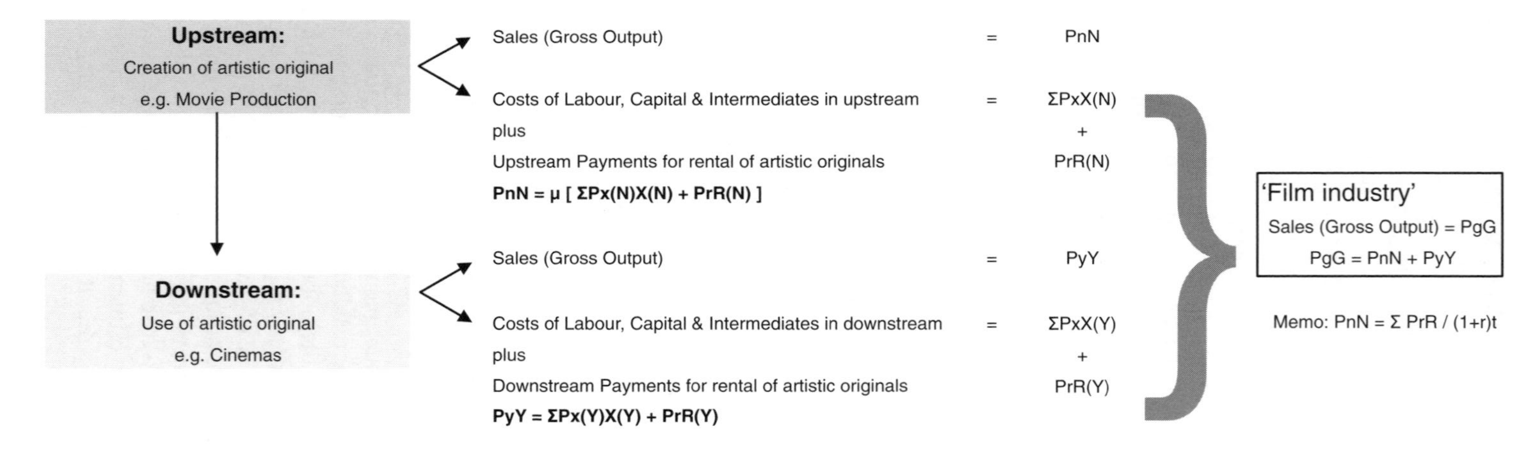

Note: To make the symmetry clear, a term for μ could also be included in the downstream. We assume that the downstream is competitive, so $\mu=1$, always. Monopoly power does however exist in the upstream, owing to the ownership of rights to a unique asset. So, in the upstream, $\mu>1$.

Figure 15.1 Theoretical framework, upstream and downstream, for the film industry

Practical Measurement of Creative Output $P^N N$

The value of creative output or activity can be estimated in a variety of ways, but broadly speaking the methods are variants of, first, an ex ante cost-based approach exploiting the relationship in equation (1) and, second, an ex post revenue-based approach exploiting the relationship in equation (2). Which method is chosen depends on the activity in question and data availability. For instance, consider 'design' and 'artistic originals'. Investment in design can be in the form of: (1) purchases from the design industry itself; and (2) in-house (or own-account) creation by firms outside the design industry. Purchases can be identified from the official Input–Output tables. Own-account creation can be estimated using data for the terms on the right-hand side of (1), that is, the labour payments to occupations that undertake design and other costs. In the case of artistic originals, production of film and TV originals can be estimated using a similar cost-based approach. For music and literary originals, explicit observation of the royalties earned from sales, performance and secondary uses means that the revenue-based approach is also applicable.

UK Creative Activity in the Context of Value Added or GDP

For reasons that are now clear, the DCMS approach to estimating creative activity might produce a highly inaccurate result. The intangibles framework offers two obvious ways to consider the contribution of creative activity to GDP. First, where the output of that creative activity has a life-length greater than one year, it is an investment and therefore a direct component of GDP expenditure. Table 15.4 sets out estimates of investment in creative assets by the UK market sector.

How do these compare to the DCMS estimates? The DCMS estimates that in 2009 the creative industries contributed £36.3 billion (2.89 per cent of UK whole economy GVA), with publishing, advertising, and TV and radio providing the greatest contributions. From Table 15.4, if we consider a very narrow range of creative capital that most closely aligns with the DCMS study (design, branding and artistic originals), we estimate UK market sector investment in creative intangibles of £33.4 billion in 2009, equivalent to 3.26 per cent of adjusted market sector GVA (MGVA),[8] or 2.66 per cent of the whole economy figure used by the DCMS. If we include investment in software, which seems reasonable, those estimates increase to £55.9 billion, 5.46 per cent and 4.45 per cent respectively. If we include a comprehensive range of intangible capital as in Table 15.3, the estimates are £124.3 billion, 12.13 per cent and 9.89 per cent respectively, almost four times higher than the DCMS estimate owing to consideration of industries outside Table 15.1 and a wider definition of creative activity. Note that there is no erroneous inclusion of non-creative output associated with use, or double-counting between production and use.

A second way to consider the role of creative assets is to estimate their contribution to growth.[9] Figure 15.2 sets out some brief results in the form of average contributions to labour productivity for the periods shown.

In 2000–09, the contribution of the most narrow definition of creative assets (design, marketing and artistic originals) was 0.08 per cent per annum, or 5.5 per cent (0.08/1.43) of annual labour productivity growth (LPG). If we consider a comprehensive range of intangible capital, which is more appropriate considering the high degrees

Table 15.4 UK market sector investment in knowledge assets, nominal

Asset	2009 (£bn)
Purchased design	11.07
Own-account design	4.39
Design	15.46
Film originals	0.26
TV (fiction) originals	1.44
TV (non-fiction) originals	0.67
Total TV originals	2.10
Literary originals	1.02
Music originals	1.33
Miscellaneous art	0.37
Total artistic originals	5.09
Advertising	10.77
Market research	2.04
Total branding	12.81
Total 'creative intangibles'	*33.36*
TOTAL INTANGIBLES	*124.25*
TOTAL TANGIBLES	*92.67*

Note: 'Creative intangibles' are defined as those assets that most closely match creative activity in Table 15.1 (design, artistic originals and branding). Total intangibles incorporate a wider definition of intangible assets, including: 'creative intangibles'; (purchased and own-account) software; (scientific and non-scientific) R&D; mineral exploration; financial product innovation; training; and organizational capital. Tangible assets are: buildings; plant and machinery (including ICT hardware); and vehicles.

Source: Goodridge, Haskel and Wallis (2012).

of complementarity between intangible assets, we estimate a contribution of 0.37 per cent per annum, or 25.6 per cent of LPG (20.1 + 5.5). Therefore, in 2000 to 2009, over a quarter of UK growth can be explained by the use of intangible assets.

Of course, some of the returns to creative assets cannot be fully appropriated by the owner(s), and some benefits probably diffuse to other agents. One element of Figure 15.2 is the contribution of the TFP residual, that is, the component of growth that is unaccounted for by the contributions of factor inputs. TFP therefore incorporates the contributions of all forms of costless advance and freely available knowledge. Whether through imitation, inspiration or some other means, part of the contribution of creative output to growth will remain in the TFP residual. The private contribution presented in Figure 15.2 therefore ought to be considered a lower bound.

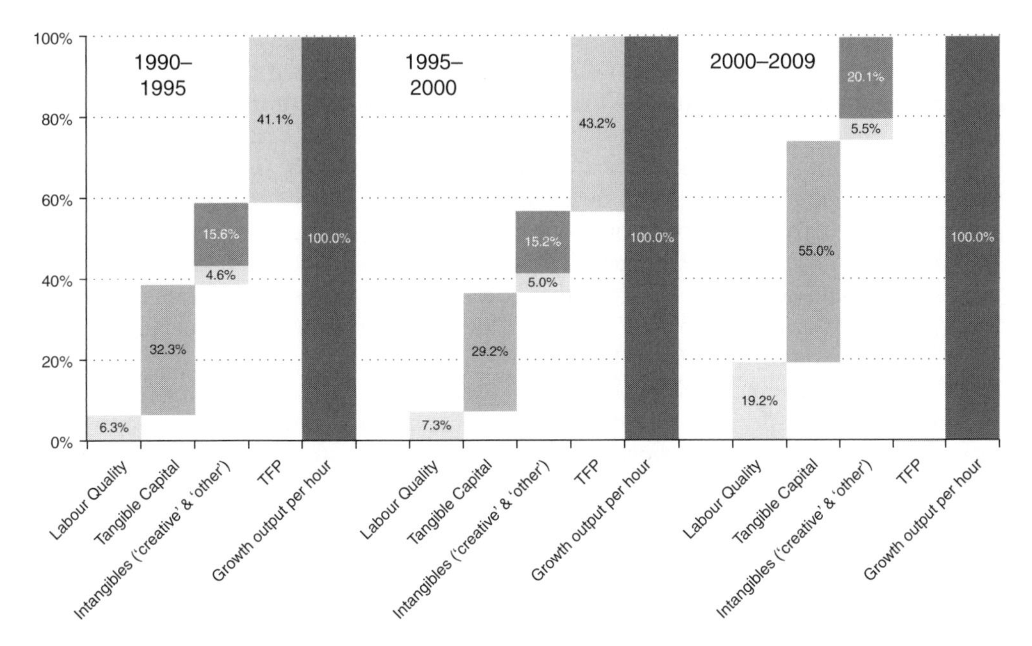

Notes: Data are average growth rates per year for intervals shown, calculated as changes in natural logs in Tornquist indices, with each contribution re-expressed as a percentage of average growth in labour productivity (LPG). The first bar is the contribution of labour composition or 'quality'. The second is the contribution of tangible capital deepening (ICT, buildings, plant and machinery, vehicles). The third bar is divided into two components. The lower part is the contribution of 'creative intangibles' (design, branding, artistic originals). The upper part is the contribution of all other intangibles defined in Table 15.3. The fourth column is the contribution of TFP, that part of growth unexplained by the contributions of factor inputs.

Source: Goodridge, Haskel and Wallis (2012).

Figure 15.2 Contribution of creative assets UK market sector labour productivity

CONCLUSION

Standard approaches for estimating the value, size or contribution of the creative sector potentially miss large chunks of creative activity and incorporate a host of other economic activity which is not in fact creative. Furthermore, failure to distinguish between production, distribution and use can result in misinterpretation of the results. This chapter has argued that a superior approach is instead to identify creative activity regardless of industry classification. In 2009: (1) 15.3 per cent of aggregate UK investment was in a narrow group of 'creative intangibles', equivalent to 3.26 per cent of adjusted MGVA; (2) for 2000–09, approximately 5.5 per cent of LPG is attributable to the use of those assets. Use of a broader definition of creative assets increases those estimates to 57.3 per cent of UK investment and 25.6 per cent of annual LPG.

Of course, these calculations do not account for any positive externalities or cultural benefits that may arise from creative activity. The contributions are solely based on the compensation that can be directly attributed from observed market transactions. Any additional social benefits or spillovers generated by creative activity remain subsumed in the TFP residual. Furthermore, creative activity may generate additional welfare beyond that measured in GDP. The estimates presented above can therefore be considered a lower bound of the true total or 'social' contribution of creative activity.

It is worth noting that one element of the contribution of creativity not considered here is that of the human capital of artists. Figure 15.2 included an estimate for the contribution of labour composition, which includes the productivity-enhancing effects of education and experience, concepts closely linked to the idea of human capital. In principle it would be possible to break that contribution down into that from artists and that from other forms of labour, and incorporate the former as an additional component of the contribution of creative activity.

Finally, one element of creative activity that receives much attention is the production of film, television, musical and literary originals, sometimes referred to as copyrighted assets. This chapter features work that sought to estimate the value of investment in those assets and their contributions to growth. The results should not be interpreted as saying anything about the value of copyright protection itself. Whether investment in those assets, or their contributions are any higher or lower than they would have been without formal IPR protection is unknown.

NOTES

* The work underlying this chapter is part of a broader research programme aimed at improving the measurement of innovation and the contribution of 'knowledge' (or 'intangible' or 'creative') capital to growth. The approach was first outlined in the seminal paper of Corrado et al. (2005). Dal Borgo et al. (2012) and Goodridge, Haskel and Wallis (2012) undertake similar studies in a UK context. The chapter also draws upon ideas and data described in Goodridge and Haskel (2011), who estimate the value of UK investment in artistic originals (film, TV and radio, music, books and art), often considered the core of the creative sector.

1. At industry level, it is only appropriate to consider the production- or income-side measures of GVA. The expenditure-side approach can be used only at aggregate levels.
2. As is the case for many forms of artistic originals, rights are split into various categories, and asset ownership is often split between multiple agents, sometimes across international borders.
3. Data and methods for estimating investment in such assets can be found in Goodridge and Haskel (2011) and Goodridge, Haskel and Mitra-Kahn (2012).
4. Two sectors are used for simplicity, with the final goods sector modelled as producing all final consumption and tangible investment goods. More detail could easily be introduced. Similarly the upstream could be disaggregated into sectors for each type or variety of good or asset.
5. For example, rental payments by cinemas are typically arranged as a percentage of box office revenues.
6. Whilst not discussed extensively here for reasons of space, it is of course important that estimation considers the residence of owners in estimation. Production or purchase of an asset owned by a UK resident is UK investment. Returns to that investment form part of UK revenues and can be earned from domestic and non-domestic sources. An asset produced in the UK but owned or purchased by a non-UK resident may be part of UK production but is *not* UK investment.
7. For a full description of such an integration, please see Goodridge, Haskel and Wallis (2012).
8. This GVA figure refers to the market sector only, defined as SIC03 sections A–K and OP, consistently adjusted for the capitalization of intangibles. The whole economy GVA figure used by the DCMS incorporates the UK non-market sector, thus including government services, and is not adjusted for the treatment of intangibles as capital goods.
9. For reasons of space a description of the growth accounting framework and methodology, and its extension to incorporating intangible assets, is not provided here. For more details please see Corrado et al. (2005), Dal Borgo et al. (2012) or Goodridge, Haskel and Wallis (2012).

BIBLIOGRAPHY

BCG (2010), *The Connected Kingdom Report: How the Internet Is Transforming the UK Economy*, available at: http://www.bcg.com/expertise_impact/publications/publicationdetails.aspx?id=tcm:12-62986.

Corrado, C., C. Hulten and D. Sichel (2005), 'Measuring capital and technology: an expanded framework', in C. Corrado, J. Haltiwanger and D. Sichel (eds), *Measuring Capital in the New Economy*, Studies in Income and Wealth, vol. 65, Chicago: University of Chicago Press, pp. 11–41.

Dal Borgo, M., P. Goodridge, J. Haskel and A. Pesole (2012), 'Productivity and growth in UK industries: an intangible investment approach', *Oxford Bulletin of Economics and Statistics*, doi: 10.1111/j.1468-0084.2012.00718.x.

DCMS (2011), *Creative Industries Economic Estimates: Full Statistical Release*, available at: http://www.culture.gov.uk/images/research/Creative-Industries-Economic-Estimates-Report-2011-update.pdf.

European Innovation Scoreboard (2007), *Comparative Analysis of Innovation Performance*, Luxembourg: Publications Office of the European Union.

Farooqui, S., P. Goodridge and J. Haskel (2011), 'The role of intellectual property rights in the UK market sector', working paper, available at: http://www.ceriba.org.uk/bin/view/CERIBA/IPRsGrowth.

Galindo-Rueda, F., J. Haskel and A. Pesole (2008), 'How much does the UK employ, spend and invest in design?', Working paper, available at: http://www.nesta.org.uk/library/documents/growth-accounting.pdf.

Goodridge, P. and J. Haskel (2011), 'Film, television and radio, books, music and art: UK investment in artistic originals', working paper, available at: http://www.ceriba.org.uk/bin/view/CERIBA/IPOArtisticOriginals.

Goodridge, P., J. Haskel and B. Mitra-Kahn (2012), *Updating the Value of UK Copyright Investment*, Newport: Intellectual Property Office, available at: http://www.ipo.gov.uk/ipresearch-ukinvestment-201206.pdf.

Goodridge, P., J. Haskel and G. Wallis (2012), *UK Innovation Index: Productivity and Growth in UK Industries*, available at http://www.nesta.org.uk/publications/working_papers/assets/features/uk_innova tion_ index_productivity_and_growth_in_uk_industries.

Hellebrandt, T. and R. Davies (2008), 'Some issues with enterprise-level industry classification: insights from the Business Structure Database', *Virtual Micro Data Laboratory Data Brief*, **5**, Spring.

Mahajan, S. (2006), *Creative Sector, 1992–2004: Extract Taken from United Kingdom Input–Output Analyses, 2006 edition*, available at: www.statistics.gov.uk/cci/article.asp?ID=1622&Pos=9&ColRank=1&Rank=1.

Neely, A. (2008), 'Exploring the financial consequences of the servitization of manufacturing', *Operations Management Research*, **1**, 103–18.

WIPO (2003), *Guide on Surveying the Economic Contribution of the Copyright-Based Industrie*s, available at: www.wipo.int/copyright/en/publications/pdf/copyright_pub_893.pdf.

FURTHER READING

Corrado et al. (2005) set out a clear exposition of how the measurement of intangible or 'creative' capital fits into wider measures of output and GDP accounting. For a recent application of this approach in a UK setting see Goodridge, Haskel and Wallis (2012), which applies the model developed in Corrado et al. (2005) in the estimation of UK innovation. For a study that fits into more conventional analysis of the creative industries, see Goodridge, Haskel and Mitra-Kahn (2012) and Goodridge and Haskel (2011), where the former provides an updated summary of a workstream described in the latter, which sought to estimate UK investment in artistic originals (film, TV and radio, music, books and miscellaneous art).

16. International trade in audiovisual products

*Gillian Doyle**

This chapter assesses the main drivers of patterns of trade in audiovisual goods. It condenses a study on audiovisual trade and policy carried out by the author for the OECD (Doyle, 2012, forthcoming). The audiovisual sector, which covers both television and film, is a significant component of the digital creative economy. Not only are television and film important in respect of employment and wealth creation, but these industries are also highly significant in cultural terms and as such are prone to a range of special policy measures and interventions designed to stimulate and support local content production in these industries. Not surprisingly then, international trade in audiovisual goods and services is affected by a range of underlying economic and market factors, by public policies and by changing technologies.

In the early days of broadcasting, television systems tended to develop very much within national territories and were shaped by national circumstances and regulations. Even into the twenty-first century, television broadcasting remains a surprisingly national phenomenon (Morris and Waisbord, 2001), but, thanks to advances in distribution technology, the television industry has become somewhat more internationalized over time (Chalaby, 2005). A major force for change has been the arrival of extra delivery channels. With the advent of cable and satellite and, more recently, with growing use of digital compression techniques and the development of the Internet, a situation of spectrum scarcity which had previously constrained market entry into broadcasting has now been transformed and replaced by one of relative abundance.

As channels have proliferated, many more services aimed at specific audience groups and niches have developed over the last 25 years, often supported by direct viewer payments or subscriptions. More channels and greater resources to support strategies of segmentation combined with the spread of trans-frontier delivery platforms have served to encourage growth in the number of thematic services aimed at transnational audience groups (for example, CNN, MTV, Eurosport, Discovery, Al Jazeera and so on). So the launch of 'international' channels is one route through which television has begun to expand across national frontiers. International language versions of news services provided by public service broadcasters (for example, BBC World) have grown in recent years. In addition, commercial services (for example, Globo Internacional offered on US pay-TV) have been seeking to extend their audiences by launching international versions aimed at diasporas (Ofcom, 2010).

Another factor that has contributed towards internationalization within the television industry has been the development of international co-production deals where partners (usually broadcasters in different territories) share production costs between them. One example would be the *Wonders of the Universe* series of science programmes transmitted both in the UK and in the US in 2011, which was a co-production between the BBC and Discovery Channel. Another is the co-production deal agreed between Beijing Zhongbie TV Arts Centre and Russian broadcaster REN TV in 2010 (Ofcom, 2010: 64).

International co-productions have increased in popularity in recent years as broadcasters have embraced opportunities to spread high 'first copy' production costs and share economies of scale (Hoskins et al., 2004).

Globalization has mostly been driven by international trade in finished programmes and formats – broadcasters acquiring transmission rights to programmes or purchasing the right to use existing tried-and-tested formats from overseas television companies, in some (though certainly not all) cases from neighbouring countries or countries that share a common language. The volume of international trade in audiovisual is large and growing. According to WTO estimates based on balance-of-payments data, global exports grew annually from 2000 at an average rate of 8 per cent and by 2007 had reached US$35 billion (WTO, 2010: 3).

The propensity for audiovisual suppliers to want to sell their wares in overseas markets is unsurprising given that, like other 'information' goods, audiovisual has public good characteristics in so far as that the act of consumption by one individual does not reduce its supply to others (Withers, 2006: 5). The main value within television content is generally to do with attributes which are immaterial (that is, its messages or meanings) and that these do not get used up or depleted in the act of consumption.

It follows that, once the 'first copy' of, say, a video or television programme has been created, it then costs little or nothing to reproduce and supply it to extra customers. As a result, the audiovisual sector is characterized by economies of scale and also by economies of scope (Doyle, 2002a). Initial production costs are typically high but then, because the content or intellectual property can be re-used again and again without additional expense, replication costs tend to be low. The wider the audience, the more profitable it will become. Extending consumption across national frontiers may well involve marginal outlays on marketing and distribution, for example costs of attendance at international content markets such as MIPCOM. In markets that have their own distinctive languages, dubbing or sub-titling may be an additional cost, although how this is shared between rights owners and distributors varies. In the case of large-budget feature films, marketing and advertising costs can be very significant. Even so, because of the public good attributes of audiovisual content, economies of scale and scope are widely available, and therefore there are great natural incentives for the makers and suppliers of audiovisual content to extend or spread the consumption of their output as widely as possible, including across national frontiers where this is feasible.

International traffic in audiovisual has increased greatly since the mid-1980s, partly because channel proliferation has triggered an explosion in demand for attractive television content (Messerlin et al., 2004:1). In the UK, for example, data from regulator Ofcom suggests that there were some 515 licensed television channels in operation in 2011 compared to just four in the early 1980s (Ofcom, 2012). At the level of the EU, it is estimated that some 9800 licensed channels had been established by 2010 compared with just 47 television channels back in 1989.[1] Albeit that the penetration of multi-channel television has been slower in developing countries, the phenomenon of channel proliferation has been a global trend, driven by changing technology and usually accompanied by an opening up of the airwaves as state broadcasting authorities have loosened regulation and issued more licences.

Although sources of data on international trade in audiovisual are somewhat patchy and inconsistent, one pattern that emerges very clearly from a variety of statistical sources is that the US is by far and away the largest exporter of television and also film content (WTO, 2010; USITC, 2011). Growth in global demand for television content has particularly benefited exporters in the United States, but it has also contributed to the development of strong regional markets for television content in Latin America, the Arab countries and the countries of East Asia, including China, South Korea and Japan. In some territories, for instance South Korea, the emphasis of policy and regulation governing the cultural sector has moved away from control and protection towards support for the development of indigenous creative industries, including broadcasting, and this has facilitated more trade and growth in regional audiovisual content markets in recent years. The term 'Korean Wave', or 'Hallyu', describes the phenomenon of an upsurge in Korean cultural exports, which came into existence around 1997 following transmission of the drama *What Is Love All About* by China Central Television which then became a hit in China and beyond (Shim, 2006: 28). A combination of strengthened demand in regional and international markets, improved quality of output and government promotional policy led to a situation where, in the early years of the twenty-first century, exports of Korean television programmes maintained 'an average increase of 40 percent each year' (Kim, 2007: 123; Shim, 2008).

While a number of popular television formats created in the US and Northern Europe have found their way into markets right around the globe (for example, production company Endemol's *Big Brother*), it remains that local cultural dynamics encourage the production and exchange of locally favoured forms of content in specific regional markets, such as in East Asia (Keane et al., 2007). Despite the widespread appeal of US-made television content for most international audiences, preference for content to which audiences experience closer 'cultural proximity' remains a factor influencing trade (Straubhaar, 2007). The benefits of cultural proximity are reflected, for example the aforementioned upsurge in the value of television and film exports from South Korea to neighbouring regional markets such as Japan and China from the mid-1990s onwards making it the so-called 'Hollywood of the East' (CMM-I, 2007; Farrar, 2010). Likewise, the success achieved by Brazil and Argentina in exporting to regional markets in South America reflects advantages of cultural proximity (WTO, 2010: 13).

Television markets in developing countries including India, China and Colombia are growing rapidly (CMM-I, 2007; WTO, 2010). In the case of India, industry growth has encouraged processes of market segmentation with more niche channels and has spurred the rapid development of indigenous television production (Awal, 2011; FICCI-KPMG, 2011). Likewise growth in the television broadcasting industry in China has supported growth in the domestic production sector albeit the industry remains subject to relatively tight state regulation (CMM-I, 2007). Russia is another growing market for television programmes and especially formats (Westcott, 2011).

However, international trade tends to be dominated by English-language content and by the output of one country in particular – the US – and this pattern is consistent over time. The UK has also achieved a notable level of success in international sales of television programmes and formats in recent years (Westcott, 2011). In 2008, UK exports reached £980 million ($1.6 billion) (PACT, 2010) and included such popular formats as the BBC's *Strictly Come Dancing* and ITV's *Who Wants to Be a Millionaire?* The Latin American countries of Venezuela and Argentina have also enjoyed increased television exports to the US in recent years (USITC, 2011: 31–4). But, as is suggested by the data in Table 16.1, the success of the US as an exporter of audiovisual is unparalleled.

Table 16.1 WTO estimates of major exporters and importers of audiovisual and related services, 2007

Rank	Major exporters	Value ($m)	Share of top 15 (%)	Annual change (%)	Rank	Major importers	Value ($m)	Share of top 15 (%)	Annual change (%)
1	United States	15043	51.5	23	1	European Union (27)	13893	63.7	1
2	European Union (27)	9962	34.1	14		Extra-EU (27) imports	6315	29.0	−16
	Extra-EU (27) exports	4063	13.9	6	2	Canada	2001	9.2	6
3	Canada	2021	6.9	−3	3	United States	1440	6.6	32
4	China	316	1.1	130	4	Japan	1044	4.8	6
5	Mexico	308	1.1	−19	5	Australia	798	3.7	13
6	Argentina	294	1.0	21	6	Russian Federation	624	2.9	36
7	Norway	272	0.9	18	7	Brazil	456	2.1	18
8	Hong Kong, China	249	0.9	−3	8	Korea, Republic of	381	1.7	66
9	Russian Federation	196	0.7	28	9	Norway	300	1.4	−21
10	Korea, Republic of	183	0.6	8	10	Mexico	259	1.2	−21
11	Australia	139	0.5	−8	11	Argentina	212	1.0	24
12	Japan	126	0.4	22	12	China	154	0.7	27
13	Albania	61	0.2	59	13	Ecuador	126	0.6	9
14	Ecuador	44	0.1	7	14	Albania	59	0.3	83
15	Colombia	21	0.1	−24	15	Croatia	55	0.3	64
	Above 15	29235	100.0	–		Above 15	21800	100.0	–

Source: WTO Secretariat estimates from 2009 based on available trade statistics (WTO, 2010: 4).

A number of factors account for this. One very key issue is the size and wealth of the domestic television market in the US. The US is home to an audience of 116 million TV households that all speak the same language. In addition, per capita expenditure on television in the US, at £265 ($425) in 2009, is far higher than in any other country (Ofcom, 2010: 132). The prospect of high returns from the home market makes it feasible for large US producers to invest in expensive scriptwriters, stars and special effects. A majority of production costs may be recouped at home in the US, and so these programmes can be made available to importing broadcasters at a relatively low cost.

Drawing on advantages of scale, wealth and language, the US television industry generates a ready supply both of broad-appeal mass entertainment programming and of the niche content needed to attract subscriptions on pay-television (such as feature films, dramas and children's programmes). As the data in Table 16.1 suggests, this material is attractive and competitive from the point of view of importing broadcasters elsewhere, and so the US has sustained a position of exceptional predominance as an exporter of audiovisual content over many years. According to the European Commission, the EU has an annual trade deficit in audiovisual with the US of some €6–7 billion, and around half of this is accounted for by television content.[2]

The statistics related to trade flows for film (as opposed to television) indicate an even stronger predominance for US-made content in international markets. The success of the US as an exporter of film contrasts with the case of India. Although India's prolific film industry generates a very significantly larger number of films than the US or any other country every year (*Screen Digest*, 2011: 324), total revenues in the order of $2–2.5 billion per annum are small compared with approximately $60 billion for the Hollywood 'majors' (Kazmin, 2011). The majors – Disney, 20th Century Fox, Paramount Pictures, Sony/Columbia Pictures, Universal Pictures and Warner Brothers – account for the bulk of the success of the US movie industry in international exports, and this reflects their corporate shape: their size and vertical structure (Doyle, 2002a: 106–10). Vertical integration enables them to use their own distribution networks to disseminate films to exhibition outlets and to reinvest a proportion of profits back into production. The scale of the US majors – the fact that each has a large production slate – is also vital in that it enables them to adopt a portfolio approach to investment in film production. According to recent estimates, over 62 per cent of global box office revenues were accounted for by these six companies alone in 2009 (USITC, 2011: 3–6). In the UK and Ireland, for example, some 90 per cent of box office receipts were accounted for by films partly or wholly made by US film companies in 2010 (BFI, 2011: 14). A similar situation exists across Europe where, in 2009, US studios accounted for 64 per cent of the box office revenues generated by the top 100 distributors (Hancock and Zhang, 2010).

'Diversity' is not a natural outcome in industries characterized by widespread economies of scale. It is typical for such sectors to gravitate towards concentrated ownership structures and a small number of providers. This is very true of the audiovisual sector, where advantages of scale are highly prevalent and influential in terms of industry structures and patterns of trade. The economics of the audiovisual industries – involving high initial production costs and the widespread availability of economies of scale – are such that content suppliers in large and wealthy markets that succeed in developing their capital bases will be at an advantage relative to less well-resourced rivals in smaller minority-language markets when it comes to exports and international trade. As a consequence, any wish to sustain a level of diversity and fragmentation over and above what free and open markets will support will call for use of special policy interventions.

On account of the perceived cultural and economic importance of audiovisual provision, many countries choose to instigate special policy initiatives to counter the risk of market dominance, particularly so by overseas suppliers. Structural interventions which limit the permitted extent of ownership (of television broadcasting in particular) are not uncommon. Curbs on ownership of audiovisual media do sometimes specifically exclude non-domestic companies and individuals and so these measures can and do sometimes act to restrict market participation and flows of international trade. Another sort of potentially discriminatory intervention commonly found in television broadcasting is the imposition of compulsory domestic content quotas (Cocq and Messerlin, 2004: 22–3; *Screen Digest*, 2010). In addition, across Europe as elsewhere, the use of subsidies and tax incentives to promote more production, distribution and exhibition of indigenous audiovisual content is relatively commonplace.

All such policy measures that favour local producers and discriminate against non-domestic suppliers have potential to restrict and distort trade and, in some cases at least, will do exactly that. For example, properly enforced quotas that limit the number of foreign films permitted entry to a national market may well impede trade and prevent local exhibitors from matching supply to underlying market demand. But, because of lax enforcement, the barriers to trade represented by measures such as tariffs and content quotas may on occasion be somewhat illusory. So in assessing the impact of discriminatory measures on audiovisual trade it is not simply the presence of measures which counts but also the extent of enforcement (Andreano and Iapadre, 2004: 111).

But the question of whether, in practice, measures that support local producers tend to choke off or widen access needs to be judged carefully and on a case-by-case basis. Evidence emerging from the available data on international trade in audiovisual as to the extent of any cause-and-effect relationship between special support measures and hindrances to trade is far from conclusive. For example, as Lange (2011) has argued, US audiovisual markets are relatively free of regulations and special support measures, but this does not appear to have stimulated or been accompanied by especially high levels of import penetration in the US. Conversely, a preponderance of special regulations, subsidies and support measures for indigenous audiovisual content production across Europe has done little to impede large inflows of film and television outputs from the US to the EU each year (Lange, 2011).

Weighing up the different measures commonly used to protect cultural diversity, it appears that some are better suited than others towards the simultaneous objective of sustaining open markets. Restrictions that prevent excessive concentrations of ownership (whether in foreign or domestic hands) are well attuned to achieving both ends. Restraints over concentrated ownership of media are frequently used by national governments as a means of promoting diversity and a plurality of voices in the media, which in turn are recognized as important for socio-political and cultural reasons. At the same time, competition-based interventions which restrict levels of ownership and cross-ownership of media can play a vital economic role in avoiding market dominance and preserving more accessible and competitive markets (Doyle, 2002b). So, in the context of reconciling the promotion of cultural diversity with preserving open markets, ownership policies can be a highly apposite instrument.

With regard to other measures commonly used to promote and protect cultural diversity, subsidies are often viewed as a less trade-distorting instrument than quotas (Voon, 2007: 20). Unlike quotas or special tariffs on imports, subsidies to protect and promote culture (for example, subsidies for local audiovisual production or special grants for public service broadcasters and/or minority-language services) will extend the range and variety of indigenous provision but without erecting trade barriers. So, again, subsidies might be regarded as a particularly appropriate measure in the context of balancing 'trade and culture' (Burri, 2011).

Copyright is another factor that shapes international trade in audiovisual. Although work on trade agreements has helped to address inconsistencies in approach to enforcement from one international territory to another, some disparities remain, and audiovisual exporters have pointed to the prevalence of piracy in some territories as a deterrent to trade flows. Digitization has added to the challenges involved in calibrating international restrictions in such a way as forges greater consensus and, in turn, support for implementation of transnational copyright protection. Despite a conviction in some quarters that, irrespective of changes in distribution technology, strict and effective enforcement of copyright is necessary to protect the interests of rights holders (Levine, 2011), the general view that the radically different nature of digital platforms means that copyright is no longer a feasible way to provide incentive and reward creation has gained ground (Brown, 2009). Protections for copyright remain vital in sustaining the development of audiovisual content supply industries, but the view that the assumptions underlying traditional copyright law are out of date and apt to stifle innovation has garnered increasing popular and political support in recent years (Lessig, 2005; Hargreaves, 2011: 1). So digitization and growth of the Internet have encouraged divided positions and, at the same time, have made enforcement of copyright more difficult and costly.

While copyright needs to be tackled and co-ordinated at international level in order to facilitate greater trade flows, it is also an area where, on account of tensions between differing ideological positions and interest groups, uncertainties caused by changing technologies, and other inherent complexities, achieving consensus is exceptionally difficult, particularly at international level. A further example of a problem has been the emergence in Europe of conflict between protection of copyright and promotion of free movement of goods and services within the 'single' market. The suggestion, raised by a recent European Court of Justice (ECJ) ruling, that territoriality may be incompatible with EU law implies that rights owners may be obliged to market their wares differently in future (Blitz and Fenton, 2011; ECJ, 2011). But, whereas the intention is to support greater cross-border trade, one possible outcome of the ECJ ruling is that sales of television rights into smaller European markets will be curtailed in future. A key lesson here is that the impetus to liberalize trade in audiovisual needs to be matched by care in assessment of how this can best be achieved and taking account of the particular circumstances of specific local and national markets.

Although measures to promote cultural diversity may sometimes be at odds with the ambition of sustaining open markets, it is by no means obvious that these two aims will always be difficult to reconcile. Indeed, the objectives of promoting diversity and of promoting more fluid trade in audiovisual content are at times highly compatible. Given the tendencies towards market dominance and monopolization in audiovisual industries discussed earlier, it follows that policy interventions which support greater production and distribution of indigenous content may sometimes serve to promote wider market access and therefore may positively facilitate rather than hamper trade. There is a risk that removal of discriminatory measures which at face value threaten to restrict trade may, in effect, serve to diminish competition and trade flows. Where frictions do occur between sustaining diversity and promoting more open trade – and it may be argued that, given the economic characteristics of audiovisual industries, such frictions are inevitable – processes of reconciliation can draw on long-standing and widespread international recognition for the legitimacy of deploying appropriate policy instruments where cultural policy objectives are at stake.

NOTES

* This chapter represents a condensed summary version of a report prepared by the author for the OECD and published at greater length elsewhere (Doyle, 2012, forthcoming).
1. Figures which draw on sources including the EAO and IDATE are cited at the website of the Association of Commercial Television in Europe (ACTE). See: www.acte.be/EPUB/easnet.dll/execreq/page?eas:dat_im=025B1B&eas:template_im=025AE9.
2. The scale of the EU deficit is referred to at the homepages for the Audiovisual Media Services Directive 2007. See: ec.europa.eu/avpolicy/reg/tvwf/promotion/index_en.htm.

REFERENCES

Andreano, S. and L. Iapadre (2004), 'Audio-visual policies and international trade: the case of Italy', in P. Guerrieri, P.L. Iapadre and G. Koopman (2004), *Cultural Diversity and International Economic Integration: The Global Governance of the Audiovisual Sector*, Cheltenham, UK and Northampton, MA, USA: Edward Elgar Publishing, pp. 96–130.

Awal, A. (2011), 'India's media sector: getting hot', *Financial Times*, available at: blogs.ft.com/beyond-brics/2011/07/28/indias-media-sector-getting-hot/#axzz1cl7HUiqM.

BFI (2011), *BFI Statistical Yearbook 2011*, London: British Film Institute.

Blitz, R. and B. Fenton (2011), 'Premier League faces TV rights shake-up', *Financial Times*, 4 November.

Brown, I. (2009), 'Can creative industries survive digital onslaught?', *Financial Times*, 2 November, available at: http://www.ft.com/cms/s/0/4f35215e-c745–11de-bb6f-00144feab49a.html#axzz1bmteI3gU.

Burri, M. (2011), 'Reconciling trade and culture: a global law perspective', *Journal of Arts Management, Law, and Society*, **41**, 1–21.

Chalaby, J. (2005), *Transnational Television Worldwide: Towards a New Media Order*, London: I.B. Tauris.

CMM-I (2007), *Trends in Audiovisual Markets: China, Mongolia and South Korea*, Beijing: UNESCO.

Cocq, E. and P. Messerlin (2004), 'French audio-visual policy: impact and compatibility with trade negotiations', in P. Guerrieri, P.L. Iapadre and G. Koopman (eds), *Cultural Diversity and International Economic Integration: The Global Governance of the Audiovisual Sector*, Cheltenham, UK and Northampton, MA, USA: Edward Elgar Publishing, pp. 21–51.

Doyle, G. (2002a), *Understanding Media Economics*, London: Sage.

Doyle, G. (2002b), *Media Ownership: The Economics and Politics of Convergence and Concentration in the UK and European Media*, London: Sage.

Doyle, G. (2012), *Audio Visual Services: International Trade and Cultural Policy*, ADBI Working Paper 355, Tokyo: Asian Development Bank Institute.

Doyle, G. (forthcoming), 'Audiovisual services: international trade and cultural policy', in C. Findlay (ed.), *International Trade in Selected Services Sectors*, Singapore: World Scientific.

ECJ (2011), *Judgement in Cases C-403/08 and C-429/08*, Press Release No. 102/11, 4 October, Luxembourg: Court of Justice of the European Union.

Farrar, L. (2010), '"Korean Wave" of pop culture sweeps across Asia', *CNN*, Turner Broadcasting System, 31 December, available at: http://edition.cnn.com/2010/WORLD/asiapcf/12/31/korea.entertainment/index.html?iref=NS1.

FICCI-KPMG (2011), *Indian Media and Entertainment Industry Report*, Mumbai: KPMG.

Guerrieri, P., P.L. Iapadre and G. Koopman (2004), *Cultural Diversity and International Economic Integration: The Global Governance of the Audiovisual Sector*, Cheltenham, UK and Northampton, MA, USA: Edward Elgar Publishing.
Hancock, D. and X. Zhang (2010), 'Europe's top 100 film distributors', *Screen Digest*, **470**, November, at p. 330.
Hargreaves, I. (2011), *Digital Opportunity: A Review of Intellectual Property and Growth*, London: TSO.
Hoskins, C., S. McFadyen and A. Finn (1997), *Global Television and Film: An Introduction to the Economics of the Business*, Oxford: Clarendon Press.
Hoskins, C., S. McFadyen and A. Finn (2004), *Media Economics*, Thousand Oaks, CA: Sage.
Kazmin, A. (2011), 'To the next level', *Financial Times*, 29 October, at p. 9.
Keane, M., A. Fung and A. Moran (2007), *New Television, Globalisation, and the East Asian Cultural Imagination*, Hong Kong: Hong Kong University Press.
Kim, Y. (2007), 'The rising East Asian "Wave": Korean media go global', in D. Thussu (ed.), *Media on the Move: Global Flow and Contra-Flow*, Abingdon: Routledge, pp. 121–35.
Lange, A. (2011), *Letter to OECD Secretariat*, 19 September, Strasbourg: European Audiovisual Observatory.
Lessig, L. (2005), *Free Culture: The Nature and Future of Creativity*, New York: Penguin Books.
Levine, R. (2011), *Free Ride: How the Internet Is Destroying the Culture Business and How the Culture Business Can Fight Back*, London: Bodley Head.
Messerlin, P., E. Cocq and S. Siwek (2004), *The Audiovisual Services Sector in GATS Negotiations*, Washington, DC: AEI Press and Paris: GEM Sciences Po.
Morris, N. and R. Waisbord (2001), *Media and Globalization: Why the State Matters*, Lanham, MD: Rowman & Littlefield.
Ofcom (2010), *International Communications Market Report*, December, London: Ofcom.
Ofcom (2012), *Communications Market Report*, July, London: Ofcom.
PACT (2010), *UK Television Exports Survey 2009*, London: PACT.
Screen Digest (2010), 'International film trade with China', *Screen Digest*, **463**, April, 107.
Screen Digest (2011), 'World film production report', *Screen Digest*, **482**, November, 323–6.
Shim, D. (2006), 'Hybridity and the rise of Korean popular culture', *Media, Culture and Society*, **28** (1), 25–44.
Shim, D. (2008), 'The growth of Korean cultural industries and the Korean Wave', in C.B. Huat and K. Iwabuchi (eds), *East Asian Popular Culture: Analysing the Korean Wave*, Hong Kong: Hong Kong University Press, pp. 15–32.
Straubhaar, J. (2007), *World Television from Global to Local*, Los Angeles: Sage.
USITC (2011), *Recent Trends in US Services Trade: 2011 Annual Report*, Investigation No. 332–345, Publication No. 4243, July, Washington, DC: United States International Trade Commission.
Voon, T. (2007), 'A new approach to audiovisual products in the WTO: rebalancing GATT and GATS', *UCLA Entertainment Law Review*, **14**, 1–32.
Westcott, T. (2011), 'International TV sales recover', *Screen Digest*, **482**, November, 315.
Withers, K. (2006), *Intellectual Property and the Knowledge Economy*, London: Institute for Public Policy Research.
WTO (2010), *Secretariat Note on Audiovisual Services*, S/C/W/310, 12 January, Geneva: World Trade Organization.

FURTHER READING

For further reading on this topic, see Hoskins et al. (1997), Cocq and Messerlin (2004) and Guerrieri et al. (2004).

17. Copyright law

Peter DiCola

Digitization has challenged copyright law in two directions, leaving a more complex set of policy problems in its wake. First, digital technology has radically enhanced the ease of copying and distributing copyrighted works, with troubling implications for creators and copyright owners. Second, digital technology has simultaneously offered copyright owners new enforcement tools and alternative exclusion strategies, with troubling implications for consumers. Legal scholars have addressed both sides of digitization's impact on copyright protection. In fact, with the challenges of digitization has come a flowering of research from multiple disciplinary perspectives, which has increased over the last 20 years. The purpose of this chapter is to offer a brief introduction and guide to recent legal scholarship on copyright. The main thesis is that copyright research by legal scholars has much to offer that is essential for economic research that is nuanced, institutionally sophisticated, and relevant to policy makers.

LEGAL RESPONSES TO TECHNOLOGICAL CHANGE

Copyright law attempted to anticipate the effects of digital technology and networked communications. In the United States, this dates back to the two-decade period of study and legislative deliberation over the most recent overhaul of the main copyright statute, the Copyright Act of 1976. The effort to deal with issues such as digital recording devices in the home, Internet radio (also known as 'webcasting'), and Internet distribution of digital files intensified during the 1990s. Congress eventually passed four major pieces of legislation during that decade, which amended the Copyright Act mostly in the direction of expanding the rights of copyright owners (Litman, 2001). But of course law cannot always fully, or even satisfactorily, anticipate technological change. And so copyright law, both in terms of statutes and in terms of judicial decisions and administrative-agency rulings, has not just anticipated but also reacted after the fact to digitization. Countries such as the UK, France, Germany, Canada, and Japan have also reacted to digitization by passing new copyright legislation and deliberating over further reforms. Despite international differences, the US experience can illustrate some of the policy issues commonly faced in these efforts. This section of the chapter offers a brief guide to the major changes to US copyright law that Congress, the courts, and administrative agencies have adopted.

Competing Theories of Copyright Law

To understand the law's anticipations and reactions, we must start at the abstract level of theory and investigate the competing philosophical justifications for copyright law's very existence. Legal scholars have spent a great deal of effort debating these theories of copyright law, with both descriptive and normative goals in mind. Understanding this debate is essential for anyone studying copyright law. One reason is descriptive in nature: many economic actors, including policy makers, believe in non-utilitarian theories of copyright law and consciously base their decisions upon these theories. Even if these actors are mistaken in their non-utilitarian beliefs – that is, even if all economic actors ultimately behave as utility maximizers – understanding their perceived motivations is an important aspect of understanding their behavior. Thus, to conduct policy-relevant research, even a strict utilitarian must be familiar with the other theories.

To be sure, utilitarianism and, in particular, economic efficiency are a central part of the scholarly discussion of copyright law's justification by legal scholars. A core theoretical frame for legal scholars is to describe copyright as an attempt to solve the public-good problem presented by intangible goods that are non-rival and (absent the strictures of copyright law) not easily excludable (Fisher, 1988). Along similar lines, one can think of copyright as an attempt to increase the production costs of would-be free-riders, that is, to consider the problem as one of competition between creators and copyists (Landes and Posner, 2003). Yet legal scholars also recognize the costs of copyright protection. Since information often has little or no marginal cost to transmit, copyright protection creates deadweight loss, which becomes the focus of neoclassical theories of intellectual property (Boldrin and Levine, 2008). More balanced theories take the deadweight loss into account as part of a tradeoff between the dynamic benefits and static costs of copyright law (Benkler, 2006). That framework can be expanded to incorporate more economic ideas, which consider a more complicated tradeoff among various dynamic benefits, dynamic costs, static benefits, and static costs (Cass and Hylton, 2013). For purposes of this section on the philosophical justifications of copyright law, my point is that legal scholars of copyright often speak in terms of tradeoffs – the language of utilitarianism.

As I cautioned above, however, many scholars, policy makers, and economic actors do not accept the notion of tradeoffs. Instead, they rely on various theories based on rights, where the notion of a right is that it trumps at least some attempt at utilitarian balancing against it. For example, many scholars advocate a philosophy of copyright law rooted in the natural-law theories of John Locke, arguing for a close analogy between intangible property and real property (Mossoff, 2005). Some of the resonance of Locke's theory comes from its invocation of moral desert on the part of creative laborers. To some – and this includes many creators themselves, if you ask them – creative effort justifies some financial reward and protection from second-comers, regardless of efficiency. The concepts of moral desert and moral disapproval of free-riders appear in many judicial opinions on copyright law. I hasten to add here that determining how Locke's philosophy should apply to intellectual property is a matter of considerable scholarly debate; for instance, a close reading might not warrant a view that Locke favored strong copyright protection (Gordon, 1993). While researchers studying copyright law may themselves favor a utilitarian approach to studying copyright law, they might well encounter moral-desert Lockeans who insist that digitization necessitates stronger copyright enforcement, irrespective of empirical studies or cost–benefit analysis.

Some rights-based theories are grounded on ideas of personality, that is, on facilitating self-actualization by citizens through their creative endeavors. One class of personality-based theory derives from Hegel's philosophy (Hughes, 1988). Another derives from

Kant's philosophy (Treiger-Bar-Am, 2008). The concept of personality rights, its role in explaining existing law, and its appeal as a normative philosophy have been subject to rigorous investigation in recent years (Hughes, 1998). This class of theories is important partly because it provides a natural motivation for actual copyright provisions that appear occasionally in the United States but even more often in the authors' rights law of continental European nations. Known as 'moral rights', these include a right to have one's work attributed to oneself, a right to maintain the physical or conceptual integrity of a work, a right to receive a share of payment when one's work of visual art is exchanged between third parties, and other such provisions (Kwall, 1985). Digitization has rekindled interest in moral rights. Rampant Internet distribution of copyrighted works, often without metadata, has encouraged many authors to seek an attribution right by making it a condition of licensing; for example, most Creative Commons licenses include an attribution right.

Other rights-based theories also play a large role in copyright scholarship and in the practice of copyright law. One strongly emerging tradition in copyright scholarship is the set of 'users' rights' theories (Patterson and Lindberg, 1991). As the name suggests, rather than focusing on the rights of authors, these theories focus on the rights of the public to use and experience creative works (Litman, 2007). Digitization presented an early situation for users' rights in the realm of software (Samuelson, 1988). This vein of copyright philosophy moves beyond thinking of users as mere consumers. Users could be second-comers who will build on existing works. They might also be something more complex: citizens who engage in a variety of activities and experiment with using and experiencing copyrighted works in different ways (Cohen, 2005). Digitization and the explosion of possible uses of copyrighted works have brought users' rights theories to the fore.

Finally, many American copyright scholars have begun investigating the relationship between copyright and the First Amendment to the US Constitution's protection of freedom of speech. One view is that the First Amendment renders certain restrictions on expression enacted by copyright invalid (Lange and Powell, 2009). Another view is that free-speech values should inform copyright policy and judicial doctrines in various ways (Netanel, 2008). For researchers interested in studying copyright and digitization, there is an immediate lesson from free-speech and the other non-utilitarian theories. Digitization has brought everyday citizens into contact, if not conflict, with copyright law. Rather than a self-contained inquiry about the industries that produce creative works, questions of copyright policy have become broad social policy touching on political issues as fundamental as the freedom of speech.

Copyright Lawmaking

With that theoretical background in mind, one can better understand the processes by which copyright law has anticipated and reacted to digitization. Perhaps the most important legislation that tried to anticipate the effects of digital technology on copyright protection was the Digital Millennium Copyright Act of 1998 (DMCA). Before the DMCA, copyright owners in the US enjoyed the exclusive rights of reproduction, distribution, adaptation, public performance, and public display. To this bundle of rights, the DMCA added the right to sue for circumventing a device that has the purpose of restricting access to a copyright work, such as password-protection software. The DMCA also added a right to sue for circumventing a device designed to prevent unauthorized exercise of one of the aforementioned exclusive rights with respect to a work, such as software that limits the number of copies one can make of a file. (Such technological-protection software is often known as digital rights management software, or DRM software.) In anticipation of increased ease of copying and Internet transmission, Congress sought to support technological means of controlling copyrighted works with legal penalties for thwarting digital rights management (Litman, 2001). Many scholars criticized these new rights for reasons related to users' rights theories of copyright law, concerns about stifling innovation, and concerns about privacy (Cohen, 1998). The DMCA anti-circumvention provisions provide an excellent case study in anticipatory legislation and unintended consequences.

Yet the DMCA also contained another interesting provision that illustrates the law in its reactive mode when faced with digitization. During the early and mid-1990s, the possibility arose that Internet service providers would face secondary liability – meaning liability stemming from the primary, wrongful actions of others – for copyright-infringing behavior on the part of their users. The DMCA included a so-called 'safe harbor' for Internet service providers who met certain criteria. But, as the 1990s gave way to the 2000s and the 2010s, other kinds of Internet companies – such as video-hosting sites and auction sites – faced the threat of secondary liability for their users' allegedly infringing activities. Faced with a succession of lawsuits against such Internet companies, the courts have often ruled that these Internet companies meet the qualifications of the DMCA safe harbor provision (Ginsburg, 2008). Congress almost certainly did not contemplate this possibility back in 1998, being unaware of the forthcoming Internet platforms and applications. But the courts had to resolve the conflicts nonetheless. It remains a matter of lively debate whether the courts are reaching efficient results and achieving the proper balance (Lemley and Reese, 2004).

Although copyright law often stems from a national statute establishing many important dimensions of the law – scope, length of term, and so on – copyright in the US also relies on courts engaged in a common law process of decision-making. Common law refers to the notion that a court deciding a case today looks to the precedent of previously decided cases, considers the analogies and distinctions, and reaches its decision about the current case with the knowledge that it will influence future courts. Scholars have recently emphasized the importance of recognizing this mode of lawmaking in addition to the legislative process (Menell, 2013). The common law process is an important way that different philosophical justifications of copyright law can come into play. The two parties to each case might each articulate opposing philosophical views, rather than just opposing interpretations of facts or legal provisions, which creates an opportunity for different philosophies to inform judicial opinions. Judges can also inject their own policy views, especially when the issue is novel or the statute is ambiguous. Common law decision-making presents a fascinating set of economic questions about the proper institutional actor to determine the policy questions presented by digitization. Which institution is in the best position to resolve the disputes between copyright owners and new technology companies?

Even the mere recognition that both legislatures and courts determine the contours of copyright law with different processes demonstrates that the institutional details are important to understanding copyright policy. The picture becomes even more complicated when one considers the role of various administrative agencies in developing copyright law and in resolving the disputes that arise between copyright owners and technology companies. In the US, the list of institutional players includes the copyright agency, the antitrust agencies, the communications agency, the trade representative, and other administrative officials (DiCola and Sag, 2012). The interplay between these different agencies, Congress, and the courts makes the question of the optimal design of institutions that much richer and more difficult. But the work of legal scholars has shown that these institutional questions matter a great deal. In the US, for example, Congress, the courts, and the communications agency combined to create decades-long delays for cable-television companies to enter the market (Wu, 2004). This aspect of copyright law, which applies more to distribution-technology companies than individual authors or creators, should receive much more study in the future.

International Harmonization

A crucial dimension to copyright law's response to digitization has been the simultaneous push for international harmonization. There has been effort to make the copyright laws of various nations more similar along almost every dimension: the length of the copyright term, the rights afforded to copyright owners, the exceptions to protection, and so on (Dreyfuss, 2004). Legal scholars have described the general logic of harmonization in intellectual property law, including: considerations of border-crossing externalities; economies of scale in governance and standard-setting; and working collectively to refrain from protectionist policies (Duffy, 2002). Okediji (2003) has modeled international intellectual property agreements as a two-stage game among countries, with treaty making as the first stage and enforcement as the second stage. The players negotiate and form coalitions to achieve their desired strength of protection, with developed nations favoring stronger provisions and developing nations favoring weaker ones. But the interaction among nations is not the only contribution of an international perspective. Some scholars have explained that the international dimension of intellectual property law must be viewed broadly to include not just international treaties but also the international dimensions of private contracts (Dinwoodie, 2000).

The push toward harmonization has shaped the law in many countries. International law also acts as a continuing constraint on reform efforts and proposal, which all scholars must keep in mind. For instance, advocates of a mandatory registration system to enhance identifiability of copyright owners for licensing purposes must explain how such a system would comply with the Berne Convention, one of the significant international treaties in copyright law (Sprigman, 2004).

ECONOMIC IMPLICATIONS

The previous section of the chapter explained the basic theories, processes, and international dimensions of copyright law. This section focuses on the economic analysis that legal scholars have conducted in the area of copyright law. In particular, this chapter will focus on research in the area of, or highly relevant to, applied microeconomic theory rather than empirical work.

The Economics of Information Goods

The legislative changes to copyright law that occurred during the 1990s, especially the DMCA anti-circumvention provisions described above, reflected a changed understanding of information goods. Jessica Litman (2001) has characterized policy makers' changing understanding of the purpose of copyright law in response to digitization as having three stages, each with its own central policy goal. I will use these three stages to illustrate what legal scholars have offered to the economic analysis of information goods.

Initially, copyright sought a balance between the interests of copyright owners and users. Balancing of interests amounts to a form of cost–benefit analysis. This accords with the standard economic framework for analyzing the provision of public goods (Gordon, 1982). The logic of government intervention stems from the prospect of underprovision, which in turn results from a good having the properties of non-rivalry and non-excludability. The standard model of public goods suggests that the government choose a mechanism to reach optimal provision of public goods based on users' preferences. Optimal provision means providing a public good up to the point that users' aggregate willingness to pay for another unit equals the cost of that additional unit. The balancing model allowed for considering the interests not only of consumers who wish to experience copyrighted works, but also of second-comer producers who wish to use existing copyrighted works as inputs in new copyrighted works (Lemley, 1997). In sum, during this stage copyright policy considered both the benefits and the costs of copyright protection.

But, in the next stage of copyright thinking, policy makers moved to a focus on maximizing incentives for creation, on the (often implicit) assumption that encouraging the most creative output would simultaneously serve the public interest. Congress had taken this tack by the time of the Copyright Act of 1976, the last major overhaul of US copyright law. That statute features a system of expansive exclusive rights and limited exceptions, arguably creating an asymmetry in favor of rights holders (Litman, 1987). This largely eliminated the cost side of the ledger. Apparently, many policy makers seek to maximize the strength of copyright protection as a shortcut to optimizing the quantity of creative works (the public good in question). But this shortcut leaves out any consideration that existing creative works are an input for new creative works. Increasing the strength of copyright protection increases the cost to creators of using copyrighted works, which tends to reduce the production of creative works and works against any enhanced incentives to create. Moreover, the ability of users and second-comer producers to realize the positive externalities of creative works depends on copyright law having certain limits and exceptions (Frischmann and Lemley, 2007). Thus, the incentive stage of copyright policy thinking reflects a narrowing of the focus to rights holders. Although this stage has an economic-sounding name, in its focus on creators and rights holders it fits better with the non-economic philosophies described above.

Finally, with the DMCA anti-circumvention provisions, increased statutory damages, and enhanced criminal penalties, copyright shifted to a metaphor of control. The thinking behind these legislative changes is that copyright owners must control all exploitations of their work. One benefit to copyright owners in this line of thinking is the ability to control derivative works – that is, adaptations, reworkings, and the like – and when those derivative works come to market. Moreover, it is important to remember that the DMCA anti-circumvention provisions allow copyright owners some control over access to copyrighted works, not just copying. Controlling access might facilitate pay-per-use business models and more fine-grained price discrimination (Boyle, 2000). Copyright owners can also bolster their control by taking advantage of contract law, for instance by using end-user licenses with terms less favorable than statutory copyright law would offer (Fisher, 1998). Whether the efficiency gains from increased control, price discrimination, and the accompanying distributive consequences are appealing will depend, again, on one's philosophy of copyright law's proper aims. My point here is only to show that the increased control that the law attempts to give copyright owners highlights some interesting economics of information goods.

The Economics of Property Rights

Not long after economists developed formal, mathematical theories of public goods and externalities, economists and legal scholars began to apply those theories to the concept of property rights. A new, economic explanation of property rights in land – having a single owner allows that owner to invest in improvements with confidence that he or she will internalize the benefits, not have them leak out as externalities – was quickly extended to intellectual property rights (Demsetz, 1969). More critical perspectives have since been offered about whether the analogy between real property and intellectual property is so close as to recommend analogous legal treatment (Boyle, 2008).

A different take on the value of intellectual property came in the form of the 'prospect theory' of patent law, which held that intellectual property rights allowed the first mover to coordinate subsequent creative activity (Kitch, 1977). Later scholars have emphasized that patents provide a signal to potential second-comer users of intellectual property that there is a single entity from whom they must obtain a license (Kieff, 2006). But a range of inefficiencies – including transaction costs, royalty stacking, and the division-of-profit problem between upstream and downstream innovators – suggest that coordination value may be more limited, at least in some settings. For example, one area in which the inefficiencies seem to dominate is the area of sample licensing, in which musicians seek to use portions of existing recordings in new recordings. Qualitative empirical evidence shows that works seeking to use more than one or two existing works are prohibitively costly to license (McLeod and DiCola, 2011).

More recent legal scholarship in the copyright field has continued to explore theoretical issues of property rights that have arisen in the wake of digitization. Digitization has contributed to interlocking trends of more copyrighted works being created, more owners of copyrights in more places, and more fragmented rights as users can easily isolate small snippets of works with widespread digital technology (Van Houweling, 2010). Also, in recent years, copyright scholars have focused on the possibilities of cooperative production that digitization affords (Benkler, 2006). Copyright law must respond to a digital environment in which creating, copying, distributing, and remixing have become easier. But copyright reforms to date have not settled the controversies over digitization. The law may have lost its sense of balance between generally applicable provisions and tailored provisions (Carroll, 2006). Seeking a new way forward, some copyright scholarship has looked to research in new institutional economics relating to commons management schemes as an alternative to private property rights (Madison et al., 2010).

All this scholarship has one major theme in common. The pressure that digitization has put on copyright law calls for new and deeper analysis of the many contours of copyright law as a kind of property regime. This analysis lies at the intersection of law and economics and would benefit from even more participation on the part of economists to study the many questions raised by copyright law scholars. Similarly, legal scholarship would benefit from more consistent and rigorous application of economics, placed in its proper context and not to the exclusion of non-economic theories. Specifically, future research in the law and economics of copyright might focus on issues like the costs of copyright protection to downstream creators and the effects of new distribution methods such as webcasting and on-demand streaming.

Public Choice Theory

Copyright law scholarship provides insights into another important area of economic interest: the sub-field known as public choice. The various philosophies of copyright law discussed above are an abstract way to understand how and why copyright policy gets made the way it does. But lobbying efforts and the peculiar incentives of various government officials are a more practical, on-the-ground way of understanding how copyright law develops. The US Copyright Act is the result of extensive lobbying efforts and negotiations, which often produced convoluted and difficult-to-interpret provisions (Litman, 1987). In response to technological change, the US Congress has often adopted a flawed process that grants disproportionate influence to particular interest groups (Litman, 1989). This approach has unfortunately carried through to the recent responses to digitization. Congress has given extensive access to interest groups representing rights holders and occasional access to interest groups representing technology companies, but little access to groups representing consumers. Interest groups can also lobby administrative agencies, making their influence even more complicated to unravel. Understanding the economic dynamics of copyright policy making remains part of the research frontier.

CONCLUSIONS

Digitization has brought major changes to the creative industries and the law that governs them. Despite early, naïve claims that activity on the Internet could not be regulated, it has become clear that copyright law, as well as telecommunications law, does in fact regulate what people do in the digital realm. The effect of law on online behavior includes explicit constraints (what lawyers call *de jure* regulation) and implicit influences on what happens in practice (*de facto* regulation). Individuals' behavior online with respect to copyrighted works is shaped by law, market forces, social norms, and even the design of software and communications technologies (Lessig, 1999). This chapter has attempted to demonstrate the many advances that copyright scholars have made in studying the impact of digitization. Ideally, this brief introduction will motivate even more economists to bring their particular tools to bear on these problems in conjunction with legal scholars and those from other disciplines. An economic approach has much to offer scholars who wish to study copyright law. At the same time, legal scholarship offers researchers insights about formal institutions and how they function in practice, a necessity for conducting policy-relevant research.

REFERENCES

Benkler, Yochai (2006), *The Wealth of Networks: How Social Production Transforms Markets and Freedom*, New Haven, CT: Yale University Press.

Boldrin, Michele and David K. Levine (2008), *Against Intellectual Monopoly*, Cambridge: Cambridge University Press.

Boyle, James (2000), 'Cruel, mean, or lavish? Economic analysis, price discrimination and digital intellectual property', *Vanderbilt Law Review*, **53** (6), 2007–39.

Boyle, James (2008), *The Public Domain: Enclosing the Commons of the Mind*, New Haven, CT: Yale University Press.

Carroll, Michael W. (2006), 'One for all: the problem of uniformity cost in intellectual property law', *American University Law Review*, **55** (4), 845–900.

Cass, Ronald and Keith Hylton (2013), *Laws of Creation: Property Rights in the World of Ideas*, Cambridge, MA: Harvard University Press.

Cohen, Julie E. (1998), '*Lochner* in cyberspace: the new economic orthodoxy of "rights management"', *Michigan Law Review*, **97** (2), 462–563.

Cohen, Julie E. (2005), 'The place of the user in copyright law', *Fordham Law Review*, **74** (2), 347–74.

Demsetz, Harold (1969), 'Information and efficiency: another viewpoint', *Journal of Law and Economics*, **12** (1), 1–22.

DiCola, Peter and Matthew Sag (2012), 'An information-gathering approach to copyright policy', *Cardozo Law Review*, **34** (1), 173–247.

Dinwoodie, Graeme B. (2000), 'A new copyright order: why national courts should create global norms', *University of Pennsylvania Law Review*, **149** (2), 469–580.

Dreyfuss, Rochelle Cooper (2004), 'TRIPS – Round II: should users strike back?', *University of Chicago Law Review*, **71** (1), 21–35.

Duffy, John F. (2002), 'Harmony and diversity in global patent law', *Berkeley Technology Law Journal*, **17** (2), 685–726.

Fisher, William W., III (1988), 'Reconstructing the fair use doctrine', *Harvard Law Review*, **101** (8), 1659–1795.

Fisher, William W., III (1998), 'Property and contract on the Internet', *Chicago-Kent Law Review*, **73** (4), 1203–56.

Frischmann, Brett M. and Mark A. Lemley (2007), 'Spillovers', *Columbia Law Review*, **107** (1), 257–301.

Ginsburg, Jane C. (2008), 'Separating the Sony sheep from the Grokster goats: reckoning the future business plans of copyright-dependent technology entrepreneurs', *Arizona Law Review*, **50** (2), 577–609.

Gordon, Wendy J. (1982), 'Fair use as market failure: a structural and economic analysis of the Betamax case and its predecessors', *Columbia Law Review*, **82** (8), 1600–1657.

Gordon, Wendy J. (1993), 'A property right in self-expression: equality and individualism in the natural law of intellectual property', *Yale Law Journal*, **102** (7), 1533–1609.

Hughes, Justin (1988), 'The philosophy of intellectual property', *Georgetown Law Journal*, **77** (2), 287–366.

Hughes, Justin (1998), 'The personality interest of artists and inventors in intellectual property', *Cardozo Arts and Entertainment Law Journal*, **16** (1), 81–181.

Kieff, F. Scott (2006), 'Coordination, property, and intellectual property: an unconventional approach to anticompetitive effects and downstream access', *Emory Law Journal*, **56** (2), 327–437.

Kitch, Edmund (1977), 'The nature and function of the patent system', *Journal of Law and Economics*, **20** (2), 265–90.

Kwall, Roberta Rosenthal (1985), 'Copyright and the moral right: is an American marriage possible?', *Vanderbilt Law Review*, **38** (1), 1–100.

Landes, William M. and Richard A. Posner (2003), *The Economic Structure of Intellectual Property Law*, Cambridge, MA: Harvard University Press.

Lange, David L. and H. Jefferson Powell (2009), *No Law: Intellectual Property in the Image of an Absolute First Amendment*, Stanford, CA: Stanford University Press.

Lemley, Mark A. (1997), 'The economics of improvement in intellectual property law', *Texas Law Review*, **75** (5), 989–1084.
Lemley, Mark A. and R. Anthony Reese (2004), 'Reducing digital copyright infringement without restricting innovation', *Stanford Law Review*, **56** (6), 1345–1434.
Lessig, Lawrence (1999), *Code and Other Laws of Cyberspace*, New York: Basic Books.
Litman, Jessica D. (1987), 'Copyright, compromise, and legislative history', *Cornell Law Review*, **72** (5), 857–904.
Litman, Jessica D. (1989), 'Copyright legislation and technological change', *Oregon Law Review*, **68** (2), 275–361.
Litman, Jessica (2001), *Digital Copyright*, Amherst, NY: Prometheus Books.
Litman, Jessica (2007), 'Lawful personal use', *Texas Law Review*, **85** (7), 1871–1920.
McLeod, Kembrew and Peter DiCola (2011), *Creative License: The Law and Culture of Digital Sampling*, Durham, NC: Duke University Press.
Madison, Michael J., Brett M. Frischmann and Katherine J. Strandburg (2010), 'Constructing commons in the cultural environment', *Cornell Law Review*, **95** (4), 657–709.
Menell, Peter S. (2013), 'The mixed heritage of federal intellectual property law and ramifications for statutory interpretation', in Shyamkrishna Balganesh (ed.), *Intellectual Property and the Common Law*, Cambridge: Cambridge University Press.
Mossoff, Adam (2005), 'Is copyright property?', *San Diego Law Review*, **42** (1), 29–43.
Netanel, Neil Weinstock (2008), *Copyright's Paradox*, Oxford: Oxford University Press.
Nimmer, Melville B. and David Nimmer (2012), *Nimmer on Copyright*, New York: Matthew Bender.
Okediji, Ruth L. (2003), 'Public welfare and the role of the WTO: reconsidering the TRIPS agreement', *Emory International Law Review*, **17** (2), 819–918.
Patry, William F. (2013), *Patry on Copyright*, New York: Thomson Reuters.
Patterson, L. Ray and Stanley W. Lindberg (1991), *Copyright: A Law of Users' Rights*, Athens: University of Georgia Press.
Samuelson, Pamela (1988), 'Modifying copyrighted software: adjusting copyright doctrine to accommodate a technology', *Jurimetrics Journal*, **28** (1), 179–221.
Sprigman, Christopher (2004), 'Reform(aliz)ing copyright', *Stanford Law Review*, **57** (2), 485–568.
Treiger-Bar-Am, Kim (2008), 'Kant on copyright: rights of transformative authorship', *Cardozo Arts and Entertainment Law Journal*, **25** (3), 1059–1103.
Van Houweling, Molly Shaffer (2010), 'Author autonomy and atomism in copyright law', *Virginia Law Review*, **96** (3), 549–642.
Wu, Timothy (2004), 'Copyright's communications policy', *Michigan Law Review*, **103** (2), 278–366.

FURTHER READING

Chapter 3 of Litman (2001) provides a good introduction to the history of copyright legislation in the US. Chapter 4 of Boyle (2008) provides a good primer on current debates in copyright law and policy. The treatises by Nimmer and Nimmer (2012) and Patry (2013) offer insightful commentary from differing perspectives, as well as reference material on many detailed topics of copyright doctrine.

18. Copyright law and royalty contracts

Richard Watt

Economic theory, and microeconomics in particular, has a strong tradition of studying incentive based decision making. In turn, one of the principal mechanisms used to give incentives to decision makers is a contract. Above all, a contract is a means under which one economic agent can give incentives to another in order that the second of the two parties carries out the decisions that best suit the first.

Contracts serve more than one fundamental purpose. Nevertheless, their principle function is that they allow for the participants to add value by *specialization and the gains from trade*, and they allow for that additional value to be shared, that is, for the contracting parties to be *compensated* for their efforts. When two parties (say an author and an intermediary) decide to enter into a contract, it is because they recognize that by cooperating together additional wealth is created. This additional wealth is the result of specialization, and is the difference between what can be charged to the next agent in the value chain (perhaps retailers, perhaps direct consumers) when the intermediary is present and what could be charged at that point without the intermediary. The contract is the mechanism that allocates that additional wealth between the two parties whose cooperation created it. Contracts allow for authors to specialize in creating new works, and for intermediaries to specialize in bringing those works to the consumers in an easily consumable format.[1] Without the ability to contract, the author would not only have to create the work in question, but also have to bring it to the consuming public.[2]

The economic theory of copyright is fundamentally a theory of incentives. It is also a theory of market failure. The basic idea is that a market failure that is founded upon the public good nature of intellectual property, which implies a significant free-user problem, may destroy the incentive of creative individuals to create intellectual products, thereby implying that the very existence of the creative good itself is endangered.

The theory goes as follows. In order for creators to want to invest their time, effort and quite possibly money in producing an item of copyrightable property, they may need to see a personal gain to them from doing so. Normally, that personal gain depends in turn upon the functioning of a market, in which access to the intellectual property is extended to users. Given that intellectual property in general, and copyrightable creations in particular, is inherently of a public good nature, without some form of regulatory intervention there is a fear that, if access is given to one user (presumably through a market transaction), then that user can subsequently provide access to others without going back to the owner of the intellectual property right in question. This, in turn, would imply the establishment of what is effectively a secondary good that competes with the original good supplied by the copyright holder, thereby diminishing the ability for the copyright holder to be compensated in the form of monetary payments from the market. Finally, the upshot may be that the original creator considers that, rather than going ahead and embarking upon the creative process in the first place, his or her time and efforts can be more effectively put to use in some other endeavour, in which case society is robbed of the creation that would otherwise have been brought into existence.

Most modern societies have deemed that the most efficient way in which to avoid such a situation is to grant copyright protection for the creation in question. Copyright protection, it is argued, imposes costs upon suppliers of the secondary good, in the form of punishment in the event that trade in those goods is detected. So long as the expected costs the copyright system places upon pirate sellers outweigh the expected gains from dealing in pirated intellectual property, then we should expect that the secondary market does not become established, and the copyright holder is free to exploit the market for the original creation to its fullest. In that way, it is often argued that the copyright system provides the incentive for creators to embark upon the creative process. Of course, the system may not totally eliminate piracy, but, so long as it reduces piracy sufficiently, the incentive for creators can be appropriately protected.

CONTRACTS IN THE COPYRIGHT VALUE CHAIN

One thing that should be apparent is that it is not the copyright system itself that provides incentives, but rather the copyright system is designed such that a market in which access to copyright content is granted can function without impediment. Within the context of that marketplace, contracts will be written along the value chain, and, as is clearly pointed out by Kretschmer (2010) and Watt (2010), the contracts are what provide the relevant incentives where they are required.

There are several places along the value chain that we can expect to see contracts. First and foremost, we might see contracts between the original creator him- or herself (the 'author') and specialist 'intermediaries'.[3] In this chapter we shall concentrate upon these kinds of contracts, above all those between an author and a publisher or distributor. Authors are specialists in creating copyright goods (writing novels, writing songs, writing movie scripts, etc.), but not in the process of embodying them in tangible media (printing physical books, recording CD Roms, directing and producing DVDs, and so on) or in the efficient promotion and distribution of the physical media once produced. Since the physical media (also known as delivery goods) are a necessary part of the process of enabling the intellectual property to be useful to users, we should expect that authors will contract out the right to affix the intellectual property to such physical media to the relevant specialists, in exchange for some sort of monetary compensation (and at this intermediary stage of the value chain there could well be several levels of contracting, for example between publishers, distributors, and retailers). The monetary compensation paid to the author as the copyright holder should be thought of as simply a way in which the economic surplus that is created by the act of affixing the content to the delivery good, and then getting it to the final consumers, is able to be shared between the content owner (the author) and the content user (distributor, producer, retailer, and so on).

Second, there is also a 'contract' when the final user (say, the consumer) purchases a copy of the delivery good containing the intellectual property. At this stage, however, the contract will most likely be implicit (although it need not be), and it will often be defined by the law rather than a specific document. Nevertheless, some delivery goods will contain the contract itself. For example, you will often find DVD movies that clearly state that by purchasing the DVD the consumer agrees not to copy and redistribute it. The act of purchasing contracts the purchaser to the terms and conditions of purchase. Likewise, books normally contain an initial page that sets out to whom the copyright belongs, and possibly also states what consumers can and cannot do,[4] which again effectively constitutes a contract that the purchaser enters into with the act of purchasing. Interestingly, the contract at the point of purchase is not between the customer and the retailer (who is the other party present when the purchase is made), but between the customer and the copyright holder.[5]

Special Aspects of Contracts along the Copyright Value Chain

There are certain aspects that differentiate the value chain of a copyright good from that of most other economic goods, and some of those aspects may lead to differences in the types of contracts that are used.

When it comes to creative goods, it is not always an easy task to determine ownership (many different rights may be bundled, and it may be the case that authorship is not registered with any public registry), and, without clear ownership, transactions for access become very difficult. Thus, one of the principal functions of copyright law is to establish exactly who owns the different rights that others would like to access. Here, I am not talking about the moral rights in a creation (that is, the correct attribution of authorship), but rather the economic rights. Ownership may become confused when there is more than one individual (or company) involved in the production and distribution chain, and copyright law should clearly establish who is the owner of what.[6]

It also happens that for copyright goods the actions of third parties (that is, persons not involved directly in the contract) are likely to be more important than in other contracting environments. This is where copyright law itself becomes a crucial element in contracting. One of the principal functions of copyright law is to attempt to limit piracy, thereby protecting the value in the contracts between the copyright holder and intermediaries.

ROYALTY CONTRACTS: RISK BEARING AND INCENTIVES

The most interesting contracts in the value chain for copyright goods, at least as far as economists are concerned, are those written between the copyright holder (here, the 'author') and an intermediary, since these are the contracts that allow for compensation to the author, and thus these contracts are a principal element in ensuring the incentives for creators. We often refer to these contracts as 'royalty' contracts, in reference to the fact that they habitually stipulate payment to the copyright holder of a royalty on each unit that is sold or, what is equivalent, a royalty for each unit of sales revenue received.[7]

Naturally, there is no overriding requirement that a contract between the original copyright holder and an intermediary stipulates a per-unit royalty. There is no reason why the contract cannot simply allow for a fixed up-front payment to the author, who then would not participate any further in the revenues from sales. More generally there are contracts that stipulate both an up-front payment and a per-unit royalty. The balance between up-front payments and royalties based on sales is determined by the need for risk sharing and provision of incentives, as we shall see below. However, before looking at risk sharing and incentives, it is worth mentioning a couple of interesting general features of royalty payments in a contract between an author and an intermediary.

There are several quite persuasive arguments against the use of royalty payments in a contract between an author and a publisher. Clearly, a royalty payment to the author is exactly equivalent to an increased marginal cost for the publisher, who now for every unit sold must pay something back to the author. This puts the publisher at a marginal cost disadvantage compared to pirate producers, who supply copies of the delivery good in question without paying a royalty to the author. In essence, the higher is the royalty in the contract the more piracy is exacerbated, and correspondingly the lower is the level of sales achieved by the legal producer (Watt, 2000). It then becomes very unclear how an increase in the royalty parameter would affect the earnings of the author.[8] Even more, it is unclear if the optimal level of the royalty from the author's point of view is even positive, or if the author would be better off using an up-front charge and no royalty. The issue of the optimal level of the royalty has been studied by Besen and Kirby (1989), Varian (2000) and Woodfield (2006).

Another argument against the use of royalties depends upon how the value of the efforts of the intermediary affects the value of the rights once the current contract has expired. Copyright law assigns ownership to the author for a specified length of time, normally life of author plus 70 years. Thus, in principle, the author can contract out certain rights to an intermediary for periods of time that are shorter than the duration of the copyright itself. This feature has led to economists considering how the actions of the parties to a contract might affect the residual value of the rights, and how this might end up conditioning the contractual terms. For example, in the context of television transmissions of a series (for example, a sit-com), Liebowitz (1987) notes that the marketing and promotion of the television station for a series have a significant effect on the value of that series after the contract with the station expires. In such a case, it is efficient that the station retains the rights in the work after it has been broadcast, and so we might expect to see contracts for outright sale, in which the author sells all of the economic rights to the TV station for a fixed up-front charge, rather than retaining the residual rights once the series has been broadcast. Such a contract would maximize the incentive of the station to appropriately market and promote the series.[9] An outright sale contract would assign all of the future risk, and all of the future value, in the work to the intermediary, and none to the original author. In reality, this argument does not imply that royalties should not be included in a contract between an author and an intermediary, since the residual value of the rights depends upon both the original value of the work and the promotional efforts of the first intermediary. But it is a persuasive argument for the intermediary also retaining some rights along with the author, perhaps also in a royalty arrangement, with successive intermediaries.[10]

On the other hand, Towse (2001) suggests that royalty contracts, which explicitly give the author ownership of some of the income generated over time by exploitation of the work, are efficient, since new works by the same author may well affect the value of existing works. When authors retain a financial interest in their works (via a royalty contract), then they have an incentive to continue to produce quality works. Take the example of a singer who when unknown writes and records a song that is contracted to a publisher for recording on to a music CD. Since the singer is unknown, perhaps this first album sells only moderately well. But then the second album becomes a huge hit and it is likely that the first CD will also begin to sell better, and the author can only participate in those new sales if a royalty contract was used. The incentives given to the singer of retaining a financial interest in the first work via a royalty contract may be the reason, at least partially, for continuing to work on new material of high quality.[11]

As we can see, whether or not royalties should be included in a contract between an author and an intermediary is a complex issue. Not only do we have to think about the effect upon piracy, the incentives for promotion by intermediaries, and the incentives for the author creating new works, but royalty payments are also very important for both risk sharing and direct incentives.

Risk Sharing

We have seen that copyright serves to enable value to be added and shared among those that create it. As we shall see shortly, it also serves as an incentive mechanism. But another fundamental purpose of copyright is to share risk among the parties. For economists, this aspect is both extremely important and extremely interesting. Once a work has been created, it will be brought to a production process in order to reach the market, and all along this process risks are present. There will be technical risks in the businesses of the intermediaries (machinery may break down, the truck carrying CDs to the retailers may crash, and so on). There is also a serious risk of piracy that may undermine the final value of the enterprise at the retail stage. And, perhaps above all, the final demand by consumers is subject to uncertainty. None of these risks can be easily insured externally, and so the risks need to be apportioned and borne by the parties working along the value chain. This is achieved through the contracts that they write among themselves. When the gains from trade are risky, and the two cooperating agents are risk averse, then it becomes efficient that the gains are shared in a way that also optimally shares the risk between the parties. Economists emphasize the benefits from sharing risk, which are at least as important as the pure compensation that is achieved by a contract.

If there were no risk involved in the cooperative endeavour, then there would be no case for a royalty payment in the contractual terms, that is, the contract would involve a one-off up-front payment only. To see this, imagine that an author supplies a book to a publisher, so that the publisher can print the book and then sell copies to consumers (via retailers). A scenario of no risk would be one in which the final demand curve of the consumers for this book were exactly known, and so a first-best optimal price can be set, and the amount of revenue that will be recovered from the market is known with certainty from the outset. The level of sales is known with certainty, and thus a fixed level of revenue is guaranteed. The contract then only needs to stipulate how much of this guaranteed level of revenue should be retained by the publisher and how much should be paid back to the author. This can be easily done with a fixed one-off payment.

Now assume that the demand curve of the consumers is uncertain, and may take any one of a number of possibilities. Aside from the difficulties this implies for price setting, clearly the level of final sales that will ensue for any given price is now a random variable, that is, revenue is risky. In this type of scenario, if the author is paid a one-off up-front payment only, then all the risk is borne by the publisher. This will only ever be efficient if the author is risk averse and the publisher is risk neutral. If the publisher is risk averse at all, then some of the risk will optimally be shared with the author,[12] and the only way to do that is to structure the contract such that the author's income from it depends upon the realization of the final sales risk. This leads directly to a contract with a royalty element.

In the sense that royalty contracts serve to apportion risk, they are effectively an insurance device.[13] One of the most celebrated results from the theoretical insurance literature is that the optimal contract for a risk averse insurance consumer is one with full insurance above a deductible (for the seminal paper on this, see Arrow, 1970). Interestingly, a deductible insurance arrangement is exactly replicated by a contract with an up-front advance payment and a royalty, which are so common in the real world of contracting between authors and publishers. In these contracts, the author is paid an up-front amount, say \$ Z, and the contract also specifies a royalty payment, of say α, but where the author will receive the royalty payment only when it exceeds the up-front payment (so the initial payment is an advance on future royalties). Thus, if we denote the final revenue that is generated by R, then the author's final income, X, is $X = \max[Z, \alpha R]$. It is important to note that, in the case in which the calculated royalty does not exceed the advance payment ($\alpha R < Z$), the author does not pay back the advance, and when the calculated royalty does exceed the advance payment ($\alpha R > Z$), since the amount Z has already been paid, the author only receives a further payment of $\alpha R - Z$. Notice then that, in such an arrangement, under no circumstances does the author's income drop below the level of the advance payment, and so very low income outcomes are totally avoided, or low income states are fully insured. On the other hand, for the states in which the royalty payment is higher than the advance payment, the author ends up getting the amount of the royalty, with all of the implied risk. For high income states, there is no insurance at all. As we can see, an up-front advance payment with a royalty gives the author full insurance for low income states and no insurance for high income states, which is exactly what a deductible insurance contract achieves by not insuring small losses (those below the deductible), and providing full (marginal) insurance for large losses (those above the deductible). Therefore it is reasonable to state that royalty contracts with an up-front advance payment are likely to be an optimal contract structure for risk averse authors.

The way a royalty contract actually shares risk is determined by the level of the royalty payment for each feasible realization of the underlying revenue variable. In order to consider this, assume for now that the revenue to be shared can take on only two possible values, say high and low, denoted by R_1 and R_2 where $R_1 > R_2$. The contract specifies that the author will be paid a fraction α_i of revenue R_i for $i = 1, 2$. To start with, any contract (α_1, α_2) such that $\alpha_1 R_1 = \alpha_2 R_2$, that is $\alpha_1 = \alpha_2 \frac{R_2}{R_1}$, gives the author a sure amount of income in each state of nature, thereby eliminating all risk for the author and correspondingly placing all risk upon the publisher.[14] On the other hand, the contract can eliminate all risk for the publisher (thereby placing all risk in the hands of the author) by setting $(1 - \alpha_1)R_1 = (1 - \alpha_2)R_2$, that is $\alpha_1 = (1 - \frac{R_2}{R_1}) + \alpha_2 \frac{R_2}{R_1}$. However, if both the author and the publisher are strictly risk averse, then in any Pareto efficient contract both must share some of the risk (i.e. neither will end up with risk free final wealth).[15]

It is also interesting to consider when the contract will stipulate $\alpha_1 = \alpha_2$, which is a very commonly observed feature of real-world contracts. These are the types of contract that stipulate that revenue will be shared in a pre-determined manner that is independent of the actual amount of revenue to be shared. So, for example, the publisher and the author might agree that the author will receive 10 per cent of sales revenue, regardless of whether revenue is high or low. This is known as constant proportional risk sharing. As it happens (see Alonso and Watt, 2003), there is only one case in which such a contract is actually Pareto efficient – both the author and the publisher must have constant relative risk averse preferences and they must both be equally risk averse. So, while constant proportional risk sharing seems to be very common, it is unlikely to be a Pareto efficient manner in which to share risk, at least for the case of an author and a publisher of a copyright work.[16]

Another reasonably common royalty structure is to offer the author a greater cut of revenue the higher is that revenue. So, for example, the author might receive 5 per cent of revenue if revenue is low, and 10 per cent if revenue is high (say, if revenue exceeds some specified threshold). Clearly, this type of royalty structure places more risk upon the author than does a constant proportional sharing structure. If the author is more risk averse than is the publisher (which is a realistic assumption), then such a royalty structure cannot be explained by risk sharing alone. In order to see why the royalties may be structured in that way, we need to consider another element of contracts, namely incentives.

Incentives

The final fundamental element of a contract is to give incentives. Incentive payments become important when one or perhaps both parties are able to take actions that affect the value of the contract to the other party, but that cannot be easily observed by the other party. Such a situation is known as 'moral hazard'.[17] Take, for example, the case of the author and the publisher. If it is known in advance that the contract will discriminate the author's royalty payment, by paying more in high revenue states than in low revenue states, then the author may exert more effort in producing a better quality work. Doing so would, presumably, enhance the probability of the high revenue state and correspondingly limit the probability of low revenue states. So, if the author anticipates that the contract will be structured such that he or she receives a greater royalty when sales are high, then he or she will have an incentive to produce the best quality work possible rather than shirking and producing low quality, which has a much greater chance of selling poorly. Notice that even if the contract stipulates constant proportional risk sharing, $\alpha_1 = \alpha_2$, the author will receive a greater monetary payment when revenue is high than when it is low,[18] and so even then there exists an incentive to produce high quality work, but when $\alpha_1 > \alpha_2$ that incentive is enhanced even further.

Interestingly, if the contract pays the author more when sales are high than when they are low, it must also be the case that the contract pays the publisher less when sales are high than when they are low. However, clearly the publisher must also exert effort, for example in marketing and promotions that will have some effect on the probabilities of high and low sales. Thus, if the author is concerned (as he or she might well be) that the publisher does in fact exert high effort in promotions, then it may be better to structure the contract such that it is the publisher, rather than the author, who gets a greater pay-off for high sales. We see here a fundamental conflict; obtaining a higher quality work in the first place may require paying the author more (and the intermediary less) when sales are high, but obtaining a higher quality of promotions may require exactly the opposite. It is interesting to conjecture how these opposing incentive effects end up being balanced.

EFFECT OF COPYRIGHT LAW UPON CONTRACTUAL TERMS

Any contract is by nature the result of a process of bargaining between two parties (or perhaps their representatives). Thus, the actual sharing rule that is stipulated in the contract is found by the two parties negotiating together. In that process of negotiation, all of the above elements (compensation, risk sharing and incentives) should be taken into account and merged together to come to an agreement on the actual sharing rule to use. However, one thing that is also quite obvious is that the general environment in which this negotiation takes place may also determine in no small part the outcome. Perhaps the most important aspect of the environment, at least for the case of contracts for copyrighted creations, is the presence of copyright law.

Copyright law establishes who actually owns the rights in question, and for how long. It also establishes important aspects that will determine the value of the enterprise to both the publisher and the author, such as how much fair use is allowed and how piracy will be punished (thereby determining to a significant degree the level of piracy that we can expect to happen). Thus, any alterations in the copyright law should lead to corresponding alterations in the contractual terms that are agreed to. The question is, then: how would we expect that a stronger copyright standard be reflected in copyright royalty contracts? This question has been considered by Michel (2006) and Muthoo (2006).

Both Michel (2006) and Muthoo (2006) study copyright royalty contracts in the context of Nash bargaining (see Nash, 1950), which in turn is one of the most commonly accepted descriptions of the outcome of a bargaining process in economics. In essence, the outcome of a Nash bargaining problem uses a constrained optimization procedure to identify a particular point on the set of Pareto efficient allocations. The point in question can be shown to be the only point that satisfies a small set of axioms, all of which are intuitively understandable and should be required of the solution to any bargaining problem. The question then is: how does an alteration in copyright law affect the parameters of that problem, and how will the change end up affecting the bargained solution?

The Nash bargaining solution depends upon several parameters that might well be affected by a strengthened copyright law. In particular, the solution depends on the bargainers' reservation utility levels, their relative bargaining powers, the size of the surplus that is to be shared, and the probabilities of the different surplus outcomes that may occur. Without going into the mathematical details of the effects, we shall consider each of them here.

First, when a bargainer's reservation utility level is increased, that bargainer should expect a more favourable outcome from the bargaining process. When the copyright standard is strengthened, we should expect that the copyright holder's reservation utility level increases. Thus for that reason an increase in the strength of the copyright standard should result in a more favourable contract for the copyright holder. On the other hand, it is also possible that the reservation utility of the intermediary also increases when the copyright standard is improved. This may be the case, for example, if the intermediary already has contracts with other copyright holders, and the stronger copyright standard increases the value of those contracts. Thus there could also be an effect such that the contract moves in favour of the intermediary. Of course, both movements are not possible (the contract cannot be more favourable to both the author and the intermediary), and so exactly how the contract changes will depend upon the relative change in the two reservation utility levels.

Second, and related to the first issue, an increase in the copyright standard might alter the relative bargaining strengths of the two parties. The most likely way in which this would happen is that under a stronger copyright standard the author (who is the copyright holder, and thus the party that copyright law protects) would enter the bargaining process in a more advantaged situation. This will unambiguously cause the final contractual terms to move in the favour of the author.

Third, the size of the surplus to be shared will probably increase owing to the implied reduction in piracy that a strengthened copyright standard should achieve. However, exactly how this increased level of surplus will be shared between the two parties will typically depend upon their utility functions, in particular their risk aversion. The way in which an increase in the wealth to be shared in any given state of nature in a Pareto optimal sharing arrangement (of which the Nash bargaining solution is an example) was considered by Wilson (1968), who shows that the additional wealth will be shared according to the ratio of their individual levels of risk tolerance[19] to the sum of the risk tolerances of them both. Thus, whether or not the increase in shareable surplus actually increases the negotiated royalty rate depends upon whether or not the original royalty rate is greater or smaller than the ratio of risk tolerance to group risk tolerance.[20]

Finally, we have the effect of a change in the probabilities of the different states of nature. If we posit that there are two states of nature, one in which piracy happens and one in which it does not,[21] then a stronger copyright standard should decrease the probability of piracy and correspondingly increase the probability of no piracy. When this happens, again it turns out that the effect upon the bargained royalty contract is ambiguous; the royalty rates for how wealth in the two states is shared may go up and they may go down. It may even be the case that one royalty rate goes up (for example, the royalty rate for the piracy state of nature) and the other goes down (the royalty rate for the no-piracy state of nature). The mathematics of this effect are complex and beyond the scope of the present chapter, but suffice it to say that the effect depends on all of the parameters of the problem, and so coming to a consensus as to the final effect will imply conditions on all of the parameters of the Nash bargaining problem.

CONCLUSIONS

The relationship between copyright and contracts is an important one, and yet one that has not received much attention in the literature. Of all the contracts along the copyright value chain, the most important are those between authors and intermediaries. These contracts serve a variety of economic purposes, of which here we have highlighted the fact that they allow for specialization and the gains from trade (that is, they increase the amount of wealth that can be shared), they provide for monetary compensation for the two parties' efforts, they share risk between the parties, and they provide incentives such that the parties exert efficient levels of effort.

We have also considered how a change (a strengthening) of the copyright standard might affect a negotiated royalty contract. In short, the overall effect is highly ambiguous, with good reasons to believe that the contract might change in favour of the author, and other good reasons to believe that it might change in favour of the intermediary. A full mathematical analysis of some of those effects would be of great value to the community of economists interested in copyright.

It is also interesting to consider if the move to a digital environment will have any effect upon the contracts between authors and intermediaries. In principle, all we would need to think about is if a move from non-digital to digital environments has any effect upon the basic underlying parameters of the bargaining process and, if it does, then to work out how that would affect the bargaining outcome. Of course, just as with a strengthening of copyright law, a switch to a digital environment will most likely involve several changes (changes in the value of the copyright good, changes in the values of reservation utilities, changes in the probability of piracy, etc.), and so the final effect would be ambiguous. It would, however, be very interesting to see some solid research into this aspect of contracting for copyright goods.

NOTES

1. It is also important to notice that entering into a contract is entirely voluntary, and so, if an incentives based contract is signed between two parties, we can be sure that it must be mutually beneficial. That is, the contract provides for greater welfare for both parties.
2. It is possible that in the digital environment this has indeed become cheaper, thereby lessening the need for contracting. However, there is also little doubt that, at least for those works with the highest value for consumers, contracting between authors and intermediaries still adds considerable value.
3. The intermediaries may be publishers, producers, distributors, retailers, and so on. For example, for the author of a musical composition, an intermediary may be the record company that records and distributes the original music, or it may be a radio station or discotheque that plays the music, or it may be a second performer who plays a cover version of the original music. Clearly there are a large number of possible intermediaries.
4. For example, a phrase such as 'All rights reserved. No part of this publication may be reproduced, stored in a retrieval system, or transmitted in any form or by any means, except as permitted by the Copyright Law, without prior permission of the publisher.'
5. The contract between purchaser and retailer is normally implicit, and involves the simple exchange of a unit of delivery good for a specified sum of money, possibly with some implied rights for the purchaser in the case of a damaged or unsuitable delivery good.
6. Along with the issue of ownership, it is also important to know exactly what it is that is owned and that can therefore be subject to market transactions. For the case of copyright goods, copyright law limits ownership both by exceptions such as fair use and by having a limited duration dimension.
7. The two are only equivalent when the unit price is constant.
8. Not only is competition from piracy important. An increase in the royalty is an increase in the marginal cost of the legal producer. Even in absence of piracy, this will lead to an increase in the price to consumers, and lower sales. Thus, by increasing the royalty parameter, the author earns a greater share of a smaller sum, and it is not clear if this is to his or her advantage. See Watt (2011a) for an analysis of the optimal royalty percentage.
9. That these types of contract are in fact not typically used, and the copyright holders often insist upon retaining the residual rights, leads Liebowitz to wonder about the 'puzzling' behaviour of the copyright holders.
10. Interestingly, this is exactly what is achieved by *droit de suite* in visual art.

11. Again, *droit de suite* comes to mind as a case in which this is actually done.
12. These results can be found in any text on the economics of risk bearing, for example Watt (2011b, chap. 5).
13. See Watt (2007) for a paper that relates copyright law itself with insurance.
14. Notice that, since $R_2 < R_1$, this contract sets $\alpha_1 < \alpha_2$, so, although the author gets the same monetary pay-off in each state of nature, this is achieved with a percentage royalty that is smaller in the good state of the world than that in the bad state of the world.
15. This result, and several others, is proved in Alonso and Watt (2003).
16. More likely these types of contract are used simply because they are easy to understand and to implement.
17. There may also exist elements of 'adverse selection', but here we shall concentrate only on the issue of moral hazard.
18. Indeed, it can even be the case that $\alpha_1 < \alpha_2$ and yet the monetary payment to the author is still greater when revenue is high compared to when it is low.
19. Risk tolerance is the inverse of absolute risk aversion.
20. Concretely, the ratio of the risk tolerance of the intermediary to the group risk tolerance.
21. Of course, there are likely to be many states of nature, possibly even a continuum of them, defined by different levels of piracy. However, the two-state example serves to capture the main points in regard to the effects of changes in the copyright standard.

REFERENCES

Alonso, J. and R. Watt (2003), 'Efficient distribution of copyright income', in W. Gordon and R. Watt (eds), *The Economics of Copyright: Developments in Research and Analysis*, Cheltenham, UK and Northampton, MA, USA: Edward Elgar Publishing.

Arrow, K. (1970), *Essays in the Theory of Risk Bearing*, Amsterdam: North-Holland.

Besen, S. and S. Kirby (1989), 'Private copying, appropriability and optimal copying royalties', *Journal of Law and Economics*, **32** (2), 255–80.

Caves, R. (2000), *Creative Industries: Contracts between Art and Commerce*, Cambridge, MA: Harvard University Press.

Kretschmer, M. (2010), 'Regulating creator contracts: the state of the art and research agenda', *Journal of Intellectual Property Law*, **18** (1), 141–72.

Liebowitz, S. (1987), 'Some puzzling behavior by owners of intellectual products: an analysis', *Contemporary Policy Issues*, **5**, 44–53.

Michel, N. (2006), 'Digital file sharing and royalty contracts in the music industry', *Review of Economic Research on Copyright Issues*, **3** (1), 29–42.

Muthoo, A. (1999), *Bargaining Theory with Applications*, Cambridge: Cambridge University Press.

Muthoo, A. (2006), 'Bargaining theory and royalty contract negotiations', *Review of Economic Research on Copyright Issues*, **3** (1), 19–27.

Nash, J. (1950), 'The bargaining problem', *Econometrica*, **18**, 155–62.

Towse, R. (2001), *Creativity, Incentive and Reward: An Economic Analysis of Copyright and Culture in the Information Age*, Cheltenham, UK and Northampton, MA, USA: Edward Elgar Publishing.

Varian, H. (2000), 'Buying, sharing and renting information goods', *Journal of Industrial Economics*, **48** (4), 473–88.

Watt, R. (2000), *Copyright and Economic Theory: Friends or Foes?*, Cheltenham, UK and Northampton, MA, USA: Edward Elgar Publishing.

Watt, R. (2007), 'What can the economics of intellectual property learn from the economics of insurance?', *Review of Law and Economics*, **3** (3), article 3.

Watt, R. (2010), 'The economic theory of copyright contracts', *Journal of Intellectual Property Law*, **18** (1), 173–206.

Watt, R. (2011a), 'Revenue sharing as compensation for copyright holders', *Review of Economic Research on Copyright Issues*, **8** (1), 51–97.

Watt, R. (2011b), *The Microeconomics of Risk and Information*, Houndmills: Palgrave Macmillan.

Wilson, R. (1968), 'The theory of syndicates', *Econometrica*, **36**, 119–32.

Woodfield, A. (2006), 'Piracy accommodation and the optimal timing of royalty payments', *Review of Economic Research on Copyright Issues*, **3** (1), 43–60.

FURTHER READING

This chapter has attempted to discuss the issues related to copyright contracts in a non-mathematical manner. Readers interested in getting to grips with the full theory could perhaps look to Alonso and Watt (2003) and Michel (2006). Much of what the chapter has discussed relies on bargaining theory. There are many great texts that deal with the bargaining problem (for example, Muthoo, 1999), although I am not aware of any that pay specific attention to bargaining and copyright contracts. The chapter has also limited itself to mainly theoretical considerations of copyright contracts, and has not really mentioned the real world. Readers interested in knowing more about the real world of copyright contracts could go to Towse (2001) and Caves (2000); see also Symposium on Copyright, Contracts and Creativity, available at: http://www.cippm.org.uk/symposia/symposium-2009.html.

19. Copyright and competition policy

Ariel Katz

Despite the tension that exists between the goals of competition law and copyright law (and other forms of intellectual property) the modern view is that ultimately these two areas of law share the same long-term goals of promoting innovation to enhance consumer welfare (Bohannan and Hovenkamp, 2011). We tolerate some static inefficiency that may result from granting exclusive rights in order to promote dynamic efficiency and long-term growth. As Frank Easterbrook, echoing Schumpeter (Schumpeter, 1994), put it, 'an anti-trust policy that reduced prices by 5 percent today at the expense of reducing by 1 percent the annual rate at which innovation lowers the cost of production would be a calamity. In the long run a continuous rate of change, compounded, swamps static losses' (Easterbrook, 1992: 119). The problem, of course, is that in formulating the optimal policies (optimal in the sense of second best) it is not clear how much static inefficiency we should tolerate in order to promote dynamic efficiency, and moreover, since today's intellectual goods are inputs for those of tomorrow, there is no magic line that allows us to know whether when we sacrifice the static we promote the dynamic or, instead, we impede both.

How the law handles this tension in the application of competition law to intellectual property has been the subject of extensive literature (for example, Lévêque and Shelanski, 2005; Boyer et al., 2009; Anderman and Ezrachi, 2011). In writing about copyright and competition policy, however, this chapter focuses on how copyright law itself applies competition policy goals from within. I will describe some features of copyright law (mostly American copyright law) that can be understood as either mitigating some of the static inefficiencies associated with copyright or ensuring that those inefficiencies do not translate into dynamic ones.

The chapter proceeds in four sections. It begins with the history and pre-history of copyright, and discusses how competition policy informed the enactment of the Statute of Anne (8 Anne c.19, 1710 (UK)), the first British copyright act, and shaped some of its key features. The second section focuses on various copyright doctrines that aim to promote competition and competitive innovation from within: it discusses the limited term of copyright, limitations on copyrightable subject matter, and doctrines such as fair use and the first sale doctrine. The third section discusses some of the challenges for the traditional allocation of rights between owners and users in the digital age, and the fourth section discusses the merit of promoting competitive goals from within as opposed to applying them externally through competition law.

THE STATUTE OF ANNE: COMPETITION POLICY AT THE CRADLE OF COPYRIGHT

The historiography of Anglo-American copyright law often begins with the Statute of Anne, which created the first statutory exclusive right over books. But, as much as that

Act resulted from the London publishers' demand for statutory exclusive rights, it also reflected Parliament's disdain of the monopoly of the Stationers' Company. This self-regulating publishers' cartel controlled the book trade and, pursuant to several licensing acts and other regulations, played a vital role in censoring publications unfavorable to the Crown or the Church (Patterson, 1968). The last of these licensing acts expired in 1695, and, after a period of 14 years during which the book trade was left entirely unregulated, Parliament enacted the Statute of Anne. Therefore, copyright law was not born as a response to a world of free copying. Quite the contrary, as much as it was responsive to publishers' demands for statutory exclusive rights, modern copyright law was also a countermeasure against an oppressive regime of press control in which censorship and exclusive print privileges were conflated. Thus, the Act contained key elements that are best understood as measures that aimed to prevent the Stationers' monopoly from reemerging. It vested initial copyright in authors rather than publishers to weaken the Stationers' stronghold on the catalogue (Nicita and Ramello, 2007). It limited the term of copyright to achieve two outcomes: maximizing access to books as they became part of the public domain and traded in a competitive market, and constraining the incumbent Stationers' market power over in-copyright books by forcing them to compete with out-of-copyright books (Nicita and Ramello, 2007). The Act removed the Stationers' Company enforcement and adjudication powers, and created a price-control mechanism to prevent 'too high and unreasonable' prices (Bracha, 2010: 1437). Lastly, while copyright depended on registration with the Stationers' Company, an alternative mechanism for securing copyright was created should the Stationers' Company refuse to register a work (Patterson, 1968).

The effect of these measures was not immediate. Notwithstanding the limited statutory term, the London publishers maintained that they held perpetual common law copyright, and it took several decades until the courts rejected this argument (Ochoa and Rose, 2001: 681–2), and, in the absence of modern competition law, some of the Stationers' anti-competitive practices, such as paying competitors to stay out of the market (Blagden, 1961), could not be directly challenged, but over time the London publishers' stronghold had weakened.

This background is useful not only for understanding how competition policy informed copyright law from its inception, but also because it provides a historical reminder of how copyright law responded to the economic, social, and political changes that the printing revolution brought about, and it also reminds us that, while limited copyright can serve as the 'engine of free expression', excessive control can serve as a highly effective break (Netanel, 2008: 3). These reminders allow us to reflect more broadly on demands for new regulations as we move from print to digital.

OWNERS' RIGHTS AND USERS' RIGHTS: COMPETITION POLICY FROM WITHIN COPYRIGHT

Although modern copyright law still retains some of the pro-competition stance that existed in its original design, over time it has omitted some measures and developed others: some minimize copyrights potential harm to static competition, while others ensure that it is not used to harm dynamic competition. Legal jargon often refers to them as 'limitations and exceptions', but they should be better understood as reflecting the law's choice to treat creative works as resources regulated as semi-commons: the state allocates usage rights over intellectual goods between different actors: some rights are allocated to creators and their assignees (copyrights), while other rights are allocated to users (users' rights) (Frischmann, 2012).

I suggest that the following three related dimensions provide a rational basis for such allocation:

- Incentive sufficiency: We would tend to allocate uses that generate marginally high incentives to owners, and otherwise to users.
- Utilizing capacity: We would allocate usage rights according to the relative capacity to utilize the work for socially desirable purposes, including innovative purposes.
- Transaction costs: We would consider the likelihood of value-maximizing voluntary exchanges. Ideally, when transaction costs, broadly understood, impede socially efficient bargaining, we would like to allocate usage rights to owners if allocating them to users would undermine the incentive to create or disseminate, and we would like to allocate usage rights to those who might have comparative advantage with respect to certain types of uses.

With these considerations in mind, let us turn to discussing some of the limitations to copyright and see how they promote competition policy goals along those three dimensions.

Limited Term

The limited term of copyright serves competitive goals, static and dynamic in different ways: It promotes static competition by allowing intra-brand competition between competing suppliers of the same work when the copyright expires. It may also increase inter-brand competition and lower prices for copyrighted books because it forces copyrighted works to compete with those in the public domain. Limited term also serves dynamic competition. If a limited term exerts competitive pressure on the prices of work, it also reduces the cost of future creation that builds on existing works. Moreover, in order to successfully compete with out-of-copyright works and command higher prices, new works have to offer something better, or at least different, than older ones (Katz, 2009c).

Let us consider limited terms along the three dimensions mentioned earlier. Discounting and depreciation imply that beyond a certain term 'the incremental incentive to create new works as a function of a longer term is likely to be very small' (Landes and Posner, 2003a: 473). Therefore, a limited term achieves efficiency gains without negatively affecting the incentive to create.

Regarding the capacity to utilize, we can fairly assume that *ceteris paribus*, and relative to any other randomly chosen firm, creators have some advantage in utilizing a work for the purpose for which it was created and around the time it was created. This advantage weakens over time, while the cost of getting permission, if required, increases as the identity of the current owner and his or her whereabouts become more difficult to discern (Katz, 2012). Moreover, at least for projects that reuse several existing works, the greater is the number of such component inputs that require permission, the higher will be the cost of getting it, as well as the likelihood of a tragedy of the anti-commons (Heller and Eisenberg, 1998). Therefore, if beyond a certain period there is no a priori reason to allocate exclusive usage rights while the cost of correcting inefficient allocation increases, it makes sense to limit the term of copyright and allow all to use the work as they see fit.

Unfortunately, the utility of time limits has diminished as the term of copyright increased through a series of copyright reforms. The public benefit arising from current copyright terms that can easily exceed a century is highly doubtful.

Offering two counter-arguments, Landes and Posner (2003a) proposed giving copyright owners an option of periodically renewing their copyrights. They maintain that, in the case of some works, continued exclusivity will reduce congestion externalities, and create better incentives to commercially exploit older works. It is not obvious, however, that congestion externalities in the consumption of creative works exist or are a matter of concern (Boldrin and Levine, 2004; Karjala, 2006). Moreover, even if the phenomenon might not be theoretically impossible, the assumption that the value of information generally increases with consumption (Gans, 2012) seems to provide a more solid basis for general policy prescriptions. Empirical work (Heald, 2007; Buccafusco and Heald, 2012) also casts doubts on the importance of the hypothesis that longer copyright terms might provide incentives for exploiting older works.

Subject Matter Limitations: Facts, Ideas, and the Merger Doctrine

Copyright may subsist only in original expressions, but not in facts (Feist v. Rural, 1991: 344), and 'in no case does copyright protection for an original work of authorship extend to any idea, procedure, process, system, method of operation, concept, principle, or discovery, regardless of the form in which it is described, explained, illustrated, or embodied in such work' (Copyright Act (US) 1976, sec.102). Even when copyright subsists in a work, copyright applies only to copying the expression or substantial part thereof, but it is not an infringement to copy any facts contained therein or any idea, process, system, and so on that the work describes. As a result, a copyright offers only a limited protection from competition. Thus, the law would allow the copyright owner to limit intra-brand/work competition, but it will not prevent others from selling competing substitutes (Breyer, 1970), even those based on ideas or information copied from the first. Consequently, copyright does not automatically result in market power, and, notwithstanding the copyright, every work is at least partially contestable. The degree of market power is a function of how many non-infringing substitutes may be available, and how close those substitutes are in the eyes of consumers.

This point has led several commentators to dismiss the concern that copyright may be a source of significant market power (for example, Landes and Posner, 2003b). But this view oversimplifies the issue, and is inconsistent not only with the easily observable fact that the price of many copyrighted works often exceeds their marginal cost substantially (Katz, 2007), but also with the fact that the purpose of granting copyright is to enable recoupment of the cost of creation by setting prices above the competitive level. If the effect of those limitations were indeed to eliminate any significant market power they would defeat copyright's very purpose to create such market power. Why then would the law adopt a set of such contradictory rules?

The three-dimensional framework described earlier provides some answers. From the incentive perspective, copyright protection over facts and ideas may not be necessary for generating sufficient incentives for three reasons. The first is that firms may resort to other means, legal, technological, and organizational, to gain some protection over facts and ideas, and the law may recognize such means, albeit reluctantly, especially when copying them might prove detrimental to the incentive to collect or create them (Katz, 2009c). Second, whether the facts or ideas that a work conveys will be repurposed, for what purposes, and how lucrative those uses might be, is entirely unpredictable. Therefore, beyond some core of predictable uses, an exclusive right to repurpose facts or ideas will provide a rapidly decreasing marginal incentive to create the work. In addition, copyrighted works are often used in ways that generate direct and indirect network effects (Katz, 2009c; Gans, 2012). As network effects intensify, competing works, even if they are similar or functionally equivalent, do not provide the same network value. The incumbent work becomes less contestable, allowing the copyright owner to recoup its investment even without the power to exclude others from copying facts and ideas. At the same time, treating facts and ideas as commons ensures that they can be used freely for many other purposes. This implies that the importance of circumscribing the scope of copyright is not so much in how it constrains the copyright owner's market power by increasing static competition, but in how it ensures that this market power – even when exists – may not be used to prevent or hinder Schumpeterian competition and innovation more broadly (Katz, 2009c).

Lastly, had protection extended over facts and ideas, reusing them would require permission. But obtaining permission is not costless, and may be plagued with accumulating transaction costs, including those arising from imperfect and asymmetric information, strategic behavior, or coordination.

In sum, since extending protection over facts and ideas is not necessary to generate sufficient incentives to create works in the first place, treating facts and ideas as commons seems preferable to exclusivity. This ensures that facts and ideas can be used at the socially efficient level (at marginal cost), which, in turn, promotes both static and dynamic competition. Statically, free facts and ideas can be used to produce competing works, constraining incumbents' market power at the margin, and dynamically promotes Schumpeterian competition, by removing barriers to the creation of other intellectual, cultural, and social goods.

The merger doctrine is another doctrine in American copyright law that limits the anti-competitive potential of copyright (Cotter, 2006). This doctrine is closely related to the idea–expression dichotomy, discussed above, but takes it one step further. Under the merger doctrine, an expression otherwise protectable will not be protected where there is only one or very few ways of expressing an idea, or when, as a result of factors such as network effects or technical compatibility, other expressions prove to be insufficient substitutes, no matter how similar they are. The doctrine's purpose is to prevent *de facto* exclusivity over the idea, which may translate into exclusivity over markets (Katz and Veel, 2013).

Fair Use

Fair use is another (and perhaps the ultimate) doctrine of American copyright law serving pro-competitive interests. Fair use extends the right of users beyond copying facts, ideas, or insubstantial parts of a work, and permits them to use substantial parts of the protected expression – even the entire work – in circumstances that otherwise would constitute infringement. Fair use mitigates some of the static losses resulting from exclusive rights in two principal ways: it constrains the copyright owner's market power by forcing it to compete, at the margin, with unauthorized but lawful copies, and it may reduce monopoly pricing deadweight loss by permitting users whose willingness to pay is higher than marginal cost but still lower than the price set by the copyright owner to use the work (effectively increasing output to the level that would occur with perfect price discrimination (Gordon, 1998). Fair use also mitigates dynamic losses by ensuring that the exercise of the copyright would not hinder downstream creativity and innovation by other authors and users (Landes and Posner, 2003b).

A standard objection to justifying fair use on these grounds is that such intervention only undermines the ability of copyright owners to set price discrimination schemes that would maximize both access and incentive (Fisher, 1987). Therefore, the early economic analysis of fair use viewed it as justified only when it responds to market failures in a narrow sense (where the cost of locating the owner, negotiating a license, and enforcing it are higher than the surplus it generates), with the implication that improvements in the technology or organization of licensing would render it unnecessary (Gordon, 1982). Under this approach, fared use would be preferable to fair use (Bell, 1998). More recent thinking accounts for a richer concept of market failure, including failures that may arise from strategic behavior, imperfect information, the negotiating parties' indifference to the negative or positive externalities of their activities, or the fact that some of these externalities are simply non-monetizable. This implies that even when licensing is feasible, fair use does not necessarily become futile (Frischmann and Lemley, 2007; Gordon, 2010).

The framework of incentives, relative capacity to utilize, and transaction costs is useful in explaining fair use. Because fair use may permit copying beyond the copying of facts, ideas, or insubstantial parts, its incentive-reducing potential is greater. Therefore, unlike the categorical allocation of facts and ideas to users, fair use analysis is more nuanced and detailed.

The US Copyright Act requires courts to consider several factors: (1) the purpose and character of the use, including whether such use is of a commercial nature or is for nonprofit educational purposes; (2) the nature of the copyrighted work; (3) the amount and substantiality of the portion used in relation to the copyrighted work as a whole; and (4) the effect of the use upon the potential market for or value of the copyrighted work (Copyright Act (US) 1976, sec.107). In their analysis, courts have to determine whether the use is merely substitutive to the work or whether it is transformative. The more transformative the use is, the more likely it will be considered fair (Sag, 2012). It is easy to see why: if the use is for another purpose, it is less likely to function as incentive-reducing substitute, and more likely to increase social utility. At the same time, the more transformative the use is, the less likely it is that the owner would have an a priori advantage in exploiting the work for that purpose, or that the prospect of using the work for that purpose would have been an important factor in the decision to create the work ex ante. Therefore, allocating that right to owners would merely increase transaction costs, impeding socially beneficial uses, without any meaningful countervailing benefits, while allocating those rights to users avoids that.

Lastly, transaction costs may be important for deciding how to allocate different uses between owners and users, and fair use analysis may factor them, albeit not always consciously. Thus, uses for research, education, criticism, and so on often generate substantial positive externalities that cannot be fully internalized by the users or may not be monetizable at all. For example, in a recent lawsuit brought by the Authors Guild against the HathiTrust, a collaborative organization of several major research libraries, the court found the defendants' activities to be fair use (Authors Guild v. HathiTrust, 2012). The scanned books were used for three purposes: (1) creating a full-text searchable inventory of the works available in the libraries to enable new opportunities for research based on metadata analysis and data-mining; (2) preservation of fragile books; and (3) access for people with certified print disabilities. All of these activities generate substantial positive externalities that cannot be fully internalized by the users or monetized at all and, as the court found, do not deprive the copyright owners of any important revenue source, because high transaction costs would prohibit the formation of a viable market for such uses.

Whether the use is for commercial or non-commercial purposes may also indicate how wide the mismatch between private and social value is. As a rough approximation, in the case of commercial uses the gap may be narrower than in non-commercial ones. But this is only a rough approximation. For example, transaction costs may impede uses that depend on reusing numerous works even if the purpose is commercial. Likewise, even if users expect to profit from the use, innovative uses might be extremely difficult to contract about because of information asymmetries, strategic behavior, problems of bilateral monopoly, and other frictions.

In sum, fair use promotes competition policy goals from within copyright because it permits some uses that mitigate the static inefficiency that copyright might create, but, at least as importantly, because it promotes dynamic efficiency by carefully allocating usage rights on the basis of how such allocation affects the various parties' incentives and capacity to innovate.

The First Sale Doctrine

One of the contentious issues in designing copyright and competition policies is the degree to which copyright owners can rely on their copyrights to impose and enforce downstream post-sale restraints. The issue arises in different contexts, for example whether copyright can be used to impose territorial restrictions and prevent parallel trade, whether copyright owners can prevent resale of used works by licensing them instead of selling them, or whether they should be permitted to maintain the retail price of their works. These debates often revolve around the doctrine of 'exhaustion', or 'first sale', which limits the exclusive rights that survive the initial authorized sale of an item protected by copyright.

Similar to fair use, the first sale doctrine can constrain the copyright owner's market power by forcing it to compete, at the margin, with secondhand copies, borrowed copies, or cheaper parallel imports, and it may reduce monopoly pricing deadweight loss by permitting those uses (Gordon, 1998). The first sale doctrine also promotes long-term dynamic efficiency because, while it may permit short-term restrictions agreed on between collaborating firms, it ensures that over the longer run intellectual goods can be freely sold and reused without being encumbered with restraints imposed by copyright owners (Katz, 2012).

The distribution of digital content has further complicated the application of the doctrine. One feature of digital content is that it can seldom be used without being reproduced. Consider the difference between a printed book and an e-book. The printing and initial distribution of a book would typically require the copyright owner's permission. But reading it does not, because copyright does not include an exclusive right to read. Likewise, reselling the book, lending it, or giving it away requires no license because of the first sale doctrine. However, the digital equivalents of such activities involve numerous reproductions, permanent or temporary. For example, buying an e-book requires downloading a copy of it, and reading it requires making temporary copies on the computer's random access memory and its screen. Being able to read the book on multiple devices and keeping it synchronized may require making additional copies on the device and on a cloud server. Copyright owners often claim that any of those acts is a reproduction for copyright purposes and therefore requires their permission. As a result, the distribution of digital content is often accompanied by contracts and licenses whose terms may permit some of those activities but not others. For example, the license may permit the user to download an e-book and install it on one or more devices for the purpose of reading it, but it may also prohibit transferring a copy to another person, or otherwise impose restrictions on how the content may be used. Digital content may be susceptible to another level of control implemented completely unilaterally through technological protection measures (TPMs), which may make it technologically impossible for users to use digital content contrary to the wishes of its sellers.

These phenomena raise several broader policy questions: Should the law treat analog uses and their digital equivalents differently, or should it be technologically neutral and focus on the economics of transactions and how they effectuate different users' experiences, irrespective of their precise technological details? Should the allocation of usage rights between owners and users change when digital technologies are involved? What is the optimal mix of public and private ordering in this field, that is, to what extent can copyright owners modify that initial allocation through contracts, license terms, and TPMs?

ALLOCATION OF RIGHTS IN THE DIGITAL AGE

Answering these questions is not an easy task, because technological developments affect each of the three dimensions discussed earlier. Obviously, digitized content can be easily reproduced and distributed, raising the specter of copyright owners losing all control and any ability to recoup their investments. Intuitively, this would suggest that to preserve incentive sufficiency copyright owners should be granted greater powers. Yet technological change has also led to 'a dramatic, and permanent fall, in the costs of production of almost all types of copyrightable subject matter' (Pollock, 2008: 52), and has facilitated the emergence of new user and open collaborative production (Benkler, 2007; Baldwin and von Hippel, 2009), suggesting that incentive sufficiency may be maintained even when rights allocated to owners are weaker.

Considering utilizing capacity also points in ambiguous directions. The digital revolution unquestionably opens new ways of utilizing works. Copyright owners may have comparative advantage in exploiting some of them (for example, commercial online distribution of newly released movies), whereas users may have comparative advantage in others (for example, commentary, fan fiction, mashups, or mass digitization projects for facilitating research and preservation). Transaction costs are also affected by technological change. A networked digital economy may facilitate transactions that may not have been possible before, but it cannot eliminate all of them. All of this suggests that what may be regarded as an optimal second best allocation of rights between owners and users in the analog world may not work the same way in the digital world.

The issue becomes more complicated because, as discussed in the previous section, whatever the initial allocation – as a matter of public ordering – is, digital technology provides copyright owners a greater opportunity to rearrange that allocation by different means of private ordering: contracts, licenses, and TPMs, potentially undermining copyright law's public policy objectives (Radin, 2003). The three-dimensional framework discussed in this chapter can be useful for drawing the limits of private ordering in this area. It helps distinguishing between circumstances where socially efficient bargaining is likely, and the initial allocation of usage rights should serve only as a default to be later rearranged through the market, and those in which the law should be less permissive of such rearrangement, because externalities and other market imperfections would impede socially efficient bargaining.

FROM WITHIN OR FROM WITHOUT?

Assuming that allocation of usage rights between copyright owners and users is desirable, let us turn to consider whether such allocation should preferably be done from within copyright law or rather by imposing external limitation through competition law. Theoretically, optimal allocation can be achieved by granting broad exclusive rights to owners initially, while relying on competition law to prevent them from being used anticompetitively. Alternatively, copyright law itself may choose to allocate narrower rights to owners and broader rights to users.

At first glance, a competition law approach might seem preferable, because generally competition law requires fact intensive rule of reason analysis of a particular challenged practice and its effects in specifically defined markets. Copyright law, even when it engages in market analysis (such as in fair use analysis), does not require the same rigor (Lemley and McKenna, 2012). This could suggest that using competition law as the preferable policy lever would achieve more precise outcomes, compared to copyright law's cruder analysis.

Nevertheless, competition law should not be seen as the preferable lever, because its methodology and procedures are ill suited for dealing with harm to dynamic competition and innovation that may result from copyright overreach. Methodologically, competition law's fundamental theory of harm is harm to static competition, not harm to innovation (at least not apart from harm to competition). Harm to innovation is inherently speculative and counterfactual: it is the harm resulting from the new ideas, products, and services that were not conceived or developed, and current competition law is ill equipped to analyze this harm in a manner sufficient to satisfy its standards of proof. Procedurally, competition law is complaint-driven. Even if its methodologies could develop to deal with harm to innovation, competition law cannot operate without a complainant. However, harm to dynamic competition and innovation can be diffuse and elusive; it may be suffered by many but no one in particular. This may lead to serious collective action problems and information gaps and asymmetries that limit competition law's capacity to effectively restrain copyright overreach.

In contrast, the main advantage of relying on internal copyright solutions is that this method regulates some uses of existing works as commons. Although this might seem to create a pro-competition bias, as discussed above, rarely will the limitation on copyright cover acts that directly undermine the incentive to create by permitting substitutive competition. The advantage of this approach is that it allows future innovators to compete freely without seeking permission or lobbying for government or judicial intervention.

CONCLUSION

This chapter discussed the tensions between copyright law and competition and some of the ways through which copyright law itself works to advance competition policy goals. It showed how competition policy goals and anti-monopoly measures have shaped the design of copyright since the Statute of Anne, and how the first British copyright act sought to encourage learning by granting limited exclusive rights operating within a competitive market system.

As copyright has expanded to cover new subject matter and new uses, so has doctrine evolved to ensure that the copyright may not result in excessive static losses resulting from unconstrained market power, and will not create dynamic losses by hindering future innovation. The chapter demonstrated how the limited term of copyright, limitations on subject matter, fair use, and the first sale doctrine attempt to achieve those goals. It also showed how a three-dimensional framework, consisting of considering incentive sufficiency, relative capacity to innovate, and transaction costs, can explain some key elements of the law, and provide guidance in its further development.

The digital revolution that we are currently experiencing alters again the ways in which knowledge is created, disseminated, and controlled, and like the print revolution may also bring about significant economic, social, and political changes. Copyright, as one of the legal institutions that regulate knowledge, may advance some changes and hinder others. So far, most recent amendments to copyright statutes aimed at adapting it to the digital age have focused on strengthening the rights of owners, without sufficient attention to how users' rights might be equally essential for achieving copyright law's ultimate purposes. Reflecting on the history of copyright and its internal mechanisms may help us to better understand the role that it should have in shaping our digital future.

REFERENCES

Anderman, S.D. and A. Ezrachi (eds) (2011), *Intellectual Property and Competition Law: New Frontiers*, New York: Oxford University Press.

Baldwin, C. and E. von Hippel (2009), 'Modeling a paradigm shift: from producer innovation to user and open collaborative innovation', available at: http://papers.ssrn.com/abstract=1502864 (accessed 22 February 2013).

Bell, T. (1998), 'Fair use vs. fared use: the impact of automated rights management on copyright's fair use doctrine', *North Carolina Law Review*, **76** (2), 557.

Benkler, Y. (2007), *The Wealth of Networks: How Social Production Transforms Markets and Freedom*, Newhaven, CT: Yale University Press.

Blagden, C. (1961), 'Thomas Carnan and the almanack monopoly', *Studies in Bibliography*, **14**, 23–43.

Bohannan, C. and H. Hovenkamp (2011), *Creation without Restraint: Promoting Liberty and Rivalry in Innovation*, New York: Oxford University Press.

Boldrin, M. and D.K. Levine (2004), 'Why Mickey Mouse is not subject to congestion: a letter on "*Eldred* and fair use"', *Economists' Voice*, **1** (2), available at: http://levine.sscnet.ucla.edu/general/intellectual/lp2.htm (accessed 2 May 2013).

Boyer, M., M.J. Trebilcock and D. Vaver (eds) (2009), *Competition Policy and Intellectual Property*, Toronto: Irwin Law.

Bracha, O. (2010), 'The adventures of the Statute of Anne in the land of unlimited possibilities: the life of a legal transplant', *Berkeley Technology Law Journal*, **25**, 1427–73.

Breyer, S. (1970), 'The uneasy case for copyright: a study of copyright in books, photocopies, and computer programs', *Harvard Law Review*, **84** (2), 281–351.

Buccafusco, C.J. and P.J. Heald (2012), 'Do bad things happen when works enter the public domain? Empirical tests of copyright term extension', available at: http://ssrn.com/abstract=2130008 (accessed 25 November 2012).

Conner, K.R. and R.P. Rumelt (1991), 'Software piracy: an analysis of protection strategies', *Management Science*, **37** (2), 125–39.

Cotter, T.F. (2006), 'The procompetitive interest in intellectual property law', *William and Mary Law Review*, **48**, 483–557.

Easterbrook, F.H. (1992), 'Ignorance and antitrust', in T.M. Jorde and D.J. Teece (eds), *Antitrust, Innovation, and Competitiveness*, Oxford: Oxford University Press, pp. 119–36.

Fisher, W.W. (1987), 'Reconstructing the fair use doctrine', *Harvard Law Review*, **101**, 1659, available at: http://cyber.law.harvard.edu/people/tfisher/IP/Fisher%20Fair%20Use%20Part%20V.pdf (accessed 2 May 2013).

Frischmann, B.M. (2012), *Infrastructure: The Social Value of Shared Resources*, New York: Oxford University Press.

Frischmann, B.M. and M.A. Lemley (2007), 'Spillovers', *Columbia Law Review*, **107**, 280.

Gans, J. (2012), *Information Wants to Be Shared*, Cambridge, MA: Harvard Business Review Press.

Gordon, W.J. (1982), 'Fair use as market failure: a structural and economic analysis of the Betamax case and its predecessors', *Columbia Law Review*, **82**, 1600–1659.

Gordon, W.J. (1998), 'Intellectual property as price discrimination: implications for contract', *Chicago–Kent Law Review*, **73**, 1367–90.

Gordon, W.J. (2010), 'Fair use markets: on weighing potential license fees', *George Washington Law Review*, **79**, 1814.

Heald, P.J. (2007), 'Property rights and the efficient exploitation of copyrighted works: an empirical analysis of public domain and copyrighted fiction bestsellers', *Minnesota Law Review*, **92**, 1031–86.

Heller, M.A. and R.S. Eisenberg (1998), 'Can patents deter innovation? The anticommons in biomedical research', *Science*, **280** (5364), 698–701.

Karjala, D.S. (2006), 'Congestion externalities and extended copyright protection', *Georgetown Law Journal*, **94** (4), 1065–86.

Katz, A. (2005a), 'A network effects perspective on software piracy', *University of Toronto Law Journal*, **55** (2), 155–216.

Katz, A. (2005b), 'The potential demise of another natural monopoly: rethinking the collective administration of performing rights', *Journal of Competition Law and Economics*, **1** (3), 541–93.

Katz, A. (2006), 'The potential demise of another natural monopoly: new technologies and the administration of performing rights', *Journal of Competition Law and Economics*, **2** (2), 245–84.

Katz, A. (2007), 'Making sense of nonsense: intellectual property, antitrust and market power', *Arizona Law Review*, **49** (4), 837–909.

Katz, A. (2009a), Commentary: is collective administration of copyrights justified by the economic literature?',

in M. Boyer, M. Trebilcock and D. Vaver (eds), *Competition Policy and Intellectual Property*, Toronto: Irwin Law, pp. 449–68, available at: http://ssrn.com/abstract=1001954 (accessed 25 August 2013).
Katz, A. (2009b), 'Copyright collectives: good solution but for which problem?', in H. First, R.C. Dreyfuss and D.L. Zimmerman (eds), *Working within the Boundaries of Intellectual Property: Innovation Policy for the Knowledge Society*, Oxford: Oxford University Press, pp. 395–430, available at: http://ssrn.com/abstract=1416798 (accessed 2 May 2013).
Katz, A. (2009c), 'Substitution and Schumpeterian effects over the life cycle of copyrighted works', *Jurimetrics*, **49** (2), 113–53.
Katz, A. (2012), 'What antitrust law can (and cannot) teach about the first sale doctrine', available at: http://papers.ssrn.com/abstract=1845842 (accessed 7 March 2013).
Katz, A. (2013), 'Fair use 2.0: the rebirth of fair dealing in Canada', in M. Geist (ed), *The Copyright Pentalogy: How the Supreme Court of Canada Shook the Foundations of Canadian Copyright Law*, Ottawa, ON: Ottawa University Press, pp. 93–156, available at: http://papers.ssrn.com/abstract=2206029 (accessed 25 August 2013).
Katz, A. and P.-E. Veel (2013), 'Beyond refusal to deal: a cross-Atlantic view of copyright, competition and innovation policies', *Antitrust Law Journal*, **79** (1), 201–46.
Landes, W.M. and R.A. Posner (2003a), 'Indefinitely renewable copyright', *University of Chicago Law Review*, **70**, 471.
Landes, W.M. and R.A. Posner (2003b), *The Economic Structure of Intellectual Property Law*, Cambridge, MA: Harvard University Press.
Lemley, M.A. and M.P. McKenna (2012), 'Is Pepsi really a substitute for Coke? Market definition in antitrust and IP', *Georgetown Law Journal*, **100**, 2055.
Lévêque, F. and H.A. Shelanski (eds) (2005), *Antitrust, Patents, and Copyright: EU and US Perspectives*, Cheltenham, UK and Northampton, MA, USA: Edward Elgar Publishing.
Liebowitz, S.J. (1985), 'Copying and indirect appropriability: photocopying of journals', *Journal of Political Economy*, **93** (5), 945–57.
Liebowitz, S.J. (2005), 'Economists' topsy-turvy view of piracy', *Review of Economic Research on Copyright Issues*, **2** (1), 5–17, available at: http://papers.ssrn.com/abstract=738864 (accessed 8 February 2013).
Netanel, N. (2008), *Copyright's Paradox*, New York: Oxford University Press.
Nicita, A. and G.B. Ramello (2007), 'Property, liability and market power: the antitrust side of copyright', *Review of Law and Economics*, **3** (3), 767–91.
Ochoa, T.T. and M. Rose (2001), 'Anti-monopoly origins of the Patent and Copyright Clause', *Journal of the Copyright Society of the U.S.A.*, **49**, 675.
Patterson, L.R. (1968), *Copyright in Historical Perspective*, Nashville, TN: Vanderbilt University Press.
Pollock, R. (2008), 'Optimal copyright over time: technological change and the stock of works', available at: http://papers.ssrn.com/abstract=1144288 (accessed 22 February 2013).
Radin, M.J. (2003), 'Regime change in intellectual property: superseding the law of the state with the law of the firm', *University of Ottawa Law and Technology Journal*, **1**, 173.
Raustiala, K. and C.J. Sprigman (2012), *The Knockoff Economy: How Imitation Sparks Innovation*, New York: Oxford University Press.
Sag, M. (2012), 'Predicting fair use', *Ohio State Law Journal*, **73** (1), 47–91.
Schumpeter, J.A. (1994), *Capitalism, Socialism, and Democracy*, New York: Routledge.
Takeyama, L.N. (1994), 'The welfare implications of unauthorized reproduction of intellectual property in the presence of demand network externalities', *Journal of Industrial Economics*, **42** (2), 155–66.

Statutes and Law Cases

Authors Guild, Inc. v. HathiTrust, 902 F.Supp.2d 445 (S.D.N.Y.) (2012).
Copyright Act, 1976 (US).
Feist Publications, Inc. v. Rural Telephone Service Co., 499 U.S. 340 (1991), available at: http://www.law.cornell.edu/copyright/cases/499_US_340.htm (accessed November 29, 2012).
Statute of Anne, An act for the encouragement of learning, by vesting the copies of printed books in the authors or purchasers of such copies, during the times therein mentioned. *8 Anne c.19, 1710 (UK)*.

FURTHER READING

I have discussed copyright and competition policy in several of my previous writings. For example, in Katz (2009c), I show how time changes the relative importance of static competition and dynamic competition over the lifetime of works, and how it may justify adjustments in the scope of copyright protection even within the term of protection. In Katz (2005a), I show how piracy may not always be detrimental to copyright owners, and how tolerating it has been a method for obtaining market dominance in the presence of network effects. This counterintuitive aspect of piracy has also been identified by Takeyama (1994), and Conner and Rumelt (1991), and is part of a larger literature on indirect appropriability (Liebowitz, 1985) that has identified that unauthorized reproduction may at times be beneficial to copyright owners, and to innovation (Raustiala and Sprigman, 2012). For a more skeptical view by one of the founding fathers of this literature see Liebowitz (2005). An area of perennial tension between copyright and competition policy that I have not discussed in this chapter is collective administration of copyright. Readers who are interested in a skeptical view about the practice are encouraged to read Katz (2005b, 2006, 2009a, 2009b, 2012). In Katz (2012), I discuss the first sale doctrine in greater detail, in Katz (2013) I discuss fair use and fair dealing, and in Katz and Veel (2013) my co-author and I discuss how the US tends to incorporate competition policy goals within copyright law while the EU has opted for greater reliance on competition law as a means for addressing copyright overreach.

20. Collective copyright management

Reto M. Hilty and Sylvie Nérisson

Collective copyright management organizations appeared in the nineteenth century. They developed an efficient way to enforce copyrights in situations in which right holders and users were too numerous for authors to do so themselves. Collective rights management organizations (CRMOs) were pioneers. During the twentieth century, notably with the appearance of new technologies – such as radio and magnetic tapes – an exponential development occurred. CRMOs extended their activities into all creative domains and increasingly enjoyed natural (sometimes even legal) and long undisputed monopolies.

The benefits of the CRMOs' ways of managing copyrights became so obvious that some legislatures established mandatory involvement of CRMOs for the exploitation of certain rights.[1] This strong position, taken together with the close link between CRMOs and the copyright-based industries, notably publishers, who became (powerful) members of most of them, over time reduced the ability of CRMOs to adapt to situations that have appeared since the emergence of digitization and the Internet. Recent initiatives of the European Commission have attempted to encourage the necessary adaptations to digitization and the Internet. Unfortunately, the contradictory approaches of two of its Directorates-General (DGs) have increased the difficulties.

The question has arisen whether digitization is leading to a decline in the collective management of copyrights in view of the emergence of individually applicable technical protection measures established by the copyright-based industries (Handke and Towse, 2007; Hansen and Schmidt-Bischoffshausen, 2007; Ricolfi, 2007). But CRMOs may yet provide for solutions to the copyright paradox (Gervais, 2010b), which remains even in the online world. By the copyright paradox is meant that copyright consists in a right to prevent the dissemination of protected works in order to incentivize this dissemination (see Netanel, 2008).

This chapter gives an overview of the impact of digitization on the current activities of CRMOs and provides perspectives on the future of collective rights management in Europe.[2] First, an introduction shows how the traditional system works, namely, by limiting the freedom of contract of right holders and users on the one hand and by reducing the exclusive rights to an entitlement to remuneration for use on the other.

The next section presents the turning point of the legal framework applicable to CRMOs on the European level, a move initiated by two of the Commission's DGs: Internal Market and Services (DG Market) and Competition (DG Comp). The following section examines the current state of collective rights management in the music and visual domains.

The chapter ends with the conclusion that the collective rights management system is currently adapting to digitization *notwithstanding* the interventions by DG Market, especially in the music sector. In fact, the availability of copyrighted works will heavily rely on collective rights management in the future as well – if not more than ever.

THE TRADITIONAL SYSTEM OF COLLECTIVE RIGHTS MANAGEMENT AND ITS MODIFICATIONS DUE TO EUROPEAN CASE LAW

'Collective rights management' is not the management of collective rights – there are no collective rights in the copyright field – but the collective management of individual copyrights, which are transferred to the CRMOs for this purpose. 'Collective management of copyright' covers the granting of non-exclusive licences for the use of works whose management has been entrusted to the CRMO (these works constitute the repertoire of the CRMO), the collection of royalties on behalf of right holders from the users of copyright-protected works of the repertoire of the CRMO, the distribution of this income among the respective right holders, the monitoring of uses licensed, and when necessary legal actions against infringement of the copyrights to which the CRMO holds title. For what are called CRMOs in this chapter, other names are common as well, such as 'collecting societies', 'authors' societies', 'copyright collectives' and so on.

If the term 'collective' refers to 'management', individual management in contrast covers the cases in which the first right holder (the author, performer or producer) or the derivative right holder (a publisher or an agent) directly contracts (licences) with a user. The motivation for individual management may be the management of scarcity, in order to increase the price of the access to specific works, or on the contrary it may be the attempt to flood a market in order to accustom the audience to a song or a performer. In contrast, such cherry-picking is not allowed by collective management. What is collective about this management is the representation, the defence and the enforcement of the rights of all its members[3] equally and via single tariffs.

How the System Works

Individual contracting in relation to copyrighted works is traditionally possible on the primary market only: the writer contracts with a publisher; the lyricist, the composer and the performer of a song deal with a music publisher or a record company, and so on. In contrast, single licensing on the secondary market for copyrighted works, related to the use of products of the primary market (for example, public performance of phonograms or DVDs, lending books or broadcasting), is generally not feasible. This is due to the quantity of works used and the sheer numbers of right holders and users concerned. Individualized contracts are not only unmanageable in practice. The bargaining power is also skewed, a single author having no leverage with established public broadcasters, public libraries or content providers. Further, from a basic cost-analysis point of view, the bundling of rights in an international repertoire via CRMOs helps to economize on transaction costs involved in contracting and monitoring.[4]

Hence collective management is worthwhile for both groups that contract with CRMOs: for right holders (whether authors, performers and producers, or derivative right holders, like publishers and heirs of initial right holders) pooling their rights in CRMOs, these act as an effective tool to bargain and enforce licences; and, for users accessing licences in all works of the respective repertoire, CRMOs act as a one-stop shop and according to established tariffs. Thus their attractiveness to both sides – taken together with the economies of scale – means CRMOs enjoy a natural monopoly on both markets in which they are active: the market for collective management services for right holders and the market for the grant of licences to users.

This position is further buttressed by the fact that CRMOs enter into reciprocal representation agreements with similar CRMOs active in their respective countries. They represent each other in licensing and monitoring their respective repertoire. This makes it possible for every CRMO in the network to license rights in works of all members of all other CRMOs, that is, on a global scale; vice versa, users obtain licences to the use of works from the members not only of the CRMO they contract with but also of all CRMOs of the reciprocal representation network. From the right holders' point of view this representation system enables them, simply by entrusting rights to one CRMO, to grant the management of their rights (that is, licensing and monitoring) in uses of their works all over the globe.

The power CRMOs possess lies in the economies of scale they generate and the immense repertoires they license. This implies a critical size. Further, the network effects help CRMOs to become indispensable and therefore to impose their bargaining power when they enter into contracts with users. The cost of this irresistible force is that the collective system does not leave any room for individual treatment: either for a customized entrustment of rights to CRMOs or for customized licences to users. In other words, the price of the CRMOs' efficiency is the limitation of the freedom of contract of right holders and users.

Limitation of Right Holders' and Users' Freedom of Contract

European case law related to antitrust explains the evolution of collective management during the past 40 years. Limitations to the freedom of contract have been challenged before the European Commission several times. In fact, according to this case law, they may be deemed as abuses of a dominant position or as distortions of competition by a cartel.[5] These cases are interesting because they help one to understand the challenges currently faced by CRMOs in adapting to the digital environment.

Let us first consider the impacts of the limitation of freedom of contract, as required by the CRMOs to strengthen their bargaining position vis-à-vis right holders.

CRMOs require from their members the assignment[6] of bundles of rights in all their works, including in their future works, for a long period. Further, if right holders withdraw their (or some of their) rights from a CRMO, they are required to entrust these rights to another CRMO.

The CRMOs justify this impossibility of selecting individual works for rights assignment in terms of the economic savings they achieve: the costs of monitoring rights individually would be much higher than the so-called blanket licence system.

The long term of the entrustment and the requirement to leave the rights under all circumstances within the frame of collective management is justified by the need for a stable repertoire, and by the paternalistic protection it offers to single authors. Owing to the network of reciprocal representation, the repertoire for which users get a licence is the same, whichever organization right holders entrust their rights to, and that offers a legal certainty to users entering a contract with a CRMO. As a result of the difficulty or the impossibility of withdrawing the rights in individual works, authors are protected against the economic pressure a user could exert on an author in order to get a rebate or a free licence.

In other words this paternalistic protection considers the position of early-stage authors, who lack strong bargaining power, and assumes therefore that the CRMO has stronger bargaining power.[7] This latter assumption relies on the global repertoire that CRMOs manage based on mutual representation. Consequently even if the tariffs appear too high and a discotheque or a radio station tries to get a cheaper licence for rights, for example for only a specific part of the repertoire, it will have to deal with the CRMO that is active on its territory.

The reverse of the coin of the strong bargaining power of CRMOs is that successful right holders give up the better contract terms they probably could negotiate themselves individually. Additionally, authors and performers seeking an audience lose the possibility of offering their works free. This solidarity – and the related limitation of the freedom of contract – is the price to pay for a certain pooling of risks of the members of a CRMO,[8] or for unionization. In the long term most members gain from economies of scale. Only multinational players, who in effect could afford their own management department, lose because of this system.

This is the model schema. We now look at the modifications that European authorities have required in the course of time.

Modifications Induced by European Case Law

The overruling principle of the European authorities is the freedom of the right holder to dispose of his or her rights, a freedom limited only by what is necessary for the efficient management of the rights.[9] The need to specify such necessary limitations led to a list of rights categories among which right holders must be free to determine which are to be collectively managed and which should remain outside the scope of the collective management.[10] Later the European Court of Justice (ECJ) recognized that the freedom of right holders may be limited for the sake of efficient management. Consequently the need for a strong position vis-à-vis users justifies the right holders' locked-in collective management and the impossibility of selecting among works.[11] Consequently for right holders the choice is as follows: entrustment to the CRMO of rights in all their works – or no collective management at all. This position still applies to the collective management of rights for analogue uses.

The Commission stated later that CRMOs may not discriminate against right holders from other member states.[12] Since then, competition has been established between CRMOs in the market for collective rights management services to right holders. In the 1989 *Tournier* and *Lucazeau* decisions of the ECJ, discotheques challenged the bundle of works and the tariffs set by the French CRMO for authors' rights in music, SACEM. Beyond that, the impossibility of choosing the scope of licences was seen as an abuse of a dominant position. However, judges of the ECJ permitted the practices of SACEM because they were the only way to 'entirely safeguard the interests of the authors, composers and publishers of music without thereby increasing the costs of managing contracts and monitoring the use of protected musical works'.[13]

The adaptation to the online and digitized environment was first discussed by the European Commission with regard to a claim raised by the electronic music band Daft Punk. The European Commission stated that technological developments had altered the balance achieved in the twentieth century between collective and individual management. Accordingly, the rule that a right holder cannot withdraw his or her rights from a CRMO in order to manage them on an individual basis should be reconsidered for the online environment. Nevertheless the Commission permitted that the withdrawal should depend on the assent of the CRMO from which the rights are to be withdrawn. This control for one thing serves the timely notification of sister organizations, which have to consider this modification of the repertoire they can license. Secondly, on the paternalistic line, this control aims at ensuring that the decision is not motivated by the economic pressure a commercial agent could exert on the concerned authors.[14] Lastly, since this decision, most rights in online uses may be entrusted to a CRMO independently of other categories of rights.

From Exclusive Rights to an Effective Remuneration

A consequence of the collective management of a right is that the right holder is deprived of the possibility of authorizing, prohibiting or negotiating uses of rights he or she has entrusted to a CRMO. Only the latter may do so. Hence creators may neither allow their works to be played at a charity concert nor make a new work available free. Even worse, they may not play their own works in public without paying royalties to the CRMO.[15]

On the other hand, collective management grants that royalties reach the original right holder (instead of derivative owners, that is, those in the copyright-based industries) without blocking access to works. This is beneficial, especially when an exclusive right would hardly be enforced.

Mandatory collective management particularly follows this remuneration-plus-access objective as regards remuneration rights. The most widespread example is the compensation fee for public lending.[16] Mandatory collective management also helps to preclude potential abuses of the exclusive feature of copyright (Hilty, 2012: 32).

In a nutshell, collective management of copyright facilitates the fundamental shift in copyright from a half-functioning system based on proprietary rules to an efficient system based on liability rules (Schovsbo, 2012; Hilty, 2012), or in other words from an excluding system to a 'business of yes' (Gervais, 2011).

THE TURNING POINT: THE EFFECT OF INTERNET AND DIGITIZATION ON THE COLLECTIVE MANAGEMENT OF COPYRIGHT ACCORDING TO THE EUROPEAN COMMISSION

An overview of the steps taken by the European Commission since 2002 illustrates its approach. The decisions of European authorities design the legal frame in which CRMOs develop their services to adapt to the emergence of a demand for uses of copyrighted works across national borders.

The *IFPI Simulcasting* Decision (2002): For Better Competition on the Users' Side

Digitization and the Internet challenge the collective management of copyright because CRMOs developed their model of management on a national basis. Until the 1990s they offered multi-repertoire licences for mono-territorial use. Since the development of the Internet, online content providers demand multi-territorial licences. In 2002 the European Commission assessed the model for reciprocal representation agreements for related rights regarding online music services (simulcasting and webcasting) that had been submitted by the International Federation of the Phonographic Industry (IFPI). The Commission exempted this agreement from the prohibition of Article 101(2) of the Treaty on the Functioning of the European Union (TFEU) on the grounds that it presented 'a number of pro-competitive elements which may significantly contribute to technical and economic progress in the field of collective management of copyright and neighbouring rights'.[17]

This agreement model was remarkable. It aimed to safeguard the traditional system of a one-stop shop for multi-repertoire licences while adding the crucial feature that online use can be multi-territorial. The amendment required by the Commission, however, was the suppression of the 'customer allocation clause'. This clause forbade CRMOs to enter into a licence contract with a user located on the territory of another CRMO. Since the Internet enables the monitoring of uses from any computer connected, the national restriction that was valid for offline uses[18] infringed the prohibition of concerted practices because it was not necessary for efficient management. This suppression enabled competition between CRMOs on the market for the grant of licences to users.

In contrast the CRMOs of the International Confederation of Societies of Authors and Composers (CISAC), the federation of CRMOs handling musical authors' rights, refused to amend the customer allocation clause of their model agreement.[19] Consequently the reciprocal representation agreements on authors' rights lapsed and have not been extended. Since then, there has been no telling whether such agreements for online uses exist and, if they do exist, what their terms are.

The Recommendation of the DG Market (2005): Forcing Competition on the Right Holders' Side

In 2005 the DG Market of the European Commission launched a Recommendation (non-binding according to EC Treaty Article 211 then applicable) that considered the progress of the *IFPI Simulcasting* model agreement as unsatisfactory and chose to promote competition between CRMOs on the market for the management of online music rights. This Recommendation raised several fundamental – and institutional – concerns and was heavily criticized.[20] The biggest substantive failure of the Recommendation was seen in the fact that the Commission ignored the delicate balances achieved by the European law based on the long practice of CRMOs: the case law acknowledged the virtues of reciprocal representation networks;[21] and certain EU directives ordered the defence of the peculiar interests of creative right holders against those of investors.[22]

The Recommendation in contrast requires 'right holders' to be free to determine which rights to entrust for which territories and uses, and therefore to be free to withdraw their rights from CRMOs without any condition. This mainly ignores the differences between categories of right holders who are members of CRMOs active in the music branch.

The consequence of this unlimited freedom was that the most powerful right holders – that is, the major multinational music publishers – withdrew from regular CRMOs their mechanical rights in the works from the UK and USA. Instead, the major music publishers entrusted these rights to 'hybrid entities' processing individual management. On the one hand, these entities include major music publishers with the biggest CRMOs of Europe; on the other hand, their management is not collective, since each entity works exclusively for pan-European licences of works of the major publisher it represents.

The beneficiaries of the Recommendation therefore were only those right holders who could afford a tailor-made management of their rights and gained the know-how and databases of CRMOs. All other right holders, however, lost out. Most authors and small publishers are not able to compare and evaluate the conditions offered by the different CRMOs in Europe, nor can they afford lawyers and business managers to do so. Small CRMOs on the other hand lost the economies of scale, since they lost the most successful part of the repertoire.

The worst effect, however, is on the users' side. Before the Commission intervened, the repertoire was stable, international and diverse, and the licensing of rights to provide access to a certain repertoire therefore was comparatively easy. After the change effected by the Recommendation, users risked infringing copyrights, since too many right holders with uncertain representatives are involved: users need licences from each 'hybrid entity' for the rights of the three major publishers in the Anglo-US repertoire and licences from a regular CRMO for the 'remaining repertoire'. With regard to the latter, the question remains whether the CRMO when approached will be able to offer a multi-repertoire and multi-territory licence, since the existence of reciprocal representation agreements today is largely unclear.

The 2012 Directive Proposal: Sticking Plaster on a Wooden Leg?

In July 2012 the DG Market published a proposal for a Directive on collective management.[23] A binding instrument as such would be very welcome. However, the proposal is based on the same mistaken assumptions as the Recommendation. Some aspects will be skimmed over below, but details will not be addressed here.[24]

At the end of the day, the DG Market with its Recommendation and its proposal succeeded only in increasing uncertainty. Hope remains in the legislative work of the EU Parliament on the proposal. The target should be that content providers are enabled to offer solid economic models and end-users to access wide repertoires.

THE CURRENT SITUATION

The Music Sector: Development of Rights Managers and Fragmentation

The music sector used to be the forerunner as regards copyright management. This situation changed abruptly with the above-mentioned Recommendation of 2005, when the music sector became the guinea pig of the DG Market. As a result, the market for licences in online use of music now consists of fragmented repertoires in which rights can be acquired.

Reciprocal representation agreements would alleviate that fragmentation, but they are established only for neighbouring rights. Consequently users can get pan-European licences from any European CRMO in the complete repertoire of recordings whose rights are administered by one CRMO that is active for the respective right holders (performers and producers of recordings). But licences in neighbouring rights are not sufficient for use of a recording unless the work is usable without authorization of the author's rights holder (that is, it is in the public domain or subject to an open licence).

Thus repertoire fragmentation and the instability that followed the Recommendation are a problem: regarding authors' rights and in contrast to neighbouring rights, the situation is uncertain and complex. What we know is that licences for online copyrights for pan-European uses of Anglo-US repertoires of the major music publishers have to be acquired from the afore-mentioned 'hybrid entities': for the repertoire of Universal Music Publishing licences are offered by DEAL, a partnership between Universal and SACEM; for EMI, the agent is CELAS, a joint venture owned by GEMA and PRS for Music (the German and English CRMOs for copyrights in music); the repertoire of Sony is offered by PAECOL, similar to CELAS, owned only by GEMA; lastly for Warner/Chappell licences are offered by one of the six following CRMOs: PRS for Music in the UK, STIM in Sweden, SACEM in France, SGAE in Spain, SABAM in Belgium and Buma/Stemra in Holland, according to the PEDL system, which relies on non-exclusive licences to these CRMOs.[25] However, it is uncertain who licenses these repertoires for mono-territorial use.

For 'the rest' one can only assume that some reciprocal representation agreements have been concluded between CRMOs (Müller, 2009), but no one knows their content. Thanks to the most probable existence of these reciprocal representation agreements between national CRMOs, the latter currently offer licences for online pan-European uses of the whole repertoire except the Anglo-US repertoire of the three majors. The validity of these licences however depends on the validity of reciprocal representation agreements, about which there is no legal certainty.

This heterogeneity is far from user-friendly. And it is fragile. The existence of CELAS for instance is questioned by a case pending in Germany. An online service provider had acquired a pan-European licence from GEMA, and provided access to copyrighted works. CELAS sued the provider for infringement of mechanical rights by using recordings of EMI – but without a licence from CELAS. The first and second instances dismissed the claim and considered that the individual licence of a mechanical right alone is not possible because it makes no sense economically.[26] Therefore the existence of CELAS, as possibly of other 'hybrid entities', depends on the final decision in this case.

This example shows how unstable the system is for the CRMOs, but from the end-user point of view the system simply does not work. Several video clips made available by their creators and/or performers themselves are not available on YouTube for users sending their request from German Internet accounts, for instance, whereas they can be viewed in France or other countries. The reason has its roots in negotiations between YouTube and right holders. But it is unclear who those right holders are that are refusing licences. As of January 2013, requests for unavailable clips led to messages blaming GEMA. GEMA has sued against this misinformation.[27] Only two things seem clear: Pan-European licences are not easy to get; and the support for cultural diversity which was a side-effect of blanket licensing systems is lost as regards the online exploitation of music, since the most rentable repertoire has flown the nest (Gilliéron, 2006; ELIAMEP, 2009; Mazziotti, 2011).

The Attempts of CRMOs to Achieve More Flexibility: The Answer of Established CRMOs to Creative Commons Licences

On 1 January 2012 SACEM and Creative Commons France launched a project[28] allowing SACEM members to apply for non-commercial licences for some of their music works. This possibility of individual treatment of single works is a big step. This is the fourth such project in the EU, following Denmark, the Netherlands and Sweden.[29] In contrast GEMA still refuses to offer such an option and focuses on the solidary and paternalistic approach instead.[30] Consequently another organization, named C3S (for Cultural Commons Collecting Society), is currently trying to capture the demand for open uses and aims to get the administrative authorization to provide collective management services in Germany.[31]

In sum, licensing in music is currently in a transition phase in which nothing is sure or predictable.

By contrast, one sector of the collective management of copyright has achieved success in the transition from analogue to digital: the visual domain.

The OnLineArt System

OnLineArt[32] offers a portal representing most CRMOs active in that area. It acts as a one-stop shop able to directly grant blanket licences for worldwide online uses of the works of 40000 painters, photographers and other visual artists, according to tariffs available on its website. The organization offers individually negotiated licences as well. The latter is not new in the visual domain: CRMOs already had to deal with issues related to *droit moral* regarding analogue uses, and worked very closely with authors, combining the bargaining power of a CRMO and personal contact with the creator.

COLLECTIVE MANAGEMENT, THE SOLUTION TO THE COPYRIGHT PARADOX

The copyright paradox is the fact that the aim of copyright is to incentivize the dissemination of protected works and that the means to reach this aim is a right to block dissemination. Collective management helps to resolve this dysfunctional situation; the involvement of CRMOs transforms the right to prohibit into a right to fairly negotiate the price of the licence, while the authorization to use works is systematically granted.

To believe that new possibilities based on digitization could bring about the end of CRMOs is to ignore the broad scope of their activities. Digitization aids data processing, and the Internet aids communication. Computers, however, do not have the know-how required to achieve a balance of all interests involved by negotiating the conditions of licences, setting tariffs, monitoring uses and enforcing rights in cases of infringement. Considering that digitization and the Internet could ever replace such skills is naive.

The legislature however should stimulate and foster efforts of CRMOs to adapt to new uses and to provide better services instead of intervening in effective systems of reciprocal representation – thereby exacerbating the fragmentation of repertoires and increasing the difficulties of rights clearance. What should be addressed in the first instance in the European law is antitrust problems specific to CRMOs. Instead the Directive proposal on collective management risks transforming CRMOs into suppliers for the copyright-based industries. This fails to address both the concerns related to competition law and the proper aim of copyright law to promote creativity and the above-mentioned balance.

Concerning business models of the dissemination of works, digitization increases the possibilities of controlling and preventing the uses of works via the application of technical protection measures. But it also allows the non-commercial dissemination of contents via the Internet, since the related costs are substantially lower compared to traditional publication systems.

Collective management may have a role to play in both situations.

Digitization enables CRMOs themselves to process individual uses. In fact, if this technology is used to enhance information on works (concerning right holders and licence terms) CRMOs can play a role in centralizing and monitoring information as well as in granting fair and efficient distribution (not neglecting the defence of creators' interests against the copyright-based industries). This is currently the case with the management of rights involved for example in YouTube. Needless to say, such business models are 'work in progress'.

Lastly, CRMOs could develop tools to enhance non-commercial uses of protected works thanks to micropayments and blanket licences. This would enable the compensation of creative people and at the same time allow the end-users sufficient access to copyrighted works (Gervais, 2011; Hilty, 2012). Extending cases in which copyrights were to be statutorily collectively managed would help in this respect, as well as possibly resolving the fragmentation issue and enhancing the legitimacy of public authorities to monitor the inner governance of CRMOs.

NOTES

1. See for example the systematically mandatory collective management for statutory claim for remuneration compensating limitations to copyright in Germany, and, on the European level: the mandatory collective management of the cable retransmission right (Satellite and Cable Directive, Art. 9) and the mandatory collective management of fees arising after the extension of the related rights of performers (Duration Directive, new Art. 3, para. 2b); the possibility given to European Union member states to impose the collective management of the unwaivable right to obtain remuneration, of the rental right (Rental and Lending Directive, Art. 5) and of the resale right (Resale Right Directive, Art. 4). Full references to all texts and decisions mentioned or quoted in this chapter are to be found in the References below.
2. For an overview of other systems, especially in the US, see Gervais (2011).
3. Licences granted by CRMOs cover the repertoire of all members of the given CRMO as long as the collective management relies only on contracts. As soon as the legislature anchors the intervention of a CRMO in the law, making the collective management of a specific right mandatory, this CRMO offers licences on the global repertoire (see, for instance, Strowel, 2011).
4. For a presentation of this rationale from a historical perspective, see Kretschmer (2002: 127). For a discussion of the primary and secondary uses, the role of CRMOs and the current switch from analogue to digital, see Ricolfi (2007).
5. Respectively banned by Arts 102 and 101 TFEU, previously Arts 86 and 85 of the Rome Treaty, then Arts 82 and 81 of its revised version of 1997.
6. 'Assign', 'mandate' or 'entrust', depending on the copyright law system of the country and of the choices made by the respective CRMOs.
7. *SABAM* decision of the ECJ (1974).

8. For a study of the risk-sharing benefits in CRMOs, and paths to improve them, see Snow and Watt (2005).
9. See *SABAM* decision at 10 and 15.
10. See *GEMA* decision of the European Commission (1971) at 23.
11. See *SABAM* decision at 7–11.
12. *GVL* case, 1981.
13. *Tournier* case, para. 3 of the operative part.
14. See *Daft Punk* decision (2002), pp. 10–12.
15. Gilliéron (2006: 947–8).
16. See Recital 12 and Art. 5(4) Rental Right Directive. Note that EU law sometimes even makes it mandatory that certain *exclusive* rights be managed by CRMOs: Art. 9(1) Satellite and Cable Directive obliges the member states to provide that the right to grant or refuse retransmission by cable can be exercised only through a CRMO. However, in this case the aim was rather to prevent single right holders from hurdling the ban on exploitation of the cable rights; see Recital 28 of the Directive.
17. *IFPI Simulcasting* decision, para. 84.
18. *Tournier* decision.
19. See the *Santiago Agreement* decision and the later *CISAC* decision. An appeal of this decision was pending at the time of writing in 2012. The General Court partially annulled on 12 April 2013 the latter Commission decision in respect of the finding of the concerted practice.
20. See Drexl (2007), Ricolfi (2007), Hilty (2010) and the comment of the Max Planck Institute for Intellectual Property (in German), available at: http://www.ip.mpg.de/files/pdf2/Stellungnahme-LizenzierungMusik1.pdf. The comment of the Max Planck Institute on the Directive proposal of 2012 sums up the main criticism; see paras 46 et seq., available at: http://www.ip.mpg.de/files/pdf2/Max_Planck_Comments_Collective_Rights_Management.pdf.
21. See *SABAM* and *IFPI Simulcasting*.
22. On mandatory collective management in Europe see note 1. The *Luksan* decision of the ECJ confirmed this approach; see para. 100.
23. COM(2012) 372.
24. See the comments of the Max Planck Institute for Intellectual Property and Competition Law.
25. For details, see http://www.warnerchappell.com/pedl/pedl.jsp and Mazziotti (2011: 13–14).
26. Munich District Court I, 25 June 2009, Case 7 O 4139/08, *ZUM* (2009: 788); Munich Court of Appeal, 29 April 2010, Case 29 U 3698/09, *ZUM* (2010: 709) (confirming the judgment of the District Court). With a more differentiated approach, Hilty (2010). An appeal is currently pending before the German Supreme Court.
27. See https://www.gema.de/presse/pressemitteilungen/presse-details/article/youtube-verhandlungen-gema-reicht-antraege-bei-schiedsstelle-ein.html.
28. See http://www.sacem.fr/cms/home/la-sacem/derniers-communiques-2011/sacem-creative-commons-signent-accord-diffusion-oeuvres.
29. See http://creativecommons.org/weblog/entry/31205.
30. See http://web.archive.org/web/20071005064532/http://www.gema.de/presse/briefe/brief59/wizards-of-os.shtml.
31. See www.c-3-s.eu.
32. See http://www.onlineart.info/.

REFERENCES

Drexl, J. (2007), 'Competition in the field of collective management: preferring "creative competition" to allocative efficiency in European copyright law', in P. Torremans (ed.), *Copyright Law: A Handbook of Contemporary Research*, Cheltenham, UK and Northampton, MA, USA: Edward Elgar Publishing, pp. 255–82.

Drexl, J., S. Nérisson, F. Trumpke and R.M. Hilty (2013), *Comments of the Max Planck Institute for Intellectual Property and Competition Law on the Proposal for a Directive of the European Parliament and of the Council on Collective Management of Copyright and Related Rights and Multi-Territorial Licensing of Rights in Musical Works for Online Uses in the Internal Market COM(2012)372*, available at: www.ip.mpg.de/files/pdf2/Max_Planck_Comments_Collective_Rights_Management.pdf.

ELIAMEP (Hellenic Foundation for European and Foreign Policy) (2009), *Collecting Societies and Cultural Diversity in the Music Sector: Study for the European Parliament*, available at: www.eliamep.gr/en/european-integration/collecting-societies-and-cultural-diversity-in-the-music-sector/cultural-diversity-imperilled-final-results-of-the-study-collecting-societies-and-cultural-diversity-in-the-music-sector/.

Gervais, D. (ed.) (2010a), *Collective Management of Copyright and Related Rights*, Alphen aan den Rijn, The Netherlands: Kluwer Law International.

Gervais, D. (2010b), 'Collective management of copyright: theory and practice in the digital age', in D. Gervais (ed.), *Collective Management of Copyright and Related Rights*, Alphen aan den Rijn, The Netherlands: Kluwer Law International, pp. 1–28.

Gervais, D. (2011), 'Keynote: the landscape of collective management schemes', *Columbia Journal of Law and the Arts*, **34** (4), Vanderbilt Law and Economics Research Paper No. 11–46, available at: http://ssrn.com/abstract=1946997 (accessed 2 May 2013).

Gilliéron, P. (2006), 'Collecting societies and the digital environment', *International Review of Intellectual Property and Competition Law*, **37** (8), 939–69.

Handke, C. and R. Towse (2007), 'Economics of copyright collecting societies', *International Review of Intellectual Property and Competition Law*, **38** (8), 937–57.

Hansen, G. and A. Schmidt-Bischoffshausen (2007), 'Economic functions of collecting societies: collective rights management in the light of transaction cost and information economics', available at: http://www.ip.mpg.de/shared/data/pdf/1_hansen_schmidt-bischoffshausen_-_economic_functions.pdf (draft version), also published in German in *GRUR Int*., 2007, 461–81.

Hilty, R.M. (2010), 'Kollektive Rechtewahrnehmung und Vergütungsregelungen: Harmonisierungsbedarf und -möglichkeiten', in M. Leistner (ed.), *Europäische Perspektiven des Geistigen Eigentums*, Tübingen: Mohr Siebeck, pp. 123–66.

Hilty, R.M. (2012), 'Individual, multiple and collective ownership: what impact on competition?', Max Planck Institute for Intellectual Property and Competition Law Research Paper No. 11–04, available at: http://papers.ssrn.com/sol3/papers.cfm?abstract_id=1774802, also published in J. Rosén (ed.) (2012), *Individualism and Collectivization in Intellectual Property Law*, Cheltenham, UK and Northampton, MA, USA: Edward Elgar Publishing, pp. 3–44.

Kretschmer, M. (2002), 'The failure of property rules in collective administration: rethinking copyright societies as regulatory instruments', *European Intellectual Property Review*, **24**, 126–37.

Mazziotti, G. (2011), 'New licensing models for online music services in the European Union: from collective to customized management', Columbia Public Law Research Paper No. 11–269, available at: http://ssrn.com/abstract=1814264.

Müller, S. (2009), 'Rechtewahrnehmung durch Verwertungsgesellschaften bei der Nutzung von Musikwerken im Internet', *ZUM*, **2009**, 121–31.

Netanel, N.W. (2008), *Copyright's Paradox*, Oxford: Oxford University Press.

Ricolfi, M. (2007), 'Individual and collective management of copyright in a digital environment', in P. Torremans (ed.), *Copyright Law: A Handbook of Contemporary Research*, Cheltenham, UK and Northampton, MA, USA: Edward Elgar Publishing, pp. 283–314.

Schovsbo, J. (2012), 'The necessity to collectivize copyright – and dangers thereof', in J. Rosén (ed.), *Individualism and Collectiveness in Intellectual Property Law*, Cheltenham, UK and Northampton, MA, USA: Edward Elgar Publishing, pp. 166–91, available at: http://papers.ssrn.com/sol3/papers.cfm?abstract_id=1632753.

Snow, A. and R. Watt (2005), 'Risk sharing and the distribution of copyright collective income', in L.N. Takeyama, W.J. Gordon and R. Towse (eds), *Developments in the Economics of Copyright: Research and Analysis*, Cheltenham, UK and Northampton, MA, USA: Edward Elgar Publishing, pp. 23–36.

Strowel, A. (2011), 'The European "extended collective licensing" model', *Columbia Journal of Law and the Arts*, **34**, 665–71.

European Decisions

CISAC: Commission decision of 16 July 2008 relating to a proceeding under Article 81 of the EC Treaty and Article 53 of the EEA Agreement (Case COMP/C2/38.698 – CISAC), available at: http://ec.europa.eu/competition/antitrust/cases/dec_docs/38698/38698_4567_1.pdf.

Daft Punk: Commission decision COMP/C2/37.219 of 12 August 2002, *Banghalter and Homem Christo v SACEM*, available at: http://ec.europa.eu/competition/antitrust/cases/dec_docs/37219/37219_11_3.pdf. This decision has been annulled by the General Court on 12 April 2013, Case T-442/08, *CISAC v European Commission*, ECR I-0000 (not yet officially reported).

GEMA: decision of the European Commission 71/224/EEC of 2 June 1971 relating to a proceeding under Article 86 of the EEC Treaty (IV/26 760 – GEMA), OJ L 134/15 of 20 June 1971.

GVL case: Commission decision 81/1030/EEC of 29 October 1981 relating to a proceeding under Article 86 of the EEC Treaty (IV/29.839 – GVL), OJ L 370/49 of 28 December 1981.

IFPI Simulcasting: Commission decision 2003/300/EC of 8 October 2002 relating to a proceeding under Article 81 of the EC Treaty and Article 53 of the EEA Agreement (Case No. COMP/C2/38.014 – IFPI

'Simulcasting'), available at: http://eur-lex.europa.eu/LexUriServ/LexUriServ.do?uri=OJ:L:2003:107:0058:0084:EN:PDF.
Luksan: ECJ decision of 9 February 2012, Case C-277/10 *Luksan v Petrus van der Let* ECR I-0000 (not yet officially reported).
SABAM: ECJ, 27 March 1974, Case 127/73, *BRT v SABAM* [1974] ECR 1974 313.
Santiago Agreement: Commission decision of 29 April 2004, Notice published pursuant to Article 27(4) of Council Regulation (EC) No. 1/2003 in Cases COMP/C2/39152 – BUMA and COMP/C2/39151 SABAM (Santiago Agreement – COMP/C2/38126).
Tournier and *Lucazeau*: ECJ, 13 July 1989, Case 395/87, *Ministère Public v. Jean-Louis Tournier*, Joined cases 110/88, 241/88, 242/88 – *François Lucazeau and others v Sociétés des Auteurs, Compositeurs et Editeurs de Musique (SACEM) and others* ECR 1989 2811.

European Legislative Acts

Duration Directive: Directive 2006/116/EC of the European Parliament and of the Council of 12 December 2006 on the term of protection of copyright and certain related rights, OJ L 372, 27.12.2006: 12–18, amended by Directive 2011/77/EU of 27 September 2011, OJ 2011 L 265/1.
Rental and Lending Directive: Rental and Lending Directive 2006/115/EC of the European Parliament and of the Council of 12 December 2006 on rental right and lending right and on certain rights related to copyright in the field of intellectual property (codified version).
Resale Right Directive: Directive 2001/84/EC of the European Parliament and of the Council of 27 September 2001 on the resale right for the benefit of the author of an original work of art.
Satellite and Cable Directive: Directive 93/83/EEC of the Council of 27 September 1993 on the coordination of certain rules concerning copyright and rights related to copyright applicable to satellite broadcasting and cable retransmission.

FURTHER READING

Gilliéron's article (2006) presents the first difficulties that CRMOs faced after the emergence of digitization and the Internet. Drexl (2007) presents the consequences of this shift for CRMOs with a competition law approach. Ricolfi (2007) describes the consequences of the shift as regards the distribution of works, and the role to be played in this regard by CRMOs. Gervais's book (2010a) gives a general overview of the legal framework applicable to CRMOs in the world and takes stock of the adaptation to the digitization. Handke and Towse (2007) present a literature review about the economics of CRMOs and identify the peculiarities of these natural monopolies that were unfortunately not considered in the Directive proposal of 2012 (Drexl et al., 2013).

21. Copyright levies

Joost Poort

Copyright levies on recording equipment or blank media exist in many countries to compensate copyright holders for the effects of private copying. In 1965, Germany was the first country to introduce a private copying levy on sound and video recording equipment, following two decisions by the German Federal Supreme Court. The German collecting society GEMA had 'asked the Supreme Court to order that producers of recording equipment be obligated, upon delivery of such recording equipment to wholesalers or retailers, to request from the latter that they communicate the identity of the purchasers to the GEMA' (Hugenholtz et al., 2003: 11). The court ruled that this would be a violation of privacy in conflict with the German Constitution. Subsequently, a levy was introduced to compensate copyright holders for the supposedly detrimental effects of copyright infringement using tape recorders.

In the ensuing decades, many countries followed suit and levies were introduced on a variety of recording or copying devices and blank media. At the present time, most countries within the European Union have copyright levies, as well as the United States, Canada, Russia and several countries in Latin America and Africa. In Asia, Japan is the only country with copyright levies (WIPO, 2012: 3).

The general idea behind copyright levies is the following. Consumers often make copies for personal use of copyright protected material such as music, films, TV shows or books. For instance, they copy a CD to play in their car or on holiday or to give to a friend. Or they record a television show to watch another time. Under strict application of copyright laws, these acts of reproduction would be an infringement. However, strict enforcement in these circumstances would be both too costly and too intrusive vis-à-vis rights of privacy and freedom of expression. Hence, legislators in many jurisdictions have introduced exceptions, which allow various forms of private copying, provided a levy is paid. This levy aims to compensate copyright holders for harm caused by the reduced sales or licensing opportunities as a result of private copying. Levies are collected by a collecting society or some general office and redistributed amongst copyright holders.

This chapter deals with the law and economics of copyright levies. It first discusses the legal framework for private copying, focusing on the European Union, where the bulk of copyright levies are collected. Second, it presents data on the amounts of levies and the sums collected in various countries. Third, this chapter analyses the economics of copyright levies. Fourth, it briefly discusses proposals to extend copyright levies to cover unauthorized file sharing over the Internet.

LEGAL BACKGROUND IN THE US AND THE EU

In the United States, home video recording was ruled to be non-infringing under the fair use doctrine by the Supreme Court in 1984. Eight years later, in 1992, the Audio Home

Recording Act (AHRA) was adopted, which imposed a levy on equipment and media primarily designed to make digital copies of music for personal use (Netanel, 2003). Levies are set by the AHRA at 2 or 3 per cent of the price of devices and media, up to a maximum of $12 (WIPO, 2012). Since the introduction of the AHRA, however, few new devices or media have been subjected to levies. Devices primarily used for copying audio-visual works, spoken word, software and databases are exempted. Moreover, CD burners used with computers, blank CDs not specifically sold for music use and most MP3 players were considered computer peripherals and as such were exempted from the levy. As a consequence, the total revenues from levies are very modest at €2.3 million in 2011 (WIPO, 2012).

The legal issues at stake in the European Union (EU) touch upon all the key economic issues with copyright levies that will be dealt with in the following sections. In the EU, the legal basis for private copies and copyright levies is found in Article 5(2)(b) of the Copyright Directive, also referred to as the Information Society Directive or InfoSoc Directive (2001/29/EC). It states that:

> Member States may provide for exceptions or limitations to the reproduction right . . . in respect of reproductions on any medium made by a natural person for private use and for ends that are neither directly nor indirectly commercial, on condition that the rightholders receive fair compensation which takes account of the application or non-application of technological measures referred to in Article 6 to the work or subject-matter concerned.

Hence, private copying can be allowed, provided authors receive fair compensation. In recital 35 of the preamble of the Directive, 'the possible harm to the rightholders resulting from the act in question' is proposed as a 'valuable criterion' for fair compensation. However, no payment may be due in the case where private copying is paid for as part of a licence fee, or if this harm is minimal (*de minimis*).

The concepts of *fair compensation* and *harm* are further explicated in the Padawan decision of the EU Court of Justice, which ruled in 2010 that:

> the concept of 'fair compensation' [. . .] is an autonomous concept of European Union law which must be interpreted uniformly in all the Member States that have introduced a private copying exception, irrespective of the power conferred on them to determine, within the limits imposed by European Union law and in particular by that directive, the form, detailed arrangements for financing and collection, and the level of that fair compensation. (EU Court of Justice, 21 October 2010, Article 37)

The Court also ruled that 'fair compensation must be calculated on the basis of the criterion of the harm caused to authors of protected works by the introduction of the private copying exception' (EU Court of Justice, 21 October 2010, Article 50). Furthermore, it ruled that 'a link is necessary between the application of the levy intended to finance fair compensation with respect to digital reproduction equipment, devices and media and the deemed use of them for the purposes of private copying' (Article 59).

Thus, according to EU law, levies should be set based on the *harm* caused to copyright holders by the introduction of the private copying exception and should take the actual use of devices and media for private copying into account. Devices and media sold to commercial users and not used for private copying should not have levies applied. This does, however, still leave much room for many differences between EU member states, as well as for lawsuits and controversy. In particular, this concerns the definition and calculation of harm, the status of copies from illegal sources and the consequences of use of encryption technology and digital rights management (DRM).

In an attempt to reach more concordance on such issues, EU Commissioner Barnier appointed António Vitorino as a mediator between stakeholders. With respect to private copies 'in the context of a service that has been licensed by rightholders' (think of iTunes for instance) Vitorino concludes that these 'do not cause any harm that would require additional remuneration in the form of private copying levies' (Vitorino, 2013). In such cases, a levy would lead to double payments. This interpretation of the somewhat enigmatic phrase 'which takes account of the application or non-application of technological measures' in the Copyright Directive was already defended by Hugenholtz et al. (2003).

On the concept of harm, Vitorino proposes

> to look at the situation which would have occurred had the exception not been in place. In particular, one needs to assess the value that consumers attach to the additional copies of lawfully acquired content that they make for their personal use. It would allow the estimate of losses incurred by rightholders due to lost licensing opportunities ('economic harm'), i.e. the additional payment they would have received for these additional copies if there were no exception. (Vitorino, 2013: 19–20)

Regrettably, Vitorino's attempt to contribute to more uniformity in the concept of harm seems to run astray when he equates 'the value that consumers attach to the additional copies', which is in principle the entire consumer surplus of private copies, to 'lost profit' which is forgone producer surplus. Moreover, to 'look at the situation which would have occurred had the exception not been in place' introduces the question whether consumer behaviour would really be much different with or without a private copying exception. The exception was introduced in the first place because enforcement was problematic.

A final issue is the position of copies from *illegal* sources. Only in a few countries – The Netherlands, Russia, Switzerland and Canada – do all copies made for private use fall within the scope of the exception, irrespective of the source (WIPO, 2012: 4). This would mean that, in all other countries, copies from illegal sources (such as file sharing on torrent networks, newsgroups, social networks or cyber-lockers) are not private copies as meant by the private copying exception. It is an important question whether it would nevertheless be lawful to levy for harm caused by copies from illegal sources, even if such copying is itself illegal. Vitorino (2013) seems to imply it is not, when he refers to 'copies of lawfully acquired content' in the quote above. Preliminary questions on the matter from courts in Denmark and The Netherlands to the EU Court of Justice are pending. In the academic literature, several proposals have been made to extend the reach of copyright levies so as to compensate for downloading from illegal sources. These are briefly discussed in the section 'Levies 2.0' in this chapter.

COPYRIGHT LEVIES IN PRACTICE IN THE EU

Following the Copyright Directive, 22 out of 27 EU member states have introduced a levy system.[1] As technology used for consuming and storing music and audio-visual

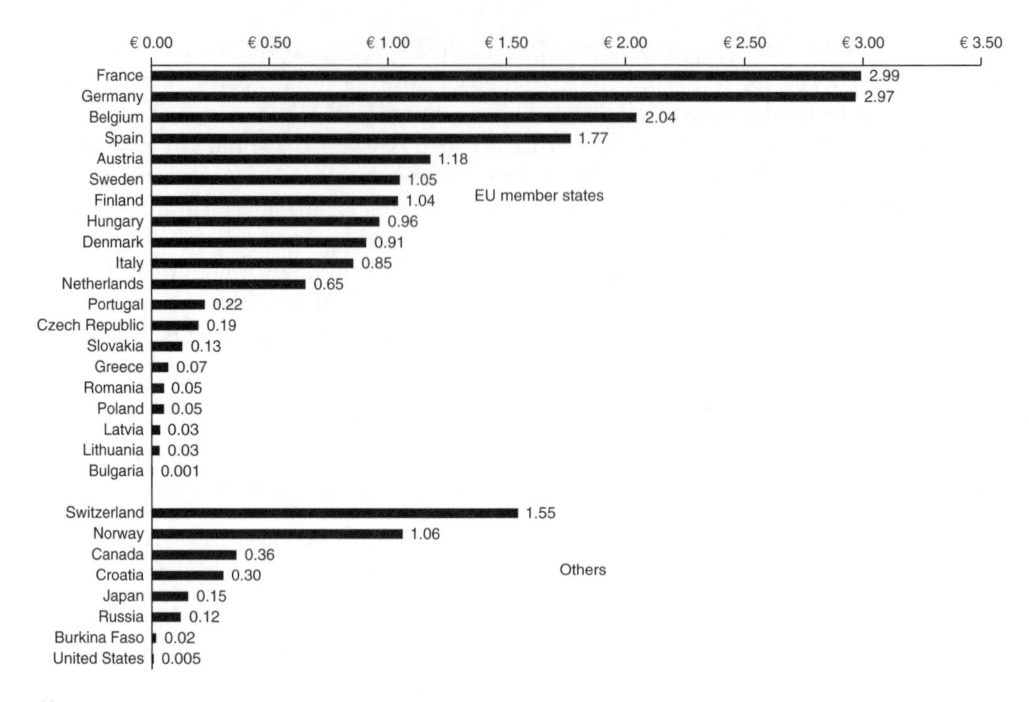

Notes:
Copyright levies are shown in € per capita.
Within the EU, no copyright levies exist in Cyprus, Ireland, Luxembourg, Malta and the UK; no data is available for Slovenia and Estonia.

Source: Based on WIPO (2012).

Figure 21.1 Copyright levies collected in 2010

material changes rapidly, so do the devices and media levied. All 22 EU countries that introduced copyright levies apply these to blank CDs and DVDs. A majority also have levies on memory cards, MP3 players and hard disc DVD recorders. A number of countries also have levies on external hard drives, PCs, tablet computers and smartphones. Game consoles are generally exempted (WIPO, 2012), as the Copyright Directive does not apply to software.

There are also massive differences in the amounts of levies. For a blank CD, for instance, nominal levies range from €0.009 in the Czech Republic to €0.35 in France. Alternatively, several countries have levies relative to the price of media, ranging from 1.25 per cent in Bulgaria to 6 per cent in Greece. For devices such as MP3 players, nominal levies often depend on the storage capacity. For a 32-GB player, copyright levies range from €1.42 in Latvia to €22.52 in Hungary. Levies for hard disc DVD recorders range as high as €50 in France (WIPO, 2012: 13–14).

Figure 21.1 gives the revenues collected per capita in 2010 in several countries. Revenues per capita range from less than €0.01 in the United States and Bulgaria to nearly €3 in France and Germany. For all EU member states taken together, revenues totalled about €648 million in 2010 (WIPO, 2012). As a result of rapidly changing technology used for storing and copying content, in combination with a tradition of litigation over the incorporation of new media and devices in levy schemes, revenues tend to vary over time. An upward driver is the rapidly increasing storage capacity of most devices, which implies that revenues based on a fixed amount per MB or GB increase rapidly in time. A downward driver is the dynamic nature of the consumer electronics market. For instance, the use of blank CDs and DVDs has plummeted as they are substituted by USB sticks and memory cards. Likewise, the market for MP3 players is cannibalized by smartphones.

THE ECONOMICS OF COPYRIGHT LEVIES

Types of Private Copying

As discussed above, copyright levies are well established in many jurisdictions, most prominently in the European Union. Nonetheless, the underlying economic rationale is not undisputed. In order to analyse the economics of copyright levies, it is useful first to distinguish various 'types' of private copying that may be different from an economic perspective:[2]

1. making copies of broadcasted material for *time shifting*, for example storing radio or TV content on a recording device to watch it another time and if desired repeatedly;
2. making '*clone copies*' of or '*format shifting*' CDs, DVDs and media files, or storing streaming audio and video for offline playback (also known as 'stream ripping' or 'stream capture');
3. making clone copies or format shifting to *share* content with members of your *household, family and friends*;
4. making clone copies or format shifting from media *rented* or *borrowed* from libraries or commercial renters;
5. making *backup copies* of content;
6. *downloading* and storing content from *unauthorized sources* on the Internet (and making ensuing copies thereof).

Typically, private copies of types 1–5 are allowed under the private copying exceptions. However, types 2–5 may be prohibited or restricted by DRM and licence or rental agreements. Downloading from unauthorized sources on the Internet (6) is prohibited in most countries.

There are economically relevant differences between all these types of copies. Copies of types 1 and 2 enhance the utility that consumers derive from their legitimate purchase or subscription. It enables them to consume this content at a more convenient time or place, on a more practical device or without carrying discs and devices around. For instance, they can keep a copy of their favourite CDs in their car or holiday home or play them on their computer, smartphone or MP3 player. Time shifting also enables them to skip advertisements in TV programmes. Copies of types 3 and 4 are different from the former in that they extend the circle of consumers that derive utility from an original unit of content.[3] Unauthorized downloading (6) resembles the former types, with the notable difference that the extended circle of consumers is anonymous and potentially unlimited. Backup copies (5) do not provide utility directly but act as an insurance against mishaps.

Harm and Indirect Appropriability

These differences are relevant in light of the *harm* that private copying may cause to copyright holders and the concept of *indirect appropriability*. If copyright levies should be understood as a compensation for harm caused by private copying, as is the case at least according to EU law, it is important to analyse the economics behind this harm more closely.

A seminal contribution to this issue is by Stan Liebowitz (1985), who studies the effect of photocopying on demand for journals. He finds that 'publishers can indirectly appropriate revenues from users who do not directly purchase journals and that photocopying has not harmed journal publishers' (Liebowitz, 1985: 945). The value consumers (or scholars) derive from copies contributes to the willingness to pay of libraries. Hence, publishers end up selling fewer originals at a higher price, which may even raise profits. If the number of copies per original differs substantially, such 'indirect appropriability' depends on the ability to price-discriminate: charging a higher price for users that are likely to enable extensive copying while preventing arbitrage. In practice, this is done by a higher price for libraries and a lower price for individuals.

In general, selling fewer originals at a higher price introduces two opposing effects which may lead to both higher and lower profits. This issue was further studied by Besen and Kirby (1989), who model private copying while distinguishing: (1) the extent to which originals and copies are perfect substitutes; and (2) whether private copying has constant or increasing marginal costs. Increasing marginal costs may stem not only from the technology itself, but also from the 'costs' of organizing the copying process within sharing groups. Besen and Kirby conclude that the effect of private copying on consumer and producer surplus and total welfare depends strongly on the assumptions about substitutability and the costs of copying.[4] When the marginal costs of copying are *constant*, copies will be distributed at marginal costs and the value of copies cannot be appropriated. The introduction of (relatively cheap) copying technology will then lower the price of originals, and profits will decline subsequently. Consumer surplus will increase, while the effect on total welfare is ambiguous. If the marginal costs of copies are *increasing*, copying leads to fewer originals sold at a higher price: indirect appropriability. Besen and Kirby (1989: 280) conclude that the effect on welfare will depend on whether or not copying is cheaper than producing originals:

> in the case where the size of the sharing group is fixed, consumer and producer welfare generally increase when copying is efficient and decline when it is not. When the costs of both originals and copies are low, however, producers will generally lose and consumers will gain from the introduction of copying.

The scenario of *constant* (near-zero) marginal costs of copying resembles the situation of unauthorized file sharing over the Internet (type 6 above), even though such file sharing did not occur when Besen and Kirby wrote their paper. Their analysis implies that indirect appropriability is not feasible in the face of online file sharing. This also follows from the fact that the number of copies generated per original copy sold may differ immensely. Some CDs will not be copied at all, while others will be ripped and uploaded to torrent sites to be seeded to millions of users (Liebowitz and Watt, 2006). This complicates price discrimination.

The scenario of *increasing* marginal costs will apply to offline private copying and may or may not cause harm, depending on the costs and value of copies in comparison to that of originals and the size of sharing groups. Copying may cause no harm at all to copyright holders, but copies may also become competitors to the originals, constraining the price the copyright holder can charge and reducing profits substantially (Varian, 2005).

Empirical testing of the net effect of private copying on profits is lacking. Surveys carried out in the context of the levy setting process in various countries typically focus on the number of private copies and the self-reported substitution rate (for example PwC, 2012; Verhue and Hilhorst, 2012) and ignore the effect of private copying on the demand for the first original and the complex dynamics of indirect appropriability. It is likely that the net effect will differ for the various types of private copying discussed in the former subsection. As mentioned there, private copies of types 1, 2 and 5 do not extend the circle of consumers that derive utility from an original unit of content. No 'copying groups' are formed in which copies become competitors to the original. Some additional sales could be forgone as a result of such copying, for instance a person buying a favourite CD twice, to play at home and in the car, but, to the extent that consumers can roughly anticipate their copying behaviour, the option to copy can be priced into the initial purchase. Put differently, the demand curve for originals will reflect the expected utility derived from such private copies. This is an important notion, as it means that in such a case a copyright levy that charges 'the value that consumers attach to the additional copies', as Vitorino (2013: 19–20) suggests, would lead to double payment.[5]

However, the demand curve will not reflect the utility of *unforeseen* copying possibilities: in the first decade after the introduction of the CD, consumers would not have expected the possibility of making perfect copies of CDs within their home, let alone ripping 500 CDs on to a portable device. Hence, the utility they derive from copying and ripping these CDs would not have been reflected in their initial purchase at the time.

Turning to copies shared within one's household, and with family and friends (type 3), the models of Besen and Kirby (1989) and Varian (2005) are more likely to apply, and private copying might be harmful to copyright holders even though some of the additional utility can be appropriated indirectly. This is partly due to the fact that the size of such copying groups is variable and price discrimination according to group size is not possible. The utility of copies of type 4 could in theory be appropriated indirectly by setting the appropriate rental prices, unless restrictions put on the rental price from a public service perspective prohibit doing so.

To summarize, the utility downloaders derive from unauthorized file sharing cannot be priced into the initial purchase. For other types of private copying, these benefits can to some extent be appropriated by using smart pricing, depending on the cost structures. For time shifting, format shifting and clone copying for personal use, this is likely to be the case. However, for copies passed on to family and friends (type 3) this may not be sufficiently possible, and harm from such private copies may well occur.

The Application of DRM

Koelman (2005) points out that digital rights management technology increases the opportunities copyright holders have to appropriate the additional utility derived from private copying. This is in line with remarks in the Copyright Directive on the 'application or non-application of technological measures' and Vitorino's (2013) recommendation that licensed copies do not require additional remuneration by a levy. However, the distinction between the 'application or non-application of technological measures' is not as binary as it may seem at first glance. DRM exists in many forms, ranging from fingerprinting or watermarking (also known as social DRM) to advanced encryption. DRM can also be used to limit the distribution of private copies among family and friends. Yet the experience so far suggests that DRM will not eradicate all unlicensed copying. The technical and privacy issues that make strict enforcement problematic and give rise to copyright levies in the first place have not disappeared with the introduction of DRM.

At the same time, now that various kinds of DRM are available to give copyright holders at least a firmer grip on copying and more opportunities for price discrimination and indirect appropriation, the decision *not* to apply DRM should also be considered. When consumers started copying and sharing CDs on a large scale, for instance, record labels introduced DRM on CDs, and until 2007 all digital music files bought from Apple's iTunes Store also contained DRM technology. However, consumers did not appreciate the way in which DRM got in the way of supposedly legitimate uses. For instance, DRM sometimes caused computers to crash, which was a nuisance to people trying to play an audio CD with their computer, even without trying to copy it. Also, the use of DRM prevented consumers from format shifting, such as ripping their own CD collection on to their MP3 player.

Thus, the use of DRM may create a disutility for consumers and have a negative effect on demand for originals. For example, consumers who play music only from a hard drive or on their phone may stop buying CDs if DRM prohibits them from ripping these. DRM may even cause consumers to revert to DRM-free content from illegal sources (Sinha et al., 2010; Vernik et al., 2011). Over the last few years, the music industry has moved away from using DRM. From an economic point of view, this should be a rational choice. Indeed, Sinha et al. (2010) find that 'the music industry can benefit from removing DRM' and that 'a DRM-free environment enhances both consumer and producer welfare by increasing the demand for legitimate products as well as consumers' willingness to pay for these products' (Sinha et al., 2010: 40). Therefore, the choice *not* to use DRM should be perceived as a rational choice in the spectrum ranging from more to less restrictive DRM technologies that are currently available. Now that right holders have these options, it no longer makes sense for copyright levies to draw a sharp line between the application and non-application of DRM as the Copyright Directive seems to suggest. The choice not to use DRM should no more entitle copyright holders to compensation than the choice to use restrictive DRM. Hugenholtz et al. (2003: 4) seem to find a legal basis for this argument, proposing that 'levies are to be phased out not in function of actual use, but of availability of technical measures on the market place'.

Finally, harm from private copying cannot be equated to harm caused by the possible introduction of a private copying exception to copyright. As pointed out in the preceding paragraphs, consumers' private copying behaviour is to a large extent independent of its legal status, since enforcement is problematic. Private copying occurs without a private copying exception. Much of any harm from private copying would thus not be the *result* of such an exception. This implies that the suggestion, both in the Padawan ruling and by Vitorino (2013), to base compensation on the harm caused by the introduction of the private copying exception could leave copyright holders empty-handed.

Other Grounds for a Levy than the Compensation of Harm

Although the concept of 'harm' is central to the legal foundations of copyright levies, harm is by itself no sufficient economic argument to introduce a levy. Kretschmer (2011: 59–66) reviews several alternative arguments.

The *transaction costs* of individual licensing are often mentioned as an economic rationale for a private copying exception. However, DRM technology has made it easier to control individual copying behaviour than was conceivable 20 years ago. More importantly, transaction costs may be an argument for an exception, but not necessarily for compensation. As Kretschmer (2011: 60) points out, 'If the current parameters of copyright law ignore "economic efficiency" (that is, if copyright law under-protects or over-protects), the transaction cost approach does not fly, as we are minimizing transaction costs towards a sub-optimal outcome.'

Another argument could lie in the extra value that devices derive from copying content. Considering that the costs of filling a 32-GB MP3 player with digitally bought MP3 files are to the tune of a few thousand euros, it is unlikely that many people would buy one if private copying were not possible. If copyright holders are not rewarded for this, they do not receive efficient incentives to create content. However, such arguments could in theory be applied to all sorts of complementary goods and generally are no cause for intervention. Effects are often bi-directional: an MP3 player enhances the utility of CDs and vice versa, and it is debatable if any compensation would enhance welfare.

A third argument understands levies as a 'tax' used to support the supply of content, which has characteristics of a public good. Like any tax, however, levies reduce demand for copies and related goods and services by increasing their price (Lunney, 2001; Fisher, 2004; Koelman, 2005; Towse, 2008). It may even reduce the incentives to introduce new copying technologies somewhat (Lunney, 2001). On the other hand, any tax system suffers from the fact that it will reduce demand for taxed products and services or reduce production. In fact, this effect may be smaller for a copyright levy than for a general tax such as VAT (Netanel, 2003).

Other Economic Concerns with Levies

Apart from the debate about the harm caused by private copying, most other concerns with levies expressed in the literature have to do with the bluntness of the levy instrument. 'Rough justice' is a term often used in the context of levy systems, which means that a levy system will never be able to make exactly the right people pay levies and the right creators receive them. The last subsection already mentioned the losses resulting from the fact that levies will be fairly uniform, which implies consumers who make many private copies will be cross-subsidized by consumers who make very few copies or no copies at all. On the other hand, it should be noted that the Padawan decision exempts professional use from levy systems, which is a step in the right direction.

Likewise, the distribution of levy revenues to creators will be 'rough', which implies creators receive imperfect price signals about the market valuation of works (Towse, 2008). In addition, the administrative costs of levy systems are stressed by some (Koelman, 2005).

LEVIES 2.0

As mentioned above, downloading and storing content from unauthorized sources (private copies of type 6) are generally not allowed under private copying exceptions. However, since the rise of peer-to-peer file sharing, several academic authors have proposed the introduction of some 'statutory licensing scheme' with a levy system to compensate for this practice (Ku, 2002; Netanel, 2003; Fisher, 2004; Gratz, 2004; Litman, 2004; Oksanen and Valimaki, 2005).[6] Such proposals are also known as for instance 'content flat rate' or 'alternative compensation scheme'.

The logic behind the introduction of a levy on such behaviour is similar to that of 'classical' levies. In return for a levy of 2 to 5 per cent on broadband access and/or computer equipment, consumers could be allowed to engage in non-commercial file sharing and creative re-use while copyright holders are compensated for the harm done. Proposals differ in what the levy should cover (music, audio-visual material, books), whether it should be mandatory for consumers and for copyright holders, and the way the revenues should be redistributed among right holders. Monitoring peer-to-peer traffic for that purpose (Ku, 2002; Netanel, 2003) would seem a reasonable option. However, such a system has privacy issues and could be manipulated. Besides, it would favour less popular content, as it would count only downloads and multiple playback would not be rewarded (Fisher, 2004). Fisher (2004) suggests sampling content consumption by consumers, while Gratz (2004) proposes using the number of peers making content available for download, and Oksanen and Valimaki (2005) argue for voting systems to democratize remuneration and involve consumers.

So far, organizations of copyright holders have not been very enthusiastic about any of these proposals. A general drawback is that such a scheme would promote file sharing, destroying what is left of the current business based on physical formats, and wiping out promising new digital models, such as iTunes, Spotify, Deezer, Netflix and the like. Indeed, as the content industry often stresses, it is hard to compete with free, but it would be even harder to compete with free when it is legalized and consumers have paid for it in advance.

CONCLUDING REMARKS

Historically, copyright levies have been introduced to compensate right holders for the harm caused by private copying. However, economic analysis has shown that the utility consumers derive from offline private copies can to a large extent be appropriated indirectly. Hence, the harm caused by private copying will be substantially smaller than the utility consumers derive from private copies or even the sales forgone by such copying. For private copies that do not lead to a proliferation of content – for example time shifting, format shifting and backup copies – there may be no harm at all, provided consumers are aware of these copying possibilities at the time of their initial purchase.

DRM and innovative pricing schemes have improved the possibilities for copyright holders to appropriate the value of private copies. Therefore, charging levies for copies that are licensed by right holders would lead to double payment. Moreover, the choice *not* to apply DRM is nowadays a rational choice that from an economic perspective should not be treated differently from the choice to apply DRM. Altogether, the case for levies to compensate for harm caused by 'classical' private copies is gradually diminishing.

The argument of indirect appropriability does not apply to unauthorized online file sharing, as there is no relation between the uploader and the downloader, and the number of copies made from an original will vary dramatically. It is an open question pending at the EU Court of Justice whether EU legislation allows levies to account for the harm caused by these copies, even if they are deemed illegal. If so, copyright levies will continue to have a sound legal basis within the EU. If not, they might still have a future as 'Levies 2.0' in combination with a statutory licensing scheme.

NOTES

1. Exceptions are: the UK and Ireland, which so far have not introduced a private copying exception other than for time-shifting TV content; and Malta, Cyprus and Luxembourg, which treated private copying as *de minimis* (Kretschmer, 2011: 10).
2. A somewhat different typology can be found in Kretschmer (2011: 9, 70), which sums up activities that users may consider private. Kretschmer also mentions user generated content, uploading and online publication, performance and distribution within networks of friends as separate categories. These are ignored here, as they are not considered to fall under the private copying exception.
3. One could also argue that making copies of legally borrowed or rented content (4) is a form of time shifting (beyond the rental period) rather than an extension of the circle of consumers.
4. Like other models discussed in this section, the models Besen and Kirby develop ignore possible positive dynamic effects of the consumption of copies on future sales (the so-called sampling effect) and negative effects of lost sales on the future production of content.
5. Time shifting of TV content is a somewhat different issue: advertisers will be aware that some of their ads will be skipped when consumers watch programmes they recorded. This will lower advertising revenues for channels, while increasing the willingness to pay consumers have for receiving TV channels and for recorders.
6. For an extensive analysis of this literature, see also Quintais (forthcoming).

REFERENCES

2001/29/EC Directive 2001/29/EC of the European Parliament and of the Council of 22 May 2001 on the harmonisation of certain aspects of copyright and related rights in the information society, *Official Journal*, L 167.

Besen, S.M. and S.N. Kirby (1989), 'Private copying, appropriability, and optimal copying royalties', *Journal of Law and Economics*, **32** (2), 255–80.

EU Court of Justice (2010), *Padawan v SGAE*, Case C-467/08, 21 October.

Fisher, William W., III (2004), 'An alternative compensation system', in *Promises to Keep: Technology, Law, and the Future of Entertainment*, Stanford, CA: Stanford University Press, pp. 199–258.

Gratz, J. (2004), 'Reform in the "Brave Kingdom": alternative compensation systems for peer-to-peer file sharing', *Minnesota Journal of Law, Science and Technology*, **6**, 399–430.
Hugenholtz, Bernt, Lucie Guibault and Sjoerd van Geffen (2003), *The Future of Levies in a Digital Environment*, Amsterdam: Institute for Information Law.
Koelman, K.L. (2005), 'The levitation of copyright: an economic view of digital home copying, levies and DRM', *Entertainment Law Review*, **4**, 75–81.
Kretschmer, Martin (2011), *Private Copying and Fair Compensation: An Empirical Study of Copyright Levies in Europe*, London: Intellectual Property Office.
Ku, R. (2002), 'The creative destruction of copyright: Napster and the new economics of digital technology', *University of Chicago Law Review*, **69** (1), 263–324.
Liebowitz, S.J. (1985), 'Copying and indirect appropriability: photocopying of journals', *Journal of Political Economy*, **93** (5), 945–57.
Liebowitz, S.J. and R. Watt (2006), 'How to best ensure remuneration for creators in the market for music? Copyright and its alternatives', *Journal of Economic Surveys*, **20** (4), 513–45.
Litman, J. (2004), 'Sharing and stealing', *Hastings Communications and Entertainment Law Journal*, **27** (1), 1–50.
Lunney, G.S. (2001), 'The death of copyright: digital technology, private copying, and the Digital Millennium Copyright Act', *Virginia Law Review*, **87** (5), 813–920.
Netanel, N.W. (2003), 'Impose a noncommercial use levy to allow free peer-to-peer file sharing', *Harvard Journal of Law and Technology*, **17** (1), 1–84.
Oksanen, V. and M. Valimaki (2005), 'Copyright levies as an alternative compensation method for recording artists and technological development', *Review of Economic Research on Copyright Issues*, **2** (2), 25–39.
PwC (2012), *Thuiskopie: onderzoek naar gederfde inkomsten door huiskopie thuiskopieën*, Amsterdam: PwC Nederland.
Quintais, J.P. (forthcoming), 'A shifting copyright Zeitgeist: alternative compensation models for digital content sharing' (draft available with the author).
Sinha, R.K., F.S. Machado and C. Sellman (2010), 'Don't think twice, it's all right: music piracy and pricing in a DRM-free environment', *Journal of Marketing*, **74** (2), 40–54.
Towse, R. (2008), 'Why has cultural economics ignored copyright?', *Journal of Cultural Economics*, **32** (4), 243–59.
Varian, H.R. (2005), 'Copying and copyright', *Journal of Economic Perspectives*, **19** (2), 121–38.
Verhue, Dieter and Marsha Hilhorst (2012), *Gebruik van opslagmedia: twintigste meting*, Amsterdam: Veldkamp.
Vernik, D.A., D. Purohit and P.S. Desai (2011), 'Music downloads and the flip side of digital rights management', *Marketing Science*, **30** (6), 1011–27.
Vitorino, António (2013), *Recommendations Resulting from the Mediation on Private Copying and Reprography Levies*, Brussels: European Commission.
WIPO (2012), *International Survey on Private Copying: Law and Practice 2012*, Geneva: World Intellectual Property Organization.

FURTHER READING

Hugenholtz et al. (2003) provide a good overview and analysis of the legal and policy background of copyright levies, focusing on the EU up until that time. A more recent and factual overview is provided by WIPO (2012). For a thorough analysis of the various proposals for alternative compensation schemes, see Quintais (forthcoming).

PART IV

COPYRIGHT AND DIGITIZATION: EMPIRICAL EVIDENCE

22. Empirical evidence on copyright

*Christian Handke**

Copyright law determines who has the right to reproduce, disseminate and modify a large share of the existing creative works, such as literary texts and news reporting, musical works, movies, video games or scientific articles. Copyright thus affects the development of digital markets as well as non-commercial use of creative works. Arguably, empirical research is essential to establish how the copyright system affects various stakeholders, and thus to develop adequate policy. This chapter reports on the empirical research regarding the socio-economic consequences of unauthorized copying and copyright.

Much empirical work has addressed the effects of unauthorized, digital copying on demand for authorized copies, in particular in markets for musical recordings and movies. That is a central question for assessing how copying affects rights holders, but it is not sufficient to inform copyright policy, because users' interests also have to be taken into account. Accordingly, this chapter reports on empirical evidence on: (1) the effect of variations in copyright protection on the supply of creative works, which is important in determining how users are affected; (2) administration and transaction costs of copyright systems; (3) unintended effects of copyright regarding competition and technological innovation; and (4) basic findings on the determinants of unauthorized copying.

Furthermore, the chapter discusses how these topics relate to one another in terms of a social welfare analysis, including the interests of all stakeholders. This is essential in moving from empirical results to an informed discussion on copyright policy.

TERMS AND THE SCOPE OF THIS CHAPTER

The chapter surveys quantitative-empirical research reported on in English and which directly relates to elements of an economic welfare analysis of copyright. The legal literature falls beyond the scope of this chapter. For a more comprehensive overview of the empirical literature, beyond economics and including qualitative research, see Goode (2012).

Throughout, 'copyright' refers to creators' exclusive rights to their original creations, as well as the related rights of performers and producers. A 'copyright system' entails laws, treaties and contracts, enforcement measures and specialized organizations such as copyright management societies that administer the copyrights of many rights holders simultaneously (Handke and Towse, 2007). 'Copyright protection' is also determined by the state of ICT. Legal entitlements are less important in practice if many people can easily access, reproduce, disseminate or modify protected works. Even though copyright laws and enforcement measures have been extended and strengthened to cover new digital ICT, effective copyright protection has decreased in practice – as indicated, for example, by the number of unauthorized copies generated in comparison to authorized copies.

A diverse range of creative works falls under the scope of copyright law, from literary works and news reporting to musical compositions and recordings, photographs, movies or computer software. What is more, 'copyright systems' have many aspects, such as: their scope (what aspects of creative works are protected); the type and intensity of enforcement measures and digital rights management (DRM) techniques; the duration of rights; the extent of fair use or other such exemptions; moral rights; and so on. Only a couple of specific aspects have been addressed in the empirical research surveyed here.

THE COSTS AND BENEFITS OF COPYRIGHT

Basic economics suggests that, with unrestricted copying, the supply of new creative works will be lower than socially desirable. Faced with competition from unauthorized sources of identical works, creators will be unable to appropriate much of the value they generate, which will diminish financial incentives to create new works.

The cost structure of the creative industries aggravates this problem. The creation of new copyright works usually entails substantial sunk costs, for example of writing a novel or producing a movie, while marginal costs of reproducing and disseminating an existing work tend to be much lower. Therefore suppliers of unauthorized copies who do not invest in the development enjoy a substantial cost advantage. With increasing use of digital ICT this cost advantage of copy-cats has probably become more pronounced. Although precise information on initial production costs is hard to come by, it is usually assumed that the costs of creating new works have decreased proportionally less than the costs of reproducing and disseminating existing works. If so, digitization will aggravate the social problem with insufficient incentives to supply new creative works.

Effective copyright establishes temporary exclusive rights to specific creative works. It enables rights holders to charge prices in excess of marginal costs. The higher revenues are supposed to help recoup development costs and motivate further investments in creativity. However, the market power of rights holders restricts the use of creative works, and managing a copyright system takes up resources. Therefore there is a trade-off between costs and benefits, and the overall improvement in social welfare has to be cast in terms of net benefits. Table 22.1 provides an overview of costs and benefits of copyright, distinguishing between short-run and long-run effects on rights holders and users.

In the short run, the supply of copyright works is taken as given. Effective copyright endows rights holders with market power, enabling them to raise prices above marginal

Table 22.1 Costs and benefits of a copyright system

	Benefits	Costs
Short run	Greater revenues to rights holders	1. Access costs to users 2. Administration costs 3. Transaction costs in trading rights
Long run	Greater incentives to supply copyright works for rights holders	Obstruction of user innovation by the costs of compliance

Source: Handke (2010a, 2012b).

cost and obtain a higher return than they would get under competition. Higher prices restrict users' access to existing works. In addition, a copyright system entails administration costs, and there are transaction costs in trading copyrights. In the long run, effective copyright is expected to increase the supply of creative new works, benefiting users. Social welfare will increase if the value of any additional works created as a result of copyright compensates for the various costs of a copyright system. This long-run assessment is the predominant economic rationalization of copyright.

As creators are usually both rights holders and users of copyrights, copyright increases the returns for creators and the costs of creating new works. Therefore it is not a given that greater copyright protection increases the supply of protected works (Landes and Posner, 1989). This argument can be extended to technological innovation. New means to reproduce, disseminate and create derivative works often require the consent of copyright holders. Copyright may boost investments of rights holders in technological innovation. It may also inhibit user innovation (Plant, 1934; Handke, 2011b). This is an important point in the context of digitization, which offers great opportunities for technological innovation in disseminating creative works.

There may also be unintended consequences beyond those covered in Table 22.1. For example, copyright may affect the contestability of markets. Specific measures to detect and inhibit copyright infringement (enforcement) may in practice conflict with other values such as freedom of expression or privacy.

In summary, copyright has to strike a complex balance between various costs and benefits. Pure theory provides the insight that copyright may make society better off under specific conditions. Where these conditions are met is an empirical question. Changes in markets – such as those resulting from digitization – may affect the case for copyright and affect the desirable extent and type of copyright protection.

In order to inform copyright policy we need a comprehensive estimate of the various costs and benefits of unauthorized copying and copyright. This is a tall order, which will hardly be accomplished perfectly. However, many sophisticated empirical studies on specific issues related to digital copying and copyright have been published over recent years.

An immediate observation from surveying the literature on digital copying to date is that it does not equally cover all relevant issues. A recent survey of the quantitative empirical work (Handke, 2012b) illustrates this point. Of all studies concerned with the impact of digital copying, 21 studies deal with short-run rights holder welfare (as indicated by revenues, sales or stock market valuation). Short-run user welfare was addressed in two papers. Work on long-run rights holder welfare after adaptation was addressed in two studies, and three studies tackled long-run user welfare as indicated by the value of new creative works supplied. As our understanding of the more immediate displacement effect of digital copying solidifies, the long-run implications require more attention.

COPYRIGHT PROTECTION AND RIGHTS HOLDER REVENUES

Many empirical studies address the impact of file-sharing on the demand for authorized copies of musical recordings. The diffusion of digital copying technology in private households coincided with very substantial reductions in sales of authorized copies in major economies in North America and Europe. Many studies imply that digital copying displaced demand, harming rights holders, even though the issue remains contentious (Smith and Telang, 2012; Liebowitz, this volume, Chapter 23). A handful of studies discuss digital copying and movies, with similar results. There has been little empirical research on other creative works covered by copyright, such as business or entertainment software, literary texts, news reporting, academic articles and so on.

The extent of unauthorized use depends on copyright law, enforcement measures and the state of ICT. Baker and Cunningham (2006) discuss the effect of numerous alterations to copyright law on the stock value of firms supplying copyright works. They find that stronger legal copyright protection has a statistically significant but minuscule positive effect.

THE BENEFITS OF UNAUTHORIZED COPYING TO USERS

Unauthorized copying occurs because users expect individually to benefit from it. Based on survey results, Rob and Waldfogel (2006) and Waldfogel (2010) estimate that, in the short run, the benefits of file-sharing to end-users far exceed the losses to rights holders. In any case, short-run benefits to users have not received much attention because the long-run effects of unauthorized copying on supply are central in the economic rationale of copyright. The question is whether benefits to users are sustainable considering any adverse effects of unauthorized copying on the supply of new creative works.

One possible extension of the short-run analysis would be the effects of unauthorized copying on retail prices of authorized services. Another is the extent to which suppliers of related ICT goods and services appropriate the value of creative works, as the demand for their products may be boosted by the availability of 'free' content online.

ADMINISTRATION AND TRANSACTION COSTS OF COPYRIGHT

The academic literature has hardly addressed the administration costs of copyright systems – how much it costs to define laws, settle conflicts and so on. In cooperation with researchers from Free University Brussels, the consultancy firm KEA (2012) has recently published estimations of the transaction costs for clearing rights by online retailers of music in Europe. These costs vary between €80 000 and €260 000 according to the scope of the repertoire, the types of services and the rights holders dealt with. This does not include the licence fee itself. A particular problem may be that it can take more than a year to strike any kind of deal.

COPYRIGHT PROTECTION AND THE SUPPLY OF CREATIVE WORKS

In the economic rationale, the fundamental issue is to what extent variations in copyright protection affect the supply of new creative works and thus user welfare in the long run.[1] There is surprisingly little empirical research on the matter. One reason may be that convincing data on the quality of supply of creative works is hard to come by. Another may be that the protracted relationship between copyright protection and content creation is harder to gauge than the more immediate effects on rights holder revenues.

Some historical-economic studies assess the impact of changes in copyright law on the supply of creative works. Khan (2004) finds that the number of full-time literary authors did not increase substantially with the US International Copyright Act of 1891. For the period between 1709 and 1850, Scherer (2008) finds no impact of copyright extensions in Europe on the number of new composers entering the market. A handful of studies address more recent copyright term extensions on the supply of creative works. None of them finds that these would have had a substantial effect on the supply of movies (Hui and Png, 2002; Png and Wang, 2009) or other types of copyright works (Landes and Posner, 2003). The impact of other changes in legal copyright protection on the supply of creative works has hardly been addressed.

Regarding the effect of digital copying, Handke (2010b; 2012a) discusses the supply of new music albums with the diffusion of file-sharing technology, which coincides with substantial sales reductions of authorized copies. Nevertheless, the number of new albums released grew in absolute terms (excluding pure digital releases), and there is no significant deviation from an upward trend in the variety of works supplied that began in the early 1990s. The consumption time of recorded music per inhabitant also expanded. This is probably driven by new ICTs that facilitate the consumption of music. More extensive music consumption also conflicts with the notion that the quality of supply would have decreased substantially.

Waldfogel (2011) assesses the quality of new music by its presence in 'best of all times' lists. While the share of newer music in such lists has decreased over time, this trend has not become more pronounced with file-sharing and sales reductions of recorded music. For the US, Waldfogel (2012) documents that the number of new albums released – including works published only as streams or downloads – 'has increased sharply' since 2000.

In short, there is no evidence that digital copying of music would have diminished the overall quality of supply. This is hardly a conclusive result. On the one hand, things might have been even better with falling costs of creating and marketing music recordings. On the other hand, the full impact of digital copying may not yet have transpired. Nevertheless, according to the limited evidence so far there has been a boom in the supply of musical works in spite of increased unauthorized copying.

The impact of digital copying and unauthorized use on the supply of other types of copyright works has hardly been studied systematically. There is incidental evidence that the sheer number of new works that become available is expanding. The quality of these works and their combined value have not been assessed. It is tricky to do that, because creative works probably give rise to substantial, positive externalities that are not reflected in their market value. One outstanding conceptual problem in the context of digitization is how to deal with user-generated content that is supplied by creators without a direct pecuniary incentive. Not only is it challenging to distinguish the professional or commercial from amateur works. Much user-generated content also draws on professional creative works, creating derivative or transformative works. Furthermore, it is unclear to what extent amateur works are substitutes for professional works, and they probably are complements in some markets.

DISTRIBUTIONAL EFFECTS AND THE IMPLICATIONS FOR COMPETITION AND INNOVATION

Much of the empirical research and theoretical work on copyright assesses the effect on rights holders or users at large. Some additional insights emerge from differentiating further.

Among private end-consumers, the direct benefits of unauthorized copying accrue to those who have access to unauthorized copies. Other end-consumers are left with any adverse effects on the supply of creative works. There may be indirect effects, when suppliers of creative works change retail prices to react to competition from unauthorized copies. This is an issue that remains largely unexplored by empirical research.

On the supply side, many studies find that, in the music industry, the effects of unauthorized copying are worse for large incumbent firms (major record companies) or superstar creators than for newcomers and fringe suppliers. This is a consistent result in a number of studies, such as Blackburn (2004), Gopal et al. (2006), Bhattacharjee et al. (2007), Mortimer et al. (2010) or Waldfogel (2012). The usual explanation is that less well-known suppliers gain relatively more from the exposure of their works as a result of unauthorized copying, whereas, for the most famous recording artists, the sales displacement is stronger than any exposure effect. The result is more frequent market entry, less stability and less concentration in the market share of leading suppliers.[2] Another possible explanation is that small enterprises or newcomers may find it easier to adapt to swift and radical changes in market conditions than large incumbents with a vested interest in the status quo (Reinganum, 1983; Handke, 2006). The difference is that long-tail effects driven by greater exposure would be permanent, as long as copying technology is widely available, whereas industry fragmentation due to more rapid adaptation from fringe suppliers will be temporary and will give way to a shake-out once the pace and scope of radical innovation diminish.

Changes in industry structure and competition due to the diffusion of digital ICT may affect innovation within creative industries in which copyright is concerned. Many creative industries are organized around narrow oligopolies of intermediary firms, such as major record companies, movie studios or publishers. If unauthorized copying undermines the position of incumbents, this may lead to greater competition and efficiency, making users and creators better off. However, the relationship between competition and innovation is complex (Gilbert, 2006a; 2006b). Archetypal competitive markets fail to allocate resources to innovation, as Schumpeter ([1942] 1975) famously observed. The predominant notion is that innovation approximates to an optimum in a situation with sufficient competition to motivate challenges to the status quo (Hicks, 1935) and some, temporary market power of successful innovators (Aghion et al., 2005). Copyright policy is the main lever with which public policy makers can influence this balance.

Copyright also influences the position of rights holders relative to suppliers of related goods and services. The relationship between suppliers of creative works (content) and suppliers of ICT goods and services seems to be very important. The obvious case in point is suppliers of file-sharing or file-hosting services that are used very extensively for copying and disseminating copyright material. There is also limited empirical evidence that other ICT suppliers benefit from the availability of unauthorized copies for their customers (on iPod sales, see Leung, 2009; on Internet traffic, see Adermon and Liang, 2010). Isolating how much of the value of copyright works is appropriated by ICT firms is tricky, as ICT goods and services are typically used for many different purposes simultaneously.

Much of the economic literature on copyright focuses on the creation of new copyright works (content creation) and on the problem that advances in copying technology may undermine incentives to create. However, digitization entails productivity increases regarding the technical and organizational means to generate, reproduce and disseminate creative works. The social value of creative works depends not only on their mere existence but also on the extent to which they are available to users. If copyright strengthens rights holders relative to users, stronger copyright protection may inhibit technological innovation by users (Plant, 1934; Handke, 2011b).[3] In functioning markets, the welfare loss would not exceed the transaction costs in trading rights. Transaction costs tend to be high in markets for intellectual property rights (Levin et al., 1987: 788; Landes and Posner, 2003: 16). Overall, there is probably a trade-off between incentives for content creation and related technological innovation – similar to the trade-off between rights holder and user innovation regarding content creation. Getting this balance right is also central for the sustainable development and competitiveness of creative industries and may influence the ICT sector as well.

INDUSTRY ADAPTATION

In the debate on digitization and copyright policy, several extensions and complications to the economic analyses have been discussed. A common theme is that creators could adapt in order to operate profitably in the presence of digital copying. There is empirical support for a couple of related arguments.

First, any adverse effects of unauthorized copying may be mitigated as rights holders adapt their operations (Varian, 2005). Regarding empirical evidence, Liebowitz (1985) documented that academic journal publishers reacted to photocopying by price discrimination. Charging higher prices to libraries, they managed to 'indirectly appropriate' much of the value of copying. See also Mortimer (2007) for movies. There is extensive evidence that positive network effects operate in markets for software, so that the market value of works increases with the number of users (Givon et al., 1995; Brynjolfsson and Kemerer, 1996). Unauthorized use of creative works may thus boost the demand for authorized copies from the same supplier, for example because more potential users become aware of these works (Bounie et al., 2006; Gopal et al., 2006). However, if network effects affect the market share of suppliers, that does not mean they necessarily affect overall demand.

Second, as unauthorized copying makes creative works available to a greater number of users, rights holders may also try to exploit positive effects on the demand for more excludable, related goods and services. Mortimer et al. (2010) show that revenues from live performances increase with file-sharing of music.[4] Suppliers of software and hardware for console based video games seem to manage their interdependence reasonably well, without full integration (Shankar and Bayus, 2002; Clements and Ohashi, 2005). An important question in all of this is how suppliers may vary prices over time in order to first develop and then exploit network effects.

Finally, there is extensive empirical evidence that creative work is motivated by non-pecuniary incentives. See for example Benhamou (2003), Lakhani and von Hippel (2003) and Towse (2006). The supply of creative works may thus be relatively insensitive to sales displacement due to unauthorized copying, and efficient copyright systems may need to cater for intrinsically motivated creators.

WHAT DETERMINES UNAUTHORIZED COPYING?

To develop efficient means to restrict unauthorized copying, and to target copyright policy, it is important to understand what determines unauthorized copying. For more comprehensive overviews of factors see Rochelandet and Le Guel (2005) and Hennig-Thurau et al. (2007).

Wealth and Education

Unauthorized copying would be less harmful if it were mainly conducted by those who would not purchase otherwise, for instance because they lack the means to pay (Bai and Waldfogel, 2010). Empirical results are mixed. On the macro-level, indicators of wealth and countries' economic development are generally found to be inversely related to piracy rates of software (for example Marron and Steel, 2000; Bezmen and Depken, 2006).[5] In micro-surveys, however, individuals' disposable income or social status rarely correlates significantly with unauthorized copying in other copyright industries (Rochelandet and Le Guel, 2005; Andersen and Frenz, 2010).

Boorstin (2004) observes that Internet access is associated with increased CD purchases for above-25-year-old US Americans, and with lower purchases for 15- to 24-year-olds. An important question that arises from this is whether unauthorized copying will remain contained to youngsters (with more spare time and less disposable income) or whether the habit sticks as the 'digital natives' grow up.

The Relative Utility of Unauthorized Copying

Unauthorized copies are typically less costly to acquire for many Internet users. The value of unauthorized copies is generally reported to be lower than that of authorized versions (for example Bezmen and Depken, 2006; Rob and Waldfogel, 2006).[6] This means that there is some scope for rights holders to sell at prices above the marginal costs of copying, however limited that may be. Some empirical work documents that end-users occasionally even purchase authorized copies after they have acquired (and consumed) an unauthorized copy if they find the work to be of high value (Gopal et al., 2006; see also Andersen and Frenz, 2010).

Besides the perceived value of copies, convenience of access and variety of available content may co-determine the choice between authorized and unauthorized copies.

Rochelandet and Le Guel (2005) found that the variety of supply motivated file-sharing of music. In Hennig-Thurau et al. (2007), file-sharing of movies is positively correlated with perceived transaction costs for accessing authorized versions. Authorized retailing services have advanced over recent years. However, covering a time period with more comprehensive authorized music retailing, Waldfogel (2010) finds that the displacement rate between unauthorized downloads and authorized downloads of recorded music is still roughly the same as that between unauthorized downloads and CDs.

Copyright Protection and Enforcement

A number of studies document that unauthorized copying is quite sensitive to public signals of greater enforcement (Blackburn, 2004; Maffioletti and Ramello, 2004; Bhattacharjee et al., 2006; Jeong and Lee, 2010; an exception is Rochelandet and Le Guel, 2005). It is not certain whether the effect is permanent, as suppliers of services related to unauthorized copying and users adapt their behaviour.

Cultural Factors

On the macro-level, Walls (2008) finds that countries with higher scores in social coordination exhibit more extensive unauthorized copying of movies. In micro-level surveys, Rochelandet and Le Guel (2005) as well as Cox et al. (2010) document that peer groups affect file-sharing of music, whereas Hennig-Thurau et al. (2007) find no such effect for movie file-sharing.

Several studies also show that moral considerations affect file-sharing. On the one hand, according to Rochelandet and Le Guel (2005) and Hennig-Thurau et al. (2007), some concern for creators reduces the intensity of unauthorized copying. On the other hand, file-sharing may also be motivated by anti-corporate attitudes (Hennig-Thurau et al., 2007), and some consider file-sharing as an act of philanthropy (Cox et al., 2010). Some users make voluntary contributions when given the opportunity (Regner and Barria, 2009; Dale and Morgan, 2010).

CONCLUDING REMARKS

It is a puzzling question whether and how the copyright system should be adapted in the course of digitization. The diffusion of digital ICT affects many relevant aspects of the creative economy. Economic theory provides a useful structure for assessing the welfare effects of unauthorized copying and copyright, and to develop testable propositions. In order to inform the copyright debate, empirical research is also needed to gauge the various costs and benefits in specific contexts and to weigh them against each other.

The existing empirical literature on the effects of unauthorized copying and copyright already provides some important insights. First, it is often found that unauthorized, digital copying displaces demand for authorized copies of recorded music and movies. There may also be some scope for rights holders to adapt, mitigating adverse effects on their profitability at least somewhat over time. Second, there is no evidence that the supply of creative works would have decreased because of unauthorized digital copying.

At least from the economic perspective, the issue is key and requires more systematic attention. Third, empirical work indicates that unauthorized copying is associated with greater contestability of markets for creative works. This should positively affect static efficiency, but the consequences for innovation and dynamic efficiency are harder to predict. Finally, it seems that many users do value authorized services and the notion that creators are rewarded, and are concerned with legal risks. This provides some opportunities for effective compensation schemes.[7]

The copyright debate often seems to be in a deadlock, while the current copyright system leaves most stakeholders dissatisfied. Part of a more sustainable solution to the copyright question may be improvements in administering and trading rights. In this sense, general purpose ICT may help to solve the very problem with unauthorized copying that it has aggravated. What is more, in the course of digitization better data on the supply and use of copyright works becomes available, enabling empirical research. Empirical work should help develop new ways to administer copyrights online that are more widely acceptable than current arrangements.

NOTES

* This chapter draws on a report to the National Academies of the Sciences (Handke, 2011a), excerpts of which were published in the *Review of Economic Research on Copyright Issues* (Handke, 2012b).

1. This section does not make the distinction from Table 22.1 regarding rights holder and user innovation. The reason is that, in practice, this distinction often does not apply. Most creators of copyright works are (prospective) rights holders. Most creators are also users of copyrights held by others at least potentially (depending on the scope of copyright protection and the originality of creations).
2. The issue is related to the discussion of the long-tail hypothesis (Anderson, 2004), and unauthorized copying may be one of the drivers of long-tail effects. For the limited evidence on long-tail effects, see for example Brynjolfsson et al. (2011) or Elberse and Oberholzer-Gee (2007).
3. This holds where intellectual property arrangements for copyright works and technological innovations are asymmetrical.
4. There have also been attempts to cover development costs through up-front payments from prospective users (crowdsourcing).
5. Bezmen and Depken (2006) do not conduct cross-country comparisons but compare US federal states.
6. If unauthorized copies are of lower value, it is wrong to calculate the harm to rights holders by multiplying the number of downloads with retail prices of authorized copies. The willingness to pay for some works accessed via unauthorized copying is lower than the retail price of 'originals'. For those works, copiers would not purchase if unauthorized copies were unavailable.
7. However, appeals to 'fairness' may not go well with coercion.

REFERENCES

Adermon, A. and C.-Y. Liang (2010), 'Piracy, music, and movies: a natural experiment', Uppsala University, Department of Economics Working Paper No. 2010:18.

Aghion, P., N. Bloom, R. Blundell, R. Griffith and P. Howitt (2005), 'Competition and innovation: an inverted U-shaped relationship', *Quarterly Journal of Economics*, **120** (2), 701–28.

Andersen, B. and M. Frenz (2010), 'Don't blame the P2P file-sharers: the impact of free music downloads on the purchase of music CDs in Canada', *Journal of Evolutionary Economics*, **20** (5), 715–40.

Anderson, C. (2004), 'The long tail', *Wired Magazine*, 12.10, available at: http://www.wired.com/wired/archive/12.10/tail.html.

Bai, J. and J. Waldfogel (2010), 'Movie piracy and sales displacement in two samples of Chinese consumers', available at: http://bpp.wharton.upenn.edu/waldfogj/pdfs/china.pdf.

Baker, M. and B. Cunningham (2006), 'Court decisions and equity markets: estimating the value of copyright protection', *Journal of Law and Economics*, **49** (2), 567–96.

Benhamou, F. (2003), 'Artists' labour markets', in R. Towse (ed.), *A Handbook of Cultural Economics*, Cheltenham, UK and Northampton, MA, USA: Edward Elgar Publishing, pp. 69–75.

Bezmen, T.L. and C.A. Depken (2006), 'Influences on software piracy: evidence from the various United States', *Economics Letters*, **90**, 356–61.

Bhattacharjee, S., R.D. Gopal, K. Lertwachara and J.R. Marsden (2006), 'Impact of legal threats on online music sharing activity: an analysis of music industry actions', *Journal of Law and Economics*, **49** (1), 91–114.

Bhattacharjee, S., R.D. Gopal, K. Lertwachara, J.R. Marsden and R. Telang (2007), 'The effect of digital sharing technologies on music markets: a survival analysis of albums on ranking charts', *Management Science*, **53** (9), 1359–74.

Blackburn, D. (2004), 'A study of online piracy and recorded music sales', Harvard University, working paper, available at: http://www.katallaxi.se/grejer/blackburn/blackburn_fs.pdf.

Boorstin, E. (2004), 'Music sales in the age of file sharing', senior thesis, Princeton University, available at: http://www.cs.princeton.edu/~felten/boorstin-thesis.pdf.

Bounie, D., M. Bourreau and P. Waelbroeck (2006), 'Piracy and the demand for films: analysis of piracy behavior in French universities', *Review of Economic Research on Copyright Issues*, **3** (2), 15–27.

Brynjolfsson, E. and C.F. Kemerer (1996), 'Network externalities in microcomputer software: an econometric analysis of the spreadsheet market', *Management Science*, **42** (12), 1627–47.

Brynjolfsson, E., Y. Hu and D. Simester (2011), 'Goodbye Pareto principle, hello long tail: the effect of search costs on the concentration of product sales', *Management Science*, **57** (8), 1373–86.

Clements, M.T. and H. Ohashi (2005), 'Indirect network effects and the product cycle: video games in the U.S., 1994–2002', *Journal of Industrial Economics*, **53** (4), 515–42.

Cox, J., A. Collins and S. Drinkwater (2010), 'Seeders, leechers and social norms: evidence from the market for illicit digital downloading', *Information Economics and Policy*, **22** (4), 299–305.

Dale, D.J. and J. Morgan (2010), 'Silence is golden – suggested donations in voluntary contribution games', available at: http://faculty.haas.berkeley.edu/rjmorgan/Silence%20is%20Golden.pdf.

Elberse, A. and F. Oberholzer-Gee (2007), 'Superstars and underdogs: an examination of the long tail phenomenon in video sales', Harvard Business School Working Paper 07–015.

Gilbert, R.J. (2006a), 'Competition and innovation', in W.D. Collins (ed.), *Issues in Competition Law and Policy*, American Bar Association Antitrust Section, available at: http://works.bepress.com/richard_gilbert/12.

Gilbert, R.J. (2006b), 'Competition and innovation', *Journal of Industrial Organization Education*, **1** (1), Article 8.

Givon, M., V. Mahajan and E. Muller (1995), 'Software piracy: estimation of lost sales and the impact on software diffusion', *Journal of Marketing*, **59** (1), 29–37.

Goode, S. (2012), 'Initial findings of a gap analysis of the digital piracy literature: undiscovered countries', *Journal of Research in Interactive Marketing*, **6** (4), 238–59.

Gopal, R.D., S. Bhattacharjee and G.L. Sanders (2006), 'Do artists benefit from online music sharing?', *Journal of Business*, **79** (3), 1503–33.

Handke, C. (2006), 'Plain destruction or creative destruction? Copyright erosion and the evolution of the record industry', *Review of Economic Research on Copyright Issues*, **3** (2), 29–51.

Handke, C. (2010a), *The Economics of Copyright and Digitisation: A Report on the Literature and the Need for Further Research*, London: Strategic Advisory Board for Intellectual Property Policy (SABIP), available at: http://www.ipo.gov.uk/ipresearch-economics-201005.pdf.

Handke, C. (2010b), 'The creative destruction of copyright: innovation in the record industry and digital copying', doctoral dissertation, Erasmus University, available at: SSRN 163034.

Handke, C. (2011a), *Economic Effects of Copyright: The Empirical Evidence So Far*, Report for the National Academies of the Sciences, Washington, DC, available at: http://sites.nationalacademies.org/PGA/step/PGA_063399.

Handke, C. (2011b), 'The innovation costs of copyright', paper presented at the Society for Economic Research on Copyright Issues Annual Congress 2011, 7 and 8 July, Bilbao, available at: http://www.serci.org/congress/papers/handke.pdf.

Handke, C. (2012a), 'Digital copying and the supply of sound recordings', *Information Economics and Policy*, **24** (1), 15–29.

Handke, C. (2012b), 'A taxonomy of empirical research on copyright: how do we inform policy?', *Review of Economic Research on Copyright Issues*, **9** (1), 47–92.

Handke, C. and R. Towse (2007), 'Economics of copyright collecting societies', *International Review of Intellectual Property and Competition Law*, **38** (8), 937–57.

Hennig-Thurau, T., V. Hennig and H. Sattler (2007), 'Consumer file-sharing of motion pictures', *Journal of Marketing*, **71** (4), 1–18.

Hicks, J.R. (1935), 'Annual survey of economic theory: the theory of monopoly', *Econometrica*, **3** (1), 1–20.

Hui, K.-L. and I.P.L. Png (2002), 'On the supply of creative work: evidence from the movies', *American Economic Review*, **92** (2), 217–20.
Jeong, G. and J. Lee (2010), 'Estimating consumer preferences for online music services', *Applied Economics*, **42** (30), 3885–93.
KEA/Vrije Universiteit Brussel (2012), *Licensing Music Works and Transaction Costs in Europe*, available at: http://www.keanet.eu/docs/music%20licensing%20and%20transaction%20costs%20-%20full.pdf.
Khan, B.Z. (2004), *Does Copyright Piracy Pay? The Effects of U.S. International Copyright Laws on the Market for Books, 1790–1920*, Working Paper 10271, Cambridge, MA: National Bureau of Economic Research.
Lakhani, K.R. and E. von Hippel (2003), 'How open source software works: "free" user-to-user assistance', *Research Policy*, **32** (6), 923–43.
Landes, W.M. and R.A. Posner (1989), 'An economic analysis of copyright law', *Journal of Legal Studies*, **18** (2), 325–63.
Landes, W.M. and R.A. Posner (2003), *The Economic Structure of Intellectual Property Law*, Cambridge, MA: Belknap Press of Harvard University Press.
Leung, T.C. (2009), *Should the Music Industry Sue Its Own Customers? Impacts of Music Piracy and Policy Suggestions*, Chinese University of Hong Kong working paper, available at: http://www.econ.cuhk.edu.hk/dept/seminar/09–10/1st-term/i-Podandpiracy.pdf.
Levin, R.C., A.K. Klevorick, R.N. Nelson, S.G. Winter, R. Gilbert and Z. Griliches (1987), 'Appropriating the returns from industrial research and development', *Brookings Papers on Economic Activity*, **1987** (3), 783–831.
Liebowitz, S.J. (1985), 'Copying and indirect appropriability: photocopying of journals', *Journal of Political Economy*, **93** (5), 945–57.
Maffioletti, A. and G.B. Ramello (2004), 'Should we put them in jail? Copyright infringement, penalties and consumer behaviour: insights from experimental data', *Review of Economic Research on Copyright Issues*, **1** (2), 81–95.
Marron, D.B. and D.G. Steel (2000), 'Which countries protect intellectual property? The case of software piracy', *Economic Inquiry*, **38** (2), 159–74.
Mortimer, J.H. (2007), 'Price discrimination, copyright law, and technological innovation: evidence from the introduction of DVDs', *Quarterly Journal of Economics*, **122** (3), 1307–50.
Mortimer, J.H., C. Nosko and A. Sorensen (2010), 'Supply responses to digital distribution: recorded music and live performances', *Information Economics and Policy*, **24** (1), 3–14.
Plant, A. (1934), 'The economic aspects of copyright in books', *Economica*, **1** (2), 167–95.
Png, I.P.L. and Q.H. Wang (2009), *Copyright Law and the Supply of Creative Work: Evidence from the Movies*, working paper, National University of Singapore, available at: http://www.comp.nus.edu.sg/~ipng/research/copyrt.pdf.
Regner, T. and J.A. Barria (2009), 'Do consumers pay voluntarily? The case of online music', *Journal of Economic Behaviour and Organization*, **71** (2), 395–406.
Reinganum, J.F. (1983), 'Uncertain innovation and the persistence of monopoly', *American Economic Review*, **73** (4), 741–8.
Rob, R. and J. Waldfogel (2006), 'Piracy on the high C's: music downloading, sales displacement, and social welfare in a sample of college students', *Journal of Law and Economics*, **49** (1), 29–62.
Rochelandet, F. and F. Le Guel (2005), 'P2P music sharing networks: why the legal fight against copies may be inefficient', *Review of Economic Research on Copyright Issues*, **2** (2), 69–82.
Scherer, F.M. (2008), 'The emergence of musical copyright in Europe from 1709 to 1850', *Review of Economic Research on Copyright Issues*, **5** (2), 3–18.
Schumpeter, J.A. ([1942] 1975), *Capitalism, Socialism and Democracy*, New York: Harper.
Shankar, V. and B.L. Bayus (2002), 'Network effects and competition: an empirical analysis of the home video game industry', *Strategic Management Journal*, **24** (4), 375–84.
Smith, M. and R. Telang (2012), 'Assessing the academic literature regarding the impact of media piracy on sales', available at: SSRN 2132153.
Towse, R. (2006), 'Copyright and artists: a view from cultural economics', *Journal of Economic Surveys*, **20** (4), 567–85.
Varian, H. (2005), 'Copying and copyright', *Journal of Economic Perspectives*, **19** (2), 121–38.
Waldfogel, J. (2010), 'Music file sharing and sales displacement in the iTunes era', *Information Economics and Policy*, **22** (4), 306–14.
Waldfogel, J. (2011), *Bye, Bye, Miss American Pie? The Supply of New Recorded Music since Napster*, Working Paper 16882, Cambridge, MA: National Bureau of Economic Research.
Waldfogel, J. (2012), 'And the bands played on: digital disintermediation and the quality of new recorded music', available at: SSRN 2117372.
Walls, W.D. (2008), 'Cross-country analysis of movie piracy', *Applied Economics*, **40** (5), 625–32.

FURTHER READING

A more extensive overview of the quantitative-empirical literature on copying and copyright is found in Handke (2011a), which includes references to several other surveys of the economic literature. Handke (2010a) discusses copyright and digitization. For coverage of other methods and disciplines than quantitative research related to economics, see Goode (2012).

23. Internet piracy: the estimated impact on sales

*Stan J. Liebowitz**

THE IMPACTS OF INTERNET PIRACY

Fifteen years ago the term 'file-sharing' was unknown. Then Napster arrived in the second half of 1999 and grew to be an international sensation during 2000. The sound recording industry experienced a dramatic swoon in sales beginning in 2000, continuing unabated (with one informative exception) through 2010. The industry has blamed this sales decline on the rapid growth of file-sharing. Although Napster was effectively shut down as an unauthorized file-sharing service within two years of its birth, its progeny live on, as do new habits developed by music listeners.

Shortly after Napster's arrival, economists began to examine the likely impacts of file-sharing on the sound recording market.[1] Further, as faster download speeds and the invention of BitTorrent allowed file-sharing to expand into movies, the impact of file-sharing on the movie industry also became a question that economists tried to answer.

Although it is clear that pirated versions of products often substitute for the purchase of an original, and this effect is unambiguously harmful to the industry, there are other, more subtle effects possibly at work, as well. Piracy could allow a consumer to discover new songs that then induce the consumer to go purchase an album that might otherwise have not been purchased, for example. This 'sampling' effect, first proposed in Liebowitz (1985), makes the theory of piracy somewhat ambiguous.

There are two separate questions that one can ask about the impact of piracy on the sales of products. First, does piracy actually harm the industries whose products are used without the permission of the copyright owner? Second, if it does harm the owners, how much does it harm them? In this chapter I intend to provide answers to these questions, to the extent that we can answer them, using the most up-to-date information available from academic studies.

MOVIE AND MUSIC SALES AFTER THE INTRODUCTION OF FILE-SHARING

The decline that has taken place in the music business is dramatic, so much so that whatever factor or factors have caused this decline must have been extremely powerful. Figure 23.1 reports the number of albums (CDs or digital) sold per person[2] (with ten digital singles being counted as an album) in the US since 1973.[3] For most of the first 30 years, sales progress upward in a somewhat irregular fashion, but then, just as Napster initiates the world into the habits of file-sharing, sales begin a fairly continuous and precipitous decline. Revenues (adjusted for inflation) follow a similar and even more precipitous path downward. A similar story is found for countries other than the US,

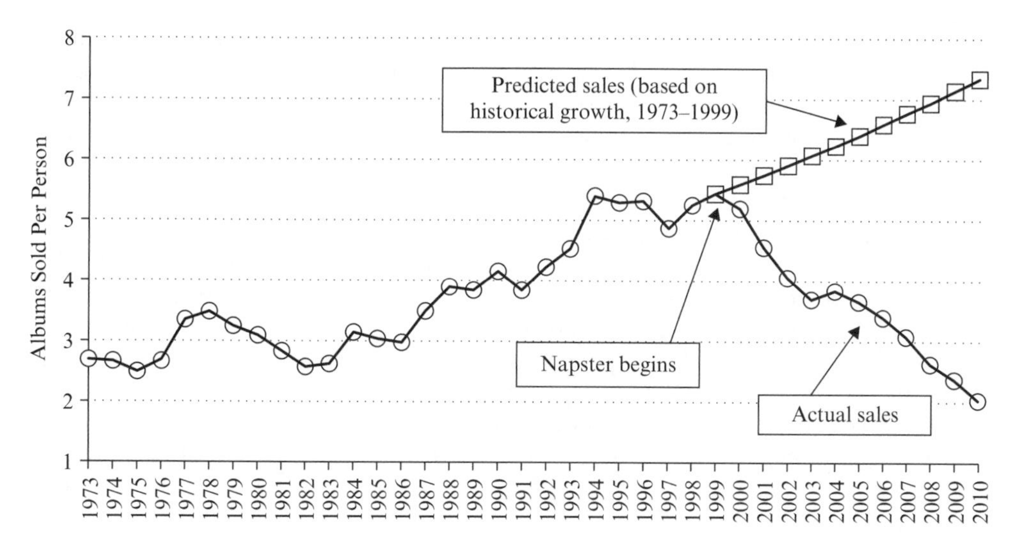

Figure 23.1 US album sales, 1973 to 2010

with every leading country experiencing a major decline in sales (see Table 23.1 in the next section).

Although the growth of Internet piracy (file-sharing) is one obvious explanation for this decline in music sales, there were many claims, particularly in the early years of file-sharing, that some other factor or factors might have been responsible for this decline in record sales. These other possible factors (e.g., the price of substitutes, recessions, format changes) were explored in Liebowitz (2004, 2006), who found no evidence to support any alternative explanation for the decline in sound recording sales.

It is also worth noting that the birth of file-sharing would not be expected to lead to a one-time drop in sales, but instead a decline that might take many years to reach its conclusion. That is because the impact of piracy would not be a function merely of the number of pirates, but also of the extent to which pirated files are substitutes for purchased files. For example, when MP3 files first appeared, they were not very good substitutes for CDs. Originally, MP3 files could only be played on computers and on MP3 players (which were quite uncommon before the iPod became available) and could not be played in cars or stereo systems. That made these files not very good substitutes for CDs, which could be played in stereo systems, in automobiles and on virtually all other music reproduction equipment. Over time, MP3 files were accepted on virtually all music reproduction devices, making pirated MP3 files much better substitutes for purchased music files than was originally the case. If piracy had a negative impact on sales, that negative impact should have increased in size over time as MP3 files were becoming better substitutes for purchased albums.

Similar factors are at work for pirated versions of movies. Movies consist of much bigger digital files than do music songs or even albums. This made the downloading of movies much more difficult and time consuming than was the case for downloading music. Thus, for an individual without broadband it was impractical to even try downloading a movie. Further, the nature of traditional file-sharing systems was that the file was downloaded from a single uploader, and uploading bandwidth was usually much less than downloading bandwidth, even with broadband. For this reason, the downloading of movies did not become practical until broadband became common and until a different download program came into existence that allowed the downloader to download the movie from multiple uploaders at the same time. This program was BitTorrent, and its arrival in 2003 played a role for movie downloading that was similar to the role that Napster's arrival in 1999 played for music downloading.

Analyzing the movie industry is complicated by the fact that the major revenue streams are generated by two quite different consumption activities: theatrical exhibitions and DVD sales. Movies that are downloaded without permission are intended to be viewed at home or on a portable device such as a laptop. In other words, movies pirated online are very similar to a DVD, except that the quality may be lower depending on the compression used in the download. For most people, a pirated movie is *not* a very good substitute for viewing the movie in a theater, because a theater almost always provides a more intense immersion in the movie than does a television at home (although some home theater systems can be very good), and attending a movie is a different type of activity than watching at home.

There may be some people seeing a movie at a theater who do not value the movie theater experience much differently from viewing the movie at home on television but only go to the theater because they wish to see a movie as soon as possible after it comes out. For these people, watching a DVD or pirated download might be a good substitute for viewing the movie in a theater. Nevertheless, if, as seems likely, most viewers in movie theaters are there because they prefer the theater experience, then pirated movie files would be a better substitute for a purchased DVD than they would for a theatrical exhibition. In this case, increased piracy would be expected to have a larger impact on DVD sales and rentals than it would on theatrical exhibitions, even though piracy might affect both markets.

The general industry evidence appears consistent with a hypothesis that piracy has hurt the movie industry. The data in Figure 23.2 are consistent with this view.

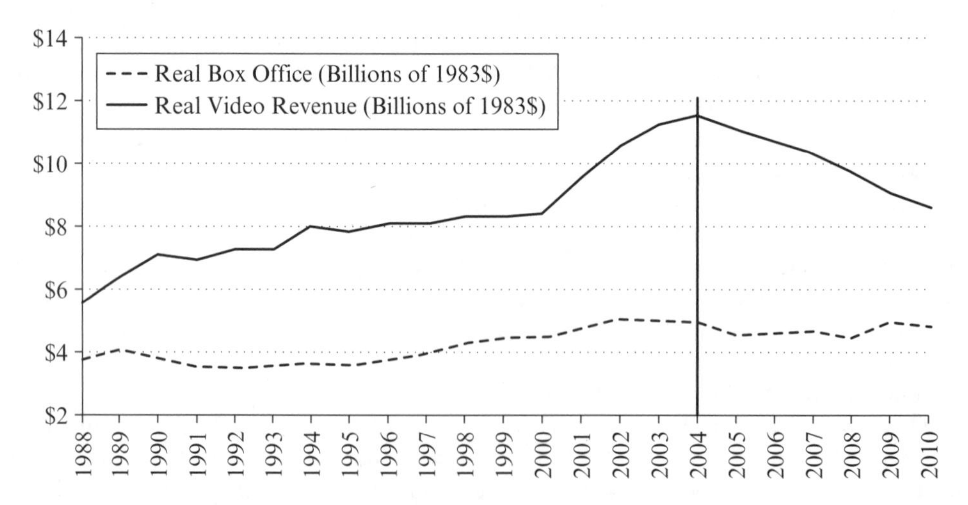

Figure 23.2 Box office and video sales and rentals, 1988 to 2010

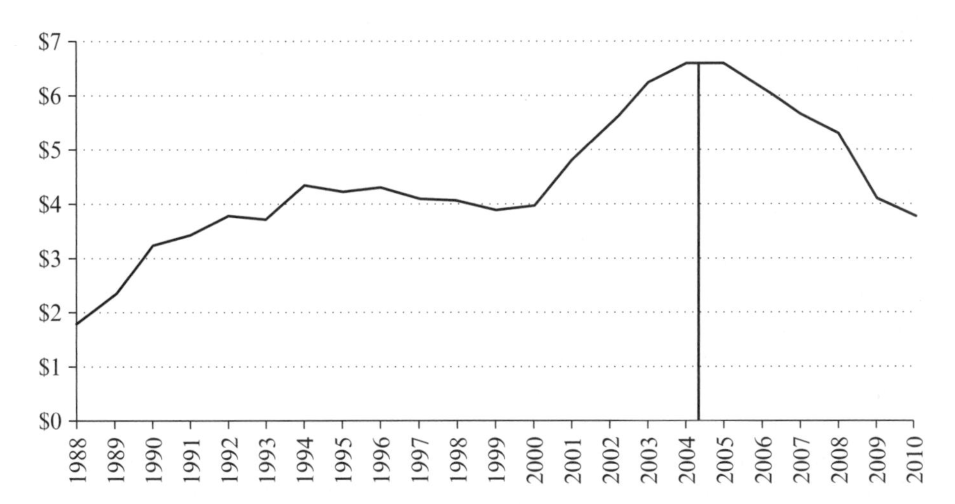

Figure 23.3 Video revenues minus box office, 1988 to 2010

Figure 23.2 reports on North American inflation adjusted revenues for exhibition of movies in theaters (the lower line) and also the sales and rentals of prerecorded movies (the upper line).[4]

Box office revenues (theatrical exhibitions) show a fairly smooth 47 percent increase from 1992 until 2002, although revenues have largely remained constant since 2002.[5] Perhaps this leveling off of box office revenues after 2002, coming as it did after a ten-year increase starting in the early 1990s, was due to piracy, but on this chart the change doesn't look particularly dramatic and the box office data in Figure 23.2 can at best be merely suggestive of causation. The box office data may merely reflect changes in the quality of movies, or changes in substitute activities.

Examining differences between box office revenues and DVD/VHS sales/rentals, however, largely controls for factors such as movie quality or the impact of movie substitutes, because these factors should affect both the theatrical and the prerecorded revenue channels. By subtracting one channel from the other the net effect should eliminate the impact of factors that affect both revenue streams.

This subtraction is performed in Figure 23.3, which has removed box office revenues from the sales and rentals of videos. Looked at in this way, video revenues from the sales, paid downloads, rentals, and paid streams of DVDs, relative to box office receipts, have fallen quite significantly after 2005, with a decline of approximately 43 percent. This is a very large decline in a short period of time. Note that this is a decline relative to box office revenues, and box office revenues themselves, as we will see below, have been estimated to have been negatively impacted by piracy as well. The fact that prerecorded movie sales and rentals have declined relative to box office receipts is consistent with the theoretical likelihood that piracy would have a larger impact on video sales and rentals than on revenues from theatrical exhibition, and thus is consistent with a view that piracy has hurt sales in this market.

Still, these are merely preliminary assessments. Economists have conducted more sophisticated studies directly testing whether file-sharing led to a decline in sound recording sales or movie revenues, and I will discuss the results of those academic studies in the next sections.

ECONOMIC ANALYSES OF THE SOUND RECORDING MARKET

In September 2011 I completed a research paper that examined all the peer-reviewed papers I could find on the subject of piracy's impact on the sound recording industry.[6] Prior to my paper, results from different studies were not easily compared, since they were measured in different ways. My main purpose was to examine the studies that found harm, since they all agreed in the direction of the impact of piracy, and to put their results into a common metric to see how consistent they were in terms of how large a negative impact was due to piracy. The common metric that I used was the share of the total industry decline that was estimated to be due to file-sharing. I also examined the few papers that did not find harm, as I discuss below. As is easily seen in Table 23.1, the size of the decline is dramatically different in different years, is different for the US versus the rest of the world, and is even different depending on whether we are measuring units (albums) or revenues. Thus the findings of the various studies of the impact of piracy on music will need to be taken in the context of the time and location of their measurements.

The standard econometric formulation for measuring the impact of file-sharing can be represented as:

$$RS = \alpha + \beta FS + \gamma Z$$

Table 23.1 Cumulative decline in music sales after 1999

	US units	US real revenues	Non-US real revenues
2000	−4.8%	−5.0%	−4.5%
2001	−13.5%	−11.4%	−7.5%
2002	−22.5%	−19.9%	−12.5%
2003	−28.3%	−26.4%	−21.0%
2004	−24.8%	−25.4%	−27.3%
2005	−27.5%	−30.7%	−32.6%
2006	−32.5%	−37.9%	−37.7%
2007	−38.4%	−47.3%	−43.1%
2008	−47.7%	−59.2%	−47.1%
2009	−52.6%	−63.9%	−49.4%
2010	−55.4%	−68.0%	−54.5%
2011	−53.9%	−68.3%	

Notes:
Units = full length albums and digital singles divided by 10.
Non-US revenues include performing rights and ringtones.

Source: US data from the RIAA database (available for a small fee), which can be found at the RIAA website (n.d.); non-US data from IFPI (2011: 7).

Table 23.2 Share of decline due to file-sharing

Published/unpublished studies	Share of decline due to file-sharing	Period	Data	Published?
Hong (2007)	>100%	2002	US	Yes
Hong (2013)	20%–40%	2002	US	Yes
Liebowitz (2006)	100%	2005	US	Yes
Liebowitz (2008)	>100%	1998–2003	US	Yes
Michel (2006)	45%	2003	US	Yes
Peitz and Waelbroeck (2004)	>100%	1998–2002	International	Yes
Rob and Waldfogel (2006)	35% or >100%	2003/04	US	Yes
Waldfogel (2010)	65%	2008/09	US	Yes
Zentner (2006)	>100%	2001	7 European	Yes
Zentner (2005)	100%	2002	International	Yes
Blackburn (2004)	>100%	2003	US	No
Zentner (2009)	75%	1997–2008	International	No

where RS stands for record (album) sales, FS stands for the amount of file-sharing and Z is an array of other covariates intended to control for factors related to record sales. Most studies use the results from the above regression to estimate the decline in sales due to file-sharing as a percentage of the sales of records $(\beta \cdot \overline{FS})$/RS, so they might conclude with a statement such as 'file-sharing led to a 10 percent decline in record sales'. Because the impact of piracy is expected to change over time and differ across regions, two studies, even if performed perfectly with ideal data, will have different results if they are based on different time periods or different geographic regions.

The metric I propose merely transposes the prior measurement by taking the measured decline in sales due to file-sharing as a percentage of the overall decline that had occurred from the start of the file-sharing era. In other words, what share of the actual decline represented in Table 23.1 appears to be due to file-sharing? It can be represented as:

$$Metric = Share\ of\ Loss \equiv \frac{(\beta . \overline{FS})}{\Delta RS}$$

I then applied this metric to all the studies that had found some degree of harm from piracy on the recording industry. There were 12 such studies, ten of them published and two of them working papers. They can be found in Table 23.2.

The first column of the table lists the author(s) of each publication and the date of publication.[7] The second column lists the share of the decline in record sales that can be attributed to piracy. The appendix of my 2011 paper goes through the details of the calculations used to derive the numbers in this table.

As seen in the second column, after the authors' names, most of the studies in Table 23.2 have results indicating that the entire enormous decline that has occurred in the record industry (see Table 23.1) is due to piracy. Results greater than 100 percent mean that sales would have increased, except for piracy.[8]

Seven of these studies have results indicating that the entire decline (or more) in sales is due to file-sharing. Another study has two results, with one of those results consistent with the full decline being due to file-sharing and the other result indicating that about a third of the decline is due to file-sharing. Three other studies indicate that file-sharing is responsible for about half, two-thirds or three-quarters of the decline, and one study finds the smallest result, that file-sharing is responsible for between 20 percent and 40 percent of the decline. It is clear that the average of these studies is not the 20 percent 'typical estimate' that has been claimed by a different survey article (Oberholzer-Gee and Strumpf, 2009), since only one of the 12 studies has a result as low as 20 percent.

There are no published articles in academic journals that find a positive impact of file-sharing on sound recording sales, although there is a study (Andersen and Frenz, 2007), conducted for a Canadian ministry, which concludes that file-sharing has a positive impact on sound recording sales.

There are two published studies that do not find that file-sharing harms sales. These are Oberholzer-Gee and Strumpf (2007) and a revised version of the Canadian ministry study using the identical data, Andersen and Frenz (2010). Obviously, when studies find no impact of file-sharing on sound recording sales there is no need to worry about making the result comparable to others, since all of the metrics will be zero. Although I discuss overall results including these two papers, I believe these two papers should be heavily discounted because I consider each to suffer from serious flaws.[9]

Although there have sometimes been claims that other published studies also find no impact of piracy on sales, Liebowitz (2011) has investigated those claims and found them to be a misreading of the literature.

Even if these two studies are added into the mix, it is still the case that the results from a majority of all studies imply that file-sharing is responsible for the entire decline in sales. This result is also consistent with the evidence from proposed alternative hypotheses to explain the decline in sound recording sales. Liebowitz (2004, 2006) carefully examined these alternative explanations and found that they were largely lacking in empirical support, and the intervening years have only strengthened that conclusion.[10] With no support for other possible explanations for the decline in record sales, logical consistency would lead to a conclusion that file-sharing was responsible for the entire decline, which is what the majority of economic studies have found, once the proper metrics are used.

ECONOMIC STUDIES OF MOVIE PIRACY

The theoretical reason that piracy might not harm creators is the possibility that piracy might be used to 'sample' a product prior to actual purchase. Any positive value of sampling, however, seems far less likely to occur for movies than for music. The reason behind this belief is that individuals watch movies only a small number of times, unlike music, where individuals may listen to a particular song dozens or hundreds of times. Because individual movies tend to be seen only a small number of times, often only one time, watching a pirated version of a movie will often be the only viewing of the movie required by an individual, thus eliminating the possibility, for movies that are only going to be seen once, of the pirated movie leading to a later purchase (the sampling effect).

There are not quite as many research papers analyzing the impact of piracy on movies,

Table 23.3 Academic studies on movie piracy

	Sales or box office	Results	Period	Data	Published?
Hennig-Thurau et al. (2007)	Both	13% negative impact on box office; 15% negative impact on DVD sales; 11% negative impact on DVD rentals	2006	German	Yes
Bounie et al. (2006)	Both	Negative impact on sales; zero on box office	2005	French	Yes
De Vany and Walls (2007)	Box office	Negative impact	Unknown	US	Yes
Rob and Waldfogel (2007)	Both	Negative impact on both	2005	US	Yes
Ma et al. (2011)	Box office	Negative impact on box office	2006–09	US	No
Zentner (2010)	Both	Decline in video sales, but not box office; large economic impact on rentals but not statistically significant	1996–2008	International	No
Danaher and Waldfogel (2012)	Box office	Negative in world; zero in US; argue that later non-US availability drives piracy	2003–06	International	No

and the analysis is somewhat more complicated than is the case for music because there are two main revenue streams that occur over different time periods: theatrical exhibition and sales/rentals of prerecorded movies. I have found seven articles, with four being published, that examine the impact of piracy on either box office receipts or prerecorded video sales/rentals, or both. These studies are listed in Table 23.3.

Every one of these studies finds a negative impact of piracy on the movie industry. The second column reports on whether the study attempted to measure the impact of piracy on box office receipts or prerecorded sales, or both. Three of the studies focused on box office only, whereas four of the studies focused on both box office and prerecorded movies. Further columns indicate the time period of the data used in the analysis, the geographic region covered by the data, the results, and whether the article has yet been published.

Because pirated videos are closer substitutes for prerecorded movies than they are for attending a movie in a theater, the impact of piracy is likely to be greater on prerecorded movies. That does not mean, however, that the impact of piracy on box office revenues would be expected to be zero. In fact, all except two of these studies find that piracy decreases box office revenues. The two studies that do not find a negative impact on box office do find a negative impact on the sale of prerecorded movies.

Note as well that, although the studies are finding a negative impact of piracy on box office revenues, box office revenues have not been falling, as seen in Figure 23.2. This means, assuming that the econometric results are correct, that box office revenues, instead of remaining relatively constant since 2002, would have increased if not for the impact of piracy. It is not necessary for revenues to decline when piracy is having a destructive impact on industry revenues.

INTERNET PIRACY IN OTHER MARKETS?

There are other markets that I have not discussed, and the reason I have not discussed them has to do with the fact that economists have not examined these markets in much detail. Computer software, for example, is a large market that is susceptible to piracy. Unlike music and movies, however, the yearly sales of software categories are expected to fluctuate as new operating systems are introduced or new forms of computing replace older forms. Additionally, there are many forms of software, such as cloud computing, monthly or yearly payments for 'seats' used by software customers, outright purchase and so forth. Getting simple statistics for these products is more difficult than for fairly straightforward products (such as movies or music).[11]

In a related manner, videogames seem similar to movies and music, but again they have large changes when new consoles are introduced, and consoles are introduced far more frequently than the format changes in music or movies (cassettes to CDs, or VHS to DVD, for example). Additionally, videogames are protected against piracy and often require a physical modification of the motherboard in order to play pirated versions, limiting the impact of piracy to the most rabid pirates.

I presume that it is for reasons such as these that economists have not investigated the impact of piracy on these markets to the same extent as for movies and music, and that is why this chapter is silent on these markets. That does not mean that piracy does not play an important role in the sales of these products but merely that we do not know the size of the role, if any, that it plays.

CONCLUSION

We have looked at music and movies, and the results are fairly unambiguous and really not all that surprising: Internet piracy harms producers of these products. The degree of harm has been quite large in music. On average, the findings for music are that the entire decline in sales since 1999 is due to piracy, and these values tend to be in the vicinity of 50–70 percent when dollars are measured in inflation adjusted units. The negative impacts of piracy on movies is not as easily put into a simple number, but all the academic studies of which I am aware find that piracy hurts sales. Given that prerecorded movie revenues have fallen by almost 45 percent relative to box office receipts in just the last few years, it is quite possible that the harm to movies is almost in the same league as the harm to music, although that is still something of a conjecture at this time. What it does seem fair to say is that the harm to these industries from Internet piracy is very large.

NOTES

* This chapter is similar to 'The impacts of Internet piracy' being published in *The Economics of Copyright: A Handbook for Students and Teachers*, edited by Richard Watt. I would like to thank the Center for the Analysis of Property Rights and Innovation for financial support.

1. The focus of much of this academic work was more on theory than empirics. For discussions of some of this earlier work see the survey by Varian (2005) or Watt (2004).
2. The number of people is defined as the population between the ages of 15 and 65, so it is not quite the normal 'per capita' based on the entire population.
3. The data for Figure 23.1 come from the RIAA database (available for a small fee), which can be found at the RIAA website (n.d.). Digital sales, including albums and singles, are included from 2004 forward. The compound growth rate from 1973 to 1999 is applied to years 2000 forward in the 'predicted sales' curve.
4. The prerecorded video data, which includes streaming revenues, come from splicing data from Digital Entertainment Group (DEG, n.d.) for 1999–2010 with *Screen Digest* (n.d.) numbers for prior years. In other words, the DEG numbers were altered for prior years based on the yearly changes that occurred in the *Screen Digest* numbers. Because *Screen Digest* numbers do not include streaming, those numbers could not be used for the entire period and DEG numbers only begin in 1999. Box office data come from *Box Office Mojo* (n.d.).
5. Box office revenues have had some large long term shifts over time, particularly after the introduction of television in the 1950s, which decimated box office revenues compared to their pre-television levels.
6. See Liebowitz (2011).
7. The full details for these articles are available in the references.
8. Values greater than 100 percent probably need a word of explanation. A value greater than 100 percent may seem odd at first glance, but such a seemingly high number is not necessarily unreasonable once properly understood. This is best illustrated through the use of a simple example. Assume that a firm sells 10 CDs in year 0, whereupon file-sharing begins in year 1. By year 5, sales have dropped to 8 CDs, a 20 percent decline. Assume as well that, in the absence of file-sharing, sales would have increased by 4 percent per year, leading to counterfactual sales in year 5 of 12.17 units. The apparent decline, the one reported by the record companies and appearing in RIAA-type statistics, is 2 units (10−8). The actual decline, determined from a perfect econometric examination, is 4.17 units (12.17−8). The conclusion would be that 208 percent of the measured decline was due to file-sharing (4.17/2). This is close to what Liebowitz (2008) found, where a counterfactual growth rate of 3.6 percent was sufficient to explain his result of 200 percent. A sound recording yearly growth rate of 3.6 percent in the US was a very typical growth rate during the three decades prior to the advent of file-sharing, indicating that a value of 200 percent is well within the realm of reasonableness.
9. I discount Oberholzer-Gee and Strumpf (2007) very heavily because there are many errors in the paper (Liebowitz, 2007), and the authors refuse to make their data available to others. Andersen and Frenz (2007, 2010) also suffer from serious flaws (see Barker and Maloney, 2012).
10. For example, the leading alternative explanation (that Liebowitz, 2006 did not find completely discredited by the evidence) was the purchase of prerecorded videos, which did show an increase after 1999 (although it seemed to begin in 1997 and was small compared to the increase in the 1980s when record sales saw robust increases). More recent evidence, however, indicates that sales of prerecorded movies fell after 2004 (see Figure 23.2), although sound recording sales showed no such pickup after 2004. Similarly, videogame sales, which had been rising since 1996, stopped rising in 2002, whereupon they remained largely unchanged until 2007, at which point there was a sudden and enormous increase, a pattern quite unrelated to the pattern of sound recording sales.
11. This is consistent with the findings of Handke (2011).

REFERENCES

Andersen, Birgitte and Marion Frenz (2007), 'The impact of music downloads and P2P file-sharing on the purchase of music: a study for Industry Canada'.

Andersen, Birgitte and Marion Frenz (2010), 'Don't blame the P2P file-sharers: the impact of free music downloads on the purchase of music CDs in Canada', *Journal of Evolutionary Economics*, **20** (5), 715–40.
Barker, George Robert and Tim John Maloney (2012), 'The impact of free music downloads on the purchase of music CDs in Canada', ANU College of Law Research Paper No. 4, available at: http://ssrn.com/abstract=2128054.
Blackburn, D. (2004), 'Online piracy and recorded music sales', working paper, Department of Economics, Harvard University.
Bounie, David, Marc Bourreau and Patrick Waelbroeck (2006), 'Piracy and the demand for films: analysis of piracy behavior in French universities', *Review of Economic Research on Copyright Issues*, **3** (2), 15–27.
Box Office Mojo (n.d.), 'Yearly box office', available at: http://boxofficemojo.com/yearly.
Danaher, Brett and Joel Waldfogel (2012), 'Reel piracy: the effect of online film piracy on international box office sales', working paper, available at: http://ssrn.com/abstract=1986299.
De Vany, Arthur S. and W. David Walls (2007), 'Estimating the effects of movie piracy on box-office revenue', *Review of Industrial Organization*, **30** (4), 291–301.
DEG (Digital Entertainment Group) (n.d.), *Data and Resources*, available at: http://www.dvdinformation.com.
Handke, Christian (2011), *Economic Effects of Copyright: The Empirical Evidence So Far*, Report, April, Washington, DC: National Academies of the Sciences.
Hennig-Thurau, Thorsten, Victor Henning and Henrik Sattler (2007), 'Consumer file sharing of motion pictures', *Journal of Marketing*, **1** (71), 1–18.
Hong, S.H. (2007), 'The recent growth of the Internet and changes in household-level demand for entertainment', *Information Economics and Policy*, **19** (3–4), 304–18.
Hong, S.H. (2013), 'Measuring the effect of Napster on recorded music sales: difference-in-differences estimates under compositional changes', *Journal of Applied Econometrics*, **28** (2), 297–324.
IFPI (International Federation of Phonographic Industries) (2011), 'Recording industry in numbers 2011', available at: www.ifpi.org.
Lessig, Laurence (2004), *Free Culture*, New York: Penguin Press.
Liebowitz, Stan J. (1985), 'Copying and indirect appropriability: photocopying of journals', *Journal of Political Economy*, **93** (5), 945–57.
Liebowitz, Stan J. (2004), 'Will MP3 downloads annihilate the record industry? The evidence so far', *Advances in the Study of Entrepreneurship, Innovation, and Economic Growth*, **15**, 229–60.
Liebowitz, Stan J. (2006), 'File-sharing: creative destruction or plain destruction?', *Journal of Law and Economics*, **49** (1), 1–28.
Liebowitz, Stan J. (2007), 'How reliable is the Oberholzer-Gee and Strumpf paper on file-sharing?', available at: http://ssrn.com/abstract=1014399.
Liebowitz, Stan J. (2008), 'Testing file-sharing's impact by examining record sales in cities', *Management Science*, **4** (54), 852–9.
Liebowitz, Stan J. (2011), 'The metric is the message: how much of the decline in sound recording sales is due to file-sharing?', working paper, available at: http://ssrn.com/abstract=1932518.
Ma, Liye, Alan Montgomery, Param Vir Singh and Michael D. Smith (2011), 'The effect of pre-release movie piracy on box-office revenue', available at: http://papers.ssrn.com/sol3/papers.cfm?abstract_id=1782924 (accessed 2 May 2013).
Michel, N. (2006), 'The impact of digital file sharing on the music industry: an empirical analysis', *Topics in Economic Analysis and Policy*, **6** (1), Article 18.
Oberholzer-Gee, Felix and Koleman Strumpf (2007), 'The effect of file sharing on record sales: an empirical analysis', *Journal of Political Economy*, **115** (1), 1–42.
Oberholzer-Gee, Felix and Koleman Strumpf (2009), 'File-sharing and copyright', in Joshua Lerner and Scott Stern (eds), *NBER's Innovation Policy and the Economy*, vol. 10, Chicago: University of Chicago Press, pp. 19–55.
Peitz, M. and P. Waelbroeck (2004), 'The effect of internet piracy on music sales: cross-section evidence', *Review of Economic Research on Copyright Issues*, **1** (2), 71–9.
RIAA (Recording Industry Association of America) (n.d.), *Key Statistics*, available at: http://riaa.com/keystatistics.php?content_selector=research-shipment-database-overview.
Rob, R. and J. Waldfogel (2006), 'Piracy on the high C's: music downloading, sales displacement and social welfare in a survey of college students', *Journal of Law and Economics*, **49** (1), 29–62.
Rob, R. and J. Waldfogel (2007), 'Piracy on the silver screen', *Journal of Industrial Economics*, **LV** (3), 379–95.
Screen Digest (n.d.), 'Video intelligence', available at: http://www.screendigest.com/intelligence/video/all_formats/video_intel_software_all_1/view.html?start_ser=vi&start_toc=1.
Varian, Hal R. (2005), 'Copying and copyright', *Journal of Economic Perspectives*, **6** (2), 121–38.
Waldfogel, Joel (2010), 'Music file sharing and sales displacement in the iTunes era', *Information Economics and Policy*, **22** (4), 306–14.

Watt, Richard (2004), 'The past and the future of the economics of copyright', *Review of Economic Research on Copyright Issues*, **1** (1), 1–11.
Zentner, A. (2005), 'File sharing and international sales of copyrighted music: an empirical analysis with a panel of countries', *Topics in Economic Analysis and Policy*, **5** (1), Article 21.
Zentner, A. (2006), 'Measuring the effect of music downloads on music purchases', *Journal of Law and Economics*, **49** (1), 63–90.
Zentner, A. (2009), 'Ten years of file sharing and its effect on international sales of copyrighted music: an empirical analysis using a panel of countries', available at: http://ssrn.com/abstract=1724444.
Zentner, A. (2010), 'Measuring the impact of file sharing on the movie industry: an empirical analysis using a panel of countries', available at: http://ssrn.com/abstract=1792615 or http://dx.doi.org/10.2139/ssrn.1792615 (accessed 3 May 2013).

FURTHER READING

There was considerable controversy at first as to whether there was any impact at all from piracy on the sales of legitimate versions of the product, and although it is generally agreed that most studies find harm there is some quasi-disagreement. For two contrasting views, it is useful to read the two surveys Oberholzer-Gee and Strumpf (2009) and Liebowitz (2011). Note, however, that the latter paper explains what are some very clear errors on the part of Oberholzer-Gee and Strumpf in simply presenting the results in the literature.

Besides the economic studies discussed in this chapter there are papers of a more philosophical legal bent, particularly from some very vocal critics who have voiced their unhappiness with copyright law and the entertainment industry. These copyright critics, sometimes associated with the concept of the 'creative commons' and the Electronic Frontier Foundation, argue that copyright laws are being used by the sound recording, movie and software industries to thwart innovative forces that would otherwise open up the market to new competition. See, for example, Laurence Lessig (2004).

Finally, there are non-academic reports that I do not discuss in this chapter. The reason, quite simply, is partly that most of these studies are self-serving on the part of the organization paying for the report but mainly that these reports are generally of low quality.

24. Artists, authors' rights and copyright

Kristín Atladottír, Martin Kretschmer and Ruth Towse

Artists, the primary creators of value-added in the chain of production in the cultural and creative industries, are the first in line for protection by authors' rights and copyright law. 'Artists' is used here as a broad, inclusive term for all kinds of initial creators of copyright works, that is, literary, musical and artistic works, and also performers of those works who have rights related to copyright. These people are always represented as beneficiaries of the rewards copyright brings, and those incentives are supposed to be essential to creativity. Artists and their representatives are often in the forefront of seeking greater protection of their rights even though economists and others have on occasion shown that that does not lead to better remuneration; nevertheless, they persist in believing in the power of artists' rights and copyright to support them (Kretschmer et al., 2011). This chapter reviews empirical research that has been done on the economic and moral rights of artists and considers the impact of digitization on them.

ARTISTS AND CREATIVITY

A great deal of the rhetoric about copyright evokes its influence on creativity as the sine qua non of the creative industries and, as those industries have come to be regarded as the source of future economic growth and welfare, the role of copyright in producing these benefits is increasingly emphasized in economic policy (Towse, 2010a). Economists and others researching empirical evidence on artists' earnings from copyright have found that evidence to conform to a well-established finding from research in cultural economics on artists' labour markets: that the winner-takes-all tendency to superstar earnings in all creative activities results in a highly skewed distribution of income, with a few high earners at the top while the majority earn incomes below the national average (Towse, 2010b). In the case of distributions of copyright royalties, a high proportion do not even attain the minimum amount that is distributed (Towse, 2001; Kretschmer, 2005; Kretschmer and Hardwick, 2007; Kretschmer et al., 2010). Moreover copyright earnings in most artistic professions not only are not the main source of income, but typically contribute only a small proportion to it. This evidence raises questions about the role economic rights play as an incentive to creativity on the part of artists and other primary creators. Copyright, and more especially authors' rights, however, offers other incentives through moral rights, and there is evidence that artists value these rights as much as if not more than the economic reward (Rajan, 2011).

Several hypotheses have been offered to explain the motivation of artists as producers of creative works and the incentive copyright and authors' rights offer. Intrinsic motivation due to the urge to create is one: adapted from social psychology by Frey, his theory of 'crowding' states that financial reward is effective as an incentive only for extrinsically motivated workers, while intrinsically motivated workers require intrinsic reward, such as their own satisfaction with their work or peer recognition as their incentive (Frey, 2000). Towse (2001) suggested that authors' rights satisfy both: moral rights protect authors' control over the use and attribution of their work, protecting their reputation and thereby offering intrinsic incentive; and economic rights offer extrinsic reward to work that is extrinsically motivated, such as that done mostly to earn money. Surveys of artists have often shown that artists divide their time between their preferred art work and other income-earning activities in order to earn a basic income, so both aspects of copyright are relevant (Towse, 2010a). Another hypothesis is that success on the market is radically uncertain and unless artists had control over their work they would not be able to benefit from later success that might come about and that would make works valuable that at the time they were produced had a low value; artists' resale right or *droit de suite* follows this logic (Solow, 1998). Atladottír (2013) suggests that artists in Iceland view their moral rights as recognition of their status and as protection for the imprint of their personality on the works they produce and, though it may not offer high rewards, they support copyright in its present form, as do many artists in other countries. Nor has digitization altered this view. These studies are reviewed below.

MARKETS FOR COPYRIGHT WORKS

Authors of copyright works (that is, literary and dramatic authors, composers, visual artists and so on – that is, 'artists') typically supply them to an intermediary who transforms them into a marketable product. They contract with the intermediary for the use of the various rights copyright affords, which may be debundled and contracted separately but mostly are not, often because the intermediary demands the use of all rights (D'Agostino, 2010; Kretschmer et al., 2010).[1] There is in fact a range of possibilities here: copyright can be transferred by the author or performer for a single flat fee, often known as a buy-out, for which the creator has effectively sold the rights. In the case of certain outright sales, though, some rights may still accrue to the creator – an artist who sells a work to a buyer retains rights to reproduction, as in a photograph of the work or its appearance in a film, and moral rights always remain with the author.[2] The more common arrangement is the royalty contract according to which the intermediary obtains the rights in exchange for a regular payment that is a percentage of the sales or licensing revenue from the work: this is the norm in publishing and in the music industry, where royalties are around 10–15 per cent of the value of sales – no sales, no royalty! (Caves, 2000). This arrangement may also involve an advance on royalties which provides the author with income before the work can be marketed and also gives the publisher an incentive to fulfil his or her part of the deal (Towse, 1999). These contractual arrangements take place in the 'primary' market for the copyright work and cover a host of possible derivative uses, such as translation, adaptation, film rights and so on.

'Secondary' use is when a published work is subsequently used in another market, say a book is read on the radio or a film is shown on television. These uses are often licensed not by the creator but by an agency or collecting society that manages copyrights on behalf of the rights holders, who are both the authors and performers and the publishers (the sound recording and film makers, literary publishers, broadcasters – that is, the entrepreneurs in the creative industries). Some secondary uses are governed by statutory licences which require 'equitable remuneration' to be paid to both parties, and those payments are made to the appropriate collecting societies and distributed by them, for example for rental, *droit de suite* and cable retransmission. That is also the case for levies on copy equipment and media (formerly cassettes and CDs, now on computers) that are mandated in many countries especially in Europe.[3]

EVIDENCE ON COPYRIGHT EARNINGS

Evidence on earnings from copyright is scant and patchy. By earnings, we mean the combination of royalties and other remuneration, such as that due from statutory licensing mentioned above. Data on copyright earnings may come from various sources, but in practice they are from questionnaire surveys of artist populations and from collecting society data. Surveys of artists have the advantage that they can obtain data on earnings from all sources as well as copyright, and they can also ask respondents questions about attitudes and experiences, which provides valuable evidence.[4] They have to rely on the accuracy of the respondents, however, and can have an unknown selection bias. Collecting society data is likely to be very accurate but relates only to secondary use of the rights that the particular society licenses, and it is therefore not possible from those data to obtain a picture of an artist's total copyright earnings, meaning both royalties from primary use and the aggregation of all secondary earnings. In some countries, however, one organization is responsible for distributing all revenues from secondary use of copyright; that is the case in Iceland, where Atladottír conducted her survey, outlined below. A further situation is where tax or insurance data are available: Finland is a country (as in all Scandinavia) where tax data are public and research on sources of income can be obtained for specific occupational groups (Hansen et al., 2003), and Kretschmer and Hardwick (2007) analysed German data from a national artists' social insurance scheme for which individuals had to provide earnings data for assessing their contribution.

For the purposes of assessing the possible contribution of copyright law to artists' earnings, two aspects are of particular interest: the level and distribution of earnings for artists compared to other professions, and earnings from the principal artistic activity compared to other sources of earnings. The median (the halfway point on the distribution) is used to report the 'typical' picture, as these earnings distributions are so uneven. Not all the research reported below provides all that information, but each makes its own contribution. Overall, what emerges from all these studies is a picture of relatively low earnings from copyright and, indeed, from all sources for the majority of artists.

The first source of data on copyright earnings in the UK was published in the report of the Monopolies and Mergers Commission (MMC), the UK's competition authority, enquiry into performing rights (MMC, 1996). It published the Performing Right Society's distribution of royalties from a range of secondary uses, which was used by researchers as evidence of the skewed nature of these earnings and the very low rewards from this source for the majority of the 'writer' members, the composers and lyricists (Kretschmer and Hardwick, 2007). Almost half the writers received less than £75, with just 3 per cent receiving £10000 or more in 1994, when the UK national median income was roughly £15 000. In other words, less than 3 per cent of composers and lyricists had a copyright income comparable to that of other workers. The typical (median) composer earned £84 in performing right income.

At more or less the same time, Taylor and Towse (1998) published what is possibly the first attempt at valuing copyright, though it dealt with only one narrow aspect. The opportunity arose for a 'natural experiment' in which a new right for musical performers was introduced into the UK under the EC Rental Directive,[5] and the authors obtained data from the Musicians' Union in the UK, and Gramex in Denmark and Sweden, these being the collecting societies for musical performers on sound recordings, which they used to estimate the value of the new right in the UK. These data also displayed a very unequal distribution of earnings from the performers' public performance rights in 1994–95: in Sweden 75 per cent of members received the minimum payment, and 32 per cent in Denmark.[6]

Such data, however, are only for one type of right and do not tell us what overall copyright earnings would be: composers might have royalties from mechanical rights, or as performers, or fees from commissions. Composers/songwriters can expect to earn a similar amount from mechanical royalties for the sale of sound recordings. The figures obtained by Kretschmer for 2000 suggest that, in the UK, about 1500 (5 per cent) composers/songwriters reach the average (mean) national wage from copyright earnings alone and, according to the German collecting society GEMA (administering both performing and mechanical rights for musical works), about 1200 German composers/songwriters (2.4 per cent) could live from their creative output (Kretschmer, 2005: 11). The study by Hansen et al. (2003) in Finland is important in providing evidence on the percentage of total income from all copyright sources compiled from the tax data of 253 composers, lyricists and arrangers: for median earners, 8 per cent of their total income came from copyright earnings (and 7 per cent from other composition sources), and even the top-quartile earners obtained only 19 per cent from royalties.

A similar picture emerges from research on collecting society data on literary authors and visual artists by Kretschmer and others (Kretschmer and Hardwick, 2007; Kretschmer et al., 2011). Kretschmer and Hardwick (2007) surveyed 25 000 authors and other professional writers (but not journalists) belonging to the Authors' Licensing Collecting Society (ALCS) in the UK and to two professional bodies of authors, Verband deutscher Schriftsteller VS and Verband deutscher Drehbuchautoren VDD, in Germany, and the results were backed up by discussions with VG Wort, the German authors' collecting society; the latter surveyed 5800 members of the Design and Artists Copyright Society (DACS) in the UK. In 2004–05, professional UK authors (defined for the purposes of the study as those spending more than half their time writing) earned a median wage of £12 330 (two-thirds of the UK national gross median wage), and in 2005 professional German authors earned a median wage of €12 000/£8280 (42 per cent of the national net median wage in Germany) (Kretschmer and Hardwick, 2007: 9). In the DACS survey, the median income of the full sample in 2010 was £12 000, of which £264 was from DACS distribution; however, there was considerable variation as between the art forms, with fine artists earning much less than designers (Kretschmer et al., 2011: 129). Careers typically were sustained by a portfolio of other activities. Close to half of visual creators (44 per cent) earned all their income from visual creation; 35 per cent had a formal second job. Besides the earnings information, both these studies asked about a range of conditions of work, such as contractual arrangements, treatment of moral rights and the impact of digitization (see below).

ARTISTS' ATTITUDE TO COPYRIGHT

Given these low levels of earnings from copyright's economic rights, one might ask why artists are so wedded to copyright. Moreover, why do they favour future royalty payments over an upfront buy-out of rights? At first blush, economists might think that a buy-out would represent the present value of expected future income from the copyright asset; however, Towse (1999) argued that artists' preference for royalty contracts over buy-outs must be explained by market imperfections and different attitudes to time preference, risk and reputation on the part of author and publisher. Asymmetric information, or indeed simply lack of information in the face of uncertainty about the reception of new works of art, means that author and publisher often have different views about the author's chances of success. As with Adam Smith's observations on risk in the *Wealth of Nations* (Book X, I), authors are likely to over-estimate their chances, while publishers, who have more experience of the market, are more realistic.

More than economic rights, though, it is argued that the moral rights embodied in the continental European law on authors' rights law play a role in the status and identity of artists, as they represent the personality interests that a creator has in his or her works. Works of art are an expression of the creator's personality, and moral rights are the embodiment of the personality interest. In line with the view that intrinsic motivation is strong for artists, moral rights provide a strong incentive to create regardless of financial reward. The Rome revisions (1928, Art. 6bis) to the 1886 Berne Convention provided for the right to claim first authorship of a work (so-called 'paternity right') and the right to object to any distortion, mutilation or other modification which would be prejudicial to the honour or reputation of the author (so-called 'integrity right'). These provisions are also known as *droit moral* from their roots in nineteenth-century French case law (Ginsburg, 1990). Moral rights are distinct from copyright as an economic (property) right in that they cannot be transferred or waived (at least in most civil law countries). The UK gave formal recognition to moral rights only with the Copyright, Designs and Patents Act 1988, sections 77–85: right to be identified as author or director (paternity right) (ss. 77–9); right to object to derogatory treatment (integrity right) (ss. 80–83). In the UK, the right to be identified as author or director has to be asserted (s. 78), and both rights can be waived by way of agreement in writing (s. 87).

In order to evaluate the perception and understanding of copyright by creators, Atladottír (2013) conducted a survey of all members of the Federation of Icelandic Artists (IFA). The ultimate purpose of the research was to empirically evaluate the role of authors' rights in the digital environment. The survey attempted to establish an empirical basis for testing two views on artist's perception and motivation: first, the incentive effect of copyright on the creation of new works; and, second, that authors' rights and the attitude to those rights held by artists rest on the deeply ingrained Romantic perception of the creator (author) and the author's personality interest in his or her works.

The research took the form of an attitudinal survey designed to ascertain artists' perceptions of the role, function and value of their rights and the level of understanding of the law on which those perceptions are based. A questionnaire was sent to the 2700 membership of the IFA, an umbrella organization representing 13 associations of professional artists. There were 838 respondents, with 526 complete questionnaires used in the statistical analysis. The study relied on the IFA and the requirements of each association for membership to determine a working definition of the population of professional artists. Although distant from the mainland of Europe and sparsely populated, Iceland has a culture that has European origins and is perceived by nationals to be firmly embedded in northern and continental European traditions. Artistic training and formal education have traditionally been sought on the mainland of Europe, and cultural theory and ideology are firmly based on the continental European school. In this sense the subjects of the study are representative of what one could term the European character and representative of the artistic population embedded in the authors' rights regime.

The survey was divided into four sections or topics, and the sections dealt respectively with income (from copyright), duration of rights, moral rights and authorship. It attempted to test artists' willingness to license or transfer copyright. In broad terms the survey corresponds to other and earlier surveys on income and employment of artists, namely that Icelandic artists rely on supplementary non-artistic work and there is little evidence of rise in income despite the growth of the creative industries. The results show that 18 per cent of respondents received 50 per cent or more of their income from copyright sources or contracts.

With questions postulating that the contractual term would be the full copyright term so as to underscore the full alienation that is contained in most licensing contracts, respondents were asked in three separate ways whether they would be willing to license for an amount that was either perceived to be fair or equal to the income they could otherwise expect; a small majority stated that they would be disinclined to do so. These replies were at odds with the expectation that most artists will license their works when offered the chance and demonstrate the respondents' adherence to the authorship ideal rather than describing accurate decision making in an actual contracting situation.

In regard to attitudes to the duration and scope of copyright, the results show that 64 per cent find the term of life of the author plus 70 years an appropriate duration and that 52 per cent considered the term to be appropriate for rights held by the author's estate or descendants. When asked about their perceptions of the function and objective of authors' rights, an overwhelming majority of respondents expressed a view that moral rights provisions provide the basis of authors' rights. When asked whether the economic rights or the moral rights served their interests better, a small majority replied that moral rights did so, and again, when asked about the abolition of either right, a majority replied that retaining moral rights was more important. A majority of respondents claim to be willing to create a similar or equal amount of work in the absence of any economic value of copyright, and a negligible number say that they would cease creation without it.

More were willing to allow their works to enter the public domain (when no further income was envisaged) if moral rights were assured. Similarly, despite a notable reluctance to license rights for the entire term, more expressed willingness to such contracting with moral rights retention. When asked to envisage the abolition of rights and a situation where other means would ensure similar economic provisions for creators, a minority replied that neither situation would affect their incentive for the creation of new works.

Overall, the survey strongly supports the proposition that authors reared in the authors' rights regime perceive of authors' rights in a similar manner to how rights were perceived when the first French and German legislation and later the Berne Convention (1886) were instituted. The idea that the relationship between the author and his or her works is a legitimate foundation for a separate set of property rights is fundamental to Icelandic artists' attitude to and perception of copyright. The highly emotive written commentaries in the survey suggest that a considerable personal interest is at stake.

In contemplating these findings, however, it is important to keep in mind that the respondents all work in the presence of authors' rights, and this fact has two significant implications: Icelandic artists are not, as findings suggest, fully aware of the purpose of copyright as an economic tool, nor do they understand sufficiently well how copyright functions as an economic tool. It is, therefore, possible to suggest that respondents did not fully understand the implications of the absence of copyright and that the economic functions of copyright were either taken for granted or expected to be performed by some other, unidentified, mechanism if copyright were absent.

What is clear is that the respondents adhere to the Romantic view of the role of the creator and that this ideology is of vital importance to both their self-perception and the incentive they have for creating. This would seem to be a potential problem for these artists in a global economy governed by international agreements such as TRIPs that do not recognize moral rights, and it is important to attempt to foresee the effect the institutional and ideological changes that both digitization and the increasing industrial structure of creative production may have on artists' attitude to copyright.

THE IMPACT OF DIGITIZATION ON ARTISTS

Research on the impact of digitization on artists is still in its infancy, and the findings of surveys that have been done show a mixed response to artists' attitudes to and/or experience of new technologies, some regarding them as a threat and others as full of promise. An early study by Madden (2004), a national survey of self-described visual artists and an online survey of 2755 musicians, found that the Internet was being widely used to make, distribute and sell their work in innovative ways and to communicate with fellow artists and fans; though they believed file-sharing should be illegal, they did not see file-sharing as a threat. 'Across the board, artists and musicians are more likely to say that the Internet has made it possible for them to make more money from their art than they are to say it has made it harder to protect their work from piracy or unlawful use' (Madden, 2004: ii). There is little evidence of the effect on earnings. The study by Kretschmer et al. (2011) referred to above provides evidence of the impact of digitization on copyright contracts and earnings. Both positive and negative effects are reported, varying between categories of creative occupations, where fine artists report many positive effects and fewer negative ones, while photographers appear to be most negatively affected, no doubt as their work can more easily be reproduced. The survey also finds that the Internet has become the main channel of exploitation for 18 per cent of the artists surveyed and the second most important medium overall.

The report by Poole (2011) for the Canadian Public Arts Funders (CPAF) found that artistic disciplines and practices have different dimensions in their relationship to digital technology: some exist because of new technologies, and others are undergoing a metamorphosis as the technology advances; many utilize opportunities that the technologies offer, for instance in production and distribution practices, but still the artistic creation itself remains largely unaffected, particularly in the performing arts. A Dutch survey on the position of authors and performing artists in the digital age (Weda et al., 2011) collected responses from 4500 authors and performing artists. It found that most creators and performers value copyright, traditional business models and collecting societies. Digital distribution and commercialization are perceived as a threat rather than an opportunity by the Dutch respondents. Most are fearful of unauthorized downloading of copyright protected works and support increased copyright enforcement measures. Few of the creators in the sample share their own work on file-sharing websites, and a small minority appreciate reuse (remixing and sampling) of their work by others. Such use is, by many, perceived as a potential threat to income. The Dutch authors applied cluster analysis to distinguish seven groups (types) of creators and performers. The analysis suggests that the perception of the effect of the digital technologies on art and the interests of the artists are influenced by age, the level of participation in new and emerging art forms, and the occupation or art form the respondent is involved in. Those who work in fields where the adoption of digital technologies is the greatest (music and film), and where it has been maintained that individual creators have lost out as a result of free riding, are more positive than others. This and other research suggests that direct participation and creation in the digital environment, as well as the level of reproducibility of the products of their creations, have a positive effect on outlook and level of optimistic acceptance as these creators become increasingly more adept in developing the skills necessary to exploit the new paradigm and developing trends in consumption.

By contrast, in their survey of artists in Australia, Throsby and Zednik (2010) found that artists use new digital technologies of various sorts in their creative practice generally and in the process of creating art where technology either enriches or changes the artwork or performance itself or enables the artist to explore new forms of creative expression. In regard specifically to the Internet, most artists across all art forms access the worldwide web frequently or occasionally for some purpose related to their creative practice, whereas a greatly smaller proportion use it frequently or occasionally in the process of creating art. The respondents were however strongly optimistic about the new technologies improving their situation, by opening up creative opportunities, improving their income-earning position, increasing exposure to their creative works and allowing for more networking and collaborating with other artists.

CONCLUSION

There is now a body of empirical work that disproves the frequently repeated view that copyright and artists' rights offer decisive financial reward to most artists and other primary creators; it has been shown time and again that that is simply not true for the vast majority of practising professional artists, though a small proportion of star artists do earn well from royalties and other copyright earnings. Though this research is partial in the sense that the whole artist population has not been tracked systematically over time (access to data often has to be negotiated on a case-by-case basis through professional bodies and collecting societies), nevertheless there is no denying the thrust of the evidence. The main incentive that copyright, especially authors' rights, law offers is likely to be in protecting moral rights. That raises the question of how this can be ensured in a digital world. It is probably too early to reach any firm conclusions about the impact of digitization on artists; survey results so far have produced mixed results which seem to depend on the experience specific types of artists have had.

NOTES

1. See also http://www.cippm.org.uk/symposia/symposium-2009.html.
2. In the UK moral rights may be waived by contract.
3. See Handke and Towse (2007) for a detailed account of the economics of copyright collecting societies and Kretschmer (2011) on levies.
4. See Kretschmer and Towse (2013) for a discussion of evidence on copyright.
5. EC Directive (1992) on rental right and lending right and on certain rights related to copyright in the field of intellectual property, 92/100/EEC, Luxembourg.
6. The predictive power of this natural experiment was, however, thwarted by the delay in the implementation of the Directive in the UK and introduction of Napster and downloading, which clouded the picture. Nevertheless, one prediction was correct: the transaction costs of a specially set up collecting society that was necessary to comply with the stipulations of the Rental Directive in the UK were greater than the benefits, and its activities were transferred elsewhere.

REFERENCES

Atladottír, Kristín (2013), 'While my taper burns: artists' perception of copyright, its function, value and incentive effect', uncompleted Ph.D. thesis, University of Iceland, Faculty of Social Science School of Business.

Caves, R. (2000), *Creative Industries: Contracts between Art and Commerce*, Cambridge, MA: Harvard University Press.

D'Agostino, G. (2010), *Copyright, Contracts and Creators*, Cheltenham, UK and Northampton, MA, USA: Edward Elgar Publishing.

Frey, B. (2000), *Arts and Economics*, Berlin: Springer.

Ginsburg, J.C. (1990), 'Moral rights in a common law system', *Entertainment Law Review*, **1** (4), 121–30.

Handke, C. and R. Towse (2007), 'Economics of copyright collecting societies', *International Review of Intellectual Property and Competition Law*, **38** (8), 937–57.

Hansen, A., V. Pönni and R. Picard (2003), *Economic Situation of Composers, Lyric Writers and Arrangers in Finland*, Turku: Turku School of Economics and Business Administration.

Kretschmer, M. (2005), 'Artists' earnings and copyright: a review of British and German music industry data in the context of digital technologies', *First Monday*, **10** (1), 1–20.

Kretschmer, M. (2011), *Private Copying and Fair Compensation: An Empirical Study of Copyright Levies in Europe*, Newport: Intellectual Property Office.

Kretschmer, M. and P. Hardwick (2007), 'Authors' earnings from copyright and non-copyright sources: a survey of 25,000 British and German writers', Centre for Intellectual Property Policy and Management, Bournemouth University, available at: http://www.cippm.org.uk/downloads/ACLS%20Full%20report.pdf (accessed 3 May 2013).

Kretschmer, M. and Towse, R. (2013), 'What constitutes evidence for copyright policy?', CREATe Working Paper No. 1, available at: http://www.copyrightevidence.org/create/esrc-evidence-symposium/.

Kretschmer, M., E. Derclaye, M. Favale and R. Watt (2010), *The Relationship between Copyright and Contract Law: A Review Commissioned by the UK Strategic Advisory Board for Intellectual Property Policy (SABIP)*, London: Intellectual Property Office, available at: http://www.ipo.gov.uk/ipresearch-relation-201007.pdf (accessed 2 May 2013).

Kretschmer, M., S. Singh, L. Bently and E. Cooper (2011), *Copyright Contracts and Earnings of Visual Creators: A Survey of 5,800 British Designers, Fine Artists, Illustrators and Photographers*, Bournemouth: Centre for Intellectual Property Policy and Management, Bournemouth University and Cambridge: Centre

for Intellectual Property and Information Law, University of Cambridge, available at: http://eprints.bournemouth.ac.uk/19521/ (accessed 2 May 2013).
Madden, M. (2004), *Artists, Musicians and the Internet*, Washington, DC: Pew Internet.
MMC (Monopolies and Mergers Commission) (1996), *Performing Rights*, cm. 3147, London: HMSO.
Poole, D. (2011), *Digital Transitions and the Impact of New Technology on the Arts*, Ottawa: Canadian Public Arts Funders, available at: http://www.cpaf-opsac.org/en/themes/documents/DigitalTransitionsReport-FINAL-EN.pdf (accessed 2 May 2013).
Rajan, M.T.S. (2011), *Moral Rights: Principles, Practice and New Technology*, Oxford: Oxford University Press.
Solow, J. (1998), 'Economic analysis of *droit de suite*', *Journal of Cultural Economics*, **22** (3), 209–26.
Taylor, M. and R. Towse (1998), 'The value of performers' rights: an economic analysis', *Media, Culture and Society*, **20** (4), 631–52.
Throsby, D. and A. Zednik (2010), *Do You Really Expect to Get Paid? An Economic Study of Professional Artists in Australia*, Strawberry Hills, NSW: Australian Council for the Arts.
Towse, R. (1999), 'Copyright, incentives and performers' earnings', *KYKLOS*, **52** (3), 369–90.
Towse, R. (2001), 'Partly for the money: incentives and rewards to artists', *KYKLOS*, **54** (2/3), 473–90.
Towse, R. (2010a), 'Creativity, copyright and the creative industries paradigm', *KYKLOS*, **63** (3), 483–500.
Towse, R. (2010b), *A Textbook of Cultural Economics*, Cambridge: Cambridge University Press.
Weda, J., I. Akker, J. Poort, P. Rutten, A. Beunen and P. Risseeuw (2011), 'Wat er speelt: de positie van makers en uitvoerend kunstenaars in de digitale omgeving', Universiteit Antwerpen, PR Onderzoek.

FURTHER READING

Towse (2010b, chaps 12–14) offers a textbook-level treatment of research on artists' labour markets and copyright.

25. New opportunities for authors

Joëlle Farchy, Mathilde Gansemer and Jessica Petrou

Legislation regarding copyright and *droit d'auteur* traditionally provides for remuneration in proportion to the success of a work, that is, all revenues earned from it. The author's remuneration is therefore linked to both the *base* remuneration, which itself depends directly on the sector's value chain, and the *rate* of remuneration as negotiated between the author and the intermediary to whom the former assigns his or her rights in a transfer agreement, the royalty copyright contract.

The purpose of this chapter is to identify to what extent authors' remuneration could benefit from the changes taking place in the digital sector using two examples: that of feature films released on video and that of the book industry in France. Digitization has indeed radically changed cost and price structures in the cultural sector. Tensions arise between authors and intermediaries in attempting to define new forms of value sharing in royalty copyright contract. While digitization offers authors new opportunities, they are nonetheless determined by changes in the market and the adaptation of forms of contracts.

A VALUE CHAIN IN COMPLETE OVERHAUL

Diversified Prices That on the Whole Are Resisting the General Decline

Like the digital transformation of the music market, the transition from material goods to digital content in the book and video industries is conducive to new forms of price setting. The paradox of the digital world is that, on the one hand, price levels are considered a handicap to the development of the new market, which indeed faces competition from an illegal, free and often more accessible supply. Given the cost of entering the digital world (the cost of digital equipment and so on) and a lesser willingness to pay for virtual content (Chartron and Moreau, 2011), consumers wait for a clear signal in terms of price before adopting these innovations. On the other hand, authors are particularly concerned about any price reduction that would diminish their base remuneration.

The Book Market

In Anglo-Saxon countries, the fall in prices is relatively widespread, even though many conflicts still simmer, especially between Apple and Amazon, the former accused of making deals with publishers to keep prices high and the latter seeking to establish the single overall price of $9.99 for all of the works of its digital catalog – 50–60 percent less than physical books in the US ($20–25). Despite the American Department of Justice (DoJ) decision allowing it to do so, the company has not yet applied these prices on its website. In contrast to the case in some European countries, in Anglo-Saxon countries book distributors have carte blanche to set their own prices (though there remains the special case of works that have become part of the public domain and that are often available free of charge on some platforms).

In Europe, contrary to consumer expectations, digital book prices are not systematically lower than those in the physical world; only new releases and bestsellers are priced lower in digital form. The simultaneous publication of paper and digital versions allows for savings due to economies of scale and therefore lowers the price of the digital version, as some of the costs are also incurred by the paper publication (see above). In France, the price of new releases is 20–30 percent lower; bestsellers can be discounted by up to 80 percent (Loncle et al., 2012). In many cases, however, the digital version of a book is more expensive than the paper version. In most European countries (Denmark, Germany, Sweden and others) this price resistance is due to the low average price of paper books (€6.21 in Germany, €7.50 in Sweden) and a much higher value added tax (VAT) on digital books.

France is a special case. There are several reasons for this low price resistance. To begin with, most publishers also have their own distribution chain that represents a large share of their total turnover: Gallimard thus owns Sodis and La Martinière de Volumen (Loncle et al., 2012), Editis owns Interforum, and Hachette owns Hachette Distribution, which accounts for 11 percent of the group's total turnover. Editors are therefore very reluctant to compete directly in the digital market, which would deprive them of this important resource.

Moreover, the consumers grant less value to digital cultural products than to tangible ones. Thus, they evaluate the pricing of digital books comparing the price of paperback format. But as Carolyn Reidy (CEO, Simon & Schuster) states: 'The right place for the e-book is after the hardcover but before the paperback.'[1] The price of digital books is based on the price of the book in its original, more expensive, hardcover version. In fact, the paperback format is in most cases the product of a second publisher, to whom the original publisher yields specific rights for this format, but not digital rights. The digital book is then the decision of the original publisher, who compares the digital version's price with that of the original publication, not that of the paperback.

The descending order of prices is therefore as follows: the original publication, the digital version and the paperback, as Table 25.1 shows for several examples.

Finally, the so-called 1981 Lang Law, which adopted the principle of a single universal price for books in order to protect a dense network of small bookstores, was extended to digital books on May 17, 2011, blocking any drastic price reductions *lex lata*. Nonetheless, one might wonder about the future of such a law whose extraterritorial clause (by which sellers, even those located abroad but who sell in France, are still bound by this single price) is contrary to the EU law that traditionally applies the law of the country (IRebs, 2011).

The Video Market

Like paper books, physical video is facing competition from a digital double: video on demand. Consumers can now access multiple platforms that allow them to download films either permanently (purchase) or temporarily (rental) via a fee per video (transactional VoD) or a subscription, or for free thanks to advertising funding.

Table 25.1 Price comparisons of different books (by format) in France, 2012

Title	Author	Date of publication	Book paper price in euros			Digital book price in euros			
			Hard-cover	Paper-back	Second-hand	Kobo	Kindle	Numilog	Gibert Jeune
La Peste	A.Camus	1947	21.3	5.95	0.9	6.49	6.49	6.49	6.49
7 ans après . . .	G.Musso	2012	21.9	not yet	10.5	13.99	14.99	13.99	13.99
Et si c'était vrai?	M. Lévy	2000	18.6	6.7	3	9.99	9.99	9.99	9.99
La Carte et le territoire	M.Houellebecq	2010	22.5	8	3	14.99	14.99	14.99	14.99

Note: Prices at May 2012.

Source: Amazon, PriceMinister, Numilog, FNAC, Gibert Jeune.

Table 25.2 iTunes pricing table

Sale (recent movies, just released)	13.99
Sale (recent movies)	9.99
Sale (older movies)	7.99
Rental (recent movies)	3.99
Rental (older movies)	2.99

Note: In euros.

Source: iTunes, 2012 rates (France).

There are significant inequalities in development of video on demand in different regions. In Europe, Germany, France and the United Kingdom are the countries that have been most receptive to this innovation. In this new market, consumers are adopting new habits that favor rental over purchase and the building of personal video libraries. Movie rental is by far the most widespread practice, representing 82 percent (CNC, 2012) of the value of the digital video market in Europe. In general, the transactional video on demand (purchase and rental) model dominates. Although movie rental fees vary greatly from one country to another, we can nonetheless establish a European average at around €5 for a recent movie and €2–3 for an older film. A video on demand purchase can meanwhile exceed €20 per movie for new releases, while the average price of a DVD in Europe is €11.50 (CNC, 2012). While comparisons of prices between the digital and physical worlds are location specific, one actor is attempting to standardize the video on demand market internationally: iTunes. iTunes offers a range of transactional video on demand in both purchase and rental form (see Table 25.2).

The French market for pay video on demand (film and otherwise) was valued at about €220 million in 2011 – an increase of 44 percent over 2010, representing 15 percent of the video market (physical and digital combined). (See Table 25.3.) By volume, movies represent 70 percent of the digital market's transactions (rentals and purchases combined).

Table 25.3 Pricing table for video on demand services

Kind of content and transaction	Average public price (euros)
Recent feature film (SD) rental	4.99
Recent feature film (HD) rental	5.99
Older feature film rental	3.99
American TV series episode rental	1.99
"Adult" content rental	9.99
Other content rental (documentary, cartoon, . . .)	1.49
Recent feature film purchase	14.99
Older feature film purchase	9.99
Subscription to a VoD offer	10

Source: IDATE (2011).

Table 25.4 Example of prices

	Year	DVD	V.O.D. rental	V.O.D. purchase	
		Amazon	Vidéofur		iTunes
Harry Potter and the Deathly Hallows – Part 2	2011	19.99	4.99	15.99	13.99
Harry Potter and the Deathly Hallows – Part 1	2010	11.99	4.99	9.99	7.99
Harry Potter and the Goblet of Fire	2006	8.32	3.99	9.99	7.99

Note: In euros.

Source: Website visits, May 2012.

In France, as for books, only recent films (for purchase) are offered in the digital world at prices lower than those in the physical world. The purchase price for recent films on digital platforms is €14.99, whereas the average price for a DVD is €18.29, even though market prices for physical videos have fallen sharply in recent years. The competition presented by both legal and illegal offers has indeed had an impact on pricing in the physical market. The average price of DVDs has dropped considerably in recent years as a result of new business practices and more flexible legislation. The average price of a DVD movie in 2004 was €22.70 (CNC, 2004), compared to €18.29 in 2011.

Regarding the purchase of older films, the conclusion is quite different. The average purchase price of older movies in digital form was higher in 2011 (€9.99) than that for the same films in physical form (€8.41). Moreover, while it accounts for the bulk of the digital market – with prices ranging from €3.99 to €5.99 per video – digital rental offers consumers rates that are comparable to or even higher than those of the physical movie rentals (€2). Video on demand rentals nonetheless offer consumers a wide choice and ease of use, allowing it to remain an attractive offer as compared with physical video rental. (See Table 25.4.) The purchase of digital films thus is a bargain for consumers when it comes to recent films, allowing them, for instance, a savings of €4 for a 2011 film.

Altogether, for books and videos, despite downward pressure on prices from consumers as well as international distribution platforms (Amazon, iTunes, etc.), which want to accelerate the development of this still fledgling market, digital content prices are resisting in France.

A CHANGE IN COST STRUCTURE AND AN OVERALL DOWNWARD TREND

While there has been a reduction in research and information costs thanks to the Internet (Bakos, 1997), as predicted by many economic analyses, the digital market is still generating new types of costs.

The Book Market

Our study involved only homothetic digital books (that is, ones identical to the paper version), without various additions. Enhanced books are still very rare and follow a different economic model, as exemplified by *L'herbier des fées*, which contains 110 interactive activities, seven short films that cost €120000 in development costs.

With the digitization of books, many of the costs inherent in the paper world, namely manufacturing and distribution costs (such as transportation costs to bookstores and from the return of unsold books, representing up to one-third of the total stock for a given title), are disappearing. These costs are estimated at between 15 percent and 30 percent depending on the source. Other costs exist (selection, proofreading and correction, and layout) but are amortized over the two types of books.

At the same time, several types of 'hidden costs' have appeared (Loncle et al., 2012). There are the costs of digitization a posteriori. It is therefore necessary to distinguish between books published in digital format as soon as they are put on the market and those digitized later, which is much more costly, as Table 25.5 shows.

Moreover, the development of a certain amount of digital rights management (DRM) based on agreements and decisions made by different actors must also be highlighted,

Table 25.5 Costs of digitization

In euros	Fiction	Essay	Guide	Collection	Comics
With digitization	1199	1572	2866	2866	502
Without digitization	665	904	1586	1586	214
New Production	151	277	513	513	70
Average cost	672	918	1685	1685	262
Illustration copyright cost			1500	1500	
Multimedia cost	750	1500	12000	5000	650
Total	1422	2418	15185	8185	1012

Note: In euros.

Source: Bienvault (2010).

as well as the cost of the bandwidth – even if the latter remains quite modest – given the size of digital book files.

Overall, we can conclude that the lower costs associated with the disappearance of production and distribution costs (15–30 percent of the overall cost, which corresponds to the drop in prices) are much greater than the burden of the new costs that have emerged.

Moreover, the cost of digital books is predominantly a fixed cost that is known once a work has been published and that will not be increased by any variable costs based on the number of works sold, unlike paper books.

The Video Market

As with books, digitization has changed the cost of distributing films on video. Digital distribution does not eliminate rights acquisition or marketing costs, which are essential to the success of a title. Production and distribution costs have, however, been shattered. The arrival of video on demand eliminates variable costs of manufacturing and logistics. The average cost of manufacturing a DVD is between €1 and €2 (CNC, 2004); the cost of logistics, including storage, inventory management and delivery to different points of sale, is estimated at between €0.5 and €1. (See Table 25.6.)

To this can be added the cost of return of unsold items. Logistics costs can therefore be as much as €1.5 per DVD for older movies. In the digital world, these publication and distribution costs are replaced by digitization, formatting and file storage costs. Preparing films for online distribution generates other specific costs. Here, the fixed costs are much greater than for those of the physical market. (See Table 25.7.)

Table 25.6 Breakdown of costs of a DVD (new releases)

Acquisition	2.25
Manufacturing	1.5
Marketing	2.5
Logistics	0.76
Total cost	7.01

Note: In euros.

Source: CNC (2004).

Table 25.7 Fixed costs for video on demand

Encoding	50
Transcoding	18
Storage	4
Metadata sorcing	between 15 and 75
Total cost	between 87 and 147

Note: In euros.

Source: CNC (2008).

Digital distribution has helped reduce certain storage and human resource costs. Distribution costs borne by video on demand platforms do not altogether disappear, however; most notably there is the cost of the bandwidth, which is substantial in this market (€0.2–€0.5 per film).

A new cost has likewise arisen – commission to IPTV (Internet Protocol Television) providers (€0.98 per transaction according to Association of Commercial Television in Europe, 2013; see also http://www.obs.coe.int). IPTV offers direct access to video on demand services on TV through the set-top boxes of Internet service providers. Internet service providers have indeed entered the film distribution sector. This type of service allows them to expand their service offerings, while the products of cultural industries become true loss leaders and means of differentiation. Ninety percent of video on demand transactions in France are made via ISP set-top boxes, not by computer. The video distributor's gross margin, however, depends on whether the video on demand service is requested by computer or via IPTV.

Thus, if we consider only production and distribution costs, selling a film in digital format (by computer) is five times less costly than selling it in physical form (CNC, 2008). These costs represent 8 percent of the value of a video on demand sale, versus 25 percent for physical sales.

All in all, for the digital market, lower costs on one hand and the relative resistance to the lowering of prices on the other result in higher base remuneration for all actors. The question is whether authors can truly benefit from this situation, given the power struggle with the other agents in the chain.

THE ROYALTY COPYRIGHT CONTRACT: THE MATERIALIZATION OF THE POWER STRUGGLE BETWEEN AUTHORS AND PUBLISHERS/PRODUCERS

In the vast majority of cases, authors themselves do not exploit their works but instead turn to economic partners – publishers in the book sector and producers in the film sector – to whom they surrender their rights of commercial use. These partners are responsible for distributing the work in return for remuneration for the author, all of which is laid out in a royalty copyright contract that formalizes a compromise between two types of agents with conflicting interests (Benhamou and Farchy, 2009).

In economic theory, this surrender can be analyzed in terms of both type of specialization of activities and transaction costs (Alonso and Watt, 2003). In the first case, the publisher or producer's professional specialization can help achieve economies of scale owing to the large number of works published or films produced. In the second case, the centralizing of books or movies within a single company responsible for all of the operational procedures of a given work allows authors to control costs as compared to multiple bilateral procedures between authors and users. Thanks to the transferability of rights, intellectual property thus facilitates the use of works by those who are able to maximize their value. Clear rules for rights assignment can facilitate these transfers by minimizing transaction costs. An efficient royalty contract adheres to the two objectives analyzed below.

Limiting the Opportunistic Behavior of Agents in Situations of Information Asymmetry

The relationship between the author and producer/publisher raises issues of lack of transparency. The author is ignorant of both the distribution efforts that will be made and the actual amounts received, the basis of his or her own remuneration. Moreover, in most cases, he or she does not know the exact legal modalities of the relationship (Caves, 2000). Issues of information asymmetry that arise follow the principal–agent type model.

Often applied in the job market, this model describes a situation where the action of an individual, the principal, is directly dependent on that of a second individual, the agent (Eaton and White, 1983), and where the principal is in a position of informational ignorance in two ways: firstly, the adverse selection that appears before the signing of the contract, as individuals do not really know the 'quality' or reliability of their partner; secondly, moral hazard *ex post facto* allowing the agent with the most information to deviate from the behavior expected by the principal. The objective of such contracts therefore is to limit opportunistic behavior (Laffont and Martimort, 2002). When the lack of information is so great that the negotiation contract terms are impossible, the royalty copyright contract follows standardized rules (Elkin-Koren, 1998).

Asymmetric information couples with a second issue, which is risk sharing between agents.

Varying Revenue Sharing Based on Risk Aversion by the Respective Agents

An efficient royalty copyright contract ratifies a risk sharing rule that is acceptable to both partners, meaning the publisher or producer and the author.

The principle of remuneration in proportion to the success of a work, officially provided for in intellectual property law, paradoxically comes back to putting most of the risk on the author, given the failure rate in cultural industries. And yet it is generally assumed that producers and publishers are less averse to risk than authors, and for several reasons. First of all, publishers and producers, by the specialization of their business and catalog effects, can pool the risks posed by certain works thanks to the success of a few blockbusters or bestsellers. For purposes of the diversity of their catalog, some may pay in advance for works they know will have very limited distribution (Loncle et al., 2012). Furthermore, with only limited finances, the author has difficulty waiting several months or even years to receive his or her remuneration; its discount rate appears higher than the market interest rate and therefore its bargaining power lower (Binmore, 1992).

To overcome this contradiction between the strict application of rewards in proportion to the success of a work and authors' strong aversion to risk, different royalty copyright contract models combining two categories of payment based on the degree of risk aversion of both agents are possible: a lump sum and an amount proportional to the revenues and thus the success of the work (Towse, 2001; Alonso and Watt, 2003). A guaranteed minimum or advance follows an insurance logic, as this advance is paid to the author regardless of the success and constitutes an incentive for the publisher or producer. If unsuccessful, the publisher or producer alone bears the loss of the guaranteed minimum payment. The greater the author's aversion to risk, the greater the chances are that he or she will want to have the greatest possible guaranteed minimum. The consequences of the latter will be a lower percentage of revenues. When risk aversion is more moderate, the author will attempt to negotiate a lower guaranteed minimum in exchange for a higher percentage of the revenues given the balance of power (Farchy et al., 2007; Robin, 2007).

THE EMERGENCE OF DIGITAL MEDIA: NEW OPPORTUNITIES FOR AUTHORS

Changes in Remuneration Sharing in Favor of Authors: The Book Market

A royalty copyright contract thus is generally composed of an advance and a variable amount, the percentage of which depends on the volume and price of sales. This percentage is estimated at 8 to 12 percent for paper books (Robin, 2007) and 15 percent for digital books (Bienvault, 2010). (See Table 25.8.) The Centre d'analyse stratégique (Center for Strategic Analysis) confirms that negotiations are under way to obtain a higher rate, which would guarantee the absolute value of authors' remuneration if prices fall.

Table 25.8 Sharing of the selling price of an e-book (without VAT) by distribution model

	Author	Publisher	Platform	Distributor	DRM	Library
Book bought on a library website	15%	35%	22%		3%	25%
Book bought on a publisher website	15%	82%			3%	
Book bought on a publishing platform	15%	35%	47%		3%	
Book bought on e-commerce website	15%	35%		50%		
Book bought on Amazon/ Apple	15%	55%		30%	3%	
Book bought on Google Books	15%	30%		55%		
Book bought on a library website (via Google)	15%	30%		18%		
Book bought on a Publienet model	50%	40%		10%		
Book bought on Amazon (direct to the author)	70%			30%		
Book bought on an author website	100%					
Paper book	8%	36% (15% manufacturing)		20% (8% distribution)		

Source: Bienvault (2010).

In France, the considerable difference between the VAT rates for the two markets (19.6 percent for digital and 5.5 percent for paper) was reduced. On April 1, 2012, the VAT for paper books was increased to 7 percent, whereas the VAT for digital works has only been 7 percent since January 1, 2012.[2] Moreover, the percentage for bookstores as regards paper books (30 percent) appears to be quite close to that of the margin for platforms like Amazon and Kobo, which have chosen to follow the model launched by Apple in the music sector (30 percent of the final price). Contrary to authors' fear of seeing their remuneration reduced, the proportional share they receive – 15 percent – is in fact clearly more favorable than in the physical world, where it is around 10 percent. Some publishers have likewise increased their margins and in higher proportions, from 15 to 40 percent (Cabinet Mazars, 2011). Paradoxically, in France, where authors' criticisms are the fiercest, their gain is even more marked, as the price of digital books in that country is higher (see above). (See Figure 25.1.)

Changes in Remuneration Sharing in Favor of Authors: The Video Market

After authors and producers, video distributors play a key role from both a business point of view and an insurance standpoint. In fact, the agreement between the producer and distributor for physical video provides for guaranteed minimum payment to producers, thereby sharing the risk between the two agents.

For video distributors – the first beneficiary of physical sales – the margin for video on demand rentals by computer is stable compared with that of the physical video rental market; this margin is nonetheless lower as regards IPTV rentals. In an economy where the IPTV rental model dominates (90 percent of the transaction volume), video distributors find themselves the losers, which seems to correspond to a reduction in their risk-taking and mission in the digital world. Their role of selector, promoter and mediator is nonetheless essential to the market (Hubac, 2010). The Internet service providers (ISPs), which themselves make an average commission per service sale equal to 24 percent of the published price, are the big winners in this transformation. Because of their leverage, ISPs are the leading players in the mass development of this new distribution channel. This type of service is also increasingly essential to these operators, who seek to differentiate themselves to ensure the perpetuity of their growth rates.

The rights holders (through both *droit d'auteur* and related rights) earn more from digital sales than physical ones in both absolute and relative value. Thus, for a film sold in digital format for €9.99, they earn €3.92 (39 percent), whereas for the same film sold on a physical medium they earn only €2.62 (13 percent). (See Figure 25.2.) Revenue sharing in the digital world thus proves favorable for rights holders in both the rental and the sales markets.

Concerning the remuneration of script writers and directors alone (who do not constitute all audiovisual authors), a memorandum of understanding was signed in France in 1999 between collective management organizations and professional associations of producers of cinematographic and audiovisual works. This memorandum has to do with authors' remuneration for the use of works on pay per view and, by extension, video on demand. It was concluded that these authors would receive 1.75 percent of the retail price (excluding tax) for on demand operating modes. This rate is higher than the average rate of remuneration of these authors in the physical video channel (1.49 percent

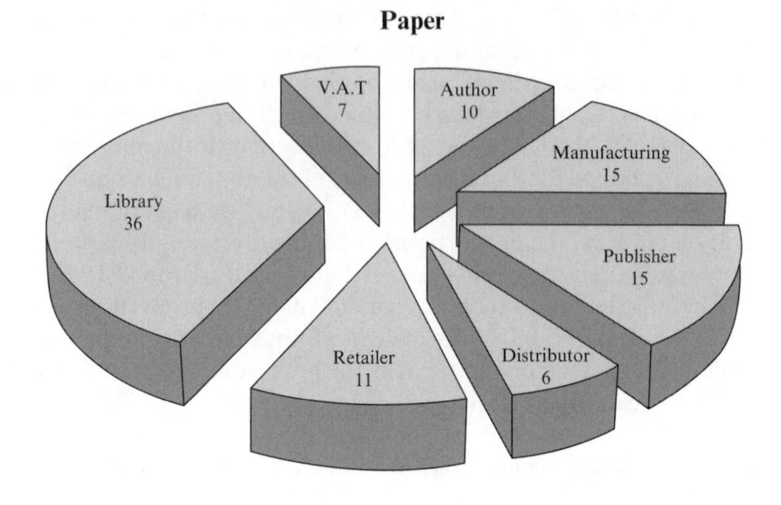

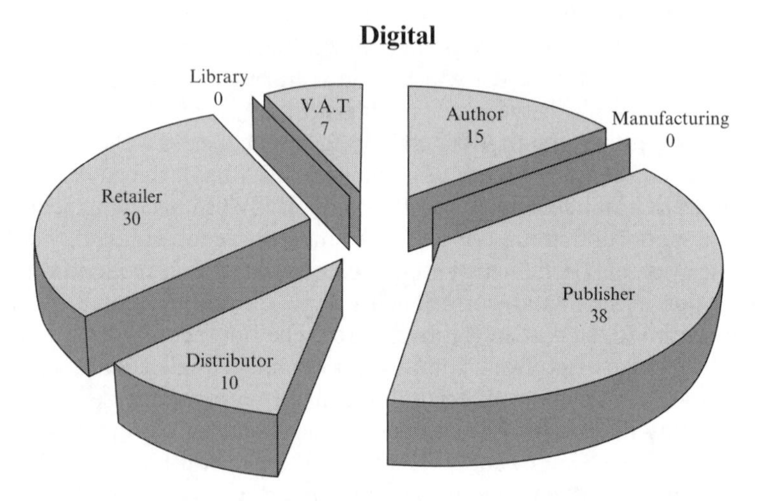

Source: Based on Bienvault (2010); Cabinet Mazars (2011); Loncle et al. (2012).

Figure 25.1 Revenue split: paper and digital books

for directors and 1.12 percent for script writers according to Farchy et al., 2007). Script writers and directors thus earn more in relative value.

Limited Opportunistic Behavior in the Digital World

The book market

In this period of transition, two major issues are under debate. First is the duration of the commitment of the author to his or her publisher. Indeed, the surrender of rights provided for in agreements is often very long, even for life. With the arrival of digital media, authors are demanding a transfer of rights for digital commercial use for a limited dura-

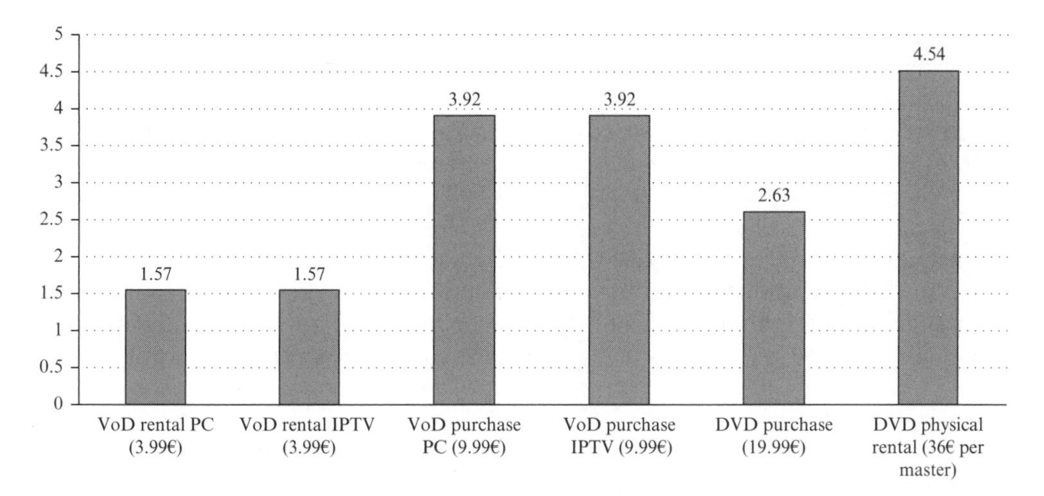

Source: CNC (2008).

Figure 25.2 Comparison of rights holders' remuneration

tion so as to adapt to the changes taking place as they occur. Second, the partners differ on the form that digital usage should take in the royalty copyright contract. Publishers want a clause regarding digital usage to be added in the same contract with identical shares of revenues and identical advances. Authors would prefer to append specific addenda where new rates proportional to sales would be established to maintain their absolute earnings. Indeed, they fear that, in the event of a future price decrease, publishers would merely pay them the same rate of proportional remuneration as in the physical world, which would automatically lower the amount of their remuneration.

Furthermore, digital opportunities could imply that authors will now be in a position to negotiate advantageously in terms of their contracts and limit the opportunistic behavior of publishers. Self-publishing is becoming more widespread, thanks in great part to methods like Amazon's Kindle Direct Publishing, which enables authors to send their manuscripts to be published without any particular terms in return for favorable distribution (70 percent for the authors, 30 percent for Amazon). Cases of success stories of unknown authors, which are still extremely rare, could grow thanks to such networks. Nonetheless, for the time being, this new avenue is primarily a weapon for renowned artists who can do without an intermediary. Thus the digital exploitation of the seven volumes of J.K. Rowling's *Harry Potter* is carried out by the author herself on the pottermore.com website.

The video market

The physical video market is considered the film industry's most opaque (Gomez, 2011). The lack of transparency between distributors and producers is due to the lack of standardized contract terms and base remuneration. Producers and rights holders lack the 'instructions' and better means for analyzing their remuneration. Negotiations between video distributors and retailers are likewise subject to suspicion by producers. The development of video on demand is in keeping with a simplification of the process and fewer gray areas between producers and distributors. This simplification is reflected in tracking forms, which are much less complex than in the physical world. The relationship between authors and producers is likewise evolving towards greater clarity, even though this market remains complex and difficult to follow over the long term.

In France, contracts linking producers and certain authors (screenwriters and directors) contain a clause referring to the October 12, 1999 protocol, which establishes a system of collective management of usage rights for video on demand. Negotiation of remuneration no longer takes place between the producer and author for this distribution channel. An additional percentage of the public price can, however, be paid directly to the author by the producer. This system of managing *droit d'auteur* in the digital world is a point of contention; whereas authors see collective management as a way of gaining transparency in the collecting of revenues, producers fear this dispossession to the benefit of societies of authors.

For actors, the growth of the digital market is therefore an opportunity for greater transparency and, as with books, limits the opportunistic behavior of intermediaries.

Towards a Renewal of Contractual Guarantees: The Video Market

Producer–distributor contracts for video on demand do not copy the model of existing contracts for physical video. Initially incorporated in the 'video' clause, video on demand is now the subject of a specific use clause in the national territory. There is almost no specific guaranteed minimum paid by the video distributor to the producer. There is nonetheless a system of guaranteed minimal remuneration per transaction (rental or purchase of a video) which protects against incentive for video distributors to price too low. The minimum the video distributor undertakes to repay the producer ranges from €1 to €1.60 per service.

The guaranteed minimum traditionally paid on the physical market has been replaced by a minimum payment per transaction in the digital world. This guarantee stands somewhere between lump sum payment and proportional remuneration, protecting the beneficiary from price decreases while remaining correlated with sales volume. This practice is also being considered for remuneration of authors. The insurance logic would be transformed for the benefit of authors.

Towards a Renewal of Contractual Guarantees: The Book Market

In this market a distinction must be made between newcomers or amateur authors and professional authors who are already established and for whom risk aversion is different.

Newcomers or amateurs who do not yet earn their living from their work are tempted to not collect insurance and rely on digital opportunities (fast distribution through social networks, the buzz phenomenon and a greater chance of being sold through the long tail) (Anderson, 2004). Conversely, established authors opt for a more insurance-based contract with a large advance and price-sharing that is more favorable to the publisher, as was already the case in the paper world.

Thus, when Amazon's Kindle Direct Publishing platform offers authors the possibility of self-publishing on favorable terms (70 percent for authors, as already mentioned), the author is not protected by any insurance in the form of an advance; if his or her book fails, he or she will not be compensated. To reassure some authors, Amazon has proposed another type of sharing whereby the author receives 35 percent and the platform commits to giving the work a certain visibility.

In addition to distinguishing between authors, considering new methods of remuneration combining a fixed share and a proportional share would boost the market while protecting authors against price fluctuations. Minimal remuneration per transaction (purchase of a book), applied among producers and distributors in video on demand markets (as well as among producers and online platforms for digital music) for royalty copyright contracts between authors and audiovisual producers, is under review in France. It would be interesting if it were likewise discussed in the book industry.

CONCLUSION

Conditions in the digital market prove paradoxical for authors. Although remuneration rates supposedly offer them more favorable opportunities, they are nonetheless very concerned about the changes taking place. Given the popularity of piracy, from which they gain nothing, the 'market' in fact accounts for only a small portion of Internet users' practices. Furthermore, authors anticipate price reductions, which would further decrease their base income. In fact, price reductions seem indispensable to the growth of the market; in 2011 in France, digital works represented only 1 percent of the total turnover for book publishing and 19 percent of that of video publishing. Only new forms of contractual guarantees can restore confidence amongst agents, allowing intermediaries to accept a higher rate of remuneration and authors a reduction in prices so as to facilitate the development of the digital market to the benefit of all.

NOTES

1. See ww.simonandschuster.biz/corporate/executive-bios.
2. Reduced again to 5.5 percent at the end of 2012 by the new government.

REFERENCES

Alonso, J. and R. Watt (2003), 'Efficient distribution of copyright income', in W.J. Gordon and R. Watt (eds), *The Economics of Copyright: Developments in Research and Analysis*, Cheltenham, UK and Northampton, MA, USA: Edward Edgar Publishing, pp. 81–103.
Anderson, C. (2004), 'The long tail', *Wired Magazine*, 12.10 (October).
Association of Commercial Television in Europe (2013), 'Facts and figures', available at: http://www.acte.be/EPUB/easnet.dll/execreq/page?eas:dat_im=025B1B&eas:template_im=025AE9.
Bakos, J.Y. (1997), 'The emerging lands for retail e-commerce', *Journal of Economic Perspectives*, **15** (1), 69–80.
Benhamou, F. and J. Farchy (2009), *Droit d'auteur et copyright*, Collection Repères, Paris: La Découverte.
Bienvault, H. (2010), 'Le coût d'un livre numérique', Study undertaken for the MOTif (Observatoire pour le livre et l'écrit en Ile-de-France), April.
Binmore, K. (1992), *Fun and Games: A Text on Game Theory*, Lexington, MA: D.C. Heath and Co.
Cabinet Mazars (2011), 'La rémunération des créateurs à l'ère numérique', Study for the Ministère de la Culture et de la Communication.

Caves, Richard E. (2000), *Creative Industries: Contracts between Art and Commerce*, Cambridge, MA: Harvard University Press.
Chartron, G. and F. Moreau (2011), 'Le "document" à l'ère de la différenciation numérique', Tendances lourdes et tensions pour les filières du document numérique, 14e colloque international sur le document électronique, INPT, Rabat, Morocco.
CNC (Centre national du cinéma et de l'image animée) (2004), *L'économie de la filière vidéo en France*, available at: http://www.cnc.fr/web/fr/etudes/-/ressources/20018;jsessionid=0505AF38E31BF51FEB0BB786445DEA40.liferay.
CNC (Centre national du cinéma et de l'image animée) (2008), *L'économie de la VoD en France*, available at: http://www.cnc.fr/web/fr/publications/-/ressources/20527;jsessionid=BF7ACCA7F78452314BAD624464E341CF.liferay.
CNC (Centre national du cinéma et de l'image animée) (2012), *Le marché de la vidéo*, Paris: CNC.
Eaton, C. and W.D. White (1983), 'The economy of high wages: an agency problem', *Economica*, New Series, **50** (198), 175–81.
Elkin-Koren, N. (1998), 'Copyrights in cyberspace – rights without law?', *Chicago–Kent Law Review*, **73**, 1155–1202.
Farchy, J. (dir.), C. Rainette and S. Poulain (2007), *Economie des droits d'auteur: le cinéma*, Culture Etudes No. 5, Paris: DEPS, Ministère de la Culture et de la Communication.
Gomez, M. (2011), 'Mission sur la transparence de la filière cinématographique – la relation entre le producteur et ses mandataires', Study realized for the Centre national du cinéma et de l'image animée.
Hubac, S. (2010), 'Mission sur le développement des services de vidéo à la demande et leur impact sur la création', Study for the Centre national du cinéma et de l'image animée.
IDATE (2011), 'Etude sur les modèles économiques des services de médias audiovisuels à la demande actifs sur le marché français', Study for the Conseil supérieur de l'audiovisuel.
IRebs (Institut de Recherche de l'European Business School) (2011), *La Revue européenne des médias*, **18–19**, Université Panthéon-Assas Paris 2.
Laffont, J.-J. and D. Martimort (2002), *The Theory of Incentives: The Principal–Agent Model*, Princeton, NJ: Princeton University Press.
Loncle, T., S. Sauneron and J. Winock (2012), 'Les acteurs de la chaîne du livre à l'ère du numérique: les auteurs et les éditeurs', Centre d'analyse stratégique, La note d'analyse No. 270, March.
NPA Conseil (2009), 'Video on demand and catch-up TV in Europe', Study realized for the Direction du développement des médias (DDM-France) and the European Audiovisual Observatory.
Robin, C. (2007), *Economie des droits d'auteur: le livre*, Culture Etudes No. 4, Paris: DEPS, Ministère de la Culture et de la Communication.
Towse, R. (2001), *Creativity, Incentive and Reward: An Economic Analysis of Copyright and Culture in the Information Age*, Cheltenham, UK and Northampton, MA, USA: Edward Elgar Publishing.

FURTHER READING

Alonso and Watt (2003) is a canonical paper about copyright royalty contracts. Towse (2001) offers an extensive overview about the issues of copyright undergoing radical transformation. Bienvault (2010), a study undertaken for the MOTif (Observatoire pour le livre et l'écrit en Ile-de-France), is a French analysis of the contemporary situation in the book sector. The European Audiovisual Observatory's study (NPA Conseil, 2009) is a complete and detailed overview of the video on demand sector in Europe.

26. Orphan works

Fabian Homberg, Marcella Favale, Martin Kretschmer, Dinusha Mendis and Davide Secchi

'Orphan works' are works in which copyright still subsists, but where the rightholder, whether that be the creator of the work or successor in title, cannot be identified and located (US Register of Copyrights, 2006). In such cases a potential user cannot be sure if the rightholder would reject a licence for this particular use or if the work has just been abandoned by its creator (Mausner, 2007). This creates a challenge for copyright law. If the envisaged use, such as making a work available online, required permission, and such permission is lacking, the user may face claims of infringement. In addition, where the work includes subject matter protected by related rights (such as performances, films, sound recordings, broadcasts and databases), often consisting of several overlapping rights, permission is required from several different rightholders.

The orphan works issue is not a minor legal difficulty. Regulatory intervention has been justified by the 'new sources of discovery' made possible by large digital libraries (Directive of the European Parliament, 2012/28/EU: Recital 1). The UK Hargreaves Review (Hargreaves, 2011: 38) states: 'The problem of orphan works – works to which access is effectively barred because the copyright holder cannot be traced – represents the starkest failure of the copyright framework to adapt.' Solutions appear to require that the copyright status be *ascertained* and *priced* with regard to each potentially orphaned artefact contained in a work, in order to ensure legal certainty within a jurisdiction.

Many institutions have large stocks of orphan works. Various European studies estimate that, for between 13 and 43 per cent of in-copyright books, the owner is unknown (Vuopala, 2010). For photographic collections of museums and archives, the numbers rise to 90 per cent of all items (Gowers, 2006). Up to 95 per cent of newspaper content held in the British Library is estimated to be orphaned; and 20–30 per cent of material in archives is also categorized as orphaned works (Stratton, 2011). The British Library calculates that it currently spends £86 248 per year on the storage and preservation of unpublished sound recordings, and £5 832 960 per year on the storage, preservation and so on of 'orphan' books (Department for Business, Innovation and Skills, 2012a: 10).

Where large amounts of collections are likely to be orphaned, archives have little incentive to preserve material that they know cannot be experienced by the public without significant legal risk. Moreover if archives are publicly funded, taxpayers are bearing the cost of preserving material they cannot experience. In the UK alone, it has been claimed that there are '2500 museums, 3393 public libraries, 3000 community archives, 979 academic libraries and approximately 3500 trust archives which might seek to use an orphan works scheme' (Department for Business, Innovation and Skills, 2012b: 9).

In this context we carried out a research programme investigating the characteristics of existing orphan works licensing schemes. First, we reviewed legislation (existing or proposed) governing orphan works in seven countries (Canada, Denmark, France, Hungary, India, Japan and the USA). Second, we conducted a rights clearance simulation. Representatives of rights clearance authorities from the countries covered in the comparative legal review were asked to provide a licence for each of six scenarios that are likely to occur in reality. Two rates for each scenario were sought, for commercial and non-commercial use. Third, we conducted a series of experiments investigating participants' attitudes towards orphan works. The following section gives an outline of suggestions on how the use of orphan works may be regulated that can be found in the literature. We then briefly present the three different parts of our research programme and summarize its initial findings.

THEORETIC SOLUTIONS

Over the past decade various solutions to the orphan works problem have been suggested. These range from making them part of the public domain, and the establishment of individual and collective licensing systems, to creating copyright exceptions, for example for academic uses (US Register of Copyrights, 2006). In some countries, proposed regulation has focused on limiting liability where an applicant has carried out a reasonably diligent search (for example Canada, the US and Japan).

Since it is not possible to force rightholders to register their works in order to obtain copyright protection as this would conflict with obligations arising under the Berne Convention (1886, Paris Act 1971, as amended in 1979), a desired side effect of any provision referring to orphan works would be to trigger the creation of voluntary registers. Various authors commented on this issue and suggested that registers would be helpful to prevent works from becoming orphans in the future.

For example, Patry and Posner (2004) make a strong case for an 'indefinite renewal' rule, that is, in order to be protected the rightholder has to fulfil certain formalities. This process can be repeated an unlimited number of times whenever the protection period expires. Mausner (2007: 398) comments: 'In addition to helping alleviate the orphan works problem, a properly indexed and searchable registry would also aid in the preservation of copyrighted works for use by subsequent generation, making this solution optimal and in line with the objectives of copyright law'. Similarly, Khong (2007: 88) argued that the creation of a statutory licensing system will provide an incentive for rightholders to build voluntary registers which help to solve the orphan works problem. The reintroduction of formalities for creators to gain strong copyright protection is likely to reduce the orphan works issue (Hansen, 2012).

Others suggest that the requirement of diligent search provides insufficient incentives for the parties involved to exert enough search effort and that classical hold-up problems arise under this regime because rightholders have monopoly power in setting the licence fee (Varian, 2006). According to Varian (2006) both issues can be mitigated by combining the diligent search requirement with a copyright registry of copyrighted works and a fixed fee catalogue that is available up front. However, as mentioned above, forcing rightholders to register their works would constitute a violation of the Berne Convention's prohibition of 'formalities'. This raises the question whether an international rather than national solution would be preferable.

Viable solutions within the current international legal framework seem to rely on two features: first, the requirement of a diligent search by the potential user trying to locate the rightholder; and, second, in case such a search does not identify a rightholder, a mechanism that establishes a licence (and a fee) that ideally matches the fee a rightholder would have charged for this particular use of his or her work. However, this counterfactual is rarely available, as there are few cases of reappearing rightholders (in the jargon, 'revenants').

LEGAL CROSS-COUNTRY COMPARISONS

In a first step we reviewed seven jurisdictions where regulatory solutions have been proposed or implemented (Canada, Denmark, France, Hungary, India, Japan and the USA). Two distinct approaches appear to be used for governing orphan works. The first may be labelled '*ex ante*', and involves rights clearing before a work is used; the second is '*ex post*', and typically involves the management of infringement risks by the user.

In the former an applicant is required to engage with an authorizing body or collecting society in order to receive a licence to make use of an orphan work. In contrast, the latter involves either the creation of a statutory copyright exception, or a limitation of liability where an applicant makes use of an orphan work after having exerted some effort to identify the potential rightholder (for example diligent search or attribution). These regimes provide different levels of protection for authors and users.

The *ex ante* approach is exemplified by Canada, Japan and India, where a potential user has to discuss terms with a copyright board. In *ex post* systems, payment is only due in cases where an author reappears. This approach is exemplified by the US. The analysis shows: a strong protection for rightholders, both *ex ante* and *ex post*, in India, Hungary and Japan; a strong protection but only *ex post* in the EU; and a relatively lower protection in the US, Canada and Denmark, the first mostly *ex ante* and the others rather balanced.

While most jurisdictions require a diligent search to be conducted by the applicant there is no uniform standard constituting a diligent search. Across jurisdictions the specifications for diligent search vary considerably. In Japan and India several steps constituting a diligent search are specified. In contrast, Canada specifies simply that 'research efforts' must be made.

The United States had proposed a 'limited liability' approach, under which the use of orphan works is possible after a reasonable search. In the case of an infringement claim orphan users are liable only for a reasonable compensation. Denmark uses an extended collective licensing system, which involves collective negotiation with users (normally for multiple licensing), valid also for non-represented authors. In turn, the EU leaves member states free to choose their regulatory approach (for example, France has chosen a central licensing system in its forthcoming legislation). All the other countries reviewed implement the central licensing system, with a central public authority granting copyright licences on orphan works. Three main regulatory approaches emerge therefore from the analysis: (1) limited liability; (2) extended collective licensing; and (3) central licensing authority.

Prices are set by central authorities in the countries that have a central licensing system, and by collecting societies in Denmark. Interestingly, national central authorities have claimed that, although no official negotiation process is provided by law, the price of licences is set on a case-by-case basis, after considering the individual circumstances of the applicant. Set prices can be challenged mostly in an ordinary court of law in the examined countries, or alternatively before the licensing authority with a quasi-judicial procedure (for example Canada). Infringement claims are handled by ordinary courts in all countries (including the US) or by licensing authorities with quasi-judicial procedures (in Hungary). In Denmark, both prices and infringement claims are under the jurisdiction of a special tribunal (the Copyright Licensing Tribunal). The above rules on price, infringement and legal remedies do not derive from EU law, which leaves these matters to member states.

In Canada, Japan, India, Denmark and France an upfront payment is required by the applicant in exchange for using orphan works. In Hungary the amount is identified but not deposited. It will be paid directly to the rightholder in the event that he or she reappears. In the US, no payment is made until a court decision is issued following an infringement claim. However, in the US, Hungary and the EU (and France) a voluntary public online register for suspected orphan works is established. In Japan, some institutions have their own register of orphan works. No register is envisaged in India, Canada or Denmark. Advertising requirements (in the national press or equivalent) are provided only in Japan and India.

The most interesting data emerged from a comparison of the requirement established in order to guarantee an *ex ante* protection to the author (for example protection of authors' rights before the work is used) and requirements established to grant an *ex post* protection to the author (for example in order to guarantee that the author receives fair compensation). The first (*ex ante*) is represented for example by provisions on accuracy of the search of the author before using the work, and steps to be taken to advertise the use of the work (through publicly accessible databases or press advertising). The second (*ex post*) is represented for example by: provisions on upfront payments for orphan works licensing; institution of an escrow system; and a simplified infringement claims system.

An inversely proportional relation might be expected between the *ex ante* and the *ex post* protection of the author. For example, a country with a limited liability system, and therefore low protection for the author before the use of the work, would be expected to have a strong compensation system, with guaranteed and efficient compensation for authors. On the contrary, countries like Hungary, India and Japan, with a strong protection for the authors before the use of the works (documented diligent search requirements, advertising, and a public registry of orphan works), would be expected to be less strict in terms of guarantees of fair compensation for the author after the work has been used.

It also needs to be considered that strong protection of rightholders (in terms of increased formalities to use orphan works) is not always consistent with the overall policy aim, which is the dissemination of orphan works and the provision of a legal framework for mass digitization projects, in the interest of the preservation of our cultural heritage. For example, Samuelson (2011) points out that extended collective licensing regimes (ECLs) have desirable properties for providing access to orphan works but that such ECLs will not be easily implemented in a US context. It is also suggested that once a work has been declared orphan a reappearing rightholder should be empowered to change this status (Samuelson, 2011).

RIGHTS CLEARANCE EXERCISE

The rights clearance exercise was designed to understand the treatment of orphan works for six scenarios in various jurisdictions that apply different regulatory regimes to orphan works. The aim of the study was to generate new data about the costs and conditions of using orphan works in different licensing systems.

We simulate a number of scenarios in which a public or private body intends to use one or more orphan works for both commercial and non-commercial purposes. Although these scenarios are hypothetical they were designed to reflect a realistic spectrum of uses that may occur. The six identified scenarios were:

1. historical geographic maps for a video game for mobile phones (up to 50 maps);
2. a vintage postcard collection for web publication and eventual sale of prints (up to 50 cards);
3. national folk tune recordings for multimedia/teaching (DVD) (up to 50);
4. re-issuing a 1960/1970s TV series as part of a digital on-demand service (one series);
5. mass digitization of photographs (archives) by a public non-profit institution, with possible sale of prints (above 100 000 items);
6. mass digitization of books by a private for-profit institution, with possible sale of books (above 100 000 items).

The subsequent analysis is exploratory in nature. It is a first attempt to collect systematic data on licences for orphan works and to develop an understanding about current international developments and best practice. Unfortunately, not all country representatives responded directly to the suggested scenarios, and only a few indications of fees were provided. We were able to gather 26 fee data points, including service charges, distributed across commercial and non-commercial uses. These conditions limit the scope for statistical analyses. Therefore we explore the data in a descriptive way and checked with the extensive qualitative comments that were submitted as response to our questionnaire.[1] Results point in the following directions.

First, the 'rights clearance exercise' revealed that the level of the fees imposed on a potential user of an orphan work is similar in all regimes we have analysed (that is, collective and individual licensing). This is an interesting finding, because it mitigates arguments related to the fact that one of the regimes (i.e. collective or individual licensing) will lead to higher fees.

Second, there is no systematic recognition of the need for permanent licences. The rights clearance exercise revealed that licensing terms varied from country to country, ranging from a monthly to a five-year licence, without the provision of a permanent licence.

Third, we find high fees that discourage mass digitization projects. Fees initially appearing very low and thus sustainable turn out to render mass digitization unviable for public and non-profit institutions.

Fourth, a limited liability system seems to have advantages for archives and other non-profit institutions exposed to orphan works, enabling those organizations to share their stock of orphaned artefacts with the public. In contrast, the upfront rights clearing seems to provide more appropriate incentives for commercial uses of orphaned artefacts, guaranteeing that a reappearing rightholder will be compensated for the exploitation of his or her work.

Overall we find very little evidence for a consistent rationale for the licensing of orphan works in the various jurisdictions. In fact, the only consistent finding is that in all cases commercial licence fees tend to exceed non-commercial ones. However, the fact that not all representatives could provide separate rates for all categories underlines the limitations of the current systems. For example, a consistent fee scheme would allow any artefact to be priced and thus permit rights clearing for the applicant. Nonetheless, in none of the jurisdictions was a fee for scenario 4, the re-issuing of an old TV series, available.

EXPERIMENTS

We conducted two experiments featuring orphan works. The first is based on music, whereas the second focuses on photographs. We employ an experimental design enabling us to investigate three major issues that are of high relevance for policy makers when designing a licensing scheme for orphan works. Thus our main research question is how performance and more information affect: (1) selection of alternatives; (2) reaction to price; and (3) reaction to copyright fees. At the time of writing, we are still conducting more fine-grained analyses based on the experimental data. The following sections present preliminary results.

Hypotheses

Accessibility and framing of information play an extremely significant role in the decision-making process (Tversky and Kahneman, 1974; Kahneman and Tversky, 1979; Kahneman, 2003). The former depends on how easily we get to information; the latter relates to individual cognitive processes and mental maps (Secchi, 2011). The theory of bounded rationality – that is, the assumptions that our decision-making skills are characterized by limited computation and access to external information – is the most natural explanation for these processes (Simon, 1955; March, 1994).

Orphan works are characterized by extremely reduced information. For some of these works, there is only the knowledge that they exist, but extremely little is known on the circumstances of their genesis, their creator(s) or anything else. Based on the assumptions above, we argue that, when more information is attached to a work of art, decision makers will be more likely to process that information and make a choice on the basis of: (1) how accessible the information is; and (2) how close it is to their mental maps. Thus we assume that individuals tend to select that artefact to which more information is attached.

Besides the quantity or the nature of information provided, the perceived *quality* has also been identified as important. A product perceived as having higher quality is also expected to be pricier than products whose quality is more difficult to grasp. Thus we assume that higher perceived quality will drive willingness to pay a fee.

Next, a potential fee for orphan works is of interest. The fee is usually embedded in the product so that its value is not disclosed to those who buy the products. A first effect is to check whether copyright fee disclosure for works that have limited information available, such as orphans, is something that decision makers are willing to support. It is also interesting to analyse to what extent this support exists. Thus we assume that individuals are more likely to accept the copyright fee for orphan works when the performance is perceived as of higher quality.

Method

The experimental design facilitates a clearer understanding of how a pricing system might be structured and how users can be incentivized to access and exploit orphan works. Two experimental designs were employed. The first experimental design focuses on choice (for example Kahneman and Tversky, 1979; Shampanier et al., 2007) and the individual reaction to a price (or fee) attached to the orphan work. The first experiment is designed to explore:

1. if individuals are affected by the medium through which the work is provided;
2. if individuals are likely to pay for works of unknown authorship;
3. if a change in fee/price may affect the choices made.

The second experimental design focuses on perceptions of a fair price (cognitive mechanisms are described in Festinger, 1957 and Kunda, 1999). The data generated allow us to make inferences about general perceptions of fairness as they might be attributed to social norms. From there we can make inferences about what authors might perceive to be a 'fair' remuneration should they reappear.

We identified an old blues song that falls under the definition of orphan works, that is, its creator remains unknown. The experimental condition features the same song performed by two artists. The first artist has been selected randomly from YouTube, while the second is a famous bluesman by the name of Doc Watson. Both videos were uploaded to the software designed for the study.

The experiment layout consists of four parts, and it is very short. The first section is to gather some demographic information and to ask questions on which form of art is preferred. The second section presents the treatment and asks participants to listen to the two artists and watch their videos, explaining that they are playing music that was composed by an unknown artist sometime in the past. Next, participants select their preferred performance. Depending on the choice, they are directed to different pages. Those opting for the YouTube artist were asked if they were willing to pay £0.49 for the performance. Then they were given the information that the price for the professional performance was £0.79 (prices were assigned after some checks on online music libraries). The next question is on the copyright fee. The text is the following: 'You know that the author of the music is unknown, hence only performance makes the music alive. Nonetheless, there is a copyright fee that is collected from a relevant collecting society that relates to author's ownership. Which fee would you see yourself more comfortable paying?' The three choices were: (1) 0p, it is unfair to pay a fee to somebody that cannot be located; (2) 1p for version 1 (YouTube); and (3) 5p for version 2 (the professional version). Those selecting the professional performance were exposed to the same questions, the only difference being the switch in the prices: £0.79 is the initial price for the song, and then the offer to change is with the YouTube performer's song that costs £0.48. The fourth section asks a question on a potential fee reduction. In the control group all pieces of information on the two performers are hidden and only music is available to participants. This removes the experimental stimulus. There were 105 individuals who participated in experiment 1.

The set-up for experiment 2 is identical. The only difference is that the artefact changes: photographs are used instead of songs. The photographs were downloaded from the Bridgeman Art Library and displayed paintings. Participants were exposed to four photographs, of which three were 'untitled' and the fourth one was titled and provided some information on where it was originally found. There were 109 individuals who participated in this experiment.

Findings

The experiments generated two major findings.[2] First, selection seems to work differently for music and visual art. Clearly, the results show that any piece of accompanying information on an orphaned photograph (such as location where it was found, or what is displayed) prompts individuals towards this particular artefact when they are asked to make a selection. In contrast, this result does not hold for music. Selection of the latter seems to be driven by the perceived quality of the performance rather than other pieces of information relating to location, origin or artist.

Second, individuals display different behaviours for music and visual art. From the fact that an individual has displayed a preference for an orphaned artefact we cannot imply that the same individual is willing to pay a licence fee for it. For example, for photographs we clearly had to reject the hypothesis that individuals are more likely to accept a copyright fee for an orphan work when it is accompanied by additional pieces of information. In contrast, in relation to music, the perceived quality of the performance is decisive, again, and is tied to an individual's willingness to pay a licence fee for an orphaned artefact. The higher the perceived quality the more willing is an individual to pay a licence fee for the orphaned music artefact.

Overall, the pattern of results indicates that if an institution wants to encourage the use of visual orphan works the inclusion of accompanying pieces of information related to the artefact may help to achieve this aim, and that users seem to consciously discriminate between the type of artefact (that is, music or photograph) according to their preferences but the 'orphan' status itself seems to be of minor importance.

CONCLUSION

Limited empirical analyses of orphan works are available, and very few datasets allowing for in-depth analyses exist. To our knowledge the data emerging from the legal review, the rights clearance simulation and the experiments are unique datasets on orphan works, and further efforts need to be made to interpret the data. This chapter demonstrates that simulation and experiments are methodologies that can build datasets suitable for the exploration of the effects of different regulatory environments. In a contested policy field where little empirical analysis has been conducted, future attempts to collect and analyse data on orphan works should be encouraged, from both user and creator perspectives. For example, the investigation of fee schemes for realizing benefits of access to culture and knowledge (while offering sufficient levels of protection to rightholders) is feasible and desirable.

NOTES

1. Details in Favale et al. (2013).
2. The complete statistical details can be found in Secchi et al. (2012).

REFERENCES

Beer, J. de and M. Bouchard (2010), 'Canada's "orphan works" regime: unlocatable copyright owners and the Copyright Board', *Oxford University Commonwealth Law Journal*, **10** (2), 215–56.

Berne Convention for the Protection of Literary and Artistic Works (1886, Paris Act 1971, as amended in 1979).

Department for Business, Innovation and Skills (2012a), *Impact Assessment BIS1054: Extended Collective Licensing*. London: BIS.

Department for Business, Innovation and Skills (2012b), *Impact Assessment BIS1063: Orphan Works*. London: BIS.

Directive of the European Parliament and of the Council on Certain Permitted Uses of Orphan Works of 25 October 2012, 2012/28/EU.

Favale, M., F. Homberg, D. Mendis, M. Kretschmer and D. Secchi (2013), 'Copyright and regulation of orphan works: A comparative review of seven jurisdictions and a rights clearance simulation', Report for UK Intellectual Property Office. Accessed at: http://www.ipo.gov.uk/ipresearch-orphan-201307.pdf.

Festinger, L. (1957), *A Theory of Cognitive Dissonance*, Stanford, CA: Stanford University Press.

Gowers, A. (2006), *Gowers Review of Intellectual Property*, London: HM Treasury.

Hansen, David R. (2012), 'Orphan works: mapping the possible solution spaces', Berkeley Digital Library Copyright Project White Paper No. 2, available at: http://ssrn.com/abstract=2019121.

Hargreaves, I. (2011), *Digital Opportunity: A Review of Intellectual Property and Growth – An Independent Report by Professor Ian Hargreaves*, London: Intellectual Property Office.

Kahneman, D. (2003), 'A perspective of judgment and choice: mapping bounded rationality', *American Psychologist*, **58** (9), 697–721.

Kahneman, D. and A. Tversky (1979), 'Prospect theory: an analysis of decision under risk', *Econometrica*, **47** (2), 263–92.

Khong, D.W. (2007), 'Orphan works, abandonware and the missing market for copyrighted goods', *International Journal of Law and Information Technology*, **15** (1), 54–89.

Kunda, Z. (1999), *Social Cognition: Making Sense of People*, Cambridge, MA: MIT Press.

March, J.G. (1994), *A Primer on Decision Making*, New York: Free Press.

Mausner, J.O. (2007), 'Copyright orphan works: a multi-pronged solution to solve a harmful market inefficiency', *Journal of Technology Law and Policy*, **12** (2), 395–426.

Patry, W.F. and R.A. Posner (2004), 'Fair use and statutory reform in the wake of Eldred', *California Law Review*, **92** (6), 1639–61.

Picker, R.C. (2009), 'The Google Book Search Settlement: a new orphan-works monopoly?', *Journal of Competition Law and Economics*, **5** (3), 83–409.

Samuelson, P. (2011), 'Legislative alternatives to the Google Book Settlement', *Columbia Journal of Law and the Arts*, **34** (4), 697–729, available at: http://ssrn.com/abstract=1818126.

Secchi, D. (2011), *Extendable Rationality: Understanding Decision Making in Organizations*, New York: Springer.

Secchi, D., F. Homberg and M. Kretschmer (2012), 'User behaviour towards orphan works: two experiments with blues and visual art', working paper, Bournemouth University.

Shampanier, K., N. Mazar and D. Ariely (2007), 'Zero as a special price: the true value of free products', *Marketing Science*, **26** (6), 742–57.
Simon, H.A. (1955), 'A behavioral theory of rational choice', *Quarterly Journal of Economics*, **69** (1), 99–118.
Stratton, S. (2011), *Seeking New Landscapes: A Rights Clearance Study in the Context of Mass Digitisation of 140 Books Published between 1870 and 2010*, London: British Library Board.
Tversky, A. and D. Kahneman (1974), 'Judgment under uncertainty: heuristics and biases', *Science*, **185**, 1124–30.
US Register of Copyrights (2006), *Report on Orphan Works: A Report of the US Register of Copyrights*, Washington, DC: United States Copyright Office, available at: http://www.copyright.gov/orphan/orphan-report.pdf.
Varian, H.R. (2006), 'Copyright term extension and orphan works', *Industrial and Corporate Change*, **15** (6), 965–80.
Vuopala, A. (2010), 'Assessment of the orphan works issue and costs for rights clearance', European Commission, DG Information Society and Media, Unit E4 (Access to Information).

FURTHER READING

The political relevance of the orphan works issue is highlighted in several government reports such as the Hargreaves Report (2011) or the *Report on Orphan Works* by the US Register of Copyrights (2006). For a detailed analysis of the role of orphan works in the Google Books Settlement case refer to Picker (2009). A detailed overview of the work and decisions of the Canadian Copyright Board is provided in de Beer and Bouchard (2010). An economic analysis of copyright and orphan works can be found in Varian (2006).

PART V

CREATIVE INDUSTRY STUDIES

27. Performing arts

Ruth Towse

The live performing arts might be thought to be an area of the creative economy in which digitization has had little impact. That, however, is far from being so, and there are many ways in which digitization is having a profound effect upon the accessibility, financing and business models of performing arts organizations, on their productions and on performers.

In cultural economics, the received view of the 'pre-digital' world was that the performing arts, such as live musical performance, musicals, opera, ballet and dance, spoken theatre, circus, cabaret and comedy shows, would inevitably have to charge prices that outstrip inflation to cover their rising costs due to the 'productivity lag' that is inherent in their labour-intensive production; or they would need subsidy or other sponsorship in order to keep prices from rising and so reducing demand. Governments in many countries give financial support to some or all of these arts to ensure their survival and to enable access to them by the public. This theory of the 'cost disease', first proposed by Baumol and Bowen (1966), and its associated fears of an 'artistic deficit' have a long history. The model is based on the view that technological progress does not increase productivity in the arts as it does in the wider economy, resulting in unbalanced growth. Digital technologies may be challenging this view in a number of ways.

So far there has been little research on the impact of digital technologies in the performing arts, and it is obviously not safe to generalize from the few published studies, the more so as it is still early days in a fast-developing area, but those studies do indicate lines of enquiry and have produced some interesting data. Moreover, casual research online produces sufficient observations on trends in the 'visible' aspects of the use of digital technologies in the performing arts, as anyone who has searched for a source of entertainment can testify! Arts funding bodies are engaged in encouraging the use of digitization in various ways, and there are a few publications on that aspect. What is more difficult to discover, though, is the impact on costs and benefits of digital technologies in performing arts organizations at the micro-level.

Overall, the most common use of digital technologies by performing arts organizations appears to be for promotion and ticketing, which for convenience are referred to here as marketing. Digital technologies are also used in several ways in the production of performances in theatres, concert halls and other venues and, what is discussed in detail here, in streamed transmissions of live performance in theatres and concert halls to cinemas or to home computers. These developments have an impact on access for audiences and ultimately on participation in the performing arts.

This chapter considers three aspects of digitization in the performing arts: its impact on marketing, on production, and on participation in and access to the live performing arts.

IMPACT ON MARKETING

The aspects of marketing that have been most affected by digital technologies are box office functions and promotion. Many venues have a website that uses images from the performances (past, present and future) to inform the potential customer about what is on offer – the pieces being performed, the style adopted in the production, the performers – and reviews by critics and others. It may well include a short audio-visual preview and a summary of the plot, history of performances and so on to whet the appetite of experienced audiences and to encourage new ones. It may be linked to social media sites and offer interactive opportunities for discussion of people's experiences, which may be a mixed blessing, however, as reactions can be both positive and negative (Thomson et al., 2013). The website also provides information on the venue – location, transport, refreshments, parking and so on – and maybe also on the history of the building and other features including its technical facilities. These services are linked to ticketing and other box office functions, such as online payment. Internet selling of tickets and bookings has made box offices more efficient for producers and consumers through enhanced personal choice of seating and record-keeping. So far, it does not seem that pricing policies have been altered by digitized box office facilities as they have with hotel and airline bookings, though some may offer preferential pricing to early bookers for season subscriptions – something that already happened before digitization. Research by MTM London (2009) for the Arts Council England (ACE, which disburses public funds to the arts) on the use of digital technologies by arts organizations in receipt of regular funding from ACE (most of which are in performing arts) found that over two-thirds had a website that offered basic marketing functions. Similarly a survey for the NEA in the US (Thomson et al., 2013) found that arts organizations of all kinds (a fifth of which were in performing arts) were increasingly using new technologies, most importantly for marketing.

Another aspect of marketing is the encouragement of new audiences and taste formation. Virtual tours of theatres, backstage facilities for the performers and stage staff, the preparation of performances in rehearsals, interviews with performers and the like have proved popular and demystify the production process, making performances and visits to venues more accessible to new audiences. The MTM London (2009: 5) survey reported that a quarter of the arts organizations had a 'rich' marketing site that promotes live 'offline' performances, while 4 per cent also offered a 'stand-alone' arts experience.

A survey of digital audiences for the arts commissioned by ACE found that Internet and other electronic device users expressed interest in learning about an art form from a five-minute clip on theatre or dance; for theatre, virtual backstage tours were most popular. However, these findings related to those who stated that they already had an interest in the art form. The most common uses of websites were finding information about a live event or artist/performer (33 per cent) and ticketing (20 per cent); 16 per cent had watched or listened to a clip of an arts performance or exhibition, and a further 8 per cent had watched or listened to a full arts performance (MTM London, 2010: 4).

Websites of arts organizations also promote their merchandise and sell other goods, such as sound recordings and DVDs. A variation on this theme is La Scala's Accenture initiative, which is a digital system for showcasing, licensing and merchandising the theatre's music and memorabilia over the Internet with the aim of raising revenue from its wealth of tangible and intangible heritage.

IMPACT ON PRODUCTION

Digital technologies have also been adopted on the production side of the performing arts both for the technical side and in relation to the artistic content of the live production. From the economic point of view, the question is how this has impacted on revenues and costs of performances and on perceived quality. Digitally controlled scenery changes and lighting have reduced the need for a considerable number of backstage staff in theatres, though making stage management a much more complex task. Digital effects are also used in performances in a variety of ways to complement or substitute for live performance, for example for music accompanying dance. In some cases, a performance has itself been created digitally between performers, such as musicians, in separate locations (Brown and Hauck, 2008). There is no reported effect on revenues or costs for these latter uses, which seem to originate for artistic purposes rather than cost-cutting ones, and in general there have been no studies of the overall effect on costs in the performing arts of adopting more capital-intensive methods of production.

Streamed Transmission of Live Performances (Live 'Broadcasts')

Live performances can be digitally streamed by satellite in real time to distant audiences in specific venues, usually to cinemas which have digital projection to HD (high-definition) screens or to a home computer. This was pioneered by New York's Metropolitan Opera (the Met) in 2006 as The Met: Live in HD programme. Though initially it made losses, the Met's programme had become profitable by 2008 and more people had seen its opera productions in cinemas than in the opera house (NESTA, 2012); by 2012 the Met Live website reported that over 3 million viewers had seen its productions worldwide.[1] It also has an 'on demand' service to iPads or computers and other digital marketing of recorded performances. Other leading opera and ballet companies (San Francisco Opera and Ballet, the Royal Opera, the Bolshoi) have followed suit. The Berliner Philharmoniker with the title 'Digital Concert Hall' pioneered transmission of live performances to a home computer. The BBC iPlayer is an app that allows listeners to download concerts and other of its broadcasts for a limited period (a week) with a highly developed website providing accompanying information. There are many such projects, but few have been studied in any detail for their economic features. One that has is the UK's NT Live.

NT Live

In 2009, NT Live was the first live streamed transmission by the UK's flagship National Theatre (NT) of two performance in the theatre in London (of Racine's *Phèdre* in a new translation by Ted Hughes) filmed by Digitaltheatre.com and digitally transmitted to 73 cinemas with HD screens throughout the UK. That was followed a couple of months later by a streamed transmission of the NT's production of Shakespeare's *All's Well That Ends Well*. There have subsequently been live transmissions of other plays in the NT's repertoire as well as by the Royal Shakespeare Company, the Young Vic, the Royal Court and other major subsidized theatre companies in the UK.

The first NT Live performance was studied in detail by Bakhshi and Throsby

(2010, 2012) and followed up by NESTA (2012), which draws on all these experiences. Bakhshi and Throsby (2010) provides a detailed analysis of audiences at both the live performance and that of the NT Live performance broadcast to cinemas in various locations in the UK, with a further audience survey for the NT transmission of *All's Well That Ends Well*. Willingness to pay by both audiences was measured, and econometric estimates made of the elasticity of demand for visits to the theatre. This was the only study of the impact of this kind of digital development in the performing arts at the time of writing.

Live screening requires the coordination of cinema and theatre performance schedules, which may be tricky (NESTA, 2012). It also necessitates coordinating video crews and those involved in the live theatrical performance – the producer (artistic director), stage and backstage staff and the actors; that is discussed in more detail in the next section.

Costs of Producing the Live Streamed Transmission ('Broadcast')

It would be more accurate, in fact, to label these the additional fixed costs of producing a joint product of the live performance and the streamed one. Any live theatrical performance is the result of long planning of the artistic production: sets, costumes and props have to be prepared, and actors and musicians (if any) hired and rehearsed off- and onstage. This requires meticulous timing in a repertory company, where actors and the stage will be used in other productions on different days. For the streamed transmission, there have to be on- and offstage camera rehearsals as well and preparation time for joint lighting and sound work for both situations; costumes and make-up will be seen in close-up, so preparation has to be made for that and built into the performance and rehearsal schedule. In general, these preparations are sunk costs, after which the marginal costs of the performance are the wages of the actors and stage staff; the filming adds to both marginal and sunk costs. A further cost is lost revenue from two sources: one is that seat prices for the theatrical audience have to be reduced, as camera operators may interfere with sight lines and, in addition, seating has to be removed to accommodate cameras.

A large team is required for the preparing the filming, for coordination with the theatrical staff (for example, a script has to be written and cues given for shots by each camera of several cameras) and for filming on the night. In principle, this is no different from a live television broadcast of a concert or theatrical performance apart from the added necessity of coordinating with cinemas. Live broadcasts are usually produced by the broadcaster; in the case of NT Live, the NT opted to be the producer (thus controlling all the rights). Further extra costs are payments due to performers for their rights in broadcast performances and royalties for the dramatist and composer of any music used. These are usually laid down in their contracts and use standard agreements with unions and copyright collecting societies.

In addition to these production costs, there are also extra costs of marketing in cinemas both domestic and foreign; a trailer must be produced for cinemas, which can also be used on the website of the theatre.

NESTA (2012: 42) reports on the split of costs for the live transmission: 48 per cent broadcast[2] production costs, 15 per cent theatre costs, 10 per cent marketing and distribution costs, 12 per cent rights and 15 per cent transmission costs.

PARTICIPATION

The National Endowment for the Arts (NEA) 2008 Survey of Public Participation in the Arts (NEA, 2010) included questions about use of electronic and digital media (TV, mobile phones, the Internet and so on) relating to the arts, showing that 21 per cent of respondents had used the Internet to view or download a performance of music, theatre or dance performance online; for those who did use it, half did so once a week. 'Digital' participants tended to be urban, higher educated, with higher incomes and somewhat older. Analysis of the data concluded: 'arts participation through media appears to encourage – rather than replace – live arts attendance. There is a strong relationship between media arts participation and live arts attendance, personal arts performance, and arts creation' (NEA, 2010: 16). Similarly, the MTM London (2010) survey reported earlier found that 8 per cent of respondents in the UK who had actually watched or listened to a full arts performance online already had an interest in the arts.

What these data seem to show is that those whose tastes for the performing arts were already established before the advent of digital access were more likely to avail themselves of digital facilities. It is probably too early to draw a strong conclusion here, as digital arts provision was in its early days when these surveys were done. An important topic that deserves further investigation is whether digital delivery simply offers new opportunities for existing audiences, broadens access to those who find it difficult to attend a live performance or develops a taste for the live arts that was not there before. That has significant implications for policy towards what are evidently costly developments.

Broadening Access

The impact on audiences has been studied in detail for the initial NT Live performances by Bakhshi and Throsby (2010, 2012). The research was conducted on audiences for the first two plays in half the cinemas by means of an online survey of both audiences, those attending the performance in the theatre and those doing so in cinemas. A third of the cinema audience reported that distance from the NT in London was the reason for choosing the NT Live screening; 40 per cent had not been to the NT; and 10 per cent had not been to any theatre in the previous year. In terms of socio-economic characteristics, the cinema audience consisted of more low-income attendees and of older people, mainly women; 13 per cent of the audience attending the NT Live in the cinema had tried but had not been able to get a ticket for the performance in the theatre. Thus the NT Live reached an altogether new audience for theatre and a significant number who had not been to the NT.

These findings relate to UK audiences, who numbered 14000; however, NT Live was also watched by another 14000 elsewhere in Europe and the US (taking into account the time difference) in a further 210 cinemas in the rest of the world. In all it was estimated that 50000 had seen the one performance either live, in cinemas by satellite transmission or on film at a later date. By comparison, the National Theatre with its three stages can accommodate audiences of only 2500 at a time.

There are limitations to the capacity of the kind of 'art' cinemas that are willing to screen these streamed transmissions, however, and indeed it is only due to an earlier policy by the UK Film Council and the Arts Council to digitize cinemas that there were enough HD screens throughout the UK that could take part. With the average seating in participating cinemas for NT Live at 200, NESTA (2012: 44) states that it is only with an international distribution to around 400 cinemas that it could be made to pay. That, of course, raises the question about prices.

Prices and Willingness to Pay

The survey of audiences for the NT Live screening in cinemas by Bakhshi and Throsby (2010) also analysed willingness to pay (wtp). Tickets to the NT Live performance of *Phèdre* in cinemas were priced at £10, which was regarded as being on the low side, a price chosen on the grounds that it was more important that the experiment worked than that revenue was raised; average cinema tickets in the UK in 2009 were £5.44. Most survey respondents thought £10 was a reasonable price. When asked what they would pay as a maximum, estimated wtp was £15 (and in the second season ticket prices were accordingly raised). Cinema goers were also asked what they would pay to see the performance they had just seen in the theatre: they regarded £25 a reasonable price for that. Prices at the NT itself are complex, with an array of concessionary prices, schemes for young attenders and so on, making a single average price almost meaningless; at the time, there was a special scheme that charged newcomers to the NT £10 (best seats were in the range of £30–40). Bakhshi and Throsby (2010, 2012) used box office data to estimate elasticity of demand (−0.24), suggesting that raising theatre ticket prices would increase revenue. By way of comparison, The Met: Live in HD, which is now profitable, was charging £25 for a provincial screening in the UK in 2012.

The first year of NT Live did not pay its way; it was expected to do so subsequently, however, and the programme was extended. One of the issues for revenue and therefore for covering costs is whether these 'broadcasts' of live performances 'cannibalize' revenues for the theatrical performances. Bakhshi and Throsby (2010) found that the cinema screening was not a substitute for but a complement to attending the theatrical show (also Bakhshi and Throsby, 2012). That is something that might change over time, however, as people begin to realize the quality of the live transmission and become more aware of the availability of these facilities. Hearsay evidence has it that attendances at the Met are falling, as audiences prefer the ease of seeing the performance in a cinema instead. Indeed, if the full cost of a visit is taken into account, the relative prices of tickets for the cinema and a downtown theatre are very different.

Quality

One of the determinants of demand for a performance is quality. Quality is achieved in a number of ways – by the work being performed, the cast, the director, sets and costumes, the facilities of the venue, its acoustics and its sightlines to the stage. Digitally streamed performances can offer top-quality sound and viewing of performances of a live show in quality cinemas or at home, where quality depends on the viewer's equipment. Filming is of high quality, with the use of specialized crews. The 'liveness' of the showing in the cinema is emphasized by the presence of an audience in the cinema, shots of the live theatre audience, and the intervals. What is missing in the cinema is the freedom to look around the stage in one's own way. In fact, some screenings are not simultaneous for geographical reasons and, moreover, the limited capacity of cinemas and excess demand have also led to deferred screenings – liveness in all but time and date (and possibly at a lower price?). This raises questions about how significant liveness actually is to the viewer. The audience survey by Bakhshi and Throsby (2010, 2012) sought to discover appreciation of 'liveness': they found that the emotional response of cinema audiences was actually greater than of those in the theatre, suggesting that there is a 'beyond live' experience of a new kind (Bakhshi and Throsby, 2010: 58; Bakhshi and Throsby, 2012: 216–17). Considering that the majority of the cinema audience had experience of live theatre, that is noteworthy. It is also the case, however, that the performances chosen for the research were of the highest quality available, at least in the UK, which is no doubt an important factor in their success.

FINANCING DIGITIZATION IN THE PERFORMING ARTS

There is some public funding available in the UK specifically to encourage digitization in the arts, and this is accompanied by specific initiatives (ACE, 2012). The ACE has a programme in conjunction with the BBC Academy offering advice and training on use of digital platforms for 'high quality arts content'. They support The Space (www.thespace.org), which offers a virtual platform for a variety of arts projects for which Arts Council England has invested £3.5 million in new commissions.

There is also the Digital R&D Fund for the Arts and Culture of £7 million provided from public funding and the National Endowment for Science, Technology and the Arts (NESTA) for development of and research on digital opportunities and business models. The BBC has created its own Digital Public Space and is digitizing its own archives, which include a considerable body of live performances of music, drama and filmed performances of opera and dance, subject to obtaining the rights to do so (see below). These initiatives were in their infancy at the time of writing.

The initial 2009 NT Live was jointly funded by Arts Council England and NESTA, which is a charitable foundation that was set up with money from the National Lottery. In other cases, sponsorship and gifts provide funding: The Met: Live in HD has several foundation and corporate sponsors. Some new projects are financed by investment from within the organization; for example, the Philharmonia Orchestra's new iPad app 'The Orchestra' utilizes the orchestra and the experience of its in-house video company along with outside expertise, essentially exploiting their 'content' in novel ways to increase revenues.

There are similar developments taking place in arts organizations great and small in many countries, but progress is uneven; for instance, Poole (2011) reports that few arts organizations in Canada have approached Canadian public arts funders for financial aid for digital initiatives. This chapter has concentrated on those high-profile ones on which some research has been done.

IMPACT ON ARTISTS

Two aspects of the impact of digitization on performing artists in theatrical performance are briefly considered here: the greater demands that digital presentation of performances of the kind discussed above makes on performers; and whether this creates new sources of income for them via their performers' rights (rights related to copyright).

Superstars and Super Looks

It was suggested earlier that the high-profile digital transmissions were successful because they employed top-quality 'talent': that may well reinforce the already present and increasing tendency to superstardom in the arts and, at the same time, sets extremely high standards that smaller-scale arts organizations cannot hope to emulate. By definition, there are few superstars and they are very expensive. Close-up filming is not familiar in some art forms or even appropriate; this may have an impact on the choice of performers based on appearance as well as their talent. For instance, opera audiences may well expect a Japanese Madam Butterfly, a naturally black Othello and a consumptive-looking Traviata. There has been some suggestion that casting and the direction of performances are being done with filming in mind.

Rights Clearance

Copyright law makes provision for digital rights for both authors and performers for the requisite media including digital usage; nevertheless, contracts may have to be specially created (or in place) for live transmissions. These will differ as between the author, conductor and featured performers and others, such as the members of an orchestra; in opera and ballet, these deals would also require remuneration to be paid to the chorus and *corps de ballet*. The Metropolitan Opera's long history of live radio transmission of opera round the world and therefore of clearing rights and negotiating contracts for recorded reproduction and distribution provided a basis for expanding into digital streaming; even before its live transmissions to cinemas were profitable it had made deals with performers and other stakeholders on all profitable future uses of its TV, DVD and online transmissions. When The Met: Live in HD was launched in 2006, fresh agreements were made with its unions offering a profit share scheme that shifted payments to a later date when revenues were assured (NESTA, 2012: 11).

The NT Live experiment was not intended to make a profit (the National Theatre is a non-profit organization with a significant government subsidy) and it did not do so. Its aim of increasing access to the NT performances in this way was supported by the various rights holders and artists involved. The NT paid a royalty to the playwrights and the directors (who have copyright in their work) and remuneration to the performers for the digital transmissions. The deal paid an upfront fee for the extra rehearsals involved and created a framework for future royalty distribution should the project become commercially successful (NESTA, 2012: 22). As reported above, the rights clearance constituted 12 per cent of the costs of the initial screening of NT Live.

We shall have to wait and see how much performers stand to gain financially from digitally transmitted performances. Authors are offered lower rates than for other types of transmission on the grounds that the market for digital is bigger so they can expect to earn as much from volume of usage; it remains to be seen if that is also the case for performers.

CONCLUSIONS

Digitization has spread quickly throughout all parts of the economy including to the arts. It is often thought of as unprecedented, and perhaps it is in terms of speed; however, it can also be viewed as another technology in a long line of technologies that have impacted on the performing arts over the last century or so – sound recording, film, radio and TV have all displaced live performance in their day while offering other forms of employment for performers. There have also been recordings and broadcasts (radio and TV) of live concerts, opera and dance performances and studio productions of plays. It remains to be seen what the long-term impact of digitization will be. The Met opera radio broadcasts have long been a feature of people's Saturdays all over the world, and that is likely to have paved the way for digital streaming to cinemas.

The received view of cultural economics is that the costs of the live performing arts are bound to increase at rates above the rate of inflation, and prices must rise accordingly unless subsidy by the state or patronage steps in to fill the gap and keep price rises down. The application of digitization surely mitigates these effects by enabling costs to be spread over a far greater audience than can be accommodated in a theatre. Those costs are higher for producing a live performance for streaming, as more technical personnel and more phases of production are involved. The Met: Live in HD reaches 1200 cinemas in 54 countries.[3] Instead of the artistic deficit forecast by Baumol and Bowen (1966), digital technologies suggest that the quality of live performance may even improve, as the profits from these transmissions can subsidize the productions; another model is La Scala's 'merchandising' of its archive, which provides financial support for its performances in times of receding subsidy. Digital offerings such as these provide an additional source of private finance for arts organizations that have previously had to rely heavily on public subsidy, which has not always been easy to justify on equity grounds and anyway is under threat from recession. In the UK and elsewhere, arts councils are accordingly actively encouraging the adoption of digital technologies in the arts.

Digitization has made box office and ticketing much more efficient, and the use of clips and the wide range of background information about productions on the arts organization's website improve consumer choice and educate non-attenders about the arts organization. Social media even do the job of advertising performances and sharing experiences with little intervention from the arts organization.

There are wider social benefits of 'live broadcasts': access is increased by removing the elitist image of the performing arts, as well as stimulating interest in these art forms; in addition, people living far from the 'original' venue, for whom the cost of travel and possibly also accommodation would make a visit to that venue too high, can experience the arts. The broadcasts offer new audiences who have not had the experience of live performance in a theatre or concert hall the opportunity to 'sample' one in another setting (at home or in a cinema). Thus live transmissions help to form tastes for performing arts. They also set quality standards and enable those who could never afford a front seat the 'best view in the house'. In addition, they offer opportunities to performing arts companies, especially smaller ones, for sharing costs and risks through co-productions which may also be 'virtual' collaborations. Bakhshi and Throsby (2012) have provided a framework for assessing innovation and for cultural value-creation that captures these cultural benefits.

There are possible downsides to the impact of digitization, as described in the chapter. As NESTA (2012) points out, these innovations are not for every arts organization. Many are too small to be able to afford the investment, and others may not feel it suits their objectives. There is the possibility that those who don't join in may lose out. A far greater danger that this author foresees is that the superstar effect will have a deleterious effect on those arts organizations that cannot compete in offering equivalent quality of star performances: the winner-takes-all tendency is very strong in the performing arts. Bakhshi and Throsby's (2013) test for 'cannibalization' – substitution between the theatrical production in the theatre and in the cinema – does not tell us about possible displacement of audiences for smaller and less prestigious arts organizations by transmissions of rich internationally known ones. The already feared predations on national cultures could also be exacerbated as globalized distribution of the world's top-quality arts organizations dominates tastes. On the other hand, there was a spin-off from the NT Live to smaller theatres which were helped to start digitally transmitting their performances.

One thing is clear, however: digitization will have a considerable impact on the performing arts in terms of access, choice, quality and taste formation as well as enabling arts organizations to develop new business models and sources of revenue. These developments also offer a whole new area for economic research.

NOTES

1. Such reporting often confuses tickets sold with number of people; the former is more likely to be correct. See http://www.metoperafamily.org/metopera/liveinhd/LiveinHD.
2. Live transmissions are called 'broadcasts' in NESTA (2012) and by Bakhshi and Throsby, though they point out that the correct term is 'narrowcast'. 'Streamed' or 'live transmission' is used in this chapter.
3. As note 1 above.

REFERENCES

Accenture, available at: http://www.accenture.com/SiteCollectionDocuments/jp-ja/PDF/industry/media-entertainment/broadcast/Accenture_industry_medi_lascala.pdf (accessed 15 April 2013).

ACE (2012), *Arts Council England's Creative Media Policy*, London: Arts Council England.

Bakhshi, H. and D. Throsby (2010), *Culture of Innovation: An Economic Analysis of Innovation in Arts and Cultural Organisations*, London: NESTA.

Bakhshi, H. and D. Throsby (2012), 'New technologies in cultural institutions: theory, evidence and policy implications', *International Journal of Cultural Policy*, **18** (2), 205–22.

Bakhshi, H. and D. Throsby (2013), 'Digital complements or substitutes? A quasi-field experiment from the Royal National Theatre', *Journal of Cultural Economics*, online 16 February.

Baumol, W. and W. Bowen (1966), *Performing Arts: The Economic Dilemma*, Hartford, CT: Twentieth Century Fund.

Brown, G. and G. Hauck (2008), 'Convergence and creativity in telematic performance: *The Adding Machine*',

Culture, Language and Representation, **VI**, 101–19, available at: http://addingmachine.bradley.edu (accessed 15 April 2013).
MTM London (2009), *Arts Council England – Digital Content Snapshot*, London: Arts Council England, available at: http://www.artscouncil.org.uk/media/uploads/downloads/MTM-snapshot.pdf (accessed 15 April 2013).
MTM London (2010), *Digital Audiences: Engagement with Arts and Culture Online*, London: Arts Council England, available at: http://www.artscouncil.org.uk/media/uploads/doc/Digital_audiences_final.pdf (accessed 15 April 2013).
NEA (2010), *How Technology Influences Arts Participation*, Washington, DC: National Endowment for the Arts.
NESTA (2012), *Digital Broadcast of Theatre: Learning from the Pilot Season: NT Live*, London: NESTA, available at: http://www.nesta.org.uk/library/documents/NTLive_web.pdf (accessed 15 April 2013).
Poole, D. (2011), *Digital Transitions and the Impact of New Technology on the Arts*, Ottawa: Canadian Public Arts Funders, available at: http://www.cpaf-opsac.org/en/themes/documents/DigitalTransitionsReport-FINAL-EN.pdf (accessed 15 April 2013).
Thomson, K., K. Purcell and L. Rainie (2013), *Arts Organizations and Digital Technologies*, Washington, DC: Pew Research Center's Internet and American Life Project, available at: http://pewinternet.org/Reports/2013/Arts-and-technology.aspx (accessed 15 April 2013).

FURTHER READING

Bakhshi and Throsby (2012, 2013) and NESTA (2012) were at the time of writing the only analytical treatments and make interesting reading. Much can be achieved by surfing the Internet looking at the websites of arts and cultural policy organizations.

28. Art markets

Payal Arora and Filip Vermeylen

The advent of digitization has had a profound impact on the art market and its institutions. In this chapter, we focus on the market for visual arts as it finds its expression in (among others) paintings, prints, drawings, photographs, sculpture and the like. These artistic disciplines claim the lion's share of the global art trade, and its objects are prominently featured by museums and galleries in both old and new art centers worldwide. Digital delivery has altered not only the content of the visual arts, but also the manner in which art is traded, consumed and valued. The various actors in the art market have embraced digitization in its many guises and forms, and its online applications, albeit at different speeds and with different intensity. The vast majority of art institutions (including artists themselves) make use of websites and incorporate databases for organizational, educational and marketing purposes. However, few have yet learned to effectively capitalize on Web 2.0 applications, even if social media offer unparalleled opportunities for community building in the art world (Castells, 2011). The largely informal and opaque character of the art market with its continued emphasis on closed dealer–collector networks and face-to-face contacts appears to pre-empt widespread use of social media for now. Still, as with other sectors of the creative economy, the art world and market are undergoing significant changes as a result of the digital revolution.

Digitization per se is not revolutionary; the transformation of analog information in the form of texts, images and sound to a digital form which can then be stored, manipulated and transmitted through a range of networks and devices has been around for decades (McQuail, 2000). The Internet has served as a vital platform to disseminate such digital content over the years. However, its contemporary Web 2.0 structure of being participatory and user-driven in the generation of content promises a revolution in how information is accessed, constructed and converged (O'Reilly, 2007).

In the art world, digitization has manifested itself through the creation of databases containing tremendous information regarding prices, the type and characteristics of a work of art, authorship, provenance and records of previous sales which are now available for professionals, art lovers and amateur art consumers. Literally millions of images of works of art have been digitized and disseminated, primarily but not exclusively in the form of websites hosted by museums, galleries, research institutions, blogs, artists and online searchable databases that have a commercial and/or educational purpose. Also, new art forms have been brought to life which have no physical presence and exist *merely* as computer images. In this brief chapter, we do not concern ourselves with digital art as such, but aim to provide a framework for understanding the art world in this information and social networking era, and reflect on the ramifications of digitization particularly with regard to knowledge construction and the valuation of works of art.

DIGITIZATION IN THE ART WORLD: CONTEXT AND HISTORY

In this section, we will survey how the various actors and gatekeepers in the art world have engaged digitization and the Internet since the late 1990s. *Art galleries* around the world have been slow in their involvement with the Internet compared to other creative industries, and it is only in recent years that gallerists have come to realize the advantages of an online presence for marketing purposes. Their websites tend to feature mostly profiles of the artists they exhibit and sell, but high-end galleries often do not advertise prices of the works for sale. The sites further include (practical) information about the gallery itself and upcoming events such as a vernissage, but seldom engage in an online dialogue with their clientele via social media. However, by the very nature of the art they present, galleries promoting digital art have been much more innovative in engaging their audiences, and have even made them part of the creative process in some instances (Bishop, 2012).

Similarly, *auction houses* have by and large been rather conservative in their engagements with the virtual realm. Certainly, the catalogues are routinely digitized and put online along with the practicalities of a live auction, but the established auction houses such as Christie's and Sotheby's have so far not been successful in capitalizing on the opportunities that new technologies offer for purely online sales. Costly attempts in the early 2000s for online transactions were soon abandoned. Most established auction houses therefore view digitization and the Internet as little more than an extra marketing tool within their existing business model, whereby one has the option to bid online on works presented at real life auctions. On the other hand, auction houses in emerging art markets have been far more cutting-edge in applying online interactive technologies in the marketing of art. For instance, India based Saffronart introduced mobile phone bidding and has been a pioneer in organizing lucrative online auctions for fine art (as opposed to the online sales organized by eBay, which offers far less valuable pieces). In some cases these virtual public sales include newly created works, and proceed without a reserve price, which precludes any buy-ins. Both strategies are considered to be detrimental by Western auction houses.

By comparison, *art museums* have made extraordinary use of the technological innovations to seek out and engage with existing and new audiences, and to make their collections available to the global community of art lovers. Shrinking national budgets for art institutions, a significant increase in the number of museums and the ensuing competition among them, and the general increase in demand for leisure-time activities have made embracing social media a necessity in public outreach (Loran, 2005). Besides, museums have been undergoing a transformation in their role from being custodians of cultural heritage to educators that engage and entertain the audience. While the 'museum without walls' idea of audience involvement has been around for decades (Bearman, 2000), new media have succeeded in materializing this through varied imaginative spaces of engagement. This works well with the bottom-up nature of social media that compels museums to listen to their audience and design exhibits and digital interfaces of art objects that appeal to a diverse and global art consumer base. There are various positive outcomes of the museums' online activity: users' increased awareness and recognition of art, reinforcement of cultural heritage and national identity, creation of a community around the museum, and the strengthening of audience trust in the institution (Kidd, 2011). Digital platforms have made interactivity possible and customization feasible through a range of creative means. For instance, we now have available the online chat option with curators on museum sites and the possibility to foster art communities around digital video hosting of new art exhibits. Another innovative feature is the online tour of the museum, where visitors can virtually navigate and educate themselves about the art collection at their own pace and through their own choosing.

Digitization of art collections has enabled audiences to access and experience museum art regardless of their physical locations. Recent advances in technology have spawned ambitious endeavors such as the Google Art Project that aim to make such experiences richer and more fulfilling to art lovers worldwide through their high-resolution images and the playful navigation affordances of top museums around the world. While art consumers enjoy browsing and learning about art through this digital edutainment approach, the museums learn about their consumers via cookies that follow visitor movements and amass a wealth of information about audience behavior and taste. A growing number of museums are opening up to co-creation with their consumers via media events that parallel the culture of Web 2.0. For instance, the Guggenheim Museum conducted a biennial of creative video called 'YouTube Play' and invited amateurs of creative and innovative videos to participate, resulting in thousands of user-generated video submissions and public visibility for the museum. Furthermore, as databases move online, there is a pressing need for more standardized and international means of coding and categorizing data for seamless inter-institutional sharing of databases. Emerging markets such as India are harnessing such possibilities by executing the most wide-scale digitization of their cultural heritage and gaining visibility by optimizing their art data for search engine algorithms. In other words, certain art forms are being privileged over others, and such politics have wide implications for what constitutes the nation's heritage online (Prasad, 2011).

An analogous development to the art repositories in the art market made possible by digitization has been the emergence of a host of *online databases* such as Art.sy, Artprice and Artnet. These databases have been populated by data related to artworks, artists and schools and are increasingly shared and made accessible through the Internet. The earliest example of a sustained effort to collect data on the art trade was compiled by Gerald Reitlinger during the 1960s, published in three volumes (Reitlinger, [1961–70] 1982). This database – now considered flawed in some respects – remains a seminal and pioneering work, since it was the first to make available longitudinal series of art prices (Guerzoni, 1995). Reitlinger's example inspired others during the dot.com revolution of the 1990s to make available large sets of data pertaining to art sales through the Internet, often with a commercial purpose in mind. These databases are sometimes exclusive, with access limited to subscribers, while in other instances they adhere to the principle of open access, where the available information is free for all to use. Either way, the wealth of these digitized datasets alerted art market professionals as well as cultural economists to the possibilities for systematic research into price histories. Econometric tools were hereby utilized to gauge the extent to which art has been a worthwhile investment compared to other financial assets. Many art indices have seen the light in the last two decades. Some are based on the repeat sales method, using prices of the same art objects traded at two or more distinct moments in time. Others are based on hedonic regression analysis, which decomposes a work of art into its constituent characteristics such as size and subject matter, and calculates the value for each of these components. Both methods have their strengths and weaknesses, but allow for more informed estimates of return on art prices and have led to a flurry of publications on art as a vehicle for investment (Ginsburgh et al., 2006 includes an extensive bibliography). Interestingly, despite the fact that great strides have been made in quantitative art market research thanks to the availability of online databases and these analytical tools, no consensus exists among cultural economists as to whether art effectively is a sound financial investment (Ashenfelter and Graddy, 2003).

Nevertheless, the growing number of investors and media covering the art trade has been relying on indices such as the Mei Moses Fine Art Index to track market trends. Recently, a new generation of relational databases is taking the characteristics model one step further. Taking their cue from the successful Music Genome Project, Art.sy has employed a staff that includes curators and art historians who deconstruct paintings into a myriad of properties ranging across formal qualities, belonging to a certain artistic movement and information concerning the artist. The aim of the company (which counts leading art gallerist Larry Gagosian and Google's Eric Schmidt among its investors) is not just to provide information, but to target potential buyers directly by offering serendipitous artistic discoveries. By taking stock of their appetite for certain types of art or artists, the site steers collectors towards comparable works that they might not be aware of, but that are in line with their revealed preferences. These computer scripts thus operate as a supply inducing demand mechanism. However, the goal of the site is not to sell directly, but to connect interested buyers to the galleries that hold these paintings in their inventory. Thus Art.sy facilitates the traditional face-to-face meeting with a dealer. Again, because of the prominence given to personal contact and direct communication, art institutions have not fully embraced e-commerce. This sets the art market somewhat apart from trends in other cultural industries such as the popular music industry, which thrives on downloadable songs and the streaming of video clips.

THE POWER OF ALGORITHMS AND THE IMPLICATIONS OF DIGITIZATION FOR THE ART MARKET

Learning about Art in the Digital Age

Contemporary digitization of art information has created the possibility of an unprecedented democratic sphere for universal access to learning that pervades across borders, regardless of geographic location, income and cultural contexts (Trant and Bearman, 2011). While indeed the Internet with its low barriers to entry has enabled widespread access to the arts, and Web 2.0 interactive features have facilitated a more personalized and immersive learning experience, this has also created a tremendous information deluge. The need to sieve through this abundance of information and be able to identify 'relevant' art knowledge has become a significant challenge. This has led corporate, state and other actors to step in and compete in the shaping of these information processes. For instance, we see well-renowned museums, galleries and auction houses invest heavily in becoming visible online, extending their expertise and authority to the virtual realm.

Interestingly, some smaller museums, commercial galleries, art lovers and amateur artists are gaining stride online as they voice their opinions and gain a mass following via creative and entrepreneurial play with digitization and social media. For instance, recent rankings of the most popular museum bloggers include the Brooklyn Art Museum (a small museum in New York), Culturegrrl (an independent professional) and Yesterday.sg (a community art blog aiming to preserve Singapore's cultural heritage) (Verboom and Arora, 2013). So it seems that who constitutes an authority in art takes on a pluralistic form within the museum blogosphere. While this digital backdoor is often the only way for smaller actors to make their mark in the contemporary artscape, this also poses a challenge in making sense of the role of conventional art experts in shaping art knowledge in the digital domain.

There is considerable concern about this popularization of art knowledge, as there is much emphasis on entertaining and educating the audience, and tailoring art information to suit the needs and moods of the audience. The art world is hardly immune to the larger social media trend of the 'filter bubble', where the web algorithm selects information for users based on their prior search behavior, location and personal profile. In other words, the art audience get trapped in their own cultural and ideological 'bubble' as alternative and novel perspectives are filtered out of their learning world (Pariser, 2011). This inward learning can perpetuate conventional ideas of art and the art market in general, irrespective of contemporary and international dynamics pervading this field. There is also a danger of peripheral learning as museums are compelled to take on more popular themes reflecting cultural tourism agendas and the new leisure-oriented demographics, with some museums potentially transforming into 'degraded cathedrals of tourism' (Davis, 2011). Lastly, it is important to keep in mind that learning of art is not just a cerebral but an emotive experience, and this affective state influences the consumer's notions on the value and meaning of the art. As digitization becomes more sophisticated in its multimodal quality, the capturing of the 'aura' of the art object to more authentically represent and encapsulate the understanding of art becomes more of a reality. However, this creates a 'digitization' divide based on those who have access to high-quality multimodal and hypertextual art experiences and those who do not, reminding us that universal learning about art in the digital age is still a distance away.

The Valuation of Art in the Digital Age

The vastly increased amount of information relative to artists and their works has had some salutary effects on the way the art market operates. In short, it has improved the transparency of a market that has long been characterized as secretive, and transaction costs have been reduced. New communication technologies help to connect dealers and collectors more efficiently, thereby lowering search costs. In addition, empowered consumers can now gather crucial information on the price histories of their favored artists without the help of an expert, allowing them to make more educated decisions about what to buy or not, and how much to spend. And, as an unintended side effect, this might reduce the opportunities for arbitrage for dealers and other intermediaries, who can theoretically be bypassed altogether.

The advantages of a digitized art market with its informed buyers – complete with art indices produced by cultural economists and professional art consultants – may have attracted new investors looking for alternatives to the usual financial assets, but this has not changed the heterogeneous nature of the visual arts market with its mostly unique and infrequently traded goods. In other words, despite the increased flow of information brought forth by technological intermediaries, investing in the art market essentially remains (in William Baumol's words) a 'floating crap game' (Baumol, 1986).

On a more fundamental level, the digitization of the art market is having an impact on the demand and valuation of works of art. After all, the aforementioned databases and other technical intermediaries associated with digitization are steered by powerful algorithms made up by sets of instructions that implicitly guide our searches. These protocols are by no means neutral and affect the valuation of art and artists in various ways. Firstly, the builders of art databases make a selection of works of art to include in their datasets, while others are left out, thereby either limiting or enhancing the exposure of browsers to particular artists and their work. More exposure leads to more visibility in the marketplace, and may increase demand for these pieces and thus result in higher prices. However, the selection criteria are often not clear. Secondly, the chosen artworks subsequently are coded to enable associations between artworks with similar characteristics. Various properties are ascribed to these works, many of them of an objective nature (size, support), while others are much more subjective (style, composition, appreciation). Again, these categorizations steer the outcome of our searches and can engender demand for previously unknown or less valued artists, but the criteria used in these coding exercises have not been made transparent. Thirdly, art databases will track and map our browsing patterns. This form of online surveillance allows them to modulate our future visits to their sites, steering us in certain directions by further enforcing the process of shaping our customized taste for art.

CONCLUDING OBSERVATIONS

The availability of a plethora of art market e-data has rendered a market that is usually seen as secretive and lacking in disclosure much more transparent. Moreover, digitization has created innumerable possibilities for research on the relationship between information access and social practice in the art world. The many inquiries by both professionals and academics have led to far greater insights into the workings of the art market than previously considered possible, and have gone a long way to resolving some of the more pressing information issues typically associated with the art world. For instance, it has become possible to track the price histories of particular works of art as they change hands on the global art market. Concretely, empirical studies have revealed that the visual art market is quite segmented, and that demand for, say, Impressionist paintings might respond differently to external shocks than the market for contemporary art (Buelens and Ginsburgh, 1993).

In general, art consumers have much more access to basic information relative to provenances, artists, exhibitions and so forth. Moreover, the emergence of countless new interactive sites has added a new layer of infrastructure to the art market, resulting in a magnifying effect, which allows for closer interactions between dealers and curators and their potential buyers and audiences. Interestingly, this avalanche of information appears to be strengthening the role of the traditional gatekeepers, rather than eroding it. It is in the face of this overwhelming availability of data that consumers are in need of trusted sources and reliable intermediaries to guide them (Arora and Vermeylen, 2013). In fact, while the Internet has certainly been instrumental in the further globalization and commercialization of the art market, the time-old structure of the market and how it operates have not been fundamentally challenged (Velthuis, 2012). Furthermore, despite the promise of a digital revolution that would render the traditional art institutions and experts obsolete, it has been suggested that the art world since the 1990s has predominantly shunned the virtual realm rather than embracing it (Bishop, 2012).

Nevertheless, it can be argued that digitization and its online manifestations are causing a shift in the art market from an object based and supply-side oriented market to a more consumer driven market. While no real empirical data exist to support this claim, it is assumed that the combination of a spectacular rise of new buyers in emerging markets and the exponential sophistication of digitization technologies outlined above (the power of big data) has started to empower buyers as well as art lovers.

This said, digital resources and overall improvements in communication technologies have not necessarily led to a dramatically more integrated global art market. Besides the strong local embeddedness of even the most outward looking art centers such as New York or Hong Kong, significant differences in transaction costs between global art centers remain, mostly as a result of diverging rates of taxation such as import and export duties, trade restrictions and value added taxes (VAT). These costs have not been offset by the leveling effect of e-commerce. Consequently, the law of one price, which would indicate a perfectly integrated market, where identical works of art (or close substitutes) are uniformly priced worldwide, does not apply (Isard, 1977; Ashenfelter and Graddy, 2003).

Finally, art is an experience good, which denotes that art lovers and buyers determine the quality of a work of art only upon consumption. Many art consumers derive pleasure from the personal contacts with artists and dealers in a gallery or at an art fair, by attending an art auction or by physically perusing the galleries of a museum admiring the tactility of the works of art displayed on the walls. While nobody seriously argues for the online reproduction of the 'aura' of material art, a sophisticated hyperrealistic platform can deeply enhance the desire to learn and to engage further with this art arena.

REFERENCES

Arora, P. and F. Vermeylen (2013), 'The end of the art connoisseur? Experts and knowledge production in the visual arts in the digital age', *Information, Communication and Society*, **16** (2), 194–214.

Ashenfelter, O. and K. Graddy (2003), 'Auctions and the price of art', *Journal of Economic Literature*, **41** (3), 767–83.

Baumol, W. (1986), 'Unnatural value: or art investment as floating crap game', *American Economic Review*, **76** (2), 10–14.

Bearman, David (2000), 'Museum strategies for success on the Internet', in Giskin Day (ed.), *Museum Collections and the Information Highway*, Proceedings of a Conference on Museums and the Internet, London: Science Museum, pp. 15–27.

Bishop, C. (2012), 'Digital divide', *Art Forum*, September.

Buelens, N. and V. Ginsburgh (1993), 'Revisiting Baumol's "Art as a floating crap game"', *European Economic Review*, **37** (7), 1351–71.

Castells, Manuel (2011), *Rise of the Network Society*, Oxford: Wiley-Blackwell.

Davis, B. (2011), 'Hype and hyperreality: zooming in on Google Art Project', ARTINFO, available at:

http://www.artinfo.com/news/story/36950/hype-andhyperreality-zooming-in-on-google-art-project/?page=2 (accessed 9 February 2013).
Ginsburgh, Victor, Jiangping Mei and Michael Moses (2006), 'The computation of price indices', in Victor Ginsburgh and David Throsby (eds), *Handbook of the Economics of Art and Culture*, Amsterdam: North-Holland, pp. 947–79.
Guerzoni, G. (1995), 'Reflections on historical series of art prices: Reitlinger's data revisited', *Journal of Cultural Economics*, **19** (3), 251–60.
Isard, P. (1977), 'How far can we push the "law of one price"?', *American Economic Review*, **67** (5), 942–8.
Kidd, J. (2011), 'Enacting engagement online: framing social media use for the museum', *Information, Technology and People*, **24** (1), 64–77.
Loran, M. (2005), 'Use of websites to increase access and develop audiences in museums: experiences in British national museums', *Digithum*, **7**, 23–8.
McQuail, Denis (2000), *McQuail's Mass Communication Theory*, 4th edn, London: Sage.
Marty, P.F. (2007), 'The changing nature of information work in museums', *Journal of the American Society for Information Science and Technology*, **58** (1), 97–101.
O'Reilly, T. (2007), 'What is Web 2.0: design patterns and business models for the next generation of software', *Communications and Strategies*, **65** (1), 17–37.
Pariser, Eli (2011), *The Filter Bubble: What the Internet Is Hiding from You*, New York: Penguin Press.
Prasad, N. (2011), 'Synergizing the collections of libraries archives and museums for better user services', *IFLA (International Federation of Library Associations) Journal*, **37** (3), 204–10.
Reitlinger, Gerald ([1961–70] 1982), *The Economics of Taste: The Rise and Fall in Picture Prices 1760–1960*, 3rd edn, 3 vols, New York: Hacker Art Books.
Trant, Jennifer and David Bearman (eds) (2011), *Museums and the Web 2011: Proceedings*, Toronto: Archives and Museum Informatics.
Velthuis, Olav (2012), 'The contemporary art between stasis and flux', in Maria Lind and Olav Velthuis (eds), *Contemporary Art and Its Commercial Markets: A Report on Current Conditions and Future Scenarios*, Berlin: Sternberg Press, pp. 17–50.
Verboom, Jessica and Payal Arora (2013), 'Museum 2.0: A study into the culture of expertise within the museum blogosphere', *First Monday*, 18(8), available at: http://firstmonday.org/ojs/index.php/fm/article/view/4538.

FURTHER READING

Comprehensive studies focusing on the wider impact of digitization on the art market as a whole are still lacking, but some important aspects have been addressed. Paul Marty (2007) provides a fine introduction to museum informatics, the study of how information science and technology affect the museum environment, while we (Arora and Vermeylen, 2013) have examined how new media have impacted the construction of art expertise, and raised doubts whether the Internet is undermining the traditional role of art experts as arbiters of taste.

29. Museums

Trilce Navarrete

Museums started exploring the use of computers in the 1960s, and by the 1990s many museums were in the process of adopting an automated work form to manage collections. By the 2000s the Internet had brought new distribution channels, and digitization became harder to avoid. During the last decade, there has been a revision in the work process, where information becomes a key asset and where the relation between the museum and its public changes to favor participatory services.

Many of the issues concerning the museum work (acquisition, preservation, exhibition, research and communication) have been thoroughly studied by cultural economists, and their insights can be applied to the digital equivalent, for instance to identify the effective use of resources for an increase in access (offline or online). There are, however, characteristic differences in the production, distribution and consumption processes as a result of digitization. These have not always been discussed.

This chapter reviews the economic literature on museums to focus on the areas relevant to digitization, applying existing theory in areas where no literature example can be found. Issues of intellectual property rights (and copyright) as a form of regulation are outside the scope of this chapter. In cultural economics, museums can be studied from three main perspectives: the museum institution (the 'firm' with inputs and outputs), the consumer of museum goods and the role of the government in supporting production of museum goods.

THE MUSEUM INSTITUTION

Museums have not changed much in their characteristics with the adoption of digitization. They are still non-profit organizations with multiple and complex functions revolving around acquisition, preservation, research, exhibition and communication of collections. Economists have looked at museums since the 1960s (see Frey and Meier, 2006 for a list of publications on the economics of museums). An important contribution to the application of cultural economics to museums can be found in the thematic edition of the *Journal of Cultural Economics* in 1998.[1] In it, museums' inputs and outputs were identified, the differences and similarities with other cultural sectors were explained and a mention or two about the Internet was made. It was still a bit early to report on the changes brought by digitization in museums. In the following, the digital aspects of museum institutions and work will be highlighted.

Inputs and Outputs

The primary input specific to the museum can be divided into capital (collection, building, hardware and data banks) and labor (Hutter, 1998: 102). Digital input includes the technology (hardware and software), the information database and the trained staff available (including registrars, curators and IT specialists, but also the entire staff). Lack of knowledgeable staff is an important deterrent in adopting digital technology (NMV and DEN, 2008).

The expectation that IT would increase output and reduce labor costs (Hutter, 1998: 104; Frey and Meier, 2006: 1025) has proved to be only partly true. Labor related to digital activities represents over 90 percent of the costs of the digital budget (DEN, 2009: 36). Registering and documenting museum objects remain manual activities, unlike library content, which is automatically catalogued upon production. Copyright clearance is another labor-intensive activity for museums wishing to publish collections online. Empirical work is needed to measure the extent to which digitization of production in museums faces the so-called cost disease, according to which a relative productivity lag drives up costs even with technological advancements.

Technological advancement can support other activities, including quality of imaging (photography, scanning, MRIs and so on), speed of scanning and processing text (with optical character recognition – OCR), cross-referencing (linking objects to themes) and increasing findability and distribution of content (onsite, online). The level of access and use of content is closely related to the degree of object registration and documentation (Zhang and Kamps, 2010). In other words, digital input strongly defines the use of output.

Output, according to Hutter (1998: 100), can be divided into (1) collecting, preserving and documenting; (2) showing (or exhibiting and communicating); and (3) selling. Output particularly related to digital technologies includes the website, digital publications (online and on CD-Roms) and virtual exhibitions (Hutter, 1998). Digital technologies have been applied to create new products and services as well as to devise new distribution channels (Bakhshi and Throsby, 2011: 1). Museums have experimented with alternative output forms, including open data sets for reuse (a machine readable version of their collections database), software applications (digital publications for smart phones and tablets), and publishing content on other websites (including portals like Europeana and social media sites like Wikipedia or Flickr). Online publication has been limited by issues of copyright leading to what has been called the black hole of the twentieth century (CEU, 2012: 3).

Interestingly, exhibiting collections in digital form requires that all other activities have taken place first: acquisition, conservation, research, documentation and interpretation (O'Hagan, 1998: 205). Selling, of course, follows display, as in the case of licensing reproduction rights for publishing purposes. Digitization of collections is not only a technology related activity, or IT cost (Stein, 2012). Publishing collections online is a type of exhibition, and as such it requires all the above-mentioned activities to take place before the actual digitization can be completed (ideally including the image, an explanation, and a list of resources). Digitization, and publishing content on the web, represents high fixed costs.

Most of the time, access by one person or many to the digitized content represents little additional cost (close to zero marginal cost). However, sometimes the high online traffic may require additional costs to improve content delivery. Such was the case for Europeana, the European Digital Library, which required additional servers after receiving 10 million hits per hour shortly after being launched, 'crashing' the site (*BBC News*, 2008).

Digitization has represented a solution to the problem of exhibiting the museums' largest capital: the collection. Not all objects in the collection can be physically exhibited; estimates of displayed objects range between 4 and 10 percent (Frey, 1994: 50; Voorthuijsen, 2009), making it a recurring source of inefficiency for economists (Frey, 1994: 50; Heilbrun and Gray, 2001: 200). Empirical research is needed to calculate digitization costs in relation to the increased visibility of objects against the opportunity cost of collections not being shown.

The value of museums is generally linked to the collection and building as well as to the services provided by these assets (Throsby, 2001: 35). Regarding digitization, the greatest asset of the museum derives from its information system: in the information contained in the collected objects, the various institutional documents (the museum's archive) and the information held individually by museum staff (Trant, 2008; Zorich, 2008).

Versioning

One important benefit of digital technology is that it has facilitated versioning of output, generally associated with information goods as proposed by Shapiro and Varian (1999). Versioning access modes to collections online, for instance, allow heritage institutions to implement price differentiation strategies. Table 29.1 shows product dimensions resulting from versioning heritage information.

Table 29.1 Versioning of access modes of digital heritage materials available online

Versioning mode	Change in content
Delay of publication	Hourly/daily/weekly delay
User interface	Personalized access/general access
Convenience	Access from home/access onsite/rush
Image resolution	High-resolution TIFF/low-resolution JPEG
Speed of operation	Use of small/large files to download Use of image rendition
Features and functions	Available on handheld device Ready-made educational kits Join/participate/add content Multilingual
Flexibility of use	Option to store/copy/print/share/search/browse/rate/buy content/reuse
Capabilities	General access/access to professionals with specialized vocabulary and hierarchy
Comprehensiveness	Partial/total access to the catalogue/collection online
Annoyance	Copyright clearance requirement/copyright free material*
Support	Possibility to ask a curator/researcher

Note: * Copyright clearance can be used as a disincentive for digitization requests, as goods are charged at a higher price or as the consumer is made responsible to clear copyright requirements. The Amsterdam City Archive declines requests to scan copyrighted materials (Holtman, 2006).

Source: Own, based on Shapiro and Varian (1999).

Versioning can take many forms. Museums can prepare digital materials related to an exhibition and offer peek previews, releasing the content selectively at different times, delaying time of publication. Content can be offered from a website and allow users to make a 'my museum' site, in a personalized user interface. Digitized content can be made accessible onsite for free but requires a charge to be accessed from the convenience of home or when delivered faster than normal (the so-called 'rush' fee).

When a digital high-resolution standard image has been produced, the museum can choose to make lower-resolution, black and white or cropped versions for distribution and sale at a different price. Higher-resolution files would be slower to download and make the system slower, though quick access can be achieved through partial download.

Features of online services can include delivering the content to be viewed in handheld devices (over the web or as an 'app'), with a language selection, and with a possibility to participate. Increasingly, services allow users to add content in the form of comments or images, or to take part in various activities (e.g. selection of works to be included in a future exhibit). Many options are possible for the permitted use of the accessible content. Users can be allowed to store content on the museum website, copy the images, download to print, share to social media sites, search the content, browse the themes, rate the quality (or popularity), purchase a copy (including a 3D print)[2] or reuse the content, for instance to program an app.[3]

Many early image banks and collections online were accessible to paying members. A common practice is to publish only part of the collection online to the general public for free and to publish a more comprehensive version with work tools (such as specialized vocabulary) for professionals, often with the use of a password. Sensitive information regarding insurance, costs and security is generally accessible to selected staff. Viewers of collections online may receive support from virtual information staff or through services like 'Ask a curator'.[4] With video collections, a clip can be made accessible online, and only after request can the entire video be seen. Copyright plays an important role in making collections available or not (Landes, 2003). Institutions may choose to publish a thumbnail but no large image of copyrighted materials or charge the interested viewer for the copyright clearance.

Digital technology has changed consumption patterns, forcing museums to adapt to the new information market. Museums are exploring the application of new business models that fit within the new forms of consumption.

Creating content that can be delivered to mobile devices onsite and offsite (that is, software applications) is favored when delivered for free. Actually, 'only blockbuster exhibitions have generated significant revenues for the museum and its tour providers' (Burnette et al., 2011: 6).

Allocation of Resources

Museums can serve different information needs for which the best allocation of resources to deliver an output is closely linked to the institutional goals. Institutions can choose to digitize the collection in great detail, but limited resources (including time) may allow covering only a small portion of the collection. On the other hand, museums can engage in mass digitization projects, covering a larger portion of the collection yet with little information about each object.

In the Netherlands, two cases can best serve to exemplify these different approaches. The Amsterdam City Archive has calculated the cost related to storage of high-resolution images from large quantities of scanned archival material and concluded it was prohibitively expensive to deliver best practice image quality. Its mass digitization program (10000 scans per week) has opted for a pragmatic approach scanning documents ensuring legibility at a basic level of imaging quality (Holtman, 2006). The opposite is true for the Amsterdam Rijksmuseum's Print Cabinet Online project. The imaging and contextual information per object is to be of the highest quality, resulting in an output of only 25000 digitized prints per year (Navarrete, forthcoming). Digitization remains an expensive activity. Identifying costs based on different types of output, while ensuring efficiency, remains a practical concern.

The discussion regarding the most effective allocation of resources is not new. Museums have struggled to balance the various core activities of acquisition, conservation, research and access, now expanded to digitization. The European Commission has recently published an overview of the current situation of digital heritage collections in 1951 institutions (including museums): 864 museums from 29 countries reported having close to 23 percent of their collections digitized; about 29 percent measure the access and use of the digital materials, and they allocate about 4.5 percent of their staff towards digital activities (Stroeker and Vogels, 2012), giving the closest indication of a digitization budget.[5] The low allocation of staff, and of resources, does not seem in accordance with the labor-intensive nature of the digitization process, involving contextualization of information and copyright clearance. The difficulty in gathering data about digitization activities (production, costs and access) may perhaps be the reason why little economic research has been done in this area.

Research on adopting digital technology, who adopts first and why (Johnson and Thomas, 1998: 81), continues to provide room for research. Bakhshi and Throsby (2011) linked innovation to technology. Digitization is, most of all, a tool to support work processes (such as information management and communication with the public). Research and development on the application of digital technologies for the heritage sector is still rare.

DIGITAL HERITAGE CONSUMPTION

According to Bakhshi and Throsby (2011: 4), digitization serves as an innovation strategy to reach audiences by publishing collections online, setting up interactive profiles in social networks and allowing consumers to participate in content creation. This would allow museums to reach a larger share of the population (broadening the audience), to attract new consumers (diversify the audience) and to intensify the engagement of audiences through interactivity (deepening the audience) (Bakhshi and Throsby, 2011: 4).

Digital content is also being consumed onsite. Museums can transfer their audio tour content to a digital base platform to give access to additional information about the objects being exhibited. Yet few consumers are 'mobile'. The San Francisco Museum of Modern Art reported only 12 percent of their visitors used the mobile phone tour (Burnette et al., 2011).

The number of visitors online is usually larger than onsite. An example is the Rijksmuseum, Amsterdam reporting just under 9 million visitors in 2010 (when only the Philips wing was open during remodeling), 1.1 million unique visits to the website and nearly 7 million search actions on the library research website (Rijksmuseum, 2010).

Digitization is believed to have resulted in an increased access to museum collections, representing a wider market (Johnson and Thomas, 1998: 80). As the cost of imaging decreased, museums enriched their information systems by adding visual representations to the object descriptions (Loebbecke and Thaller, 2011: 359). The distribution of digitized content, however, is still rare. For most museums, the main purpose of their websites is to stimulate *onsite* visits. In the Netherlands, only 31 percent of museums reported publishing part of their collections information online (DEN, 2009: 25), while 99 percent of museums reported having 'at least one website' (NMV and DEN, 2008).

The low level of publication of collections compared to the high online presence of the museum may be explained by the fear of cannibalization, or losing onsite visitors if collections were to be made available online. However, the opposite seems to be true: the online visit complements the onsite visit (D. Peacock, 2002; Marty, 2008; Peereboom et al., 2010; Bakhski and Throsby, 2011: 10). Taking again the case of the Netherlands, only four museums had a website in 1998, growing to 81 percent by 2002 and again to 99 percent by 2007 (NMV and DEN, 2008). As the digital medium is slowly adopted and understood, it is to be expected that collection publication will take time to catch on.

Another argument against the publication of collections online has to do with the loss of the special sensation of experiencing objects onsite compared to viewing a reproduction online. The experience of copies of paintings and prints throughout history shows that valuation of reproductions has changed in time (for example, we now value engravings, which used to be seen as reproductions, as originals). One important element is that reproductions result in an increased access to images, making collections better known (Benhamou, 2006: 280). Clearly, digitization of museum collections is about the distribution of object information and not of the objects themselves. For digital art, being 'born digital' and by nature multiple (it can be seen at the same time in two or more different locations), issues of originality or copying are irrelevant.

For some, cultural information systems available online are seen as add-ons to the museum institution, while for others they are seen as a replacement of the onsite visit (Loebbecke and Thaller, 2011). But what exactly is the perception of the consumer? In the Netherlands, the number of visits to museums was much higher than the visits to museum websites in 2006, with 38 percent and 15 percent of the surveyed population respectively (at least one visit in the last year) (Wubs and Huysmans, 2006). A follow-up survey may deliver a different result.

Criteria for Selection and Consumption

Heritage consumption is known to return individual benefits (pleasure of viewing the pieces and impressing others by having seen them or having purchased a piece) as well as providing investment opportunities for the future (A. Peacock, 1997: 397). Consumption of digital heritage has other characteristics. It includes a selection process of a fundamentally different nature. Museums are used to selecting the content that is available to the consumer through the museum visit. The choice of a museum visit may follow criteria based on proximity, current exhibition or holiday destination. Online, it is the consumer who selects the content to be viewed and used from the available information space. The consumer also selects the distribution channel (such as a museum website, a search engine or a 'virtual' collection such as the Google Art Project) to best fit his or her needs (Mackenzie Owen, 2007). It is about 'the right information, in the right amount, and in the right package . . . provided at the right time' (Yakel, 2000: 28).

Navarrete (2013) has proposed that digital cultural consumption follows a new set of principles shifting from cultural and economic values to information values. Consumers select content based on reliability, validity, completeness, actuality, verifiability, correctness, integrity, relevance and access. These criteria for selection are closer to information use than to cultural consumption (see Shapiro and Varian, 1999 for an overview of the economics of information).

Content on the Internet is expected to be free of charge (Farchy, 2003) and to suit the users' needs. These needs are dependent on the consumer's personal characteristics (age, gender, interests, needs), social role and general context (availability and knowledge of technology) (Boekhorst et al., 2005: 66–7). The consumer's community context has been identified to play an important role in the consumer's behavior on the Internet (Hutter, 2003). The Internet has only magnified the frame of reference to select consumption, a phenomenon that has been previously observed with radio, film, television and recorded music (Frey, 1998: 116).

The general consumer more often selects digital heritage material from what Frey (1998) has described as 'superstar' museums. These institutions hold the 'world famous painters and world famous paintings' (Frey, 1998: 114). That is partially because selection for distribution favors famous works. Google first approached 17 superstar museums to be part of the Google Art Project, including the Van Gogh Museum and the Rijksmuseum in Amsterdam, the Museum of Modern Art and the Metropolitan Museum of Art in New York, the Museo Reina Sofia and Museo Thyssen-Bornemisza in Madrid, and Tate Britain and the National Gallery in London. The project has expanded to showcase 151 museums from 40 countries.[6]

Onsite, visitors are willing to pay a higher entrance fee to see superstar museums and temporary exhibitions in order to see the world-famous painters and paintings (Frey and Meier, 2006: 1036–40). Online, consumers are willing to substitute quality for free access, high resolution for speed, and branding for convenience. It no longer matters where the content comes from as long as it satisfies the information need. Many users are more interested in finding information (images, background documentation) than in visiting a specific museum online. This is in conflict with the tendency of museums to use the Internet as a promotional device for themselves as institutions.

Users' Data

Digital content from museums can be found at the museum websites, but also in the Google Art Project, Wikipedia, image banks, iTunes and Europeana, as well as in a number of video games, blogs and software applications. Little empirical work is found on the economic use and reuse of digital heritage materials because of the complexity in forms and channels for distribution. Research has been conducted on the information needs of the first users, the cultural heritage experts or museum staff, revealing that systems for information management are too simple. Users must combine information from several sources (e.g. the information about the object and its image from two different databases) and use relatively simple tools made for fact finding to do their research (Amin et al., 2008). Systems are not meeting the needs of the museum staff. It is therefore no surprise that experts (curators) have been reluctant to adopt the digital medium.

Similarly, users of the Metropolitan Museum of Art website 'want to search across data, on multiple dimensions, and find other artworks that interest them' (Rainbow et al., 2012: 1). Consumers expect relevant results to their search actions as much as experts do. However, this is not easy to achieve, since technical and procedural solutions generally require a change in organizational practice (Camarero et al., 2011: 253), often with financial implications. Observing the behavior of the general user has revealed that individuals benefit from the Internet's ease of searching for digital cultural heritage (accessible overall and any time) (Wubs and Huysmans, 2006). Searching usually begins from a keyword search using a search engine (Nordbotten, 2000), and selected content includes context (for example, image and explanation) (Zhang and Kamps, 2010). This means that institutions benefit from following the findability best practice to ensure their content is found, from positioning it also outside of their website (say on Wikipedia) and from presenting images linked to text (instead of a list of images with little context information). Empirical research is needed to measure the level of investment in online publication in relation to the actual use. Data from past technologies used (such as microfilm, photography, slides and videos) may provide an interesting contrast. Demographic information of online users specific to museums would be beneficial. Visitors onsite are generally valued for the income they may represent (through ticket sales, shop, restaurant or parking fees). Visitors online are content users who require new measuring and valuation methods.

Digital technology allows the automatic documentation of user behavior, for example through web logs. This data holds quite interesting and mostly not exploited information about digital cultural heritage users. Data includes the number of people visiting a website, the number of times a file was downloaded, the frequency of mentions in tweets, or the type of question words used in the search engine. D. Peacock (2002) discusses the potential of log files to produce quantitative data, and Voorbij (2009) analyses the current web analytics methods. Museums use web data for simple indicators, yet complex efficiency indicators could easily be devised, such as following the growth of access after digitization (frequency of access to a work before and after digitization). Twenty-nine percent of European museums reported gathering some form of consumers' data regarding use of digital content (Stroeker and Vogels, 2011), suggesting data may become increasingly available to better understand users' behavior. However, the use of performance indicators based on users' digital data is yet to be institutionalized in museums.

Production as a Service

Curiously, digitization has opened up the production process to include the consumer becoming a 'prosumer'. In a way this is not new. Johnson and Thomas (1998: 78) wrote about volunteers being a group who 'receive utility from the production process itself: they are both consumers and producers'. Digitally, consumers have an active stand and expect to participate as part of the museum experience.

Participation has many forms and can include sharing (participating in distribution), ranking (participating in ordering), commenting (supporting contextualization), correcting (improving the quality of information), adding new information (including text, images or video to expand the digital collection), reworking the data (for example, programming an app) and financing (known as 'crowdfunding'). Social media has served as a platform to satisfy participation without having to change much inside the institution. Contributing content, however, has raised questions of authority. If a museum publishes a catalogue that includes information provided by the public, who owns the content, who is responsible for it and who can profit from it?

Many art history students support data entry during their museum internships. Yet opening up public participation online has not been fully accepted. Selected museums are seeking presentation solutions for the two types of data (the authorized and the public one). Options include reviewing the consumers' data before merging it with the rest, or presenting multiple sources of information within the museum website (including text by curators, the public, Wikipedia and the like).

A number of Dutch museums are 'freeing' their data (images and metadata) to allow the public to rework collection information into new products. The Amsterdam Museum and the Rijksmuseum, Amsterdam are two examples of Dutch museums offering open data as a service. Users have visualized collections in a timeline and have applied face recognition software to all available images, to name a few examples.[7]

THE ROLE OF THE GOVERNMENT

The government has always played a key role in supporting adoption of new technologies (for a historical account of the role of the government in media adoption see Crowley and Heyer, 2003 and Briggs and Burke, 2009). Adoption of digital means has been no exception. Governments have supported digitization efforts through subsidies and through regulation. One study in Italy found that adoption of laser technology for the restoration of heritage depends on government intervention. While benefits were well known and the technology had been established, restoration firms presented a 'significant resistance to adopting new technologies' unless the State provided the explicit request or the funds for the use of laser technology (Verbano et al., 2008: 7).

In the Netherlands, the government first supported the use of computers in museums by making mainframe computers available for data processing in the 1960s and 1970s. Subsidies were given to selected projects throughout the years, but it was not until the 1990s that structural funds were earmarked, part of the e-government plan, for R&D and for advancing online services (Navarrete, forthcoming). The government also supported the establishment of a basic standard for object valuation, which was adopted nationwide in the 1990s through a comprehensive subsidy plan to inventory collections. From the mid-2000s, digitization projects were funded with increasingly strict guidelines to augment the use of standards and best practice, including earmarking funds for digital activities (Navarrete, forthcoming). Recently, the Dutch government has been supporting the commercialization of digital heritage content by supporting collaboration between museums and the private sector to develop new market products (Navarrete, forthcoming). This appears to be an international trend (Bakhshi and Throsby, 2011: 2).

Reasons for Supporting Digitization of Museum Collections

The government has three reasons for supporting digital activities in museums: efficiency, capitalization on past investment, and the welfare of present and future generations (public value). First, supporting the adoption of digital technology and the use of standards can assist productivity improvement. Data management using paper cards limits the use of information, not to mention that it represents redundant work and has a higher chance of errors. Harmonizing standard use in the work field assists the gathering of data to inform decision-making also at government level.

Dutch museums are what Schuster (1998) calls hybrid institutions. They are (for the most part) private independent organizations that care for collections (partly) owned by the government, they are housed in buildings (generally) owned by the state and they receive structural funding from the national, regional or local government (OCW, 2006). Digitization of objects has been (mostly) financed by the government. That makes digitized collections a type of government data. The Dutch government has an open data policy, meaning that government data should be freely available to the public. However, the heritage collections are yet to be considered (Donner, 2011).

Having a digitized collection paid for by the government, of objects that are property of the state, cared for by the museum, and in need of sustainable management has raised issues of long-term financial responsibility. The Dutch National Library has taken responsibility for storing and ensuring long term access to digitized collections paid for by government grants (i.e. part of the Memory of the Netherlands) (Navarrete, forthcoming).

A second reason for supporting digital activities has to do with the preservation of past investments and their future use. In the Netherlands, the largest government subsidies have been given for digitization aimed at the preservation of collections, first of all museum collections (the so-called Delta Plan for the Preservation of Collections running from 1990 until 1995), then for paper collections (under the Metamorfoze project started in 1997) and recently for audiovisual materials (the Images for the Future project planned from 2009 until 2013) (Navarrete, forthcoming), this last project having a huge budget of €154 million for seven years (TNO, 2010).

Digitized heritage content represents the repositioning of a past investment involving the acquisition, preservation and research of the objects. Allowing use and reuse of the digitized materials is seen by the Dutch government as key to stimulating innovation and production (Donner, 2011). Recent Dutch R&D funding schemes involve collaboration between heritage institutions and the private sector to bring new heritage products and services to the market (Navarrete, 2013).

Another reason to fund digitization relates to the safeguarding of access to the national cultural heritage by present and future generations (Camarero et al., 2011). Museum collections have been digitized to increase access to content and to preserve the original object, so that future generations will have access to it. This is because cultural heritage is a merit good, because it has public good characteristics and because of the externalities brought by production and consumption (D. Peacock, 2002).

Digital preservation has also been on the government's agenda. The Dutch Department of Internal Affairs has been part of the Netherlands Coalition for Digital Sustainability in charge of developing an infrastructure to ensure sustainable access to content.[8] Preserving museum content is part of the Coalition's strategy.

Similarly, Europeana, the European Digital Library, was formed to increase access to European heritage materials. It has further adopted an open data policy, making 20 million objects (images and metadata) available for free reuse (with a creative commons public domain license).[9]

CONCLUSIONS

Economic research of museums has focused on descriptive applications of theory and on development of theoretic models for traditional activities. Digital activities have gained recent attention as to their role in innovation. Research that takes advantage of large digital data sets (such as web statistics) and that considers the museum as part of an information market (not only cultural or leisure) is on the future agenda.

Consumption onsite has been greatly analyzed. New forms of use and reuse have emerged. A better understanding of access and new forms to measure use are still an open field. Public funding of museums has had a great impact on the adoption of computers and in museums' digital practice. Curiously, funding for building a digital infrastructure (digitization of collections) remains limited, if existent at all.

There is still much room for research on the economics of digitization and museums. Future analysis would benefit from an interdisciplinary approach combining economic theory with information science.

NOTES

1. The publication resulted from a selection of papers presented at the Economics of Museums Conference held at Durham University in March 1998 (Johnson and Thomas, 1998).
2. The San Francisco Asian Art Museum and the New York Metropolitan Museum of Art are exploring the use of 3D technology to increase access to collections (http://www.wired.com/design/2012/10/scanathon/).
3. An example can be found in the Rijksmuseum, Amsterdam API (application programming interface) offered on their website for people to develop application software (see http://www.rijksmuseum.nl/en/api?lang=en).
4. The 'Ask a curator day' involves asking questions via Twitter, and the 'Ask a curator' website showcases curators answering questions. In 2010, 340 museums participated, and in 2012 it grew to more than 600. For an initial tweet analysis from 2012, see http://blog.lamagnetica.com/2012/09/27/el-askacurator-day-en-numeros-y-valoraciones/.
5. Data was gathered as part of the ENUMERATE project (2011–13); for information about the project see http://www.enumerate.eu.
6. http://en.wikipedia.org/wiki/Google_Art_Project.
7. http://www.rijksmuseum.nl/en/api?lang=en.
8. http://www.ncdd.nl.
9. http://pro.europeana.eu/web/guest/news.

REFERENCES

Amin, Alia, Jacco van Ossenbruggen and Lynda Hardman (2008), 'Understanding cultural heritage experts' information seeking needs', in *Proceedings of the 8th ACM/IEEE-CS Joint Conference on Digital Libraries JCDL '08*, New York: ACM, pp. 39–47.

Bakhshi, Hasan and David Throsby (2011), 'New technologies in cultural institutions: theory, evidence and policy implications', *International Journal of Cultural Policy*, **18** (2), 1–18.

BBC News (2008), 'European online library crashes', Europe News, 21 November, available at: http://news.bbc.co.uk/2/hi/europe/7742390.stm.

Benhamou, Françoise (2006), 'Copies of artworks: the case of paintings and prints', in Victor A. Ginsburgh and David Throsby (eds), *Handbook of the Economics of Art and Culture*, vol. 1, Amsterdam: North-Holland, pp. 253–83.

Beunen, Annemarie and Tjeerd Schiphof (2006), *Juridische Wegwijzer*, The Hague: Taskforce Archiven, available at: http://www.taskforce-archieven.nl/nieuws/symposium6december2006.

Boekhorst, Albert, Inge Kwast and Diane Wevers (2005), *Informatievaardigheden*, Utrecht: Lemma.

Briggs, Asa and Peter Burke (2009), *A Social History of the Media: From Gutenberg to the Internet*, Cambridge: Polity Press.

Burnette, Allegra, Nancy Proctor and Peter Samis (2011), 'Getting on (not under) the Mobile 2.0 bus: emerging issues in the mobile business model', in *Museums and the Web 2011*, available at: http://www.museumsandtheweb.com/mw2011/papers/getting_on_not_under_the_mobile_20_bus.

Camarero, Carmen, M.J. Garrido and E. Vicente (2011), 'How cultural organizations' size and funding influence innovation and performance: the case of museums', *Journal of Cultural Economics*, **35** (4), 247–66.

CEU (Council of the European Union) (2012), *Council Conclusions on the Digitisation and Online Accessibility of Cultural Material and Digital Preservation*, Brussels: Council of the European Union, available at: http://sca.jiscinvolve.org/wp/2012/05/25/europeana-council-of-europe-conclusions/raadsconclusies-europeana/.

Crowley, David and Paul Heyer (2003), *Communication in History: Technology, Culture, Society*, Boston, MA: Allyn & Bacon.

DEN (2009), *De Bijlage bij de Digitale Feiten*, Zoetermeer: Digitaal Erfgoed Nederland.

Donner, Jean Piet Hendrik (2011), Letter to the Chair of the Parliament regarding the reuse of open data, 'Naar betere vindbaarheid en herbruikbaarheid can overheidsinformatie', ref. 2011–20000224638.

Farchy, Joëlle (2003), 'Internet: culture', in R. Towse (ed.), *A Handbook of Cultural Economics*, Cheltenham, UK and Northampton, MA, USA: Edward Elgar Publishing, pp. 276–80.

Frey, Bruno (1994), *La economia del arte*, Barcelona: La Caixa.

Frey, Bruno (1998), 'Superstar museums: an economic analysis', *Journal of Cultural Economics*, **22** (2–3), 113–25.

Frey, Bruno and Stephan Meier (2006), 'The economics of museums', in David Throsby and Victor Ginsburgh (eds), *Handbook of the Economics of Art and Culture*, vol. 1, Amsterdam: North-Holland, pp. 1017–47.

Heilbrun, James and Charles M. Gray (2001), *The Economics of Art and Culture*, Cambridge: Cambridge University Press.

Holtman, Marc (2006), *Digitization Simplified: Large-Scale Digitizing for Archive Research*, Amsterdam: Stadsarchief Amsterdam.

Hutter, Michael (1998), 'Communication productivity: a major cause for the changing output of art museums', *Journal of Cultural Economics*, **22** (2–3), 99–112.

Hutter, Michael (2003), 'Information goods', in R. Towse (ed.), *A Handbook of Cultural Economics*, Cheltenham, UK and Northampton, MA, USA: Edward Elgar Publishing, pp. 263–8.

Johnson, Peter and Barry Thomas (1998), 'The economics of museums: A research perspective', *Journal of Cultural Economics*, **22** (2–3), 75–85.

Landes, William (2003), 'Copyright', in R. Towse (ed.), *A Handbook of Cultural Economics*, Cheltenham, UK and Northampton, MA, USA: Edward Elgar Publishing, pp. 132–42.

Loebbecke, Claudia and Manfred Thaller (2011), 'Digitization as an IT response to the preservation of Europe's cultural heritage', in A. Carugati and C. Rossignoli (eds), *Emerging Themes in Information Systems and Organization Studies*, Heidelberg: Springer, pp. 359–72.

Mackenzie Owen, J. (2007), *The Scientific Article in the Age of Digitization*, Dordrecht: Springer.

Marty, Paul (2008), 'Interactive technologies', in Paul Marty and Katherine Burton Jones (eds), *Museum Informatics: People, Information and Technology in Museums*, New York: Routledge, pp. 131–7.

Navarrete, Trilce (2013), 'Digital cultural heritage', in Ilde Rizzo and Anna Mignosa (eds), *Handbook on the Economics of Cultural Heritage*, Cheltenham, UK and Northampton, MA, USA: Edward Elgar Publishing, pp. 251–71.

Navarrete, Trilce (forthcoming), 'A history of digitization: Dutch museums', dissertation, University of Amsterdam.

NMV and DEN (2008), *ICT gebruik in musea*, Groningen: Reekx Advies.

Nordbotten, Joan (2000), 'Entering through the side door – a usage analysis of a web presentation', in *Museums and the Web 2000*, available at: http://www.museumsandtheweb.com/mw2000/papers/nordbotten/nordbotten.html.

O'Hagan, John (1998), 'Art museums: collections, deaccessioning and donations', *Journal of Cultural Economics*, **22** (2–3), 197–207.

OCW (2006) *Cultural Policy in the Netherlands*, Boekmanstudies, Amsterdam: Boekman.

Peacock, Alan (1997), 'A future for the past: the political economy of heritage', in R. Towse (ed.), *Cultural Economics: The Arts, Heritage and the Media Industries*, Cheltenham, UK and Lyme, NH, USA: Edward Elgar Publishing, pp. 189–243.

Peacock, Darren (2002), 'Statistics, structure and satisfied customers: using web log data to improve site performance', in J. Trant and D. Bearman (eds), *Museums and the Web 2002: Proceedings*, Toronto: Archives and Museum Informatics, available at: http://www.museumsandtheweb.com/mw2002/papers/peacock/peacock.html.

Peereboom, Marianne, Edith Schreurs and Marthe de Vet (2010), 'Van Gogh's letters: or how to make the results of 15 years of research widely accessible for various audiences and how to involve them', in J. Trant and D. Bearman (eds), *Museums and the Web 2010: Proceedings*, 31 March, Toronto: Archives and Museum Informatics, available at: http://www.archimuse.com/mw2010/papers/peereboom/peereboom.html (accessed 1 October 2012).

Rainbow, Rachel, Alex Morrison and Matt Morgan (2012), 'Providing accessible online collections', in *Museums and the Web 2012: Proceedings*, available at: http://www.museumsandtheweb.com/mw2012/papers/providing_accessible_online_collections.

Rijksmuseum (2010), *Jaarverslag 2009*, Amsterdam: Rijksmuseum.

Schuster, Mark (1998), 'Neither public nor private: the hybridization of museums', *Journal of Cultural Economics*, **22** (1–2), 127–50.

Shapiro, Carl and Hal Varian (1999), *Information Rules: A Strategic Guide to the Network Economy*, Boston, MA: Harvard Business School Press.

Stein, Robert (2012), 'Blow up your digital strategy: changing the conversation about museums and technology', in *Museums and the Web 2012: Proceedings*, available at: http://www.museumsandtheweb.com/mw2012/papers/blow_up_your_digital_strategy_changing_the_c_1.

Stroeker, Natasha and Rene Vogels (2012), *Survey Report on Digitization in European Cultural Heritage Institutions 2012*, report prepared for ENUMERATE, Zoetermeer: Panteia, available at: http://www.enumerate.eu/en/statistics/.

Throsby, David (2001), *Economics and Culture*, Cambridge: Cambridge University Press.

TNO (2010), *Tussentijdse Evaluatie Beelden voor de Toekomst*, report, Delft: TNO.

Trant, Jennifer (2008), 'Curating collections knowledge: museums on the cyberinfrastructure', in Paul Marty and Katherine Burton Jones (eds), *Museum Informatics: People, Information and Technology in Museums*, New York: Routledge, pp. 275–92.

Verbano, Chiara, Karen Venturini, Giorgio Petroni and Anna Nosella (2008), 'Characteristics of Italian art restoration firms and factors influencing their adoption of laser technology', *Journal of Cultural Economics*, **32** (1), 3–34.

Voorbij, Henk (2009), *Webstatistieken: achtergronden, mogelijkheden en valkuilen*, report, part of the research project More Digital Facts commissioned by the Den Foundation, available at: http://www.den.nl/kennis/thema/webstatistieken.

Voorthuijsen, Anka van (2009), 'Museum 3.0 wordt community', *Binnenlands Bestuur Magazine*, 4 December, 42–5.

Wubs, Henrieke and Frank Huysmans (2006), *Klik naar het verleden: een onderzoek naar gebruikers van digitaal erfgoed: hun profielen en zoekstrategieen*, The Hague: SCP.

Yakel, Elizabeth (2000), 'Knowledge management: the archivist's and records manager's perspective', *Information Management Journal*, **34** (3), 24–30.

Zhang, Junte and Jaap Kamps (2010), 'Search log analysis of user stereotypes, information seeking behavior, and contextual evaluation', *Proceedings of the Third Symposium on Information Interaction in Context*, New York: ACM, pp. 245–54.

Zorich, Diane (2008), 'Information policy in museums', in Paul Marty and Katherine Burton Jones (eds), *Museum Informatics: People, Information and Technology in Museums*, New York: Routledge, pp. 85–106.

FURTHER READING

The basics of information economy can be found in Shapiro and Varian (1999). Crowley and Heyer (2003) and Briggs and Burke (2009) present a thorough history of issues related to the adoption of new technologies, including changes in the production, distribution and consumption of information goods. Navarrete (forthcoming) presents a detailed account of the digitization history of Dutch museums, including a chapter on government policy and financing. Navarrete (2013) gives an account of the supply and demand of digital heritage information. Beunen and Schiphof (2006) wrote a legal guide for online publication of museum content (in Dutch), including the basics of copyright. The *WIPO Guide on Managing Intellectual Property for Museums* is available at http://www.wipo.int/copyright/en/museums_ip/guide.html. The Museums and the Web international conference is a good reference for the application of digital technology in museum work.

30. Publishing

Patrik Wikström and Anette Johansson

Publishing is no doubt one of the oldest and most diverse sectors in the creative economy. While publishing originally was associated with print and paper, the term is nowadays also commonly used to represent organizations that control, administer and license intellectual properties in other sectors of the creative economy such as videogames and music. While the title of this chapter is 'Publishing', we have no intention of covering all publishing related activities, but will focus on the economic consequences of digitization on two traditional and important print media sectors, namely books and magazines. Within these sectors we will specifically focus on consumer magazines and trade books, in other words books and magazines that are sold via commercial retailers to consumers.

It is relevant to study these two publishing industries, since they share a number of very fundamental characteristics and have experienced similar economic consequences caused by the digitization of the creative economy. Both industries have undergone a gradual shift from print to digital and increasingly rely on revenues based on digital content carriers such as e-books, tablet magazine applications, special interest websites, blogs and so on.

THE CHANGING STRUCTURE OF THE PUBLISHING INDUSTRIES

National publishing markets are typically dominated by a handful of large publishing houses that operate a portfolio of magazine titles and book imprints. These publishing houses follow a general trend in the media industries and increasingly diversify into other media related areas such as radio and TV broadcasting and different online content products (Johansson et al., 2012). While major publishing houses may dominate a large share of the market, most publishers are very small businesses that run only a single magazine title or release a handful of books per year. These publishers are often primarily driven by an artistic, ideological or literary passion rather than by economic goals – actual profitability is relatively far down on their list of priorities.

The book and the magazine industry value chains follow the same principles as most other media industries and can be structured into five sections: content acquisition and creation; content selection and processing; transformation of content into distributable form; distribution; and marketing and promotion (Picard, 2011: 46). While both industries have this common basic structure there are at the same time significant differences between how magazines and books are produced. For instance, while the main actors in the book sector traditionally include authors, agents, publishers, wholesalers and retailers, the actors in the magazine industry include journalists, publishers, distributors and retail stores.

Digital technologies and innovations break up some of these structures, redefine established roles and introduce entirely new actors into the system. They enable more cost efficient ways to produce, distribute, and communicate with readers and users, and thereby they radically lower the barriers to entering the publishing industry (Rightscom, 2004). During the last decade the emergence of innovations such as online self-publishing book services and social media platforms has not only simplified processes and lowered entry barriers within the established national markets; it has also made it easier for both incumbent and entrepreneurial publishers to reach international markets (Rønning and Slaatta, 2012). Although social media services and other digital innovations are used by publishers in all shapes and sizes, taken together it particularly means that smaller players – independent authors or publishers – can market their products without having to rely on the resources and networks provided by traditional multinational publishers. This structural transformation is particularly apparent in the book sector, where it increases the opportunities of self-management and independence.

As digital innovations create opportunities for entrants into the publishing industries, they at the same time significantly intensify the competition within the industry. Some of the entrants into the magazine and book industries take advantage of the lower production costs and compete directly against the incumbent publishers with their printed products. During the last few decades the number of book and magazine titles has significantly increased in many markets and, since the total circulation has not had a parallel development, the average revenues per title have decreased. Other entrants, primarily in the magazine sectors, choose to compete with the established actors by offering their content and services online, without any ties to a traditional print product. There is a significant turbulence among these start-ups as they try to shape value propositions that are able to appeal to both the reader market and the advertiser market. While most start-ups fail, some of them have indeed been able to set up sustainable businesses. By abandoning the print product altogether, these actors are able to offer targeted advertising products that are more cost efficient and provide better feedback to advertisers compared to the advertising products offered by magazines operating in the print world. In addition, the online magazines' advantageous cost structure enables them to offer their editorial content to their readers at a low cost – or sometimes even free. The competition from these and other new actors in the magazine industry requires the incumbent magazine publishers to transform their value proposition and to find new sources of revenues, which we will discuss in the next section of this chapter. The traditional magazine publishers have had varying success with their attempts to adapt to the new business conditions. While some have been able to transform their businesses to remain competitive, other magazine publishers have decided that the only way to improve their cost structures and achieve economies of scale is by merging with other publishers. The digitization of the publishing industries can thereby be seen as one of the most significant drivers of acquisitions and mergers between major and mid-sized magazine and book publishers during the last decade (for example, Hultén et al., 2010).

In addition to new entrants that are competing with the incumbents, the digitization of the publishing industries has also enabled the establishment of entirely new types of actors. These actors contribute to the transformation of the publishing industries, as they provide innovative services such as content aggregation, social network platforms or publishing tools for aspiring writers. In the next sections we will examine how such novel products and services contribute to the transformation of the publishing industries.

FROM PAPER TO SCREEN

Many entrepreneurs, inventors and digital visionaries have dreamed about one day being able to replace the paper used for magazines and books with a new medium that would allow for a new kind of literature. For instance, in the early 1970s, Alan Kay speculated about the Dynabook, 'a personal computer for children of all ages' (Kay, 1972) that conceptualized many of the features of contemporary laptops, tablets and e-book readers. A decade later, in 1983, Steve Jobs explained in a speech that the strategy of Apple Computer was to 'put an incredibly great computer in a book that you can carry around with you that you can learn how to use in twenty minutes' (Brown, 2012). In 1990, Tim Berners-Lee, at the time at the research institute CERN in Geneva, proposed a world wide web of 'hypertext' documents that would be stored on computer servers and accessible from web browsers via data networks (Berners-Lee and Cailliau, 1990). These key technologies have been around for decades, but it was not until November 2007 that they were fused into a product that was reasonably accessible and appealing to a mainstream audience. The company that achieved this feat was the online retailer Amazon, and the product was called the Kindle (Levy, 2007). The Kindle was by no means the first e-book reader, but it was the first integrated *platform* that allowed readers to conveniently buy books and subscribe to magazines and newspapers that could be wirelessly downloaded to a reading device. After the launch of the Kindle in 2007, a plethora of other e-book readers, tablet computers and media platforms entered the market. The market is still far from any kind of equilibrium, and a considerable number of actors are struggling to get a share of the growing market. Amazon's early entrance into the market has enabled it to remain ahead of the competition, and currently merely two other players – Google and Apple – are recognized as forces in the global market of electronic magazines, books and newspapers.

It is interesting to note that these actors – Amazon, Apple and Google – are three US-based technology companies without any deeper commitments to the publishing industries. It is too early to judge the consequences of the total American dominance in this part of the creative economy, but it is already clear that the US e-book market is by far the most advanced in the world. In 2011, US trade publishers' e-book revenues were $1.97 billion, which is 16 per cent of the total trade industry (Harrison, 2012). Pew Internet Research (Rainie and Duggan, 2012) has also reported that 33 per cent of all Americans own some kind of e-book reading device, and approximately 23 per cent read e-books on a regular basis. In order to put the US statistics in some kind of perspective it may be useful to compare them with the figures for an average European book market, Sweden – a small but otherwise fairly technologically sophisticated market. Less than 8 per cent of the Swedish population owned an e-book reading device during 2011 (Facht, 2012), and the publishers' revenues from e-books still were not even close to 1 per cent of the total trade book market (SvF, 2012). There are several explanations of the differences between the American and non-Anglophone European countries. Regardless of the explanations, the rapid development in the US and the dominance of the three US-based retailers of online culture ought to create some sense of urgency in publishing minds in other parts of the world.

Retailers have always been gatekeepers in the publishing industries, but the emergence of these global industry leaders has wrung even more power out of the hands of the publishers. One example of this power shift is the weakening of the traditional industry practice that once allowed book publishers to dictate the retail price of a book and merely use the retailer as an agent for the sale. When Amazon launched its e-book business in 2010 it wanted to sell its e-books for $9.99 even if that meant it lost money on every unit it sold. Amazon wanted to quickly build a locked-in user base that had spent their monies on building a substantial e-book library that could be read only using an Amazon Kindle device. That way, the readers' costs of switching to another competing platform, such as Google or Apple, would be discouragingly high and they would stay loyal to the Amazon Kindle. Many publishers were concerned by 'the $9.99 problem' and the potential monopoly of the Amazon platform, and they refused to do business with Amazon as long as the retailer sold their books at a discounted price. After lengthy processes between publishers, retailers and competition authorities both in Europe and in the US, most major publishers eventually had to succumb and abandon their traditional 'agency model' in favour of the 'wholesale model' suggested by Amazon. It remains to be seen how this will shape the continued development of the e-books market, but it is likely that the major retailers will continue to strengthen their market dominance and that it will be very difficult for small or mid-sized online retailers to stay in the market.

Amazon is one of the primary driving forces during the shaping of e-book pricing practices, and Apple plays a corresponding role for electronically distributed magazines. The agency model remains robust in the e-magazine sector, but the three leading online retailers' market dominance enables them nevertheless to demand a 30 per cent commission on all subscription fees and any purchase that takes place within applications on their platforms. When Apple announced this commission-based model to magazine publishers in 2011, most publishers found the notion of paying 30 per cent of their subscription fees to Apple outrageous. However, as time goes by, Apple's magazine storefront grows ever stronger. As a consequence it becomes increasingly difficult not to participate, and even some of the most sceptical magazine publishers have eventually decided to accept the model.

Both these examples from the magazine and the book industries illustrate how the shift from paper to screen also means a shift of power from the stage in the value chain where content is created, selected and processed to the stage where content is packaged and distributed (Picard, 2011: 46). Even though this shift of power may perceived as a threat to publishers, many publishers, particularly in the magazine sector, have actually been relatively eager to embrace the prospects offered by tablet computers in general and particularly Apple's iPad device. Since the launch of the first iPad in April 2010, magazine publishers have aggressively developed products for the platform. One way to explain this technological enthusiasm in an industry where Luddite tendencies otherwise are fairly common is by looking at the even less appealing alternatives. As already mentioned in this chapter, magazine publishers perceive a growing competition from a range of new online as well as print media outlets. Many magazine publishers have already experimented with online publications during the last decades – most often without being able to reach sustainable profitability. In the light of these disappointing experiences, the iPad appeared to offer a controlled and safe haven where traditional publishing business rules could be upheld – even though a considerable share of the revenues had to be surrendered to the governor of the ecosystem, that is, Apple.

Many large magazine publishers launched their first iPad magazines as early as 2010. In order not to add to the competition to their print media products, and since they controlled the pricing of their iPad products, most publishers chose to set the price of their tablet magazines at the same level as the printed single-copy sales. It is difficult to make any conclusions whether their pricing strategies have been successful or not. Studies show that the circulation of tablet magazines has been growing rapidly. For example, digital magazine circulation in the USA went from 1.4 million in 2010 to 3.3 million in 2011 (AAM, 2012), and according to a study made by VDZ, the trade organization for German magazine publishers (VDZ, 2011), 68 per cent of iPad owners in Germany read magazines on their iPads. On the other hand, a number of ambitious tablet magazines, including News Corp's *The Daily*, have failed miserably, while paper as the main medium for magazine products slowly but surely continues to deteriorate.

BRANDS AND NICHES

As the traditional print magazine business continues to crumble, publishers search for ways to expand their businesses into other safer ground. One successful route ahead has been to extend strong print media brands into other product areas such as events, custom publishing, merchandising, broadcasting, videogames and so on. As the demand for traditional print products decreases among readers and advertisers, brand extension has been a critical route forward for the magazine sector. Brand extension in the book industry is somewhat different from brand extension in the magazine industry. Magazine publishers usually have their own brands as a starting point, for example *Vogue* or *Cosmopolitan*, while book publishers instead build on literary characters such as Harry Potter, or authors such as J.K. Rowling. Book publishers are also in a somewhat different situation from magazine publishers as a result of the growing independence of authors that has been enabled by digital technology. Authors increasingly use the opportunity to digitally produce, market, distribute and communicate their own books and build their own brand, independently from a publisher (for example, Harrison, 2012). This independence can also be manifested as collaborative networks across different media actors and regions. One such example is the collaboration between J.K. Rowling and Sony in the extension of the Harry Potter brand to the Pottermore website and storefront, where e-books and audio downloads are sold in partnership with J.K. Rowling's publishers worldwide. This use of collaborative partnerships to create new value networks is seen in several industries that are affected by digital technologies (Chesbrough and Schwartz, 2007). For major multimedia actors, brand extensions open up considerable opportunities of integrating their brands across different media platforms. A recent example includes the European reality television franchise *The Best Singers* (controlled by Zodiak Rights in the Netherlands), where nationally famous singers meet and present interpretations of each other's songs. In one of the countries where the format was licensed, the licensee combined an iPad magazine and the television series into what commonly is referred to as a transmedia production (for example, Jenkins, 2003). The publication schedule of the magazine was synchronized with the television shows and introduced the show before it was aired, and followed up with interviews and extra material afterwards.

The extension of brands into new product categories and distribution platforms has contributed to the blurring of boundaries between industries and markets and poses interesting questions regarding what actually defines a book or a magazine. One manifestation of this convergence is the emergence of the Kindle Single format developed by Amazon. Kindle Singles are simply self-contained stories that are too long for the traditional magazine article format but not long enough for the traditional book format. Authors of a Single submit their stories to Amazon via its Kindle Direct Publishing platform without requiring any involvement from other agents or publishers. The writers set the price anywhere between $0.99 and $4.99, they keep 70 per cent of the revenues, and they retain all rights to their work. It should be noted that, even though a traditional publisher is not part of this process, the Kindle Single format is not self-publishing, as Amazon reviews each submission and decides what is accepted and what is not.

The Kindle Single is only one example of the convergence between two already closely connected industries. As publishers of magazines and books gradually learn how to take advantage of the interactive features that advanced tablet computers offer there is no real need to restrict the magazine format to static texts and images. It is likely that the tablet computer is the platform that finally will be able to bring text, video, audio and images together into a multimedia magazine that has a closer kinship with videogames than with the traditional products that were distributed on paper. While these so-called 'enhanced' e-books and e-magazines have been expected to arrive for a number of years, there are still a number of obstacles that seem to be hampering their development. One problem is the fact that the costs of producing enhanced e-books and e-magazines are significantly higher compared to traditional equivalents. A second problem is that many publishers simply lack the competences, and many writers lack the creative inclination, required to produce 'enhanced' publications that are able to bring something new to the market.

Regardless of whether the new book- and magazine-related products are Singles, interactive applications or any other kind of brand extension, they all add to the plethora of titles that compete for the readers' media spending. The competition between all these titles is certainly fierce, but it may actually be possible for a large number of them to be economically viable, since digitization radically reduces the cost of creating or finding a suitable market niche. The improved possibility of establishing economically viable micro-niches is an essential feature of the contemporary publishing industry. Brynjolfsson et al. (2003) have made considerable contributions to the theoretical understanding of this phenomenon, and Anderson (2006) has introduced the concept to the mainstream audience as 'the long tail'. It deserves to be noted that, while 'the long tail' predicts that digitization makes smaller niches economically viable, it has perhaps even greater consequences on aggregator level, primarily for retailers of magazines and books. As books and magazines become digital products, the costs for storing and distributing these products are very close to zero. This means that it is cost neutral to a retailer whether it has 10 000 titles or 10 million titles in its portfolio. Also, assuming that there are, for example, 10 million titles in the portfolio and that they all have the same price, it does not matter if the retailer sells 10 million copies of a single title or a single copy of every title in the portfolio. It is important to note that in the latter case the logic of the long tail is not very profitable to the owners of each individual title – in that case, the logic of the long tail is primarily benefiting the retailer.

READERS AS WRITERS

One of the most fundamental consequences of the digitization of the creative economy is how digital technologies strengthen the audiences' ability to create and distribute digital content to their friends and the world. This phenomenon has spurred a new type of audience behaviour where the readers' role as producers of content is just as relevant as their role as consumers of content. In order to emphasize this new audience behaviour, they are sometimes referred to as 'produsers', rather than users or producers (Bruns, 2008).

The emergence of the produser has significant consequences for most sectors of the creative economy, including magazines and books. Certainly, readers of for instance special interest consumer magazines have always been produsers, and have constituted a tight community where the magazine has served not only as a tool for one-way communication but as a forum where the members of the community are able to discuss the topics that tie them together. While in the pre-digital days this communication was made via mail and telephone, it now primarily takes place in online chat rooms, on message boards and so on. The contemporary magazine publisher that wishes to stay competitive is simply required to provide efficient tools and routines to its readers that facilitate this communication between readers. The experience of reading a book is certainly very different from that of reading a magazine, but the consequences of the new audience behaviour are similar for both sectors. While the community aspects are emphasized in the case of magazines, the social aspects related to book reading are more concerned with books as markers of identity. When books are traded as physical objects, the living-room bookshelf is the place to put the books on display in order to tell the world about one's interests and literary achievements. In a world of e-books and social media, the living-room bookshelf is replaced by social media services that allow readers to share their thoughts about books they have read. These tools may also be integrated in the real-time reading experience so that readers who are reading the same book may comment and highlight passages in the book that they feel are particularly interesting for various reasons.

The magazine and book readers' desire to use digital technologies to comment on the stories they read, to discuss others' comments, to share their experiences with their friends and so on goes beyond being a curious consumer behaviour merely of academic importance. The readers' comments and discussions actually add to the value that is generated by a particular literary service such as Amazon Kindle, Apple iBooks or Google Play. For instance, the value of reading Paul Auster's novel *City of Glass* may be significantly amplified if it is experienced via a service provider that offers features that allow users to interact and discuss the meaning of the novel, and perhaps share photos from the New York streets that are mentioned in the book. Put differently, the service providers are able to significantly differentiate the products they offer by offering 'contextual' features and services that allow the readers to do things with the stories they experience. Actually, it is possible to envision a future where *all* books are available from *all* service providers, and when the competition between the service providers is entirely based on what kind of features they offer and the quality of the user-generated content that has been accumulated during the operation of their service.

Studies show that only a small fraction of the members in a community are genuinely interested in getting involved in being produsers and in creating and uploading text, images or other types of content (for example, Olin-Scheller and Wikström, 2010).

While this may be the case, it should not be interpreted as if user-generated content is relevant only to other produsers. Readers may read and appreciate the user-generated content even though they themselves do not contribute to the conversation. The readers' level of involvement in the content generation is a continuum that ranges from merely reading the professionally produced content all the way to considering the professionally generated content merely as a seed that should be used as a starting point for one's own creative expressions. Some readers at the latter end of the continuum are still not satisfied with being dependent on the works of others – they want to write their own stories and they dream of one day being writers who are able to earn a living from their craft. The same principles that lower the barriers to entering the publishing industries for new publishing companies also enable individuals with access to a computer and the Internet to reach an audience and to be published. Such self-publishing has traditionally been looked down upon as a last resort for writers who are unable to secure a proper publishing contract. However, the emergence of e-books and self-publishing solutions such as Lulu, Xlibris and formats such as the aforementioned Kindle Single opens up new venues for the aspiring writer and strengthens the position of the self-published writer in the publishing ecosystem. During 2011 more than 235 000 self-published electronic and printed books were released in the United States (Harrison, 2012). Most of these titles do not sell very well at all, but there are nevertheless a large enough number of very successful self-published titles that serve as motivational success stories for other aspiring writers. The most successful self-publishing story of all is no doubt Erika Mitchell's (pen name E.L. James) *Fifty Shades of Grey*, which initially was posted in 2010 on a *Twilight* fan website before it was revised and released as an e-book in 2011 by a small independent Australian publisher. By the end of 2012 the *Fifty Shades of Grey* series had sold more than 65 million copies worldwide, and E.L. James was named the Publishing Person of the Year (Deahl, 2012).

LOOKING FORWARD

In this chapter we have explored three of the most significant consequences for these sectors. First, we have noted how electronic distribution of traditional stories continues to take market shares from physical distribution. Second, the emergence of transmedia productions that extend entertainment brands across multiple platforms questions the very notion of what is a book or a magazine. Third, we have discussed how the role of the audience as active producers of content and co-creators of value becomes increasingly difficult to ignore. The economic consequences of digitization on the publishing of books and magazines may already be perceived as transformative, but there are no indications that these sectors are even close to settling into some kind of new normal.

REFERENCES

AAM (2012), *How Media Companies Are Innovating and Investing in Cross-Platform Opportunities*, Arlington Heights, IL: Alliance for Audited Media.

Anderson, C. (2006), *The Long Tail: Why the Future of Business Is Selling Less of More*, New York: Hyperion.

Berners-Lee, T. and R. Cailliau (1990), 'WorldWideWeb: proposal for a hypertext project', unpublished CERN document, 12 November, available at: http://www.w3.org/Proposal.html (accessed 5 November 2012).
Brown, M. (2012), 'The "lost" Steve Jobs speech from 1983; foreshadowing wireless networking, the iPad, and the App Store', *Life, Liberty, and Technology*, 2 October, available at: http://lifelibertytech.com/2012/10/02/the-lost-steve-jobs-speech-from-1983-foreshadowing-wireless-networking-the-ipad-and-the-app-store/ (accessed 2 November 2012).
Bruns, A. (2008), *Blogs, Wikipedia, Second Life, and Beyond: From Production to Produsage*, New York: Peter Lang.
Brynjolfsson, E., M.D. Smith and Y. Hu (2003), 'Consumer surplus in the digital economy: estimating the value of increased product variety at online booksellers', *Management Science*, **49**, 1580–96.
Chesbrough, H. and K. Schwartz (2007), 'Innovating business models with co-development partnerships', *Research Technology Management*, **50**, 55–9.
Daly, C., P. Henry and E. Ryder (1997), *The Magazine Publishing Industry*, Boston, MA: Allyn & Bacon.
Deahl, R. (2012), 'E.L. James: PW's Publishing Person of the Year', *Publishers Weekly*, 30 November, available at: http://www.publishersweekly.com/pw/by-topic/industry-news/people/article/54956-e-l-james-pw-s-publishing-person-of-the-year.html (accessed 28 December 2012).
Facht, U. (2012), *Aktuell statistik om e-böcker*, Göteborg: Nordicom, available at: http://www.nordicom.gu.se/common/stat_xls/2189_3090_E-bokstatistik_antologi_2012.pdf (accessed 27 December 2012).
Harrison, C. (2012), 'Self-publishing industry explodes, brings rewards, challenges', *Miami Herald*, 11 November, available at: http://www.miamiherald.com/2012/11/11/v-fullstory/3092294/self-publishing-industry-explodes.html (accessed 20 November 2012).
Hultén, O., S. Tjernström and S. Melesko (eds) (2010), *Media Mergers and the Defence of Pluralism*, Göteborg: Nordicom.
Jenkins, H. (2003), 'Transmedia storytelling', *MIT Technology Review*, 15 January, available at: http://www.technologyreview.com/news/401760/transmedia-storytelling/ (accessed 28 December 2012).
Johansson, A., H.-K. Ellonen and A. Jantunen (2012), 'Magazine publishers embracing new media: exploring their capabilities and decision making logic', *Journal of Media Business Studies*, **9** (2), 97–114.
Kay, A. (1972), *A Personal Computer for Children of All Ages*, Palo Alto, CA: Xerox Palo Alto Research Center.
Levy, S. (2007), 'The future of reading', *Newsweek*, 17 November, available at: http://www.newsweek.com/id/70983 (accessed 15 November 2012).
Olin-Scheller, C. and P. Wikström (2010), *Författande fans*, Lund: Studentlitteratur.
Picard, R. (2011), *The Economics and Financing of Media Companies*, 2nd edn, New York: Fordham University Press.
Rainie, L. and M. Duggan (2012), 'E-book reading jumps; print book reading declines', Pew Internet and American Life Project, 27 December, available at: http://libraries.pewinternet.org/2012/12/27/e-book-reading-jumps-print-book-reading-declines/ (accessed 27 December 2012).
Rightscom (2004), *Publishing Market Watch, Sector Report 3: The European Magazine and Journal Market*, Brussels: European Commission, Enterprise Directorate-General.
Rønning, H. and T. Slaatta (2012), 'Regional and national structures in the publishing industry with particular reference to the Nordic situation', *Journal of Media Business Studies*, **9** (1), 101–22.
Striphas, T. (2009), *The Late Age of Print: Everyday Book Culture from Consumerism to Control*, New York: Columbia University Press.
SvF (2012), *Branschstatistik 2011*, Stockholm: Svenska Förläggareföreningen, available at: http://forlaggare.se/media/47293/svfstat_2011_web.pdf (accessed 30 November 2012).
Thompson, J.B. (2010), *Merchants of Culture*, Cambridge: Polity Press.
VDZ (2011), *Apple Landscape Analyse: Tiefenanalyse zu Apple und exklusive iPad-Nutzerbefragung*, Berlin: Verband Deutscher Zeitschriftenverleger.

FURTHER READING

The Magazine Publishing Industry (Daly et al., 1997) gives a general overview of the magazine industry, and the impact from the Internet and digital technology is discussed in 'Magazine Publishers Embracing New Media' (Johansson et al., 2012). Thompson's *Merchants of Culture* (2010) and Striphas's *The Late Age of Print* (2009) analyse the transformation of the book publishing industry.

31. E-book and book publishing

Joëlle Farchy, Mathilde Gansemer and Jessica Petrou

Like the music and film industries, the publishing industry is now facing the digital wave. Turnover losses in the physical markets for the former two have been impressive. The global recorded music market lost 41 percent of its value between 2003 and 2011 (SNEP, 2012) – a drop that was significantly more marked in the United States than in other major markets. In the US, the DVD market, estimated at $20 million in 2006 (DEG, 2010), had fallen to $14 million in 2010. Losses in the physical markets are only partially offset by the development of digital markets. In 2011, digital sales represented one-third of global revenues for the recording industry (SNEP, 2012). In Europe, the virtual film market represented 8 percent of the video market (physical and digital) in 2010 (CNC, 2012).

Sales from literary publications in Europe remained stable over recent years (€22.8 billion in 2011 versus €23.5 billion in 2010 and €23 billion in 2009) (FEP-FEE, 2012); with the exception of the United Kingdom, digital book sales were still marginal until 2011.[1] In the US, the Association of American Publishers (2011) announced that, in 2012 for the first time, e-book sales surpassed paper book sales ($282.3 versus $229.6 million). The development of the digital book market is very unequal depending on the region. For instance, 98 percent of the e-book sales in America are in the United States; 45 percent of those in Asia are in South Korea. The UK leads Europe with 52 percent of sales (A.T. Kearney and Book Republic, 2012). This new market likewise has its own hardware requirements. One such support, the touch pad, was not designed specifically for reading e-books. The bestselling model, the iPad, marketed by Apple in 2010, sold 7 million units in 2010 versus nearly 12 million in the first quarter of 2012 alone, making for an increase in sales of approximately $20 million ($5 million in 2010 to almost $25 million in 2012). A second, Amazon's Kindle, a support specifically designed for reading books in homothetic format, made its appearance in 2007. According to the *Kindle Nation Daily*, approximately 35 million units would be in circulation by the end of 2012.

Faced with this new competition – which, in its infancy, was largely illegal and free – cultural industries have had to change their strategy not only by developing attractive and legal digital offers, but by adapting their physical offer as well. Downward price adjustments have already been made in the video and recorded music industries. However, producers and editors have recognized the advantage of proposing a more sophisticated, more expensive offer to consumers who are willing to pay more for works (collectors' editions, for instance), thereby excluding those consumers who tend to pay less and turn to piracy through classic price discrimination (Shapiro, 1988).

This chapter analyzes changes in the book industry resulting from the development of digital technology in three areas: (1) the categories of works available; (2) consumer uses; and (3) the potential consequences of these changes on cultural diversity.

THE DIGITAL OFFER: MORE APPROPRIATE CATEGORIES OF WORKS

Digital technology opens up a great many opportunities for all types of works. However, only some appear to be overwhelmingly affected by this phenomenon.

Preferred Genres

Publishing is not the only cultural industry where substitutability between physical and digital products is correlated to genre. In the film industry, action and science fiction films tend to be more conducive to digital format (and digital pirating as well), which has an important financial impact on theatrical releases and DVD sales (Danaher and Waldfogel, 2008; Martikainen, 2011).

Sales in the digital book market likewise tend to be concentrated in certain genres. While fiction is the leading genre for paper books, it is even more so for digital books: fiction represents 27 percent of the market share for paper books versus 70 percent of that for e-books. In the physical market, after fiction, the two bestselling genres are 'practical' books (self-help, hobbies, etc. with 24 percent) and non-fiction (science, biographies, etc. with 23 percent). Sales for these genres in the digital book market, however, are much lower: non-fiction represents 12 percent of the market share and 'practical' books only 8 percent. Of the 70 percent that is fiction, sentimental works such as Harlequin works prove especially suited to digital format (Bounie et al., 2012).

Enhanced Books, or the Birth of a New Genre

Digital technology has paved the way for the birth of a new genre – enhanced books – which can be distinguished from so-called homothetic format books in that they not only are composed of text but also incorporate other media, such as music, videos and so on. This burgeoning genre has fostered new types of reading and literary products, which *L'herbier des fées*, a children's book by Benjamin Lacombe and Sébastien Perez with 110 interactive features and seven short films woven into the text, exemplifies. Its production model was entirely different, with development fees of up to €120 000 (Loncle et al., 2012) – significantly higher than simple scans of paper versions of *beaux livres* or illustrated books (approximately €9000), or digital books of the same kind produced directly from digital files (approximately €1586) (Bienvault, 2010).

Four genres prove particularly suited to enhanced content:

- Children's books. The shift to digital turns the drawings typically found in paper versions into interactive features and richer content. However, a study has shown that enhanced books do not have the same educational value for children, who pay less attention to the text and are distracted by the slightest media interaction (Chiong et al., 2012). Some authors, like Julie Donaldson for her work *The Gruffalo*, have refused the addition of interactive applications (Solym, 2012).
- School textbooks. A sweeping digital education policy in France aims to equip schools with a 'digital blackboard' or 'interactive whiteboard' (IWB), digital

notebooks and tablets to save money and space and to replace quickly outdated textbooks with programs.
- Comic books. These include particularly the popular mangas of Japan, as well as 'augmented reality' works like François Schuiten's book *La Douce*, about the mythical locomotive of the 1930s, that recently came out in France. The book received funding and support (in the form of access to its drawing programs) from Dassault Systems.
- Travel guides. An example is the popular *Lonely Planet*.

Paradoxically, hardware supports made exclusively for digital books (such as readers) allow only for the reading of homothetic books, as e-ink is not compatible with enhanced media. Enhanced books can therefore be read only on tablets; after the Kindle, Amazon launched its own tablet, the Kindle Fire (September 2012), for this other market.

Successful Works Are Those Most Affected

The digital offer is also defined by works' success. Nowadays, editors are choosing to publish the superstars of the paper world in digital format, thereby guaranteeing their superstar status in the latter form as well (Bounie et al., 2012). Hu and Smith (2011) did a study on Amazon based on the true experience of an editor who decided to stagger the digital and paper version publication of works (for new publications, both formats usually appear at the same time). The researchers showed that: (1) staggering had a negative impact on e-book sales, and the positive effects on physical sales were not present for the whole market; and (2) the positive effects of this staggering on physical sales worked only for bestsellers, which consumers prefer not to wait for – regardless of the support.

Nonetheless, with the exception of a few bestsellers (the top 10), successful books are not necessarily equally so in both formats. For instance, 48.9 percent of the top 100 digital works were *not* on the list of top 100 paper books (Bounie et al., 2012). Similarly, in the online used book sector, the bestselling works are not the same as those for paper works, thus limiting competition between the formats (Ghose et al., 2006).

Successful works are also those most frequently pirated – a trend that has also been observed for the film (Elberse and Eliashberg, 2003; Danaher and Waldfogel, 2008) and music industries. In the latter, successful albums are highly substitutable with illegal digital versions, which can affect physical sales (Tanaka, 2004; Blackburn, 2007).

New Outlets for Niche Products

Unlike successful products, niche products more readily benefit from the existence of free and illegal digital versions, which can boost physical sales via a penetration effect (Tanaka, 2004; Blackburn, 2007). Information about these products can stimulate paid consumption by calling attention to their value and increasing consumer awareness, and through buzz phenomena (Shapiro and Varian, 1999) that act as a kind of product promotion. In the music industry, pirating lets consumers discover and enjoy artists whose albums they would not necessarily have bought without having heard them first. Digital technology can be an opportunity for fragile firms, facilitating the visibility of their works using cheaper, more flexible promotional tools to demonstrate their innovativeness (Spellman, 2006). Bhattacharjee et al. (2007) thus demonstrated that the downloading of music files has reduced the performance gap (in terms of length of time in the top 100) between artists on independent labels and those on major ones.

In the book industry, some books – called 'digital outsiders' by Bounie et al. (2012) – would have remained unknown had it not been for their digital versions. Thus, 'not all unpaid consumption displaces paid consumption' (Danaher and Waldfogel, 2008). However, the implications of sampling are not the same for all cultural industries (Rob and Waldfogel, 2007); while it helps promote marginal works and increase consumers' willingness to pay and appreciate the utility in doing so (Liebowitz, 2005; Peitz and Waelbroeck, 2006; Bhattacharjee et al., 2007), it must be relativized for the film (Bounie et al., 2006) and, to an even greater extent, literary publishing industries, where reading of works is not necessarily a repeat action and requires more attention than listening to a song.

CHANGES IN READING PRACTICES

The birth of digital technology has dramatically changed reading habits, albeit without completely eradicating traditional print reading. In fact, what have emerged are hybrid reading behaviors (different supports, different media). Thus reader profiles have also changed.

A Decline in Paper Reading That Is Unrelated to the Birth of Digital Technology

The decline observed in reading and book sales is not a direct consequence of the digital revolution. In fact, the decline in paper reading started before the emergence of digital versions and continues today, as one can see from the statistics beginning with the first surveys on cultural practices in the 1970s in France. 'Screen culture' (*la culture de l'écran*) (Chambat and Ehrenberg, 1988) is the first phase of a transformation that started in the 1960s. The spread of the Internet was the beginning of a new chapter in paper reading's decline.

Simultaneous with this trend is the symbolic loss of power of the paper book object, resulting most notably in the emergence of reading practices that are closer to reality than they were 30 years ago. Thus the French claim less often than in the past to be readers when, in fact, they are not. Moreover, paper reading faces competition not only from digital books, but from a whole set of new practices; new media and entertainment, digital learning tools that directly compete with paper books, social networks and so on all accentuate the decline of a solitary, time-consuming practice. These different trends leave the debate between those who insist on a medium's resistance to the birth of its 'successors' and impossibility of flawless substitution in the medium term (Darnton, 2011) and those who, on the contrary, point out that these past decades have witnessed the extinction of supports like VHS and cassette tapes (Benhamou, 2012).

New Reading Practices

The proliferation of ways of reading increases the polymorphous nature of the activity; how can one compare reading a traditional paper book with reading hyperlinked, media-enhanced texts on the Internet? Consultative or piecemeal reading, 'gleaning' (Leclerc, 2012) and 'flitting' (Assouline, 2012) are all practices that are on the rise. Paradoxically, however, digital formats do not facilitate non-linear reading (Bounie et al., 2012). Some are concerned about such changes, stressing the loss of concentration and memory caused by non-linear reading (Baccino, 2011) or reading enhanced books (Solym, 2012); others denounce the myth that paper is 'more conducive to understanding a text than are screens' and propose actual training in digital reading (Assouline, 2012) and claim that classic-format e-books have the same virtues as paper ones (Chiong et al., 2012).

Nonetheless, the categorizing of types of readers according to the support is a debate that already seems to belong to the past. Much as the astute reader was once one who knew how to read certain passages of Proust, Balzac or *War and Peace* diagonally to fully enjoy such masters and masterpieces (Barthes, 1973), the astute reader of the twenty-first century must master all types of reading skills – from traditional linear reading to piecemeal reading (Donnat, 2012).

Complementary Uses

The importance of the reading format

Buying behaviors and dynamics are different in the physical and digital worlds (Bounie et al., 2010). Thus, the ranking of bestselling paper books on the Internet changes much faster than the same ranking in physical stores. A similar trend can be seen in the music market, where the ranking of the most frequently downloaded tracks is more volatile.

The two markets, however, are closely linked. Users of digital offers do not limit themselves to this format and are likewise consumers of paper books. Fan fiction books available for free online arouse such enthusiasm that some fans use print-on-demand services to have them printed out on paper! In the US, the increase in the number of digital books has gone hand in hand with the decline of the paper book market in recent years. Some see digital books as an opportunity to revive not only the book market overall, but the paper book market in particular (Bosman, 2011; Darnton, 2011). In fact, some consumers who choose format over content (Hu and Smith, 2011) have become readers thanks to the emergence of digital technology.

Furthermore, reading has become a 'transmedia' practice (Donnat, 2012), meaning that consumers are led to read via other media. In this way, TV series (*Game of Thrones* and so on) and films (such as *Twilight*) have prompted many fans to read the books that inspired the audiovisual works. Finally, the multitude of supports for acquiring knowledge suggests that new cognitive skills will continue to be born, as the task of synthesizing a myriad of resources requires a certain amount of dexterity when it comes to digital technology.

Once used to digital format, consumers have a hard time returning to a purely paper offer. Danaher et al. (2010) showed that the presence of video content in legal downloading had an impact on piracy, but no conclusive effect on the physical market. Similar findings have been observed for the literary publishing market. For most books, delaying the release of their digital version relative to the paper one had little positive effect on the market for the latter, but a negative effect on that of the former (Hu and Smith, 2011; see above). This primacy of distribution channels is also marked by consumer choice, which tends to be more driven by whether the offer is physical or digital than by its legality or illegality (Danaher et al., 2010).

Consumer profiles

Consumption therefore tends to become 'hybrid', with the digital portion varying depending on the type of consumer. Preference for the format appears to be stronger than preference for the work itself for individuals who are more at ease using the Internet (Frambach et al., 2007). Consumer profiles indeed explain the favored uses:

- Given the choice between a digital and a physical offer, the user's age often determines his or her choice. Internet access has quite a negative impact on younger people's CD consumption and a rather positive one on that of their elders (Boorstin, 2004).
- We likewise see strong similarities between user profiles (demographics, lifestyle, tastes, etc.) by observing legal versus illegal digital offers (McKenzie, 2009; Danaher et al., 2010). One of the key variables is the level of consumption. Michel (2006) demonstrates that the negative impact of music file sharing on CD sales is only true for the biggest music consumers. The biggest users of digital versions of works consume more, and often more eclectic, cultural works than do average users. Waldfogel (2009) thus showed that the people who watch the most TV series also watch the most series on the Internet. Similarly, the viewers who watch the most series on legal sites are also those who watch the most series on illegal sites.
- Finally, within the strictly illegal digital supply chain, we can notably distinguish two types of users whose motives and buying behaviors differ greatly (Bounie et al., 2007): 'explorers', who use file sharing to discover new works, which can have a positive impact on sales, and 'pirates', who substitute legal offerings with illegal ones.

DIGITAL TECHNOLOGY AND CULTURAL DIVERSITY

Digital technology is conducive to a more diverse offer. The positive impact on consumer well-being generated by this increase in variety is greater than that which results from the increase in competition (Brynjolfsson et al., 2003). However, a more diverse supply does not necessarily mean more diversity is consumed.

A Broader Offer

Digital technology creates a broader offer (Brynjolfsson et al., 2006):

- Niche works that are not profitable in the physical world can become so in digital format, as distribution costs are considerably reduced, thus offering the potential for a wider range of works. Moreover, digital technology allows for easier access to works that are hard or even impossible to find in paper format, which explains why works sold in digital format tend to be older than works sold on paper (Bounie et al., 2012).

- Online distribution platforms offer many more titles than a store ever could, thereby increasing the offer available to each consumer. This relative volume makes it easier for consumers to locate works.

Nevertheless, all paper works are not available in digital format. According to MOTif (2011), the digital book supply in France (80000 digital titles in 2011, 70000 in 2010) in 2011 represented only 15 percent of all the titles available in physical format. The development of this offer has galvanized, and the rate of digital publications from paper works is now sustained. It is worth noting that, in France, the number of digital video titles still trails behind that of physical titles; large specialty stores carry more than 61 000 titles, while digital video platforms offer only 41 680 (CNC, 2012). The rate of development of digital offers likewise varies greatly according to the area of the world. In the US, Barnes & Noble boasted more than 2 million digital book titles in 2011; only one other market, the UK's, where nearly a million digital works were listed in 2010, came close (Wischenbart, 2011).

The digital technology often associated with the decline in turnover observed in physical markets of cultural industries (see the introduction to the chapter) upsets both distribution and creation. For the time being, the vitality of creation – if measured in terms of the number of new works produced each year – paradoxically does not seem to be affected by sales losses (Oberholzer-Gee and Strumpf, 2007); the three main cultural industries touched by digital technology (recorded music, video and literary publishing) have not seen a decline in the number of new works produced each year. On the contrary, between 2000 and 2007, a period when the music industry was strongly hit by piracy, the number of new releases doubled in North America; of the 79695 albums that came out in 2007, 25159 were digital. The same observation can be made for videos. Between 2003 and 2007, the number of feature-length films produced each year worldwide went up by 30 percent. In the US, the number of new literary works increased by 66 percent between 2003 and 2007, largely because of self-publishing. Self-publishing has boomed thanks to platforms like Lulu, XinXii and Amazon's online bookstore, which allows users to bypass the editor filter. The 'classic' offer is also on the rise, with the number of new publications jumping from 215138 in 2002 to 328259 in 2010 – an increase of 52 percent (Bowker International, 2011).

A Less Certain Impact on Diversity Consumed

The impact of an abundant supply on the diversity of the demand has given rise to two contradictory theories in economic literature: that of the 'superstars', according to which digital technology will remain in line with a the rising volume of works consumed, and that of the long tail, which, on the contrary, sees digital technology as increasing the diversity of works consumed.

The 'superstar' effect was theorized long before the birth of digital technology (Rosen, 1981; Adler 1985). When applied to it, these hypotheses point to a strengthening in the weight of superstars as a result of a decrease in the marginal costs of distribution. Rosen posits that, for consumers, there is low substitutability between different talents, so that a slight advantage in terms of talent results in a notable advantage in terms of sales. As digital technology greatly reduces congestion costs for the consumption of superstar works, the success of such works is even more marked. Adler speaks of what is, in his opinion, 'the need on the part of consumers to consume the same art that others do', as social interaction draws on a common cultural background. Thus, as successful works already have an ever-increasing advantage, digital technology will only amplify this network effect.

According to Anderson (2004), the lowering of production and distribution costs on the contrary represents an opportunity for more personal works. He describes the 'long tail' phenomenon by a relative drop in the importance of bestsellers combined with a growing number of works with a small number of consumers. According to Anderson, the Internet, in fact, diversifies the offer by eliminating the problem of rarity which exists in the physical world and lowering distribution costs. Meanwhile, new tools reduce information costs for consumers, especially for products about which information is less readily available (that is, more restricted). Digital technology thereby unifies all the conditions necessary for diversifying the demand – if, indeed, it is diversity the consumer wants.

One variable in particular plays a pivotal role in the relationship between the long tail and superstar effects: how Internet users find their bearings. As the volume of network content increases, consumers' ability to identify the information most pertinent to them becomes more difficult. Hyperchoice can prove counterproductive if the consumer does not have the tools necessary to make a choice (Gourville and Soman, 1995). In a seminal article written long before the birth of the Internet, Herbert Simon (1971) developed the idea that, in a world where information is abundant, it cannot be a source of value (recall that the economy is often defined as the allocation of scarce resources). In this new attention economy, consumers' time and attention have become rare key resources (Goldhaber, 1997; Lanham, 2006).

In this context, recommendation tools, specialized websites, social networks and file sharing between Internet users allow consumers to find the works that best suit their desires or discover new works that they would not necessarily have found otherwise. The effects of digital tools on the diversity of the demand are, however, subject to debate. In fact, recommendation tools can have a negative impact in terms of the diversity of the demand (Fleder and Hosanagar, 2009) when they are skewed in favor of successful works. This same bias toward star works can be seen in the comments made by users (Dellarocas and Narayan, 2007), who are more likely to share their opinion when the work has already been commented on. All works are not equal – not even on the Internet.

Empirical studies are struggling to divvy up the consequences of digital technology between the long tail and superstar effects (see Table 31.1). Regarding the literary publishing industry, Peltier and Moreau (2012) have highlighted that, in France, there has been a shift in book consumption from bestsellers to more moderately successful works. The study also highlights a veritable change in consumer habits that can be expected to last.

Table 31.1 Summary of research findings on shifts in demand with digitization

	Data	Results	
Brynjolfsson et al. (2011)	Comparison of digital versus catalogue sales	Sales less concentrated in the digital channel. Positive impact of search tools on parts of the niche product market.	Long tail
Tan and Netessine (2009)	Netflix data (film and TV series rentals) 2000–05	Increased demand for bestsellers. New film releases too rapid for consumers to have time to discover them.	Superstar
Elberse and Oberholzer-Gee (2008)	Physical sales (video) in the US (2000–05)	Lengthening of the tail/higher volume of unsold titles. Decrease in the relative weight of superstar products; higher sales for an increasingly small number of works.	Long tail and superstar
Benghozi (2008)	Physical CD and video sales in France (2002–05)	Elongating of the tail in general. Superstar effect less pronounced in the digital circuit. Seasonal results.	Effects too weak to be called long tail
Brynjolfsson et al. (2003)	Weekly book sales on Amazon (2001)	Nearly 40 percent of sales on Amazon were niche works that are not available in stores.	Long tail

CONCLUSION

The emergence of e-books has upset the paper book market, yet without threatening it entirely or destroying it. The success of digital technology depends on genre, and consumers' preference for specific channels depends on the work, thereby creating even greater nuance. Reader behaviors have been changing for many years now and have become much more hybrid, but are not fundamentally oriented to digital technology.

By offering diversity, digital technology broadens the range of possibilities but also increases tensions between and the anxieties of the various agents in the chain. Authors are concerned about the future of their rights and the optimal form of transfer agreement in this new world. Professional editors are seeking their place in response to the emergence of amateur practices while noting the reactions of readers on the Internet in view of obtaining the rights to works that have proven successful. After self-publishing her work *Fifty Shades of Grey* on the Internet in 2011 and receiving heavy promotion on social networks, British author E.L. James was spotted by Random House, a classic American publishing company, and had sold over 40 million copies of her book by late 2012. The work has broken the Kindle record, with more than a million e-books sold.

NOTE

1. Digital book sales were 11 percent of revenue from publishing in the UK versus 1 percent in Germany, 2 percent in France and 0.3 percent in the rest of Europe (Enders Analysis, 2012).

REFERENCES

Adler, Moshe (1985), 'Stardom and talent', *American Economic Review*, **75**, 208–12.

Anderson, Chris (2004), 'The long tail', *Wired Magazine*, 12.10 (October).

Association of American Publishers (2011), *Bookstat*, available at: http://www.publishers.org/press/74/ (accessed 6 May 2013).

Assouline, Pierre (2012), 'La métamorphose du lecteur', *Le Débat* (Gallimard), **170**, May–August, 78–89.

A.T. Kearney and Bookrepublic (2012), *Do Readers Dream of Electronic Books*?, available at: http://www.slideshare.net/IfBookThen/bonfantiferrario-at-kearney-bookrepublic (accessed 6 May 2013).

Baccino, Thierry (2011), 'Lire sur internet, est-ce toujours lire?', *BBF*, **5**, 63–6.

Barthes, Roland (1973), *Le Plaisir du texte*, Collection Points Essais, Paris: Éditions du Seuil.

Benghozi, Pierre-Jean (2008), *Effet long tail ou effet podium: une analyse empirique des ventes de produits culturels en France*, Ministère de la Culture et de la Communication, March, Paris: PREG-CRG.

Benhamou, Françoise (2012), 'Le livre et son double: reflexions sur le livre numérique', *Le Débat* (Gallimard), **170**, May–August, 90–102.

Bhattacharjee, S., R. Gopal, K. Lertwachara, J. Marsden and R. Telang (2007), 'The effect of digital sharing technologies on music markets: a survival analysis of albums on ranking charts', *Management Science*, **53** (9), 1359–74.

Bienvault, H. (2010), 'Le coût d'un livre numérique', Study for the MOTif (Observatory for books and writing in Ile-de-France), April.

Blackburn, David (2007), 'The heterogenous effects of copying: the case of recorded music', working paper, Harvard University.

Boorstin, Eric (2004), 'Music sales in the age of file-sharing', thesis, University of Princeton.

Bosman, Julie (2011), 'Using e-books to sell more print versions', *New York Times*, 26 June.

Bounie, David, Patrick Waelbroeck and Marc Bourreau (2006), 'Piracy and the demand for films: analysis of piracy behavior in French universities', *Review of Economic Research on Copyright Issues*, **3** (2), 15–27.

Bounie, David, Marc Bourreau and Patrick Waelbroeck (2007), 'Pirates or explorers? Analysis of music consumption in French graduate schools', *Brussels Economic Review*, **50** (2), 73–98.

Bounie, David, Bora Eang and Patrick Waelbroeck (2010), 'Marché Internet et réseaux physiques: comparaison des ventes de livres en France', *Revue d'économie politique*, **120**, 141–62.

Bounie, David, Bora Eang, Marvin Sirbu and Patrick Waelbroeck (2012), 'Superstars and outsiders in online markets: an empirical analysis of electronic books', working paper, Telecom ParisTech.

Bowker International (2011), 'Annual report on US print book publishing', available at: http://www.bowker.com/en-US/aboutus/press_room/2012/pr_06052012.shtml (accessed 6 May 2013).

Brynjolfsson, Erik, Yu Hu and Michael Smith (2003), 'Consumer surplus in the digital economy: estimating the value of increased product variety at online booksellers', *Management Science*, **49**, 1580–96.

Brynjolfsson, Erik, Yu Hu and Michael Smith (2006), 'From niches to riches: anatomy of the long tail', *Sloan Management Review*, **47** (4), 67–71.

Brynjolfsson, Erik, Yu Hu and Duncan Simester (2011), 'Goodbye Pareto principle, hello long tail: the effect of search costs on the concentration of product sales', *Management Science*, **57**, 1373–86.

Chambat, Pierre and Alain Ehrenberg (1988), 'De la télévision à la culture de l'écran', *Le Débat* (Gallimard), **52**, November–December, 107.

Chiong, Cynthia, Jinny Ree and Lori Takeuchi (2012), 'Print books vs. e-books', Report of the Joan Ganz Cooney Center.

CNC (2012), *Le marché de la video*, Paris: CNC.

Danaher, Brett and Joel Waldfogel (2008), 'Reel piracy: the effect of online film piracy on international box office sales', working paper.

Danaher, Brett, Samita Dhanasobhon and Michael Smith (2010), 'Converting pirates without cannibalizing purchasers: the impact of digital distribution on physical sales and internet piracy', *Marketing Science*, **29**, 1138–51.

Darnton, Robert (2011), 'Le numérique, une chance pour l'édition!', *Télérama*, **3192**, 16 March.

DEG (Digital Entertainment Group) (2010), *Year-End Home Entertainment Report*, available at: http://www.degonline.org/pressreleases/2011/f_Q410.pdf (accessed 6 May 2013).

Dellarocas, Chrysanthos N. and Ritu Narayan (2007), 'Tall heads vs. long tails: do consumer reviews increase the informational inequality between hit and niche products?', Robert H. Smith School of Business Research Paper No. 06–056.

Donnat, Olivier (2012), 'La lecture régulière de livres: un recul ancien et général', *Le Débat* (Gallimard), **170**, May–August, 42–51.

Elberse, Anita and Jehoshua Eliashberg (2003), 'Demand and supply dynamics for sequentially released products in international markets: the case of motion pictures', *Marketing Science*, **22**, 329–54.

Elberse, Anita and Felix Oberholzer-Gee (2008), 'Superstars and underdogs: an examination of the long tail phenomenon in video sales', Working Paper 07–015, Harvard Business School.
Enders Analysis (2012), 'Digital Europe: diversity and opportunity', available at: http://www.letsgoconnected.eu/files/Lets_go_connected-Full_report.pdf (accessed 6 May 2013).
FEP-FEE (2011), *European Book Publishing Statistics*, available at: http://www.fep-fee.eu/FEP-Statistics-for-the-year-2011 (accessed 6 May 2013).
Fleder, Dan and Kartik Hosanagar (2009), 'Blockbuster culture's next rise or fall: the impact of recommender systems on sales diversity', *Management Science*, **55**, 697–712.
Frambach, Ruud T., Henk Roest and Trichy V. Krishnan (2007), 'The impact of consumer Internet experience on channel preference and usage intentions across the different stages of the buying process', *Journal of Interactive Marketing*, **21**, 26–41.
Ghose, Anindya, Michael D. Smith and Rahul Telang (2006), 'Internet exchanges for used books: an empirical analysis of product cannibalization and welfare impact', *Information Systems Research*, **17**, 3–19.
Goldhaber, Michael H. (1997), 'The attention economy and the Net', *First Monday*, **2** (4).
Gourville, John T. and Dilip Soman (2005), 'Overchoice and assortment type: when and why variety backfires', *Marketing Science*, **24**, 382–95.
Hu, Yu and Michael D. Smith (2011), 'The impact of ebook distribution on print sales: analysis of a natural experiment', working paper.
Lanham, Richard (2006), *The Economics of Attention: Style and Substance in the Age of Information*, Chicago: University of Chicago Press.
Leclerc, Caroline (2012), 'Remarques sur l'évolution des pratiques de lectures étudiantes', *Le Débat* (Gallimard), **170**, May–August, 70–72.
Liebowitz, Stanley (2005), 'Pitfalls in measuring the impact of file-sharing on the sound recording market', *CESifo Economic Studies*, **51**, 435–73.
Loncle, Thomas, Sarah Sauneron, François Vielliard and Julien Winock (2012), *Les acteurs de la chaîne du livre à l'ère du numérique: les auteurs et les éditeurs*, La note d'analyse 270, March, Paris: Centre d'analyse stratégique.
McKenzie, Jordy (2009), 'Illegal music downloading and its impact on legitimate sales: Australian empirical evidence', *Australian Economic Papers*, **48** (4), 296–307.
Martikainen, Emmi (2011), 'Does file-sharing reduce DVD sales?', working paper, University of Turku.
Michel, Norbert (2006), 'The impact of digital file sharing on the music industry: an empirical analysis', *B.E. Journal of Economic Analysis and Policy*, **6**, 1–22.
MOTif (2011), 'EBookZ: l'offre légale et illégale de livres numériques', May, Paris, available at: http://www.lemotif.fr/fichier/motif_fichier/171/fichier_fichier_.l.offre.de.livres.numa.riques.en.france.le.motif.pdfhttp://www.lemotif.fr/fichier/motif_fichier/171/fichier_fichier_.l.offre.de.livres.numa.riques.en.france.le.motif.pdf (accessed 6 May 2013).
Oberholzer-Gee, Felix and Koleman Strumpf (2007), 'The effect of file sharing on record sales: an empirical analysis', *Journal of Political Economy*, **115** (1), 1–42.
Peitz, Martin and Patrick Waelbroeck (2006), 'Why the music industry may gain from free downloading: the role of sampling', *International Journal of Industrial Organization*, **24** (5), 907–13.
Peltier, Stéphanie and François Moreau (2012), 'Internet and the "long tail vs. superstar effect" debate: evidence from the French book market', *Applied Economic Letters*, **9** (8).
Rob, Rafael and Joel Waldfogel (2007), 'Piracy on the silver screen', *Journal of Industrial Economics*, **55** (3), 379–95.
Rosen, Sherwin (1981), 'The economics of superstars', *American Economic Review*, **71**, 845–58.
Shapiro, Carl (1988), 'Economic effects of home copying', mimeo, Princeton University.
Shapiro, Carl and Hal Varian (1999), *Information Rules: A Strategic Guide to the Networked Economy*, Boston, MA: Harvard Business School Press.
Simon, Herbert (1971), 'Designing organizations for an information-rich world', in M. Greenberger (ed.), *Computers, Communications and the Public Interest*, Baltimore, MD: Johns Hopkins Press, pp. 40–41.
SNEP (2012), *Economie de la production musicale*, Paris: Irma, available at: http://www.irma.asso.fr/Economie-de-la-production-musicale?lang=fr (accessed 5 May 2013).
Solym, Clément (2012), 'Les ebooks enrichis, trop divertissants pour les enfants', *Actualité*, 7 June.
Spellman, Peter (2006), *Indie Power: A Business-Building Guide for Record Labels, Music Production Houses, and Merchant Musicians*, 2nd edn, Boston, MA: MBS Business Media.
Tan, Tom and Serguei Netessine (2009), 'Is Tom Cruise threatened? Using Netflix prize data to examine the long tail of electronic commerce', working paper, University of Pennsylvania.
Tanaka, Tatsuo (2004), 'Does file-sharing reduce CD sales? A case of Japan', conference paper prepared for Conference on IT Innovation, Hitotsubashi University, Tokyo.
Waldfogel, Joel (2009), 'Lost on the web: does web distribution stimulate or depress television viewing?', *Information Economics and Policy*, **21** (2), 158–68.

Wischenbart, Rüdiger (2011), 'The global ebook market, 2011', available at: http://www.publishersweekly.com/binary-data/ARTICLE_ATTACHMENT/file/000/000/522–1.pdf (accessed 6 May 2013).

FURTHER READING

For a global statistical database on books, see Rüdiger Wischenbart (2011) or FEP-FEE (2011). Hu and Smith (2011) offer a natural experiment in the e-book sector on Amazon and evaluate the potential substitution between paper books and e-books. Frambach et al. (2007) study the intrinsic preference for a channel more than for contents. Brynjolfsson et al. (2003) estimate the economic impact of increased product variety made available through electronic markets. Peltier and Moreau (2012) give a comparison of two adversarial economic theses in the book sector: the long tail and the superstar effect.

32. Academic publishing and open access

*Frank Mueller-Langer and Marc Scheufen**

With the digital era and the spread of the Internet, the academic publishing market is currently facing another revolution after the invention of the Xerox copier in 1959. While copyright was broadened through a series of significant reforms after the Xerox copier had been introduced, new business models, especially open access (OA), seem to have recently put copyright and its role in academia into debate.[1]

Two developments motivate this 'OA debate'. First, subscription prices for academic journals have increased, which has forced (university) libraries to significantly cut their journal portfolios. Second, copyright as an incentive mechanism seems negligible in academia, as researchers are motivated by reputation gains and CV effects rather than direct financial returns from publishing their works. Consequently, the OA publishing model may be seen as a superior alternative for the conventional closed access (CA) publishing model.

This chapter critically reviews the OA debate by discussing theoretical and empirical arguments on the role of copyright in academic publishing. A brief historical examination introduces the altering conditions for scholarly publishing and highlights the new trade-off in the digital age. By locating the debate within a broader stream of current research, we provide alleys for further research and a glimpse of possible future scenarios. It is shown that copyright may be both a blessing and a curse in establishing an effective framework for scientific progress.

COPYRIGHT IN ACADEMIC PUBLISHING

From an economics perspective, copyright is a simple means to correct the 'free-rider' environment that surrounds information goods. It does so by granting exclusive rights and hence by providing monetary incentives for creative endeavor (Arrow, 1962). Stated differently, copyright enforces a temporary monopoly creating both benefits (incentives for creative works) and costs (deadweight loss) from a social welfare perspective.[2]

In the academic publishing market, copyright did not play a pivotal role until the mid-twentieth century. In fact, the relationship of copyright and scientific journals was 'merely occasional; because many of the earliest journal publishers were learned societies and then academic institutions, copyright was licensed explicitly or implicitly to them, though it did not have a central role in the business' (Ramello, 2010: 13). In contrast, the presence of pirated copies may have even increased the popularity and reputation of particular journal articles and journal titles.[3]

This changed dramatically once commercial publishers started to enter the journal publishing market by launching new or acquiring existing titles in the second half of the twentieth century. After the introduction of the Xerox 914 copier in 1959, academic publishers induced significant revisions in copyright law. As a result, a series of court cases tackling the practice of copying journal articles en masse from library collections forwarded an era that somewhat revolutionized copyright law. Photocopying changed the trade-off for balancing the interests in copyright law in two respects. First, the Xerox technology dramatically eased the process of copying printed material. Second, copying en masse from library collections significantly dropped article unit costs. As a consequence, copy and original entered into competition (Liebowitz, 1985).

The advent of the Internet and the development of technologies to digitize information goods had at least three far-reaching implications for copyright legislation: (1) digitization supersedes the need for any physical media such as paper or CDs; (2) a digital copy is a perfect substitute for the original work; (3) digitization has reduced the marginal cost of copying to virtually zero. Several legislative steps have been trying to adapt copyright to the new conditions. Arguably, the most significant change is the introduction of technological measures to control the access and use of electronic content – so-called digital rights management (DRM) technologies.[4] The implications of the steps taken are far-reaching and go beyond the question of who should own intellectual assets. Public protest against recent proposals for a reform of copyright, such as the Stop Online Piracy Act (SOPA), shows the intensity in this debate. Besides, there has recently been a movement in the academic publishing market that questioned whether the traditional copyright model solves the trade-off between creating incentives for scholars to create high quality research output on the one hand and allowing the public fast access to those works on the other (Eger and Scheufen, 2012a: 49 ff.). In fact, the Internet fostered the emergence of an alternative business model for publishing academic works – not only by scientific associations, but also by scholars themselves. The OA model seeks free online access of academic works:

> permitting any user to read, download, copy, distribute, print, search or link to the full text of these articles, crawl them for indexing, pass them as data to software, or use them for any other lawful purpose, without financial, legal, or technical barriers other than those inseparable from gaining access to the internet itself. (Budapest Open Access Initiative, 2002)

In this regard, OA to scholarly literature can be achieved by means of two complementary strategies: (1) self-archiving ('the green road'); and (2) OA journals ('the gold road'). The differentiation between the green and gold roads towards OA was introduced by Harnad et al. (2004). Self-archiving means that mostly preprint versions of papers are made available online without extensive quality control by third parties. OA journals make papers freely available and provide the general services of journal publishing: peer review to ensure quality, editing and typesetting. However, Bergstrom (2001) and Mueller-Langer and Watt (2010) suggest that, in practice, the academic community rather than publishers fulfill many of the tasks of filtering (i.e. refereeing) the articles for quality, editing and sometimes even typesetting.[5] Consequently, it may be questionable to what extent publishers actually add value to the production process and whether profit margins of 25 percent or more seem reasonable for incentivizing publishers to publish (Ramello, 2010).

THE OA MOVEMENT

The OA movement gained momentum especially throughout the last decade. Starting in 2002, several initiatives have fostered OA publishing in academia.[6] In addition, scientific associations like the National Institutes of Health started to promote OA publishing by fostering the development of OA journals as well as self-archiving platforms. The *Directory of Open Access Journals* (*DOAJ*)[7] listed more than 8500 OA journals as of December 2012.[8] Despite this vast number of journals, OA still seems to play a minor role in academic publishing. As Table 32.1 documents, OA journals often lack in reputation as compared to well-established CA journals. Only 7.5 percent of all leading journals with an impact factor greater than 1 operate on the basis of the OA principle.[9] In 2011, each of the major academic publishers under study had more academic journals by themselves in this category than all OA journals combined. While OA journals have an average impact factor (IF) of 1.44, established commercial publishers like Elsevier (2.67) or Wiley-Blackwell (2.58) show substantially higher reputation measures and hence market power. Similar findings are reported when focusing on the average Eigenfactor as a measurement for journal reputation. Here, the figures for established publishers such as Elsevier (0.015) or Wiley-Blackwell (0.013) reveal a reputation advantage that exceeds the level in the group of all leading OA. The analysis of other performance measures, such as the maximum IF or the average yearly number of articles per journal, leads to similar results. Table 32.1 gives an overview on some market structure characteristics, comparing OA and CA journal publishers.[10]

There is considerable variation in the impact factors of OA journals by academic discipline (see Figure 32.1). OA journals seem to be most established in biology, physics

Table 32.1 Quality indicators of leading closed access journals administered by major publishers and leading open access journals

	Closed access publishers			All leading open access journals
	Elsevier	Wiley-Blackwell	Springer	
No. of journals with an impact factor ≥ 1	609	639	546	373
Market share of journals with an impact factor ≥ 1 (%)	12.2	12.8	11.0	7.5
Average impact factor per journal	2.67	2.58	1.64	1.44
Maximum impact factor	38.3	23.5	11.5	17.5
Average Eigenfactor	0.015	0.013	0.006	0.005
Average number of articles per journal	206	171	109	128

Notes:
All journals with IF ≥ 1.
We generated the matching data via metadata harvesting. See OAI (2008).

Source: Authors' calculations based on data by ISI Web of Science (Thomson Reuters Journal Citation Reports, 2011).

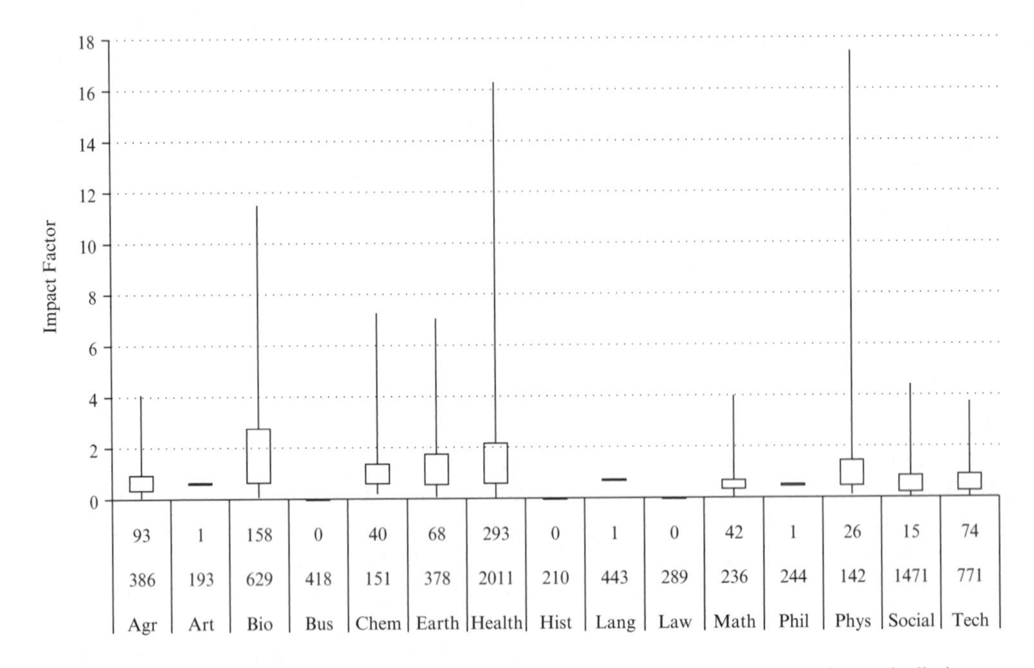

Note: The vertical line illustrates the minimum and maximum. The height of the rectangle sets the limits for the first and third quartile of the distribution. The first row of the horizontal axis illustrates the number of all ISI listed OA journals, while the second row shows the number of all OA journals in the respective disciplines. Row three denotes the disciplines, where Agr = agriculture and food sciences, Art = arts and architecture, Bio = biology and life sciences, Bus = business and economics, Chem = chemistry, Earth = earth and environmental sciences, Health = health sciences, Hist = history and archeology, Lang = languages and literature, Law = law and political sciences, Math = mathematics and statistics, Phil = philosophy and religion, Phys = physics and astronomy, Social = social sciences and Tech = technology and engineering.

Source: Authors' calculations based on data from ISI Web of Science (Thomson Reuters Journal Citation Reports, 2011).

Figure 32.1 Boxplot of OA journals' impact factors by academic discipline

and health sciences, while in most other disciplines OA journals are hardly ever listed by ISI Web of Science.

DIGITIZATION AND ACADEMIC PUBLISHING

Digitization provides new opportunities in the context of the traditional publishing model. It offers new marketing strategies for commercial journal publishers, making bundling of different versions (electronic and print version) and different journals (so-called 'big deals') predominant price discriminating strategies in the academic journal market (Edlin and Rubinfeld, 2004). As a consequence, journal expenditures increased by 273 percent between 1986 and 2004 (Ramello, 2010). Journal unit costs increased by 188 percent as compared to an increase in the consumer price index of 73 percent for the same time period. In some disciplines, for example physics and chemistry, journal prices even increased by more than 600 percent from 1984 to 2001 (Edlin and Rubinfeld, 2004). As a consequence of this vast increase in journal subscription prices, together with budget cuts in several countries, libraries were forced to significantly change their subscription portfolios.[11] This development is a driving force behind the OA debate.

Moreover, the role of copyright – necessary for the possibility of exclusion and hence such pricing strategies – may be questioned. The reason is that the rationale of a primarily monetary reward as an incentive for an author's creative endeavor seems negligible in scientific research. Authors hardly ever receive royalties for publishing an academic work. Other motivational factors, like cites and reputation (peer recognition) or labor market signals, may be more important to motivate academic work. This is one reason why scholars in law and economics have discussed whether academic articles should fall under copyright at all (Shavell, 2010; Mueller-Langer and Watt, 2012). Accordingly, forced OA by abolishing copyright for academic works may be a possible policy alternative in shaping the future of academic publishing.[12]

THE OPEN ACCESS DEBATE IN SCIENCE

Recently, most attention in the OA debate has been directed at a seminal paper by Steven Shavell (2010), which asked whether copyright for academic works should be abolished. In a nutshell, Shavell argues as follows: (1) scholars' driving motivation is the accumulation of reputation, which is increasing readership; (2) OA will most likely increase readership and hence scholarly esteem; (3) most universities will have an incentive to cover the publishing costs when moving to an 'author pays' principle.[13]

However, McCabe and Snyder (2005) suggest that OA is more likely to be a feature of lower quality journals. They argue that, under OA, profit maximizing publishers may accept more articles than would be socially efficient. A less strict review process would increase publishers' profits and reduce paper quality. Similarly, Jeon and Rochet (2010) find that OA forces publishers to set socially inefficient quality thresholds for paper acceptance. This challenges Shavell's (2010) notion that OA publishing would positively affect authors' reputation. In addition, Mueller-Langer and Watt (2010) are skeptical regarding Shavell's modeling assumption that scholarly esteem can be proxied by readership alone, that is, by the number of reads. The authors argue that this holds true only if reputation as a function of readership would be strictly increasing for all values of readership. However, a journal's impact factor or reputation may be more important in the scholar's decision making than readership. Arguably, an author would more likely submit a paper to a well-esteemed journal with only a few readers than to a low-esteemed journal with a larger audience. Mueller-Langer and Watt (2012) indirectly account for the importance of a journal's reputation in deciding where to publish. They model both author's and reader's perspectives in a two-sided model where journals act as intermediaries linking authors and readers. Mueller-Langer and Watt (2012) identify countervailing effects of abolishing copyright for academic works and find scenarios in which quality for journal articles may increase under OA. The overall welfare effects of a removal of copyright for academic works are ambiguous. In the light of this finding, we suggest that further empirical research on the academic publishing market may make an important contribution to the OA debate.

Several papers have investigated the influence of online or free online access on readership and citations. The literature provides a rather differentiated picture on a possible citation advantage of an OA regime, ranging from an OA citation advantage by a factor of 3 (Lawrence and Giles, 2001) to the conclusion that OA does not generate a significantly higher citation rate (Davis et al., 2008; McCabe and Snyder, 2011) or one that is declining by 7 percent per year (Davis, 2009). Despite some doubt about the degree to which OA may induce higher citation rates, a broad literature stream gives confidence in believing that readership and citations may be at least weakly higher in an OA regime (Harnad, 2012). Accordingly, Eysenbach (2006) finds significantly higher citation rates for OA journal articles in the fields of biology, physics and social sciences. Similar findings are recorded by Norris et al. (2008) in ecology, applied mathematics, sociology and economics. Hajjem et al. (2005) find a citation advantage ranging between 25 and 250 percent by discipline and year for ten different disciplines. Furthermore, Bernius and Hanauske (2009) suggest that a scholar may increase peer recognition and hence scholarly esteem when switching to OA.

Feess and Scheufen (2011) consider possible distortion effects if not all universities cover the publication costs when moving towards an 'author pays' principle. The authors consider publication as a contest for reputation, and thereby for scarce positions at universities. Assuming that researchers differ in talent and top universities are more likely to cover publication costs, Feess and Scheufen (2011) find that researchers' rent-seeking motives may contradict some of the conclusions in Shavell (2010). While in Shavell (2010) private research incentives can never be too high (assuming no negative externalities between authors), Feess and Scheufen (2011) emphasize that social welfare may not be strictly increasing in research activity, which has important implications for the superiority of either regime. Accordingly, OA is always superior if and only if we believe researchers' private effort levels to be already too high, as a larger readership and the asymmetry in publishing costs will correct some of the distortions in the traditional publishing model. In the other case, it will depend on the model's parameters which of the two regimes might produce a better outcome.

Mueller-Langer and Watt (2010) are interested in the possible effects of charging authors a publication fee. A universal OA regime may be particularly detrimental for research institutions exhibiting a relatively large publication output. Accordingly, the best institutions would have to bear relatively higher publication costs as compared to mediocre institutions with a lower publication output. As a result, Mueller-Langer and Watt (2010) emphasize the need to assess the pricing scheme within an OA regime more carefully, especially considering possible distribution effects across institutions.

Last but not least, an emerging literature has been investigating whether the market mechanisms will bring about a more widespread adoption of OA regimes (assuming a universal OA regime to be superior) or whether some coordination failure might prevent this transition. Several authors argue that authors may be locked in owing to a reputation advantage of established CA journals. Cavaleri et al. (2009) invoke the picture of a 'chicken and egg' problem, where newly launched OA journals are restricted in accumulating a decent level of reputation. Good papers that would attract the interest of readers and thus increase the reputation of an OA journal will not be submitted to an OA journal with a low reputation. The dominant strategy of submitting to well-esteemed closed access journals leaves the authors locked in to the weak Nash equilibrium and a 'wait and see' attitude regarding OA (Mann et al., 2008). A survey by Eger et al. (2013) among 2151 scholars in Germany finds that this attitude may, however, differ considerably between disciplines. A positive attitude towards both OA journals and self-archiving depends on the reputation and the widespread adoption of OA, but also on the specific characteristics in the reward structure of the discipline.

HYBRID OPEN ACCESS

Publishers such as Springer and Oxford University Press, among others, have recently introduced the hybrid open access (HOA) business model for academic publications in peer-reviewed journals (Davis, 2009; Björk, 2012). In contrast to the traditional subscription-based CA business model, the HOA publication format gives authors the option of paying an HOA publication fee (up to $3000) to make their paper immediately and freely available online without any embargo period. Under HOA, the copyright remains with the authors. Mueller-Langer and Watt (2013) empirically analyze the effect of HOA at the paper level by comparing citation rates and quality factors for HOA papers to those of CA papers that appear in the same journal. They find that HOA papers generate slightly higher cites than CA papers. Commercial publishers may use the HOA format, which is basically the 'author pays' principle as discussed above, as a second source for revenue on top of the revenue generated from subscription prices (double dipping).

OPEN ACCESS TO DATA

The technological revolution ushered in by the Internet and the increase of possibilities in the digital environment have not only changed the business model of commercial publishers but also facilitated and spurred the creation and use of data sets for scientific purposes. For instance, in economics, the number of empirical articles has significantly increased in recent years as a result of greater data availability and lower costs of data creation. However, even though data availability is an essential feature for the scientific principle of replication and further research (Dewald et al., 1986; McCullough, 2009), Anderson et al. (2008) suggest that authors generally hesitate to share their data and code. For instance, Andreoli-Versbach and Mueller-Langer (2013a) provide evidence for the status quo in economics with respect to data sharing using a data set with 488 hand-collected observations randomly taken from researchers' academic webpages. Out of the sample, 435 researchers (89.14 percent) neither have a data-and-code section nor indicate whether and where their data is available; 8.81 percent of researchers share some of their data, whereas only 2.05 percent fully share.

In a follow-up paper, Andreoli-Versbach and Mueller-Langer (2013b) analyze the incentives of researchers to voluntarily make their data publicly available and thus to provide the scientific community with voluntary OA to data. The analysis suggests that a trade-off arises between the ex post benefits associated with OA to data and reduced incentives to create the data ex ante. It shows that forced OA to data, for example mandatory data availability policies of journals, may lead to welfare-reducing strategic delays of submission. Finally, Andreoli-Versbach and Mueller-Langer (2013b) find that forced OA to data is welfare enhancing if and only if researchers have no incentives to postpone the date of submission and if the positive effect of data availability outweighs the negative effect associated with reduced efforts to create data.

RETRO-DIGITIZATION: GOOGLE BOOK SEARCH

We have so far focused our analysis on OA to journal articles and data. In the following, we address the electronic availability of books, including academic books. A recent strand of literature studies the Google Book Search (GBS) Project (Lichtman, 2008; Grimmelmann, 2009; Bechtold, 2010; Samuelson, 2010a). GBS aims at maximizing the accessibility to books by making digitized books publicly available and searchable worldwide via an Internet book search engine. Beginning in 2004, Google has pursued the retro-digitization of millions of books en masse from library collections with the vision of creating a digital library that allows worldwide and free access to books. While the supporters of GBS conceive this as a first reasonable step towards the largest online body of human knowledge and as a means to promote the democratization of knowledge, its opponents fear negative effects as a result of an erosion of copyright protection, as it may undermine excludability and appropriability (Samuelson, 2010b, 2010c). Mueller-Langer and Scheufen (2011) analyze the GBS Project, focusing on a possible fair-use argumentation against the claims of copyright infringement. This question arises, as the GBS search engine offers its users a short excerpt of the book containing the search term, allowing users to browse sample pages in a limited preview. Copyrighted books are not generally excluded from the digitization and the preview, as Google does not ask right holders for explicit permission. The right holders may, however, exclude their books from GBS by choosing to opt out. Mueller-Langer and Scheufen (2011) suggest that GBS may provide a solution for the unsolved dilemma of orphan works. However, the authors also claim that Google's pricing algorithm for dividing the revenues from integrated advertisements and e-commerce activities for orphan works between the Book Rights Registry and Google should replicate a competitive market outcome under third-party oversight.[14]

In fall 2005, the US Authors Guild and five publishers initiated litigation over GBS, suing Google for copyright infringement. Google reverted to the fair-use argument to legitimate GBS. In particular, Google argued that the limited preview was a fair and thus non-infringing use of copyrighted book content and that GBS improved the accessibility to knowledge. The proceedings of the litigation resulted in a $125 million class action settlement. For instance, Google agreed to make a payment of $60 to the settlement fund per book that was digitized without permission of the right holders. In November 2009, the parties filed an amended settlement agreement which has consequences regarding the international scope of the settlement in particular. It includes only books that were either registered with the US Copyright Office or published in the UK, Australia or Canada up to January 5, 2009. However, Judge Denny Chin rejected the proposed settlement (US District Court Southern District of New York, 2011). In particular, Chin suggested that many of the concerns from a copyright law perspective would be ameliorated if the settlement moved from an 'opt-out' to an 'opt-in' approach.[15] As of February 2013, the parties were considering their next steps.

Finally, it is important to note that, from an academics' perspective, the retro-digitization of academic books increases the accessibility to knowledge, particularly for scientific purposes, and thus facilitates research. Academic books that are out of copyright will be in the public domain and appear in a full text version with an option to browse a book online or directly download a PDF copy. For in-copyright books, GBS allows access to snippets of a book containing the search term, which also facilitates academic research and scientific progress.

CONCLUSIONS, CAVEATS AND QUESTIONS FOR FURTHER RESEARCH

Even though other topics such as GBS are fiercely debated among academics and regulators, OA to academic works appears to be the most important issue for academic publishing in the digital era. Despite the recent flood of research investigating the impact of a shift towards OA in academic publishing, many questions remain unresolved. Three future scenarios seem possible: (1) a universal OA regime; (2) a coexistence of CA and OA business models; (3) the continued predominance of CA. In the first scenario, the question arises of whether a change in copyright or alternative legislative steps (for example an inalienable right of secondary publication) are reasonable and/or necessary for promoting OA. In addition, copyright may have important implications for academic publishing in the first place (for example journal reputation). In contrast, the current coexistence of both regimes may raise doubt as to the ability of OA journals to compete with well-established CA journals. Accordingly, the question arises whether OA journals will be able to successfully increase reputation and hence their impact factor in the long run. From a social welfare perspective, a possible downside for other stakeholders associated with the 'reputation advantage' of established CA publishers may be 'double dipping' strategies by using HOA models for discriminating prices and maximizing profits. Consequently, future studies should investigate which approaches, tools or strategies would provide research institutions and research funders with a counterbalance against the market power of well-known commercial publishers.

As for OA to data, it will have to be answered which approaches, tools or strategies research institutions or external funders of research may choose to increase the (career) incentives of affiliated academics to share their data with the academic community. In addition, new standards of data citations and concepts of data co-authorship have been established (Altman and King, 2007). The analysis of the specific effects of these new standards and concepts on authors and journals appears to be a promising path for further empirical research. Besides, further research on the impact of GBS on scholarly communication is needed. One may argue that GBS fills a gap with respect to the accessibility of knowledge, as it provides a retro-digitization in contrast to the OA movement. Finally, the ongoing debate on HOA, OA to data and the benefits and cost of GBS may induce a substantial impulse for the future scientific discussion on the role of copyright law in the information age and particularly its effects on academic publishing.

NOTES

* The authors thank Jonas Rathfelder for excellent research assistance.

1. Note that there are several other important topics in the context of academic publishing and digitization, such as plagiarism (Collins et al., 2007) or preservation of original content (Astle and Muir, 2002). Nevertheless, the increasing attention OA publishing has gained through recent initiatives, commitments by academic societies and the introduction of policies to promote OA shows its relevance for the future of academic publishing and motivates the focus of the present chapter.
2. See Towse et al. (2008) on the economics of copyright and copying.
3. A famous example is the *Journal des Savants*, which was first published on 5 January 1665 and is known as the first scholarly publication. Ramello (2010) points out that the popularity of the journal was heavily influenced by 'the flourishing number of pirated copies that were widely distributed in France' (Ramello, 2010: 13) and which were clearly infringing the predominant statutory printing privileges.
4. See Bechtold (2004) for a thorough analysis of DRM in the United States and Europe.
5. On the latter point see Hilty (2006).
6. Three initiatives laid the foundations of the OA principle: (1) the Budapest Open Access Initiative in 2002; (2) the Bethesda Statement on Open Access Publishing in 2003; and (3) the Berlin Declaration on Open Access to Knowledge in the Sciences and Humanities in 2003.
7. The *DOAJ* lists all OA journals that follow the lines of the definition by the Budapest Open Access Initiative; that is, only pure OA journals are listed. Not subject are so-called hybrid OA models, that is, business models that provide delayed, partial or retrospective OA or offer an additional open choice option to authors subject to the payment of an author fee. Bernius et al. (2009) provide an overview of academic publishing models.
8. Leading OA disciplines are the health sciences with more than 24 percent of all OA journals, besides social sciences (over 17 percent) and technological engineering (over 9 percent). The standardized Gini coefficient is 0.49 (author's calculations based on the relative distribution of OA journals by discipline). Most OA journals were launched in the US (1272), followed by Brazil (806), the UK (575) and India (473); see *DOAJ* (2013).
9. The impact factor measures the average number of citations of a journal in a particular year or period (Garfield, 1955). The Eigenfactor rates the influence of journals, taking into account whether a citation comes from a journal with a high or low reputation. For further reading see for example Bergstrom et al. (2008).
10. Data restrictions, in particular with respect to the history of journal subscription prices of commercial publishers, did not allow us to provide a broader overview of the academic journal market.
11. A website by Bergstrom (2013) provides an extensive overview and links for further reading on journal pricing.
12. Note that the abolishment of copyright is only one instrument that is currently discussed in the context of OA. However, OA does not necessarily imply that authors lose all of their rights, but the author may in fact retain certain rights, for example the right to attribution (Depoorter et al., 2009) or to fair compensation or modification rights. In this regard, policy recommendations may range from voluntary to statutory limitation or exceptions to copyright (especially for publicly funded research, for example inalienable right of secondary publication).
13. Note that the most important change when moving towards a universal OA regime will be that journals will impose an author's fee for covering the publication costs. The reason is that journal content in an OA world is by definition free of charge. Nevertheless, there are also other income sources that are used in practice. Here major OA publishers have reverted to grants (Public Library of Science, PLoS), print subscriptions (Hindawi Publishing Corporation) or advertising (Medknow Publications) as additional income source.
14. See Picker (2009) for a divergent view.
15. See Mueller-Langer and Scheufen (2011) for a critical investigation of Chin's judgment.

REFERENCES

Altman, Micah and Gary King (2007), 'A proposed standard for the scholarly citation of quantitative data', *D-Lib Magazine*, **13** (3/4), available at: http://www.dlib.org/dlib/march07/altman/03altman.html (accessed 6 May 2013).

Anderson, Richard G., William H. Greene, Bruce D. McCullough and Hrishikesh D. Vinod (2008), 'The role of data/code archives in the future of economic research', *Journal of Economic Methodology*, **15**, 99–119.

Andreoli-Versbach, Patrick and Frank Mueller-Langer (2013a), 'Open access to data: an ideal professed but not practised', RatSWD (German Data Forum) Working Paper Series No. 215, pp. 1–10.

Andreoli-Versbach, Patrick and Frank Mueller-Langer (2013b), 'Climbing the shoulders of giants: open access to data', European Datawatch Extended Project working paper, mimeo.

Arrow, Kenneth (1962), 'Economic welfare and the allocation of resources for invention', in Richard Nelson (ed.), *The Rate and Direction of Inventive Activity*, Cambridge, MA: Harvard University Press, pp. 609–24.

Astle, Peter J. and Adrienne Muir (2002), 'Digitization and preservation in public libraries and archives', *Journal of Librarianship and Information Science*, **34** (2), 67–79.

Bechtold, Stefan (2004), 'Digital rights management in the United States and Europe', *American Journal of Comparative Law*, **52**, 323–82.

Bechtold, Stefan (2010), 'Google Book Search – a rich field of scholarship', *International Review of Intellectual Property and Competition Law*, **41** (3), 1–2.

Bergstrom, Theodore C. (2001), 'Free labor for costly journals?', *Journal of Economic Perspectives*, **15** (3), 183–98.

Bergstrom, Theodore C. (2013), 'Journal pricing page', available at: http://www.econ.ucsb.edu/~tedb/Journals/jpricing.html (accessed 21 January 2013).

Bergstrom, Theodore C., Jevin D. West and Marc A. Wiseman (2008), 'The Eigenfactor™ metrics', *Journal of Neuroscience*, **28** (45), 11433–4.

Bernius, Steffen and Matthias Hanauske (2009), 'Open access to scientific literature – increasing citations as an incentive for authors to make their publications freely accessible', Hawaii International Conference on Systems Sciences (HICSS-42), Waikoloa, Hawaii.

Bernius, Steffen, Matthias Hanauske, Wolfgang König and Berndt Dugall (2009), 'Open access models and their implications for the players on the scientific publishing market', *Economic Analysis and Policy*, **39** (1), 103–15.

Björk, Bo-Christer (2012), 'The hybrid model for open access publication of scholarly articles – a failed experiment?', *Journal of the American Society of Information Sciences and Technology*, **63** (8), 1496–1504.

Bosch, Xavier (2009), 'A reflection on open-access, citation counts, and the future of scientific publishing', *Archivum Immunologiae et Therapiae Experimentalis*, **57**, 91–3.

Brody, Tim, Stevan Harnad and Les Carr (2006), 'Earlier web usage statistics as predictors of later citation impact', *Journal of the Association for Information Science and Technology*, **57** (8), 1060–72.

Budapest Open Access Initiative (2002), available at: http://www.budapestopenaccessinitiative.org/read (accessed 6 August 2013).

Cavaleri, Michael, Giovanni B. Ramello and Vittorio Valli (2009), 'Publishing an e-journal on a shoe string: is it a sustainable project?', *Economic Analysis and Policy*, **39** (1), 89–101.

Collins, Alan, Guy Judge and Neil Rickman (2007), 'On the economics of plagiarism', *European Journal of Law and Economics*, **24**, 93–107.

Davis, Philip M. (2009), 'Author-choice open access publishing in the biological and medical literature', *Journal of the American Society for Information Science and Technology*, **60** (1), 3–8.

Davis, Philip M., Bruce V. Lewenstein, Daniel H. Simon, James G. Booth and Mathew J.L. Connolly (2008), 'Open access publishing, article downloads, and citations: randomised controlled trial', *BMJ*, **337**, a568.

Depoorter, Ben, Adam Holland and Elisabeth Somerstein (2009), 'Copyright abolition and attribution', *Review of Law and Economics*, **5** (3), 1063–80.

Dewald, William G., Jerry G. Thursby and Richard G. Anderson (1986), 'Replication in empirical economics: the *Journal of Money, Credit and Banking* Project', *American Economic Review*, **76**, 587–603.

DOAJ (*Directory of Open Access Journals*) (2013), 'DOAJ by country', available at: http://www.doaj.org/doaj?func=byCountry&uiLanguage=en (accessed 21 January 2013).

Edlin, Aaron S. and Daniel L. Rubinfeld (2004), 'Exclusion or efficient pricing? The "big deal" bundling of academic journals', *Antitrust Law Journal*, **72**, 128–59.

Eger, Thomas and Marc Scheufen (2012a), 'The past and the future of copyright law: technological change and beyond', in Jef De Mot (ed.), *Liber Amicorum Boudewijn Bouckaert*, Brugge: die Keure, pp. 37–65.

Eger, Thomas and Marc Scheufen (2012b), 'Das Urheberrecht im Zeitenwandel: Von Gutenberg zum Cyberspace', in Christian Mueller, Frank Trosky and Marion Weber (eds), *Ökonomik als allgemeine Theorie menschlichen Verhaltens*, Schriften zu Ordnungsfragen der Wirtschaft, vol. 94, Stuttgart: Lucius & Lucius, pp. 151–79.

Eger, Thomas, Marc Scheufen and Daniel Meierrieks (2013), 'The determinants of open access publishing: survey evidence from Germany', SSRN working paper, available at: http://papers.ssrn.com/sol3/papers.cfm?abstract_id=2232675.

Eysenbach, Gunther (2006), 'Citation advantage of open access articles', *PLoS Biology*, **4** (5), e157, available at doi: 10.1371/journal.pbio.0040157 (accessed 21 January 2013).

Feess, Eberhard and Marc Scheufen (2011), 'Academic copyright in the publishing game: a contest perspective', SSRN working paper, available at: http://papers.ssrn.com/sol3/papers%20.cfm?abstract_id=1793867 (accessed 21 January 2013).
Gans, Joshua (ed.) (2000), *Publishing Economics: Analyses of the Academic Journal Market in Economics*, Cheltenham, UK and Northampton, MA, USA: Edward Elgar Publishing.
Garfield, Eugene (1955), 'The meaning of the impact factor', *International Journal of Clinical and Health Psychology*, **3** (2), 363–9.
Gaulé, Patrick and Nicolas Maystreb (2011), 'Getting cited: does open access help?', *Research Policy*, **40**, 1332–8.
Grimmelmann, J. (2009), 'How to fix the Google Book Search Settlement', *Journal of Internet Law*, **12** (10), 10–20.
Hajjem, Chawki, Stevan Harnad and Yves Gingras (2005), 'Ten-year cross-disciplinary comparison of the growth of open access and how it increases research citation impact', *IEEE Data Engineering Bulletin*, **18** (4), 39–47.
Harnad, Stevan (2012), 'The effect of open access and downloads ("hits") on citation impact: a bibliography of studies', available at: http://opcit.eprints.org/oacitation-biblio.html (accessed 21 January 2013).
Harnad, Stevan, Tim Brody, François Vallières, Les Carr, Steve Hitchcock, Yves Gingras, Charles Oppenheim, Heinrich Stamerjohanns and Eberhard R. Hilf (2004), 'The access/impact problem and the green and gold roads to open access', *Serials Review*, **30** (4), 310–14.
Hilty, Reto M. (2006), 'Five lessons about copyright in the information society: reaction of the scientific community to over-protection and what policy-makers should learn', *Journal of the Copyright Society of the USA*, **53** (1), 103–38.
Jeon, Doh-Shin and Jean-Charles Rochet (2010), 'The pricing of academic journals: a two-sided market perspective', *American Economic Journal: Microeconomics*, **2**, 222–55.
Lawrence, Steve and C. Lee Giles (2001), 'Accessibility of information on the web', *Intelligence*, **11** (1), 32–9.
Lemley, Mark (2011), 'Is the sky falling on the content industries?', *Journal of Telecommunications and High Technology Law*, **9**, 125–35.
Lichtman, Douglas (2008), 'Copyright as innovation policy: Google Book Search from a law and economics perspective', in Josh Lerner and Scott Stern (eds), *Innovation Policy and the Economy*, vol. 9, Cambridge, MA: National Bureau of Economic Research, pp. 55–77.
Liebowitz, Stan L. (1985), 'Copying and indirect appropriability: photocopying of journals', *Journal of Political Economy*, **93** (5), 945–57.
McCabe, Mark J. and Christopher M. Snyder (2005), 'Open access and academic journal quality', *American Economic Review Papers and Proceedings*, **95** (2), 453–8.
McCabe, Mark J. and Christopher M. Snyder (2007), 'Academic journal prices in a digital age: a two-sided market model', *B.E. Journal of Economic Analysis and Policy*, **7** (1), Article 2.
McCabe, Mark J. and Christopher M. Snyder (2011), 'Did online access to journals change the economics literature?', SSRN working paper, available at: http://papers.ssrn.com/sol3/papers.cfm?abstract_id=1746243 (accessed 21 January 2013).
McCullough, Bruce D. (2009), 'Open access economics journals and the market for reproducible economic research', *Economic Analysis and Policy*, **39** (1), 118–26.
McGuigan, Glenn S. (2004), 'Publishing perils in academe: the serials crisis and the economics of the academic journal publishing industry', *Journal of Business and Finance Librarianship*, **10** (1), 13–26.
Mann, Florian, Benedikt von Walter, Thomas Hess and Rolf T. Wigand (2008), 'Open access publishing in science: why it is highly appreciated but rarely used', *Communications of the ACM*, **51**.
Mueller-Langer, Frank and Marc Scheufen (2011), 'The Google Book Search Settlement: a law and economics analysis', *Review of Economic Research on Copyright Issues*, **8** (1), 7–50.
Mueller-Langer, Frank and Richard Watt (2010), 'Copyright and open access for academic works', *Review of Economic Research on Copyright Issues*, **7** (1), 45–65.
Mueller-Langer, Frank and Richard Watt (2012), 'Optimal pricing and quality of academic journals and the ambiguous welfare effects of forced open access: a two-sided market model', TILEC Discussion Paper No. 2012–019, available at: http://papers.ssrn.com/sol3/papers.cfm?abstract_id=2045956 (accessed 21 January 2013).
Mueller-Langer, Frank and Richard Watt (2013), 'Analysis of the impact of hybrid open access on journals and authors', mimeo, University of Canterbury.
Norris, Michael, Charles Oppenheim and Fytton Rowland (2008), 'The citation advantage of open access articles', *Journal of the American Society for Information Science and Technology*, **59** (12), 1963–72.
OAI (2008), 'The Open Archives Initiatives Protocol for Metadata Harvesting', available at: http://www.openarchives.org/OAI/openarchivesprotocol.html (accessed 25 February 2013).
Picker, Randal C. (2009), 'The Google Book Search Settlement: a new orphan-works monopoly?', *Journal of Competition Law and Economics*, **5** (3), 383–409.

Pitsoulis, Athanassios and Jan Schnellenbach (2012), 'On property rights and incentives in academic publishing', *Research Policy*, **41** (8), 1440–47.
Ramello, Giovanni B. (2010), 'Copyright and endogenous market structure: a glimpse from the journal-publishing market', *Review of Economic Research on Copyright Issues*, **7** (1), 7–29.
Samuelson, Pamela (2010a), 'Google Book Search and the future of books in cyberspace', *Minnesota Law Review*, **94** (5), 1308–74.
Samuelson, Pamela (2010b), 'Legally speaking: the dead souls of the Google Book Search Settlement', *Communications of the ACM*, **53** (7), 32–4.
Samuelson, Pamela (2010c), 'Academic author objections to the Google Book Search Settlement', *Journal of Telecommunications and High Technology Law*, **8**, 491–522.
Shavell, Steven (2010), 'Should copyright of academic works be abolished?', *Journal of Legal Analysis*, **2** (1), 301–58.
Suber, Peter (2012), *Open Access*, Cambridge, MA: MIT Press.
Thomson Reuters Journal Citation Reports (2011), available at: http://thomsonreuters.com/products_services/science/science_products/a-z/journal_citation_reports/ (accessed 21 January 2013).
Towse, Ruth, Christian Handke and Paul Stepan (2008), 'The economics of copyright law: a stocktake of the literature', *Review of Economic Research on Copyright Issues*, **5** (1), 1–22.
US District Court Southern District of New York (2011), 'Authors Guild et al. vs. Google Inc., Opinion 05 Civ. 8136 (DC)', available at: http://amlawdaily.typepad.com/googlebooksopinion.pdf (accessed 21 January 2013).
Zimmermann, Christian (2009), 'The economics of open access publishing', *Economic Analysis and Policy*, **39** (1), 49–52.

FURTHER READING

For an overview on the economics of academic publishing and journal prices, see Gans (2000), Bergstrom (2001), McGuigan (2004), McCabe and Snyder (2007), Zimmermann (2009) and Pitsoulis and Schnellenbach (2012). See Bernius et al. (2009) for an analysis of the implications of OA for publishers, libraries and funding organizations. For an overview on the impact of OA on citation counts, see Brody et al. (2006), Bosch (2009) and Gaulé and Maystreb (2011). Suber (2012) provides a thorough overview on the benefits of OA for authors and readers.

For a more general review on how new technologies have affected the content industries, see Lemley (2011). Many useful sources on the history of copyright and its relationship with the diffusion of new technologies are presented in Eger and Scheufen (2012a, 2012b).

33. News

Piet Bakker and Richard van der Wurff

The dominant discourse on the future of newspapers is build around threat. Circulation, readership and advertising revenues are dropping at a fast rate, presumably because of the Internet. Audiences – especially younger groups – and advertisers move online, which leaves papers with a declining and ageing readership, less revenue and the threat of extinction. Meyer (2004) estimated that in 2043 no American would read a daily newspaper. At the same time, but less prominent, we also find a discourse of hope. In this discourse, digital technologies enable journalistic innovation, product customization and customer involvement.

Whether out of fear or out of hope, newspaper publishers feel the urgent need to go online, following readers and advertisers. 'Even the most confident of newspaper bosses', wrote the *Economist* (2006, n.p.), 'now agree that they will survive in the long term only if . . . they can reinvent themselves on the Internet and on other new-media platforms such as mobile phones and portable electronic devices.' But, so far, online strategies of news providers turn out to be commercially far less successful than print strategies. Moreover, online successes tend to be developed by enterprises without a stake in traditional newspapers.

In this chapter we first review the economic characteristics of news. Next, we track developments in the newspaper industry in some detail. Finally, we review some solutions and discuss the main strategic approaches that publishers adopt in the digital age.

THE PECULIARITIES OF NEWS AS ECONOMIC PRODUCT

News is an atypical product. One distinguishing aspect concerns the political and social significance of news in democratic societies. But news is also different from 'normal' products in economic terms. It has peculiar economic characteristics that explain why it is difficult, if not impossible, to produce and sell general news as a stand-alone product on a commercial basis.

The *news value chain* includes five technologically and strategically distinct activities (Porter, [1985] 1998). These are: the *production* of news stories by journalists; the *bundling* of news stories (and other types of information, including advertising) into the newspaper master copy; the *reproduction* on printing presses; the physical *delivery*; and *marketing and sales* to audiences and advertisers.

High First Copy Costs and Non-Rivalry in Consumption

A crucial characteristic of the newspaper industry in the first two stages – production and bundling – is that these involve *high first copy costs*. This is because news is a labour-intensive product, produced by highly skilled journalists. For example, editorial costs made up 28 per cent of total costs of Dutch newspapers in 2011 (NDP Nieuwsmedia, 2012). These costs are independent of the number of actual readers. The third and fourth stages of news production – reproduction and delivery – involve considerable high fixed costs as well, for example of printing presses. For Dutch newspapers, the costs of reproduction and distribution amount to 41 per cent of total costs. The remaining 32 per cent are for marketing (18 per cent) and overhead (14 per cent) (NDP Nieuwsmedia, 2012).

The high levels of fixed and first copy costs imply that newspaper publishers can profit from considerable *economies of scale*. These explain the strong tendency towards editorial cooperation and consolidation. One might even argue that news production is to some extent a natural monopoly. Why send two journalists to cover an event when one can do the job for a larger audience? The answer is to be found in the social importance of news media: more journalists may cover an event from more different angles and thus contribute to a diverse supply of news.

Partly reflecting the high first copy costs, the marginal costs of serving one additional customer with news are low to non-existent. News, as virtual product, tends to be *non-rival* in consumption. In practical terms this holds more strongly for digital than for print news. But even printed newspapers are easily passed from household to household. For newspaper companies, this translates into the problem of *excluding non-payers* from using the product. Why would someone pay if you can get the same news free?

High first copy costs, non-rival consumption and non-excludability imply that news – once produced – is not really scarce. This explains why news cannot easily be sold like any other product on commercial markets. One additional issue is that news generates *positive externalities*. When citizens are well informed, society as a whole benefits. But, for the newspaper business, the problem is that society does not pay for these positive externalities. Another issue is the *merit good* character of news: citizens are not willing to pay its 'true' value.

Print Newspaper Business Models

Traditionally, publishers dealt with these problems by bundling news with advertising (Sparks, 2003). They circumvented the problem of generating sufficient revenues to pay the costs of news production from *users*, by selling 'access to audiences' to advertisers (Picard, 1989).

The success of this strategy depended, in the final analysis, on scale economies that publishers could realize, in combination with a lack of competing communication channels. Important social actors – governments, politicians, advertisers and NGOs – depended on newspapers to reach their audiences; and citizens depended on newspapers to learn about the world around them. That is why New Yorkers felt 'awfully lost' when their favourite newspapers were not available for 17 days in 1945 (Berelson, 1949: 125). That is also why publishers could offer exclusive value to advertisers and earn enough to pay for news production and make a decent profit as well. Ultimately, the newspaper business was a profitable business because and to the extent that newspapers were dominant communication channels in society.

This comfortable position was destroyed by technological change. Terrestrial TV and cable TV opened up alternative channels that reduced the dependency of society on newspapers and their publishers. Next, any remaining dependencies were ended with the advent of the Internet. These technologies enable any organization or individual that wishes to do so to reproduce and distribute content to audiences at virtually no cost.

Digitization

Thus, nowadays, sources and advertisers bypass news media and communicate directly with audiences. There is no need to bundle various news stories with other information and advertising in print to make reproduction and distribution technically and economically feasible. Instead, companies may offer targeted services directly to customers. Examples are classified ads, job vacancies, weather forecasts and stock market information. Other companies pass on news free as part of their marketing strategy. They buy, copy and aggregate news from newspaper publishers and news agencies and add it to their websites to attract audiences.

The new scarcities are not in reproduction and distribution, but in making the match between content senders and receivers. This explains the economic successes of search engine providers (for example Google) and other players that make online content accessible and connect users (for example Facebook). For newspaper publishers, these developments mean that they have to find new ways to produce and sell news as a profitable commercial product.

Fortunately for newspaper publishers, print and digital news are not perfect substitutes. If that had been the case, printed newspapers would have been extinct for a decade, as online news is cheaper, faster and available almost everywhere. Younger audiences usually do not have much appetite for print media. But older users stick with the medium they have grown up with, while adding new media to their media diet. This habitual use of printed news could explain why the decline in print circulation, though persistent, is still modest.

WHAT IS ACTUALLY GOING ON?

How urgent is the situation for traditional publishers? How fast is circulation dropping, and how quickly are online revenues increasing? We provide data on print circulations for mature newspaper markets where newspapers operate on a for-profit business model. This model can be found in Western Europe, Northern America and some Asian countries. For online revenues – a sensitive business area – we have to rely on anecdotal information.

Print Circulation

In 20 European markets for which we have complete data over the last 15 years (see Table 33.1), total paid circulation – the number of copies sold – dropped 24 per cent from 97 million copies in 1995 to 73 million in 2010. Between 1995 and 2004, the average drop in circulation was 1.4 per cent per year. Between 2005 and 2010, the average annual decline almost doubled to 2.7 per cent. Newspaper penetration – the number of newspapers distributed per 100 inhabitants – decreased from 24 in 2000 to 19 in 2010.

Figure 33.1 shows the development of paid newspaper circulation in 1995–2010 for

Table 33.1 Changes in newspaper circulation with and without free papers, 2000 to 2010

	2000–10 paid %	2000–10 paid and free %
Major newspaper markets		
Germany	−20	−21
UK	−20	−12
France	−14	16
Italy	−25	21
Spain	−12	30
Smaller newspaper markets		
Netherlands	−20	−7
Sweden	−13	−12
Poland	5	13
Switzerland	−22	−10
Austria	−20	11
Finland	−14	−16
Norway	−28	−28
Belgium	−12	−6
Czech Republic	−27	−17
Hungary	−30	−22
Denmark	−33	−3
Ireland	22	33
Portugal	−4	43
Luxembourg	−8	128
Iceland	−47	52

Source: *World Press Trends* (1996–2011) (paid papers) and own research (free circulation).

five major European newspaper markets. The other markets show the same pattern. We find a decline in every country, except for Ireland – and this is mainly because the method of measurement was changed in that country.

When free circulation is included the picture is somewhat different, with major differences between countries. Norway and Germany have no free newspapers, whereas free newspapers are important in the UK, France and Italy. Countries like Spain, the Netherlands, Sweden and Denmark have experienced a rapid growth followed by a fast decline within a few years.

Overall, circulation of free newspapers increased rapidly to 24 million in 2007 and then dropped again to 16 million in 2010 (and to 14 million in 2012). This slowed down the overall decline in circulation. When we consider both paid and free newspapers, circulation in the European markets dropped by 8 per cent between 1995 (97 million) and 2010 (89 million) and penetration declined from 26 (2000) to 24 (2010).

Outside Europe, the only 'mature' newspaper market with complete circulation data from 1995 onwards is the USA. Between 1995 and 2010 circulation declined by 24 per cent from 58 to 44 million. Free newspaper circulation in the US was 2.5 million in 2010. Canadian data for 2010 are not available, but between 1995 and 2009 paid circulation dropped by 22 per cent. Free circulation of 1.5 million, however, resulted in a growth of total Canadian circulation. Japan has a circulation of 49 million on a 15-plus population

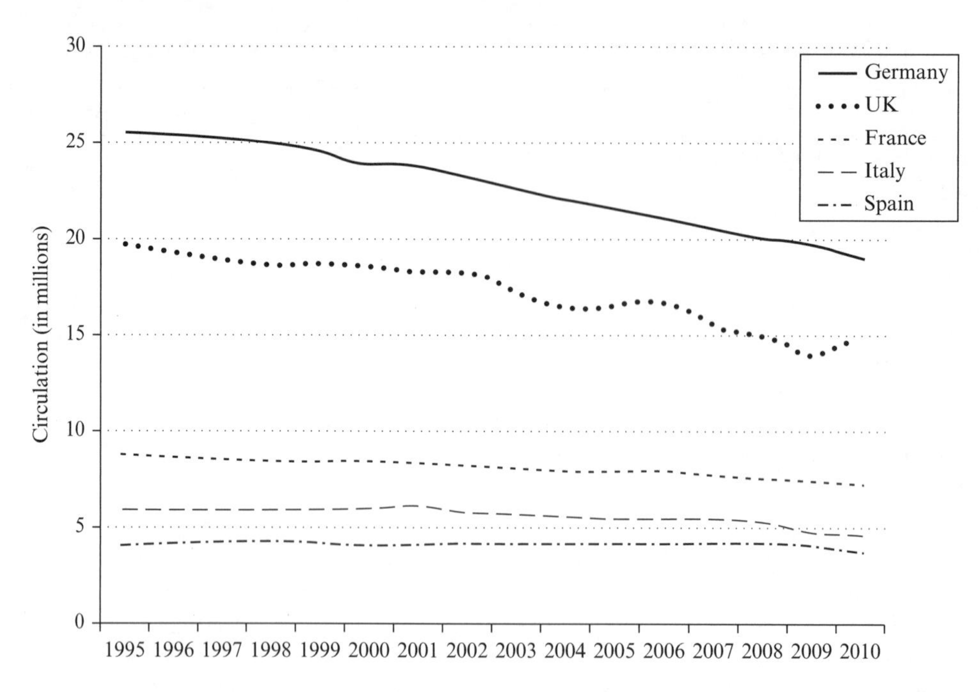

Source: *World Press Trends* (1996–2011).

Figure 33.1 Paid newspaper circulation in the five largest European markets, 1995 to 2010

of 110 million. Between 2005 (no earlier comparable data are available) and 2010 circulation dropped by 6 per cent.

A Double Bind

The decline in paid newspaper circulation might not yet be labelled 'dramatic', but the rate of decline is increasing: over the five years 2005–10, circulation decreased 2.7 per cent per annum in Europe and 3.5 per cent in Canada and the US. It is hard to substantiate that this decline in paid circulation is directly caused by digitization, but it is very probable that digitization affects newspaper circulation. The decline in circulation increased over the last few years, when Internet penetration also showed substantial growth, and evidence is growing that younger and more highly educated groups in developed markets are indeed substituting online for print news media (Mitchelstein and Boczkowski, 2010; Taipale, 2013). Furthermore, it is not only circulation decline that publishers are worried about – the related decline in advertising income is even more damaging.

Whatever the cause of the decline might be, publishers aim for online revenues to increase at least as quickly as print circulations and revenues decline, so that in the years to come online editions will start to pay their own way, print editions can be phased out and printing and distribution costs can be saved (Thurman and Myllylahti, 2009).

However, the promise of growing digital revenues from newspaper websites has so far proved to be unjustified. Online experiments and endeavours did not result in substantial income. Instead of delivering revenues they offered users the possibility to access online content free and stop reading a printed newspaper.

Publishers still generate a substantial part of their income from their print newspaper operations (Picard and Dal Zotto, 2006). Earning money online is an art that only a few publishers have mastered. One strategy, charging money for online news, has been discussed for some time, but remains a loss-making endeavour unless providers have unique content to offer. The archetypical example is the *Wall Street Journal*, which has been charging its online subscribers since 1996. Another strategy is to earn money with online advertising, but this strategy brings publishers into a very competitive market where they face strong and innovative challengers. Players such as Google, eBay and Monsterboard have taken important parts of newspapers' advertising business.

In combination, free content and more intense competition on advertising markets mean that publishers would need at least ten online readers to replace the revenues brought by one print reader, as the *Economist* (2006) estimated some years ago. Publishers thus face a difficult economic choice: investing in new online ventures with lower revenues and the additional risk of undermining the remaining print revenues even faster; or harvesting the print market and running the risk that the company will have to close its doors when the last remaining print readers cannot pay for the costs of news production any more. This explains the double bind publishers are experiencing.

SOLUTIONS?

A common approach for publishers, especially in the early days of the World Wide Web, was to complement their print newspaper with an online edition. These online newspapers allowed publishers to experiment with online content and possible revenue streams. A typical online version offered news stories from the print edition, content interactivity (hyperlinks), personal interactivity (user contributions) and perhaps some multimedia (van der Wurff and Lauf, 2005). Most of these models were offered free. They relied heavily on the print counterpart for news content and generated some additional advertising revenues for the company. Later, publishers introduced paid services, such as access to archives and news on mobile platforms (Herbert and Thurman, 2007).

Integration

The next step towards developing a working digital revenue model seems to be the integration of online and offline operations. In this view, online is no longer a different, subordinate department but fully integrated into the organization. In theory this saves costs, as it ends the practice of having different reporters covering one subject and results in more output – on different platforms – as all staff members are expected to produce material for all platforms. It usually also involves publishing online first.

This integrated 'digital first' strategy is employed by media like the *Guardian* and the *Daily Telegraph* (UK), *Le Monde* (France), *Blick* (Switzerland) and *El País* (Spain), yet other media actually do the opposite. Austrian quality newspaper *Der Standard* totally separated the online and offline operation. And the Dutch Persgroep (*de Volkskrant*, *AD*, *Trouw*, *Parool*) reversed its integration strategy. In these organizations, online journalists work at their own offices, with their own editor-in-chief.

Rethinking Free

Publishers increasingly rethink the strategy of free online content. When Rupert Murdoch announced in 2009 that 'quality journalism is not cheap' and erected a pay wall at his UK flagship *The Times*, other media followed suit. They preferred a targeted – and probably smaller – paying audience over a mass audience that offered low revenues because of low advertising rates. Media that cater for a high-end audience, offering proprietary need-to-know content, usually employ this strategy. This includes papers like the *Financial Times* (UK), *Wall Street Journal* and *New York Times* (US), as well as *Handelsblatt* (Germany).

In the Netherlands, financial paper *Financieele Dagblad* charges for access to its website and offers an e-paper as well. In 2012, 20 per cent of its subscriptions were digital. Two smaller Christian newspapers also employ pay walls and sell between 5 per cent and 10 per cent of their subscriptions online. Other papers sell online subscriptions but keep the website free. This includes quality paper *NRC Handelsblad* and its sister paper *NRC.NEXT*, which sell 9 per cent and 7 per cent, respectively, of their circulation in digital copies. But quality paper *de Volkskrant* sells only 2 per cent online, and all other Dutch newspapers (including the two largest national dailies and all regional papers) sold not more than 1 per cent in digital copies in 2012. This illustrates that financial or otherwise specialized papers, and national papers with a high-income audience, have a much better chance of selling digital news than mass-oriented newspapers, traditional tabloids and local papers.

Tablets and Mobile

The introduction of the iPad in 2010 stimulated the shift towards paid news. Online users may have a low inclination to pay for online content, but on the iPad the behaviour seems to be different. Because of the easy-to-use and trusted iTunes Store payment model, users are more apt to buy digital versions. Most newspapers have introduced iPad editions: enriched replicas of the printed newspaper, but for a lower price. It saves costs by avoiding printing and offline distribution. However, Apple takes a substantial cut from the sales made through its store.

Publishers also target other tablet users by launching editions for Android and Windows platforms, platform-independent html5 versions, and editions for different smartphones. Although some content – usually for phones – is still free, the dominant business model on tablets is paid content.

In the Netherlands, the regional newspapers of the Mecom group introduced paid iPad apps for their editions in 2012 for €16 per month. *NRC Handelsblad* launched a reader app with a limited number of stories from the paid newspaper (at €4.49 per month), specially formatted for online users. Former free daily *De Pers*, which did not survive print competition, reinvented itself online with a paid app for €4.99 per month.

Users can also subscribe to individual journalists for €1.79 per month. *De Telegraaf*, the Netherlands' largest newspaper, introduced a paid iPad app, too. The trend is clear: there is more paid-for access on tablets and mobile phones than on traditional personal computers.

Low(er)-Cost News

Sometimes some content is offered free, after which registration and subscription are required (the freemium model). A 'hidden' pay wall is erected when media offer only a limited amount of content online and direct readers to the newspaper or the paid edition for the full story. Alternatively, a newspaper that cannot offer exclusive need-to-know content and caters for a more general audience can expand the free-for-all strategy – going for the mass market.

A strategy that complements the free-for-all strategy could be to reduce costs by sharing editorial information with other companies. Sharing content on the local and national level is a type of cooperation in which newspapers companies frequently engage, next to cooperation in the advertising and distribution businesses (Picard and Dal Zotto, 2006).

Another strategy that companies adopt to reduce cost is to lay off personnel and invest less in resources for news production. Publishers of free newspapers in particular excel in filling news pages with a fraction (one-tenth) of the journalists that a paid-for paper would need (*Economist*, 2006). Digital news operations likewise cut costs by employing fewer or less expensive people (freelancers and non-union personnel), using stock photos instead of original work, relying more on PR material, press agencies and syndicated content, and using technology on websites for aggregation, promotion via social media, and even automatic article writing (Bakker, 2012). These strategic choices clearly have negative repercussions for the quality and scope of reporting, threatening local and foreign coverage and journalistic independence (Lewis et al., 2008; Siles and Boczkowski, 2012).

A third option to reduce costs, with more uncertain implications for news quality, is for news media to rely on free content provided by volunteers, experts, citizen journalists and users. One example is the *Huffington Post*, a successful and profitable news and commentary blog that relies for its content on more than 6000 volunteers. But many traditional news media, too, rely on user contributions and participatory journalism, to reduce costs and produce content that readers might find attractive (Deuze et al., 2007).

Other Sources of Revenues

Rather than reducing costs, another approach is to look for third parties to finance news production. Individual journalists and publishers have been seeking the support of foundations, angel investors and citizens to pay the costs of investigative reporting (crowd funding) (Drew, 2010: 23). Some commentators also advocate stronger government support for the press (Nordenson, 2007: 41).

Government, foundation and community support for the production of quality news makes economic sense when we consider the public and merit good character of news. If quality news is important for society and should be made available to all, public-interest organizations could ensure sufficient incentives by compensating for the revenues that the market fails to offer owing to the economic peculiarities of the news product.

Online news providers may also choose to develop the commercial side of their news services. The underlying idea is that news services capitalize on their visibility and the trust that they command to facilitate a wide range of commercial transactions (Ihlström and Palmer, 2002). One particular model that is frequently discussed in this respect is the (hyper)local platform model: the bundling of all kinds of locally relevant content – news, business information, blogs, opinions, user-generated content, advertising – into one local platform, together with commercial and, perhaps, transaction services. Example are the AOL-owned Patch Media network of hyperlocal sites in the USA, the Examiner network (USA), My Heimat (Germany) and Dichtbij (the Netherlands). Not all models are successful. NBC-owned EveryBlock closed in February 2013.

MULTIPLE MEDIA, MULTIPLE STRATEGIES

There is no single solution that would reliably deal with all challenges that newspaper publishers face. Media are active on all available channels and platforms to serve customers. But adding and shifting across distribution channels and platforms do not by themselves solve the fundamental economic problem that newspaper publishers and other news providers face: how to turn news, with its peculiar economic characteristics, into a marketable and profitable product. What value do news providers actually add? Relying on Porter's ([1985] 1998) typology of business strategies and reviewing the strategies and experiments in the newspaper business, two main approaches emerge.

The first is to serve news as a premium product to affluent professionals with a high need for up-to-date quality information and a concomitant willingness to pay for that content. Providers adopting this *differentiation* strategy add *journalistic* value by selecting, investigating, relating and explaining news events. Additional value might be added by customizing news services, providing investigative journalism on demand and facilitating user interactions and networking. This strategy brings back the original newspaper as a means of communication for the political-economic and cultural elites, but in a modern-day electronic version (Picard, 2002: 30–31). Examples might be the *Economist*, the *Washington Post* and the *Financial Times*, which all use print, online and mobile platforms to serve their customers with a combination of real-time factual news, online opportunities to browse and interact, and multiple channels with which to consult and appreciate in-depth background information and analysis.

The second approach is to serve basic news as a commodity to mass audiences in return for advertising revenues. This is a digital variant of the print strategy that most publishers of free newspapers have been following for some time. In economic terms, this is a *low-cost* strategy. It depends on the realization of economies of scale and scope to reduce costs, and on the generation of advertising revenues (and perhaps third-party payments) to cover these costs. Providers of such low-cost news services may add additional value by selling brand and line extensions and facilitating user contributions, thereby increasing the attractiveness of their services.

The End of Mass Quality News?

In retrospect, the second half of the twentieth century might very well be classified as the golden age of newspapers. The newspaper businesses yielded high profits, and newspapers provided news to large audiences. But this period also witnessed important and potentially destructive technological changes.

In the future, news suppliers need to adapt their strategies. Most likely, we will see a diversification of strategies – with some providers offering expensive and high-quality topical information to interested elites and others serving low-cost newsy bits of information to mass audiences. For local publishers this will be problematic, as they usually employed an 'in-between' model for a general audience. Given the high level of competition online it is quite possible that not all models will survive. We will also witness, within each strategy, a diverse use of different media channels to serve appropriate content and services to customers at acceptable prices – taking into consideration that payment is more acceptable to audiences on some platforms (print, mobile and tablets) than on others (the traditional World Wide Web or television). Major changes in these strategies are only to be expected when governments, social actors or customers become considerably more convinced that quality news is a valuable product that they should also pay for in an online environment.

REFERENCES

Bakker, Piet (2012), 'Aggregation, content farms and Huffinization: the rise of low-pay and no-pay journalism', *Journalism Practice*, **6** (5–6), 627–37.

Berelson, Bernard (1949), 'What "missing the newspaper" means', in Paul F. Lazarsfeld and Frank N. Stanton (eds), *Communications Research 1948–1949*, New York: Harper & Brothers, pp. 111–29.

Deuze, Mark, Axel Bruns and Christoph Neuberger (2007), 'Preparing for an age of participatory news', *Journalism Practice*, **1** (3), 322–38.

Drew, Jill (2010), 'The new investigators', *Columbia Journalism Review*, **49** (1), 20–27.

Economist (2006), 'The newspaper industry: more media, less news', 24 August.

Herbert, Jack and Neil Thurman (2007), 'Paid content strategies for news websites', *Journalism Practice*, **1** (2), 208–26.

Ihlström, Carina and Jonathan Palmer (2002), 'Revenues for online newspapers: owner and user perceptions', *Electronic Markets*, **12** (4), 228–36.

Küng, Lucy, Robert G. Picard and Ruth Towse (eds) (2008), *The Internet and the Mass Media*, London: Sage.

Lewis, Justin, Andrew Williams and Bob Franklin (2008), 'Four rumours and an explanation', *Journalism Practice*, **2** (1), 27–45.

Meyer, Philip E. (2004), *The Vanishing Newspaper: Saving Journalism in the Information Age*, Columbia: University of Missouri Press.

Mitchelstein, Eugenia and Pablo J. Boczkowski (2010), 'Online news consumption research: an assessment of past work and an agenda for the future', *New Media Society*, **12**, 1085–1102.

NDP Nieuwsmedia (2012), *Jaarverslag 2011*, Amsterdam: NDP Nieuwsmedia.

Nordenson, Bree (2007), 'The Uncle Sam solution', *Columbia Journalism Review*, **46** (3), 37–41.

Picard, Robert G. (1989), *Media Economics*, Newbury Park, CA: Sage.

Picard, Robert G. (2002), *The Economics and Financing of Media Companies*, New York: Fordham University Press.

Picard, Robert G. and Jeffrey H. Brody (1997), *The Newspaper Publishing Industry*, Boston, MA: Allyn & Bacon.

Picard, Robert G. and Cinzia Dal Zotto (2006), *Business Models of Newspaper Publishing Companies*, Darmstadt: IFRA.

Porter, Michael ([1985] 1998), *Competitive Advantage*, New York: Free Press.

Siapera, Eugenia and Andreas Veglis (eds) (2012), *The Handbook of Global Online Journalism*, Chichester: Wiley-Blackwell.
Siles, Ignacio and Pablo J. Boczkowski (2012), 'Making sense of the newspaper crisis: a critical assessment of existing research and an agenda for future work', *New Media and Society*, **14** (8), 1375–94.
Sparks, Colin (2003), 'The contribution of online newspapers to the public sphere', *Trends in Communication*, **11** (2), 111–26.
Taipale, Sakari (2013), 'The relationship between Internet use, online and printed newspaper reading in Finland: investigating the direct and moderating effects of gender', *European Journal of Communication*, **28**, 5–18.
Thurman, Neil and Merja Myllylahti (2009), 'Taking the paper out of news: a case study of *Taloussanomat*, Europe's first online-only newspaper', *Journalism Studies*, **10** (5), 691–708.
World Press Trends (1996–2011), Paris: World Association of Newspapers.
Wurff, Richard van der and Edmund Lauf (eds) (2005), *Print and Online Newspapers in Europe: A Comparative Analysis in 16 Countries*, Amsterdam: Het Spinhuis.

FURTHER READING

Robert G. Picard and Jeffrey H. Brody's book on *The Newspaper Publishing Industry* (1997) offers a good introduction to newspaper economics and management. The impact of digitization on journalism is discussed in *The Handbook of Global Online Journalism*, edited by Siapera and Veglis (2012). The impact of digitization on media companies, including newspaper publishers, is covered by *The Internet and the Mass Media*, edited by Küng et al. (2008). A good source of quantitative data is *World Press Trends*, published every year by the World Association of Newspapers.

34. Digital music

Patrick Waelbroeck

The music industry is facing the development of online peer-to-peer (P2P) networks, direct download services and other forms of digital piracy, while at the same time the format and promotion of pre-recorded music are undergoing major transformations with the advent of MP3 and new forms of collective promotion, particularly through online communities and social networks. This chapter analyzes the main effects of digital distribution on demand and supply in the music sector and draws their implications for copyright policies and new business models. In a related work, Waldfogel (2012a) compares the digital music transition to what happened to travel agencies facing Internet intermediaries such as Expedia or Orbitz. Traditional travel agencies have gradually disappeared, but one can hardly argue that this is a social loss, because Internet intermediaries are more cost efficient and can propose better and customized offers. It is therefore not socially optimal to prevent this digital transition, because new digital distribution channels allow consumers to listen to music in new ways but also because digital music increases the long tail of electronic commerce by increasing the variety of titles available online. While the pre-recorded music industry was the first to experience the challenges of digital distribution, other cultural industries such as the movies and book industries are currently facing the same challenges.

In this chapter, I will review how the transition to digital music is affecting both demand and supply. On the demand side, new ways of consuming digital music (streaming, download à la carte, cloud, transfer and synchronization with smartphones) have attracted new customers, therefore expanding demand. In addition, digital distribution has extended the long tail of music sales, increasing variety and demand at the same time. However, digital piracy can reduce demand, most likely for best-selling albums and titles.[1] On the supply side, digital music has two distinct effects on producers of music (labels and/or artists) and retailers including online intermediaries. First, for labels and artists, both marginal and fixed costs of production of digital music have decreased with new affordable home studios, as well as online storage and cloud services. This can benefit artists with less media exposure or a smaller potential audience and thus favor entry into the market. Secondly, retailers and online intermediaries can benefit from two types of positive informational externalities associated with digital music distribution. On the one hand, retailers, such as Amazon, can benefit from comments and recommendations posted by users on their online platform, which generates additional sales to customers with similar taste and increase the knowledge on their customers. On the other hand, some online platforms act as intermediaries between consumers and advertisers. They can exploit indirect positive externalities between different sides of the market. Indeed, advertisers are looking for a platform where there are many active consumers, while consumers benefit from a large number of advertisers when they search for products and services. This is especially true for online streaming services that offer free music financed by ads. Weyl (2010) discusses multi-sided platforms. A well-known result from this literature is that the platform in two-sided markets should subsidize the side of the market that is most important for the development of the platform, by supplying it for free. Indeed, digital music distribution platforms, such as YouTube, Dailymotion, Deezer or Spotify, offer music for free to users and charge advertisers. Moreover, there are externalities of digital music on producers of equipment (for instance Apple) and operating systems (Google Android, Apple iOS and Microsoft Windows phone), as they can supply music at low prices and generate profits with complementary products. These externalities together with economies of scale can result in the concentration of digital distribution among a small number of large retailers. Overall, digital music distribution is improving the negotiating power of retailers relative to producers of recorded music (Alain and Waelbroeck, 2006). This is especially true for online intermediaries, who can collect users' information and gain better knowledge of demand conditions.

Before starting the analysis, it is important to remember that the music industry is divided into several markets of various sizes, covered by different types of actors and professions. For instance, in the United Kingdom, the value of recorded music was £1112 million in 2011, compared to £1624 million for the estimated value of live concerts and to £448 million for PRS for Music gross collections (PRS for Music, 2012).

I will not analyze cross-market effects, such as positive externalities from digital music (especially free music from P2P networks or from free online streaming services, such as Spotify or Deezer) to live shows. I will concentrate on the market for recorded music and the challenges that digital distribution brings to this market.

The remainder of the chapter is organized as follows. First, I analyze how digital piracy and digital music distribution are changing the demand for recorded music. Next, I review new business models and give perspectives on how cloud services will challenge existing offers. I conclude the chapter by discussing the policy implications of digital music distribution.

DIGITAL PIRACY AND THE LONG TAIL OF DIGITAL MUSIC DISTRIBUTION

In this section, I analyze how illegal distribution of music is affecting the legal market for recorded music and how digital music is expanding the variety of products available to the consumer.

Digital Piracy

There are two opposing arguments to analyze the effect of digital piracy on the music industry. Peitz and Waelbroeck (2006a) and Belleflamme and Peitz (2012) review this literature. The majors defend the first argument: an album that is downloaded is not purchased. In other words, the MP3 files are perfect substitutes for legal sales. The other argument is based on the idea that music is a good that consumers need to 'taste' before they can determine its value. Thus, Internet downloads allow consumers to discover new talents. New artists can benefit from this exposure to enter the market and sell albums to people with high availability to pay for their music as well as tickets for their concerts (Duchêne and Waelbroeck, 2006). In a sense, digital copies can be a form of cheap advertising. Peitz and Waelbroeck (2006b) use a model where products are differentiated to establish conditions under which a company benefits from giving away free samples in the form of MP3 files. The main economic trade-off balances the negative effect of competition of free copies and the positive matching effect by which consumers discover which products suit their tastes best, increasing their willingness to pay for music.

New business models offer market solutions to exploit the segment of 'free' users who have a low willingness to pay for music, including pirates. Halmenschlager and Waelbroeck (2012) analyze a freemium model in which a firm sells a free version and a premium version to consumers who have the option to download copies from the Internet. The key parameter of the model is the restrictions imposed on the free version (compared to the premium version): increasing restrictions improves the conversion of free users to premium users but also diverts some consumers to digital piracy. We show that freemium is optimal when the cost of copying is small. New business models that address all segments of the demand, including the segment of consumers with a low willingness to pay for music, are promising and are developing very fast. I will present the main offers below.

This brings us to the empirical studies on the effect of P2P networks on music sales. This literature is reviewed by Dejean (2009) and Waldfogel (2012b). It is difficult to study illegal behavior such as downloading music files on P2P networks. Peitz and Waelbroeck (2004) used macro-economic data on the downloading of MP3s and estimate that illegal downloading was responsible for 7 percent of the decline in CD sales between 2000 and 2001. The magnitude of this effect has been confirmed by Zentner (2006), also using consumer data. It is safe to assume that this is an upper bound of the effect of digital piracy on sales, since these studies were carried out at the beginning of the file-sharing era, at the peak of Napster fame. The effect is probably lower now because the number of people downloading on a daily basis has remained constant for several years. What is more, new business models based on digital music are expanding very fast, and digital sales outperform CD sales in most industrial countries.

To fight online piracy, the French government created Hadopi (Haute autorité pour la diffusion des oeuvres et la protection des droits sur internet), an independent institution with the power to cut the connection of Internet users caught downloading copyrighted files from P2P networks (after several written warnings).[2] The effectiveness of this law is questionable. First, the law will not stop those who download a lot, because there are other ways to get music files for free, such as direct file transfers from portable hard drives, direct downloads, private networks and intranets. Moreover, many people who download are reluctant to pay for pre-recorded music and will not buy more CDs with strengthened protection and repression by the Hadopi law. They can turn to free alternatives such as streaming and web radios or substitute music with new forms of digital entertainment such as online social networks and online games. In the United States, the Record Industry of America Association (RIAA) started more than 20000 lawsuits (the state does not intervene directly, unlike the situation in France). Most were settled out of court. This did not stop CD sales from declining in the US or elsewhere in the world (see IFPI, 2013).

Long Tail Effects of E-Commerce

Communities and online stores, as well as TV channels, allow consumers to search for new talents and promote new and old releases. This role was previously exclusively reserved to record labels that use mass media such as TV and radio for promotion. This development can partly be attributed to the superiority of online communities over traditional stores for everything related to meta-information, or information on how new products match consumer tastes. CDs were traditionally sold in bricks-and-mortar stores with limited space on the shelves. New releases were mainly promoted on mass media such as radio and television. Since the cost of promoting and distributing a CD does not depend on the size of the potential audience, producers and retailers focused their marketing efforts on one or two artists rather than share their marketing campaign among many artists. This factor combined with bandwagon and network effects contributed to a situation of 'winner take all' in which a small number of superstars generated the vast majority of CD sales.

New technologies offer new ways for consumers to discover new songs and artists (through the development of social networks, file-sharing networks, Internet radio, podcast or streaming). Consumers also have new formats to listen to music (individual titles, albums or playlists on iTunes, subscription to streaming services such as Spotify or Deezer), and MP3 players in particular have increased consumer exposure to music.

Online communities are great tools for exploring new music and getting personal recommendations. They generate what some people have called the phenomenon of the 'long tail': a business model based on the exploitation of niche products hard to find in physical stores. These communities include file-sharing networks, against which the record companies went to court, as well as social networks like MySpace or Facebook. Thousands of musicians, amateurs or semi-professionals, regularly connect to these sites to try to generate some buzz. Thus, the promotion is done by the consumer or by the artists themselves. Some artists such as Radiohead have bypassed their record company to promote their album online. Only a decade ago, the promotion of new releases was almost exclusively done by record labels. In addition, many retail stores have opened online community spaces to their customers to enable them to read reviews posted by other users of the site and receive personalized recommendations. Amazon offers such community tools as well as an online marketplace that brings together professionals and individuals. Bounie et al. (2010) showed that the best-selling cultural products on these platforms are different from the best-selling products in physical stores: among all titles that entered the monthly top 100 list in the digital marketplace, only 30 percent were among the top 100 best-sellers offline. What is more, the demand in the online marketplace was less concentrated on the top sellers (having a longer and fatter tail) and less concentrated on new releases.

This evolution of the recommendation and promotion system has an impact on online retailers: the entry of new players aggregating content and offering music playlists customized by users themselves. It is a value proposition very different from that of radio or TV, since each user can have his or her own program and prescriptions are largely controlled by users. Firms using these new forms of collective promotions are new and entered the market less than five years ago.

FOCUS: DIGITAL MUSIC IN FRANCE

This section draws on an expert report for the French Hadopi on how profits are shared in the digital music sector (Waelbroeck et al., 2011). The report was based on interviews with main actors of the digital music sector in France and matched their accounts with data from business intelligence companies.

In 2005, the legal market for digital music was characterized as follows: (1) lack of legal offers to compete with online P2P networks; (2) top sellers in physical stores were top sellers online, and sales were concentrated on new commercial releases; (3) online retailers were not exploiting the potential of the Internet to tailor product lines to consumer needs; and (4) the long tail phenomenon was more a theory than an economic reality.

In the meantime, we have seen a gradual increase in online product variety: (1) between 2008 and 2010 the number of titles or albums sold through online platforms went from 2 million to 4 million; and (2) some online platforms such as Amazon.fr achieve 66 percent of their turnover with titles that are not in the top 1000 best-selling list, using a mix of advertising campaigns and low prices to attract new buyers.

However, there are 10 million titles (out of 14 million available on legal online platforms) that are still not generating revenues. The main reason is the following: the Internet is not yet a good way to trigger interest in new music. On the radio, playlists with a limited number of songs often triggered a purchase through repetition and memorization. This mechanism has no equivalent on the web yet. Most industry experts consider on-demand streaming services as tools for 'trying' music, but the potential of these services to turn repetition into sales is still limited. There is a need to develop new online tools and the know-how to improve the strength of Internet prescription and to increase the variety of products sold to consumers.

Online sales of digital music are indeed less dominated by the top 50 best-selling titles or albums than physical sales, but some titles are streamed only a couple of times. This is very costly for collecting societies, such as SACEM in France, which have to redistribute a few cents per title to copyright owners. This is why SACEM refers to the long tail as the 'long nightmare'.

DIGITAL DISTRIBUTION AND THE SUPPLY OF DIGITAL MUSIC

Business Models for Digital Music

While in 2005 the legal offers were limited to two digital music formats (single and album) with a unique price, online music services now combine access to content with innovative services (playlists, recommendations, mobility, metadata and so on). This creates value for users. There has been a proliferation of offers, covering a broad range of prices (for titles and albums) and subscription fees (from free to €9.99 per month). Further product differentiation is based on: the quality and comfort of listening (with or without advertising); repeated listening to the same title or album; degrees of flexibility with listening on demand, interactive or semi-linear streaming; and portability, with playlists being transferable to smartphone.

With flexible prices and a large catalog of albums in digital format, digital albums sales have doubled between 2008 and 2010. Music fans who cannot or do not wish to pay for music can access a wide range of free legal services financed by ads, giving record labels the opportunity to monetize large web audiences. The development of price discrimination strategies has led to the creation of a virtuous circle of value creation where the customer is central. This is a fundamental transformation of the music market that requires all players to change their business models. This section describes how new business models try to segment demands for digital music.

Mobile Customization

This market corresponds to the marketing of music clips for ringtones and logos on mobile phones. This market is declining with the gradual introduction of smartphones.

A la Carte Downloads (iTunes)

In France, à la carte downloads accounted for 54 percent of digital revenues in 2010, and grew at a rate of 20 percent in 2010 compared to 2009. In addition, there is a changing balance between sales of singles and albums on digital platforms. Album sales have driven growth of paid downloads over recent years, increasing by 46 percent in 2009 compared to 2008 and by 38 percent in 2010 compared to 2009. The share of album sales in trade value increased from 38 percent in 2008 to 51 percent in 2010. Digital downloads have also increased elsewhere, and digital sales (including subscriptions) are now greater than physical sales in the top five markets for music (IFPI, 2012).

Subscriptions

Subscription services offer either unlimited listening per month or a pre-determined number of downloads for a monthly fee. In France, streaming services with unlimited listening represent the majority of the offers. Monthly subscriptions to services that offer unlimited downloads are no longer on offer. Streaming services, such as Spotify or Deezer, are generally available in three types of subscriptions: with access from a fixed PC only; with access from both a PC and a mobile phone; or bundled with a telecommunications service, such as a mobile plan or a fixed broadband plan.

Online streaming services are often based on a 'freemium' model where a free service is available with a number of restrictions. The conversion of free users to paying subscribers is made possible by adding a dimension of 'mobility' to the basic offer, which allows premium users to access their playlists anytime on their smartphone. Another lever to convert free users to paying subscribers is to improve the quality of music streams to premium subscribers (relative to the sound quality available to free users) or to give them access to more titles from the catalog. Finally, the cost to the user can be partially supported by an industrial partner, for example 'bundled' to a subscription to a mobile operator. In the case of the partnership between Orange and Deezer, Deezer is activated on the 'Origami' mobile subscription. Orange pays a monthly fee per subscriber to Deezer, even if the user has not activated this service.[3] At the end of this period, if the user wants to continue to use Deezer, he or she has to pay €5.00 per month instead of the regular € 9.99 Deezer price. This has an effect on churning (the share of customers lost to competitors), since Orange customers who want to switch to another Internet service provider have to pay an opportunity cost. Some Origami subscriptions even include Deezer premium in the regular subscription fee at no additional cost.

Bundling and other forms of versioning are a promising way to reach new consumers. Globally, subscription services are also the fastest growing segment of digital sales (IFPI, 2012).

Free Services

Web radios supply programs from simultaneously broadcasted radio. Most players in this segment are traditional radio stations. According to Médiamétrie, interviewed in Waelbroeck et al. (2011), the average monthly audience of these services in 2010 was 9.4 million unique visitors in France.[4] If a person wants more flexibility to listen to a radio program (for instance time-shifting), he or she can turn to a smartradio.

Smartradios are generally run by 'pure players' of the Internet that offer semi-interactive programs for each listener (the Digital Millennium Copyright Act defines smartradios in the US). Some smartradios offer programs designed by algorithms that take into account user preferences (for example Pandora in the US uses algorithms to break down a song into components such as rhythm and harmonic schemes and rates its similarity to songs a user enjoys). Other smartradios offer customized programs based on a user's activity on online social networks (for example, lastfm).

On-demand streaming allows users even more flexibility in the way they listen to music. Users of these free streaming services, such as the free versions of Spotify or Deezer, can access a large catalog of music, but can listen to music only for a limited number of hours (typically five hours) per month or to a specific title a limited number of times (typically five times). Users of these services can sometimes share playlists with other members of their community. It is also likely that these systems may gradually become tools used by music producers and online aggregators to better target users with personal recommendations.

Video-sharing sites like YouTube and Dailymotion are also increasingly used for listening to music on demand, but are generally not subject to restrictions, as music songs are mainly provided by users themselves or correspond to official videos paid for by advertising.

Music in the Cloud

New services based on cloud services are becoming available in the US and in Europe. Most offers are based on a freemium model: the user can store and stream content for free on a desktop but has to pay in order to play the content on a smartphone. I will only review cloud services where the user can upload his or her own content. Services such as Rdio are similar to Spotify.

Cloud services raise both piracy and privacy concerns. First, cloud storage allows users not only to upload files, but also to share some files with friends or other users. Since the capacity of these services is limited only by the storage capacities of servers, we might expect offers of several hundreds of GB quite soon. Secondly, there are privacy concerns, as firms operating cloud services can exploit personal data. At the same time, this is another component of value for these firms.

Music cloud services usually exist for desktops and for smartphones. Given our previous discussion of the value of streaming services on portable devices such as smartphones, the key component of the value chain lies in the ability to become a major supplier of these types of devices. It is therefore not surprising that, at this stage of early development of the technology, most cloud offers are restricted to a single technical platform (Armstrong, 2006). For example, Google Music can stream only to Android devices. Amazon Cloud Player can both store and stream content for an annual fee; content can be played on Android devices. Finally, Apple's upcoming iCloud service syncs to iTunes and is compatible only with devices running Apple's operating system. At the time of writing, the iCloud service did not yet stream to mobile devices.

THE IMPACT OF DIGITAL DISTRIBUTION TECHNOLOGIES ON ARTISTS AND RECORD LABELS

Digital technologies can reduce the costs of creating music. Indeed, a large number of recording artists and small record labels exploit opportunities to cheaply produce multimedia works. On the one hand, the home studio allows recording artists to better prepare studio recording, which reduces recording costs. On the other hand, these tools allow artists who have signed with small record labels to produce a greater diversity of genres and styles and to address additional or new micro-markets. Digital tools have also significantly helped to reduce the production costs of music video clips.

Digital distribution brings new forms of collaboration between artists, producers and music fans. First, online technologies bring artists closer to their fans. Artists, producers and labels use social media such as Twitter or Facebook to reach their fans, who generate cheap word-of-mouth promotion. Artists can also directly sell merchandising to Twitter followers.

This trend requires creators to produce some audiovisual content to capture the attention of the public via websites. These new practices also require editorial know-how on 'customer relations'. Furthermore, these new tools have enabled the emergence of new sources of funding (crowd-funding on mymajorcompany.com for instance) and give the opportunity to create a database of fans around new projects of artists. These tools can also trigger 'buzz': the advice of friends counts often more than signals from traditional media. However, exploiting this situation requires the implementation of new knowledge about the management of fan communities. Scouting and crowd-sourcing tools such as the French Noomiz can help the 'discovery' and financing of new genres and new artists. This evolution also favors the emergence of new players or digital labels, which can more effectively apply this type of expertise.

Actors in the independent music scene[5] interviewed in Waelbroeck et al. (2011) were rather positive about the new business opportunities brought by digital services. The hope is that they can reach a larger audience than in physical markets. Premium streaming subscriptions seem to attract an audience of people very engaged with music, who consume a greater diversity of repertoire than mainstream or 'free' consumers. Their success may promote the market share of 'indies'.

Independent suppliers of music see download platforms as new ways to develop business opportunities in music markets that have been hard to enter through traditional physical channels. Digital distribution platforms have allowed numerous independent artists (who often have no international distribution offline) to sell outside of their home country. All players develop strategies to monetize audiences in emerging markets (in Asia and Africa, for example) where physical distribution does not exist. Emerging countries have become an engine of growth for many artists (Waelbroeck et al., 2011).

CONCLUSIONS AND POLICY IMPLICATIONS

The music industry is witnessing a shift from a one-size-fits-all product (the CD) to products and services designed to address all segments of demand for music. Consumers can listen to digital music by using a free service like YouTube or Dailymotion, by paying a monthly fee to a streaming music service such as Spotify or Deezer, or by downloading music files from iTunes. They can also choose whether they want to listen to a single title, an album or a playlist. The versioning of digital music has led to significant growth of digital sales. However, there are still many challenges facing digital music. First, free online services monetize their audiences rather than content and are funded by advertising. The two main types of advertising formats are 'display', where one or more graphic banners appear on the current page of the browser while listening, and 'in-stream audio', where an advertisement is played between two audio songs. Streaming video sites like YouTube or Dailymotion earn money by selling both 'display' and 'in-stream video' ads, similar to 'ad clips' on TV. While the video advertising market is growing rapidly, there are uncertainties about the long-term development of prices of display advertising. 'In-stream audio' is still at an early stage and suffers from the lack of certified tools to measure audiences.

Secondly, promoting artistic projects requires that new artists are visible in an online environment where there is already a deficit of attention. There is a multiplicity of formats available on a growing number of media. Refining customer segmentation implies complex pricing policies, as suppliers seek to attract consumers through a growing number of intermediaries. But prices of digital music are constrained by competing formats. All these factors require platforms to deliver digital contents in sophisticated ways. One of the greatest promises in the market for music is probably that the data gathered by major retailers of digital music will help to predict the notoriously fickle demand for music somewhat better.

NOTES

1. See, for instance, the study of Rob and Waldfogel (2006), which analyzes how college students acquire best-selling CDs. There is evidence that less well-known artists benefit from the exposure on P2P networks. See Peitz and Waelbroeck (2006b) on this point.
2. Only P2P networks are monitored, not direct download websites such as the defunct Megaupload, now Mega.
3. To compensate, the monthly fee paid by Orange is lower than the value of a subscription paid by a

premium Deezer subscriber. In this way, the larger corporation (Orange) takes on some of the risk of the smaller partner (Deezer).
4. A unique visitor is usually identified by a cookie that combines the computer ID and the active browser.
5. Independent labels and musicians are defined as those who are not owned or under contract with one of the four major record labels.

REFERENCES

Alain, M.L. and P. Waelbroeck (2006), 'Music variety and retail concentration', in P. Waelbroeck (ed.), *Proceedings of the 2006 Telecom Paris conference on the economics of ICT*, Paris: Telecom Paris.
Armstrong, M. (2006), 'Competition in two-sided markets', *Rand Journal of Economics*, **37**, 325–66.
Belleflamme, P. and M. Peitz (2012), 'Digital piracy: theory', in M. Peitz and J. Waldfogel (eds), *The Oxford Handbook of the Digital Economy*, Oxford: Oxford University Press.
Bounie, D., B. Eang and P. Waelbroeck (2010), 'Les plateformes de vente en ligne: une opportunité pour les industries culturelles?', *Revue économique*, **62** (1), 101–12.
Dejean, S. (2009), 'What can we learn from empirical studies about piracy?', *CESifo Economic Studies*, **55** (2), 326–52.
Duchêne, A. and Waelbroeck, P. (2006), 'The legal and technological battle in the music industry: information-push versus information-pull technologies', *International Review of Law and Economics*, **26** (4), 565–80.
Halmenschlager, C. and P. Waelbroeck (2012), 'Fighting free with free: streaming vs. piracy', mimeo.
IFPI (International Federation of the Phonographic Industry) (2012), *Digital Music Report 2012*, available at http://www.ifpi.org/content/library/DMR2012.pdf (accessed 21 April 2013).
IFPI (International Federation of the Phonographic Industry) (2013), *Recording Industry in Numbers*, available at http://www.ifpi.org/content/section_resources/rin/rin.html (accessed 21 April 2013).
Peitz, M. and P. Waelbroeck (2004), 'The effect of Internet piracy on music sales: cross-section evidence', *Review of Economic Research on Copyright Issues*, **1** (2), 71–9.
Peitz, M. and P. Waelbroeck (2006a), 'Piracy of digital products: a critical review of the theoretical literature', *Information Economics and Policy*, **18** (4), 449–76.
Peitz, M. and P. Waelbroeck (2006b), 'Why the music industry may gain from free downloading – the role of sampling', *International Journal of Industrial Organization*, **24** (5), 907–13.
PRS for Music (2012), *Adding Up the UK Music Industry 2011*, available at: http://www.prsformusic.com/aboutus/corporateresources/reportsandpublications/addinguptheindustry2011/Documents/Economic%20Insight%2011%20Dec.pdf (accessed 21 April 2013).
Rob, R. and J. Waldfogel (2006), 'Piracy on the high C's: music downloading, sales displacement, and social welfare in a sample of college students', *Journal of Law and Economics*, **49** (1), 29–62.
Waelbroeck, P., P. Astor and C. Waigner (2011), 'Rapport "Engagement 8 – Partage des données relatives à l'économie du secteur et état actuel du partage de la valeur"', available at: http://www.hadopi.fr/sites/default/files/page/pdf/Rapport_Engagement8.pdf (accessed 21 April 2013).
Waldfogel, J. (2012a), 'Copyright research in the digital age: moving from piracy to the supply of new products', *American Economic Review*, **102** (3), 337–42.
Waldfogel, J. (2012b), 'Digital piracy: empirics', in M. Peitz and J. Waldfogel (eds), *The Oxford Handbook of the Digital Economy*, Oxford: Oxford University Press.
Weyl, G.E. (2010), 'A price theory of multi-sided platforms', *American Economic Review*, **100** (4), 1642–72.
Zentner, A. (2006), 'Measuring the effect of file sharing on music purchases', *Journal of Law and Economics*, **49** (1), 63–90.

FURTHER READING

Belleflamme and Peitz (2012) and Waldfogel (2012b) provide recent surveys on digital piracy. The Digital Music Report of the International Federation of the Phonographic Industry (IFPI, 2012) discusses new business models in the market for recorded music.

35. Film

Paul Stepan

The effects of digitization on the film sector are massive, and we are only just beginning to understand the extent to which changes have taken place and will take place in the years to come. There are many aspects to the topic of film and digitization that cannot be discussed in depth or even hinted at in a single chapter. First, there is a range of different film industries in various parts of the world that differ in organization, financial volume, culture and customs. Writing about 'the' film industry is often equated with analysing Hollywood and the North American audiovisual sector but neglects the Indian film industry (Bollywood) and the less well-known but for the African continent very important video industry in Nigeria (Nollywood), whose history is very closely connected with the relatively cheap yet professional-standard equipment for video production. In 2010 Nollywood produced more titles than any other national film industry in the world. It is an industry that caters for the African market and owing to its peculiarities in distribution is largely unconnected to the global film industry (Lobato, 2010). Furthermore there is a very vibrant Japanese film industry and another one in South Korea, one in South America and, of course, the one producing the most feature films for theatrical use every year, the European film industry. According to the Focus report (Marché du Film, 2010), there were almost twice as many feature films produced in the European Union (EU) as in the US. All these industries work differently regarding the organization, role of the producer, investment, copyright, neighbouring rights and so on. Assessing the impact of digitization, apart from technical issues, on the production side of the film industries is consequently very difficult and needs to be narrowed down to a specific case.

The relatively easy accessibility of digital equipment has increased diversity in film productions over and above geocultural differences. While TV series such as *Mad Men* are produced on 35-mm film, a technology mostly used for cinema films, many movies for theatrical use are filmed with digital equipment. In some productions the use of digital cameras is a creative decision, while in other productions it is due to budget constraints. The same thing happened in the 1960s when famous Western movies such as *A Fistful of Dollars*, *The Good, the Bad and the Ugly* and *Once Upon a Time in the West* were produced in Techniscope instead of the state-of-the-art format Cinemascope, which was too expensive. Many of the world-famous close-up scenes result from the technical limitations rather than from a creative decision. Nevertheless the technology also allowed for cheaper productions in the same way as digitization does nowadays, for example in Nollywood and many other parts of the world. Small cameras and digital equipment have often been used to create a private atmosphere or to emulate the authenticity of a true story. Digitization in the film industries enables both new creative possibilities and lower financial entry barriers for movie production.

Another angle from which to look at digitization and the film industry is to focus not on production but on distribution and changes on the business side and to business models. How are movies provided for the Internet? How do consumption, authorized and unauthorized distribution and the supply in general change? To answer these questions it is legitimate to write about 'the' film industry, since in principle it has much less to do with country-specific industries and policies or financial restrictions than with digital copying and the possibilities and risks deriving from digital technologies and the Internet. Of course, regional differences do also play an important role in the current distribution system, but this is one of the main challenges to overcome in order to adjust to digital business models, as I explain below.

Once a master copy is produced and available in any digital format the process of distribution is in principle the same for all movies. This chapter deals with changes in the distribution of movies, legal and illegal, business models, voluntary sharing and changes in theatrical distribution.

CINEMA, TV AND VIDEO

Film distribution consists mainly of three areas: theatrical distribution; broadcasting including pay TV; and video distribution, which includes an end consumer market, video on demand and video rental. In the past decade video distribution has been adopted for online applications. The three distribution channels are often sequenced in chronological order (especially in Europe), beginning with theatrical distribution, followed by selling and renting videos and finishing with broadcasting; video on demand and pay TV come in at different stages depending on regions and laws (see Figure 35.1).

> Sequential distribution patterns are determined by the principle of the second-best alternative – a corollary of the price-discriminating market-segmentation strategies . . . Films are normally first distributed to the market that generates the highest marginal revenue over the least amount

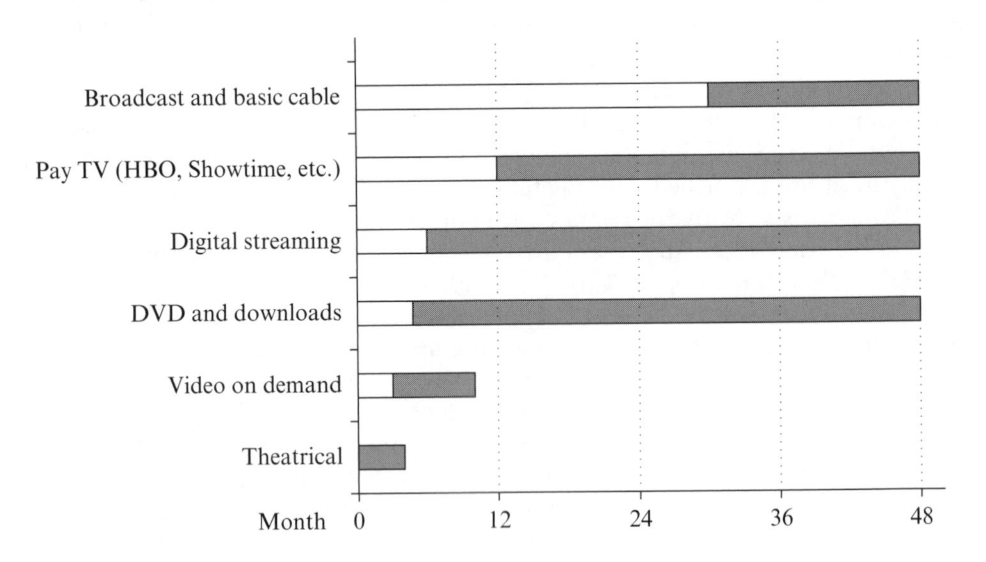

Source: Vogel, 2011: 127.

Figure 35.1 Typical market windows from release date, circa *2010*

> of time. They then 'cascade' in order of marginal-revenue contribution down to markets that return the lowest revenues per unit time. This has historically meant theatrical release, followed by licensing to pay cable program distributors, home video, television networks, and finally local television syndicators. (Vogel, 2011: 126)

In some countries – Austria, Greece, Germany, France, Luxembourg, Portugal and many others – in addition to the economically driven chronology there is also a legal retention period that regulates the time span between theatrical release and video release and between video release and broadcasting in order to protect cinema revenues and video distribution. These regulations are an additional hurdle for adjusting the distribution chronology to the requirements of the new technologies (Kuhr, 2008).

For most subsidized European movies and for successful commercial movies from all over the world, business plans include all the available channels, while less successful commercially produced movies are distributed only on a video market. In Europe almost all feature films are publicly subsidized to some extent and, if they are, theatrical screening is compulsory in most countries according to the subsidy guidelines or associated laws.

Compared to the music industry or book publishing, the film industry faces an additional problem, namely the difficulties deriving from global demand and the traditional regional markets. While CDs are usually released globally, movies are released and licensed country by country, which also affects theatrical distribution as well as broadcasting and DVD distribution. The territorial, rights-based business structure is challenged by the Internet (Vogel, 2011), which makes unauthorized copies available worldwide, while authorized copies are available only in certain countries.

Be it economic reasoning or legal obligations, chronological distribution is in question, since the Internet tends to synchronize geographically as well as between technologies. Both price discriminatory models and legal obligations rely on excludability and controllability, features that cannot be taken for granted on the Internet.

Cinema Distribution

The underlying established business models for theatrical distribution are still in place and working, though digitization has led to some adjustments. Cinemas are equivalent to live performances, and hence admission can be controlled and seating is scarce. Therefore selling tickets for admission works in the digital age in the same way as it did at the beginning of the twentieth century. Whether the unauthorized distribution of film copies has a direct influence on cinema attendance remains ambiguous. While Bounie et al. (2006) argue that file sharing has no direct impact on attendance, De Vany and Walls (2007) suggest a significant and strong negative correlation. However, there are two factors that have changed through digitization. First, the appearance of peer-to-peer technologies puts pressure on the timeline of the release strategies and, second, there is a drastic reduction of 'shipping' costs between distributor and movie theatres.

In analogue times the decision on the number of prints for release in a particular country was an important issue, since celluloid copying was expensive and sending physical copies around much more difficult. The distributor was responsible for overseeing worldwide shipping, retrieving copies, sending advertising materials and merchandise in advance, ensuring that the movies were played in the right cinema with the agreed technical equipment and negotiating the lease for the movie (Kill and Taylor, 2009). Some of these very complex tasks are made easier through digitization. In order to satisfy the demand in a region the numbers of copies needed had to be estimated, which of course was associated with a risk to recouping the outlay. Digital technologies open the possibility of satisfying any demand worldwide within hours. Even though the business models are not in place for this kind of flexible programming it is not restricted by technology any more. Having sufficient copies is no longer the bottleneck in the industry, and so global releases are much easier and overall cheaper than they used to be when expensive celluloid copies had to be shipped around the world. Since the number of copies and especially the various versions (dubbed, audio and subtitles) are not an issue any more, the optimization problem has become significantly easier and flexibility has increased. This might change the business plans of theatres in the future if they can pick and choose movies at any time without being limited by weekly changes of the programme (Husak, 2004).

The second event that amplified the trend towards global release strategies is competition launched through file sharing. Since the introduction of file sharing the distribution companies have lost control over global availability and access to their films. They are left with the control over legal distribution but not accessibility in general. Since very early on, it has been understood that one of the main motives for the illegal distribution of films is the lack of availability (German Federal Film Board, 2002). The Brennerstudie shows that, as early as 2002, as well as in subsequent annual reports, the main motive for unauthorized downloading in Germany was to watch movies as close to the world premiere as possible. Since the early days of illegal supply, technologies for copying and bandwidth have dramatically improved, and so very quickly, if not even before the global cinema release, unauthorized copies are available. Often these early copies are of very poor quality, if, for example, they were made with a video camera from a screen, but nevertheless they are available and hence they are a substitute of some sort for theatrical screening, and that has induced competition. This led the major studios to release the most expensive films in production and expected global blockbusters within a week worldwide (Culkin and Randle, 2003; Vogel, 2011). However, this is true for only a small number of films, while the vast majority are still released country by country.

TV Distribution

Similarly to theatrical distribution, broadcasting is merely indirectly affected by digitization. Again, the business models are still intact and working, both for pay TV, in advertising-based commercial TV's two-sided markets, and for the European case of semi-publicly financed broadcasting. Digitization, however, has an impact on TV shows that have been increasingly sold as DVDs in the past. If shows are live and the live element is important, as in talk shows, news and topical debates, Internet distribution is hardly a competitor and is more valuable as providing an archive. This is true especially for the broadcasting of sports events, interviews of all kinds, news in an audiovisual format and many other shows. Very different is the impact of the Internet on pre-recorded shows and especially TV series. File sharing and streaming portals test commercial broadcasting stations on the extent to which their programming meets the consumer's interests, since shows can be downloaded or streamed hours after their world premiere. In other words, 'Internet-based technology already provides viewers with unprecedented control over when and where entertainment may be enjoyed. Such technology has already appreciably lowered the price per view and further diffuses the economic power of the more traditional suppliers of programming' (Vogel, 2011: 99).

Since the mid-2000s high-quality TV shows featuring international superstars have boomed. New series such as *Mad Men*, *Boardwalk Empire*, *Downton Abbey*, *The Sopranos* or the fantasy series *Game of Thrones* are expensive, high-quality productions that differ in many respects from the relatively cheap sitcoms that dominated the previous decades. These series, mostly from American broadcasting companies such as HBO, AMC or NBC, are as popular in file sharing networks and streaming portals as on TV. *Game of Thrones*, an HBO series that can only be legally seen in countries where HBO offers its services and then only in combination with a particular subscription, very quickly evolved to be the most downloaded series of all time up to the time of writing. Nevertheless there seems to be a willingness to pay in other territories, as initiatives such as 'Take my money, HBO'[1] suggest, although there is no firm evidence. Again regional segmentation of a global demand makes unauthorized copying attractive in many parts of the world. Although new business plans are evolving on iTunes, Netflix and many others, it seems that the industry has not yet found the right tool to compete in usability and speed with unauthorized copying.

Video and DVD Distribution

In business-to-business deals for cinema distribution and broadcasting the changes due to digitization have been relatively minor compared to the ones in the end consumer market, that of video distribution. Theatre admission and broadcasting are still based on the traditional models, while the Internet can often provide only a weak substitute. By contrast, for video distribution unauthorized file sharing and streaming often are a very close substitute or can even provide consumers with added value. The principal problem is the transformation of content bound to a carrier into an intangible product that is essentially a true public good, while in the analogue distribution chain it was not film, music or literature that was sold but the physical carriers attached to the content, that is, DVDs, CDs and books. These carriers were scarce private goods and are subject to the market mechanisms of private goods. What digitization achieved was the separation of content and carrier, which turned content, as Kenneth Arrow had already predicted in 1962, into a public good (Arrow, 1962). Non-rivalry in consumption is not new, since there has been private copying before, but the non-excludability on a global scale led to the loss of control by the seller. As already stated above, the industry is still in charge of the legal distribution, but illegal distribution now appears as a competitor (Varian, 2005), and in the case of video distribution there is direct competition between legal and illegal supply. How important usability is in this respect was shown in a study by Danaher et al. (2010). Their analysis focused on a controversy between NBC and iTunes: NBC asked for a higher remuneration for its content provided on iTunes, which was rejected. NBC set a cut-off date for iTunes to reconsider and, after unsuccessful negotiations, took down its whole catalogue, mostly major TV series. The study followed the illegal traffic for these NBC series two weeks prior to and two weeks after the cut-off date and showed that legal demand was to a great extent replaced by illegal demand. The authors concluded that a lack of legal supply results immediately in an increase in illegal demand. This shows two very important issues. First, online customers would rather go from legal to illegal than from online purchasing back to a store. Second, there is, or at least was, a willingness to pay even if there is an unauthorized free alternative.

Although there are many similarities between the film and music industries in their approach to coping with file sharing, there is one notable difference. The use of digital rights management (DRM) has been abandoned to a large extent in the music industry after a series of trials, but not in the film industry.

Even though the film on a DVD was digital, the distribution of DVDs was in analogue form. As explained above, films were released not globally but territory by territory, starting with theatre exhibition and followed by video and TV, and theatrical distribution moved from one country to another, since the celluloid copies were costly and management of the logistics of shipping copies around was a difficult task. Through the scheme of sequential market distribution where films are first released on the market with the highest marginal revenue, the country-by-country chronology was kept intact (Vogel, 2011). In addition in some countries it is even a legal obligation to wait with video distribution for a certain retention period. Out of this chronology of events video distribution, like theatrical distribution, is regional, and accordingly the film industry introduced regional codes for DVDs, which restricted customers from playing videos in other regions. There are six different regional codes, which determine the areas where the movies can be screened, and in addition there is one for all territories, one for future use of DVDs and media copies and one for aircraft and cruise ships. The hardware, the DVD player, also corresponds to these regional codes, so that it should be impossible to play a European copy in Australia or to buy a copy through an online seller in the USA if the film is not yet available somewhere else. Thus regional codes segregate the world, a restriction that has no significance for unauthorized copies on the World Wide Web.

However, the traditional chronology of releases caused problems beyond those experienced by the music industry. As already mentioned above it is known that the main motivation for illegal downloading of movies derives from a lack of a legal supply between the world premiere and the regional release. DRM and technological protection measures (TPM) cannot control supply any more. The same structures go for online services such as iTunes, Netflix or Amazon. iTunes does not offer its services for film in many European and other countries. To be able to buy movies legally online an IP address and a credit card registered in a particular country are requested. In many countries in the world there is no legal online supply for movies at all and, even if private copying is legal in principle, bypassing DRM is not. Therefore there is no legal way for movies to be seen on tablets or other mobile devices. In many countries around the world the only legal way of watching a movie privately is to play a DVD on a DVD player. For all the hardware devices of recent years such as tablets, movie players, mobile phones, notebooks without a DVD drive and so on there is no legal means of viewing a DVD.

The Internet triggered two opposing trends: first, the legal possibility of tracking the use of every single legal licence; and, second, the impossibility of controlling unauthorized copying and streaming. Therefore to analyse the effects of digitization on film distribution, two very different scenarios need to be discussed. In the changes in legal distribution one can pin down two important developments. First is the use of DRM and

TPM. Almost every single DVD is nowadays legally protected from private copying not only through regional coding but also through DRM and TPM. The technical protection is very weak and can be bypassed with many programmes that do not require any advanced knowledge. Regardless of the technological possibilities and the insignificance of the technological hurdle, it is illegal to copy or rip a protected DVD. This implies that, when buying a DVD, one does not buy the right to privately use the film for one's own convenience but only for a very particular and restricted use, namely to play the film on a DVD player.

This practice of protection leads to the second noteworthy change in legal film distribution for end consumers. It is the change in the product itself. If one buys a DVD or a download on one of the platforms, the product bought is merely a licence to access that product. Various limitations to use leave the customer with a licence to view the movie in a particular way and, if downloaded, the licence can even be revoked, as in the case of Amazon's Kindle, iTunes and many other services that associate the use of the content with a piece of up-to-date software and hardware. These problems are increasing with cloud computing, since customers are granted access to film libraries, but they can easily be cut off by the company providing the cloud. Even the change of an operating system on a computer can be the end of access to one's playlists and libraries, as in the case of iTunes when migrating, say, to Linux. There is no iTunes application for Linux, and hence customers of iTunes lose all their legally bought access when choosing a different operating system. Translated to the analogue world it means that one would have to buy every book in the library again every time one buys a new shelf or replacing all LPs when buying a new record player of a different brand. Therefore the value of the licences is different from buying content on an unprotected carrier.

The change to the products supplied is still ongoing. It is very difficult to predict in which direction the trends will go, but it is of great importance to have a close look at what the commodity sold turns into. A video on VHS was comparatively easy to define, since once it was sold the element of control through the distributor had ended, not even mentioning the right of first sale, another characteristic that decreases the value of digital licences (Hinkes, 2007; Rotstein et al., 2010).

At the same time illegal distribution took a very different turn. An illegal copy has no limitations to its use, concerning either regional coding or DRM and TPM. Files can be stored and played on any device regardless of the hardware company or the operating systems. One can say that in terms of usability illegal copies beat the legal ones. This is especially remarkable as ripping movies adds consumers' value, which means not only that illegal copies are a substitute for legal copies but that they are improved versions. Varian (2005) states that the new technology needs to be treated as a competitor. Consequently the discussions on how legal distribution can compete with file sharing headed in the direction of adding value. Fourteen years after Napster, the first global download client, it appears that illegal copies are still more valuable for consumers than legal ones, which puts the business strategies in perspective.

Simultaneously, there is a development in cloud services and film streaming that is closer to a rental market than to purchasing movies. Currently, there is both legal and illegal supply in this area. In the music sector legal streaming services such as Spotify are increasing in importance. For a monthly fee any piece of music from the supplier's catalogue can be streamed at any time and with unlimited repetition. While such services begin to appear in some countries, for example Netflix, Hulu or Amazon Prime, there is no global provision yet, especially not for film.

NEW BUSINESS MODELS

The film industry tends towards vertical integration, which historically has been an issue in the US antitrust Paramount case of 1948, when the Supreme Court ruled against the major studios and separated the movie theatres from the production studios, thereby ending the so-called Hollywood studio system. Although the effects of the subsequent disintegration were mixed for the public – less danger of abuse of power and a better bargaining position for competitors on the one hand, but higher prices on the other (Whitney, 1955) – the urge to vertical control has persisted while the consent decree is still in place. However, the bargaining power of new retailers towards the right holders of the big catalogues is very limited, which gives them a great deal of influence on future business models.

New business models not only for the film industry but for digital distribution in general have been called for and suggested even before peer-to-peer file sharing was invented. Takeyama (1994) emphasized the importance of network externalities in the presence of widespread copying technologies, arguing that the harm done through unauthorized copying is overstated if demand network externalities are in place. Varian (2005) made the point that a new technology should be treated as a new competitor entering the market and triggering competition. If there is a difference in price between legal and unauthorized copies there must be something else that makes the product attractive for consumers. Shapiro and Varian (1998) suggested that versioning and bundling information goods would add value to online products. On the same lines Bakos and Brynjolfsson (1999) found that bundling large numbers of information goods and selling them for a fixed price result in an increase in profits in total as well as for a single item.

In spite of all the problems stated above and the suggested new business models and analysis of economists from very early on, little has changed in business models in the film industry since peer-to-peer file sharing was introduced. It is the regional markets and DRM that add to the problems in the film industry, but these are issues that cannot be solved through new business models.

Be that as it may, new business models in the film industry are hardly in place yet, so there is little evidence for evaluating them. The power granted to the copyright holders of back catalogues creates a sense of inertia that limits the flexibility of the film industry to adjust to the new technological environment, so that, 14 years after the introduction of file sharing, the discussion of new business models remains merely theoretical.

CONCLUSIONS

The music industry was the first industry to be hit by new technologies for a number of reasons. First there was hardly any language barrier, the application of copyright law is different for music than for film and much more suited to tackle a global market, and the size of the MP3 files was so small that sharing was easily feasible given the bandwidth and storage capacity in the early 2000s. Therefore the film industry had a relatively comfortable position and enough time to prepare for the refinement of the new technologies that could manage large audiovisual files just as easily as audio files.

Surprisingly this preparation did not take place, and instead the film industry visited all the stops that the music industry had passed by a couple of years earlier: suing users for downloading; DRM and TPM and therefore reducing the value of a legally purchased copy instead of, as Shapiro and Varian (1998) suggested, adding value in order to compete with unauthorized supply; and in general trying to persist with the analogue way of distribution. Consequently the biggest changes are still to come for theatrical screening as well as for TV and also for the various forms of the video market.

The biggest changes are likely to be a shift from regional releases towards global releases for all channels. This has already begun for theatrical screening for major box office hits and might be intensified in the near future. The second shift will most likely be towards synchronizing not only regions but also technologies, since it is not the supply in general that can be controlled but only the authorized part of it. This means rethinking the handed-down sequential distribution patterns and therefore the release chronology from theatres to video, video on demand and pay TV, to free TV.

NOTE

1. www.takemymoneyhbo.com (accessed 28 March 2013).

REFERENCES

Arrow, J. Kenneth (1962), 'Economic welfare and the allocation of resources for invention', in Universities–National Bureau (ed.), *The Rate and Direction of Inventive Activity*, Princeton, NJ: Princeton University Press, pp. 609–26.

Bakos, Yannis and Erik Brynjolfsson (1999), 'Bundling information goods: pricing, profits and efficiency', *Management Science*, **45** (12), 1613–30.

Bounie, David, M. Bourreau and P. Waelbroeck (2006), 'Piracy and the demand for films: analysis of piracy behavior in French universities', *RERCI*, **3** (2), 15–27.

Culkin, Nigel and Keith Randle (2003), 'Digital cinema: opportunities and challenges', *Convergence*, **9** (4), 79–98.

Danaher, Brett, Samita Dhanasobhon, Michael D. Smith and Rahul Telang (2010), 'Converting pirates without cannibalizing purchasers: the impact of digital distribution on physical sales and Internet piracy', *Marketing Science*, **29** (6), 1138–51.

De Vany, Arthur S. and W. David Walls (2007), 'Estimating the effects of movie piracy on box-office', *Revenue Industrial Organisation*, **30** (4), 291–301.

German Federal Film Board (2002), *Brennerstudie 1: Studie über das Kopieren und Downloaden von Spielfilmen*, Berlin: German Federal Film Board.

Hinkes, Eric Matthew (2007), 'Access controls in the digital era and the fair use/first sale doctrines', *Santa Clara Computer and High Tech Law Journal*, **23** (4), 685–726.

Husak, Walt (2004), 'Economic and other considerations for digital cinema', *Signal Processing: Image Communication*, **19** (9), 921–36.

Kill, Rebekka and Laura Taylor (2009), 'Cinema and film in the entertainment industry', in Stuart Moss (ed.), *The Entertainment Industry*, CABI Tourism Texts, Wallingford, Oxfordshire: CABI, pp. 78–94.

Kuhr, Martin (2008), 'Verwertungsfenster im Wandel Herausforderungen für die Chronologie audiovisueller Medien', Supplement to *Rechtliche Rundschau der europäischen audivisuellen Informationsstelle*, **14** (4), 1–8.

Lobato, Ramon (2010), 'Creative industries and informal economies: lessons from Nollywood', *International Journal of Cultural Studies*, **13** (4), 337–54.

Marché du Film (2010), *Focus 2010: World Film Market Trends*, Paris: Festival de Cannes.
Rotstein, Robert H., Emily F. Evitt and Matthew Williams (2010), 'The first sale doctrine in the digital age', *Intellectual Property and Technology Law Journal*, **22** (3), 23–8.
Shapiro, Carl and Hal Varian (1998), 'Versioning: the smart way to sell information', *Harvard Business Review*, **42** (6), 106–14.
Takeyama, Lisa (1994), 'The welfare implications of unauthorized reproduction of intellectual property in the presence of demand network externalities', *Journal of Industrial Economics*, **42** (2), 155–65.
Varian, Hal (2005), 'Copying and copyright', *Journal of Economic Perspectives*, **19** (2), 121–38.
Vogel, Harold L. (ed.) (2011), *Entertainment Industry Economics*, 8th edn, New York: Cambridge University Press.
Whitney, Simon N. (1955), 'Vertical disintegration in the motion picture industry', *American Economic Review, Papers and Proceedings of the Sixty-Seventh Annual Meeting of the American Economic Association*, **45** (2), 491–8.

FURTHER READING

Vogel (2011), Chapters 3, 4, 5, 7 and 8 give a good overview of film economics. Owing to regular updates, the 8th edition covers digital issues if need be.

36. Broadcasting

Glenn Withers

Media has been at the forefront of the impact of the changes wrought by digital technology. Schumpeter's 'gale of creative destruction' is perfectly well illustrated in this landscape. Its impact is greatest in print media, but electronic media have their own adjustments to make, and borders between media are increasingly convergent or blurred.

In this chapter the nature of digitization and its implications for broadcasting are examined. The consequent changes in the analysis of broadcasting, and the nature of policy changes required are also presented.

THE TECHNOLOGY TRAJECTORY

For radio and television, technological change has long been part of its nature. From the introduction of these media via airwave signals from terrestrial transmitters to receivers, there has been the emergence of satellite transmission and cable, which helped transcend the limitations of geography. Technological enhancements, such as movement from black and white to colour transmission within terrestrial analogue technology or the emergence of recording technologies such as sound and video recorders and players all the way through to iPods, have continued to buffet the sector.

Most important though is how digitization through compression has allowed each of airwave, satellite and cable analogue transmission platforms to be enhanced and transformed, diminishing frequency spectrum limits, allowing substitute or complementary service provision, and interactivity and customization, and facilitating user-pays charging arrangements all in real time.

To listeners and viewers the dominant media players in broadcasting have in the past been advertising-funded private broadcasters and taxpayer-funded public broadcasters. Their operations have changed substantially with the new technology. Direct user payments for access have instead grown and advertising has fallen, and the growth of an abundance of new outlets for information and entertainment has challenged the traditional justifications for public service provision of broadcasting services.

'Glocalization' has also transcended the national pre-eminence of more traditional broadcasting, and broadcasting in the digital age in many ways looks more like a conventional industry and less like the special case that has allowed it a quite distinctive status in national affairs. Nevertheless, while the previously distinctive spectrum access limitations of the broadcasting industry have diminished, continuing issues of market power, property rights protection and political and social implications of broadcasting activity persist even as their form changes.

Alongside the direct digitization of extant broadcasting transmission, the digitization of telecommunications including the Internet also is transforming the broadcasting and other media, as it blurs the boundaries for each. The constellation of change, from

Internet Protocol television through wireless fastweb and personal mobile transmission and new reception technology, to fibre to the home telecommunication and including the new social media, is now profoundly changing broadcasting too.

Some have seen these developments and the contemporary pace of ongoing change to be so great as to constitute a paradigm shift and not simply further evolution. The extent of changes in train as regards proliferation of technologies capable of carrying combinations of video, written and voice communication has been characterized as 'a portmanteau bursting at the seams as more and more activities are stuffed into it' (Inglis, 2006: 582).

Others also see a downside and speak of the risk of disruption and loss of service from malfunctions and deliberate denials of service by terrorists, hackers and others (including governments) under digital integration. This latter however has tended to be a specialist IT security issue rather than a concern of economists.

ANALOGUE BROADCASTING AND POLICIES

The basic economic propositions about broadcasting evolved for analogue broadcasting. They derived from the fact of a limited frequency spectrum that had to be allocated between competing uses, including many beyond broadcasting, such as for police and defence purposes. They also acknowledged the fact that broadcasting had important political and social implications for the government and culture of a country. The control of spectrum by government meant that regulatory power to make such allocation, often reflected in a high concentration in broadcasting media, also extended readily into public service requirements of the kind reflected in the role and responsibilities of public broadcasting.

With free-to-air terrestrial analogue technology for broadcasting, funding derived from programme sales wholesale to advertisers, for private broadcasters, and from tax revenues or licence payment equivalents, for public broadcasters. In some countries community subscription through voluntary payments for free-to-air broadcasting added a community element to the broadcasting scene, but 'free rider' behaviour meant that this approach rarely came to be a major component of the broadcasting revenue arrangements.

Monopoly power therefore was one policy concern resulting from the limited availability of frequency spectrum. Limited spectrum restricted competition, as did economies of scale and scope both vertically and horizontally in the broadcasting value chain. Problems of congestion or signal interference exacerbate the problems emerging from a limited frequency spectrum to reduce competition further.

The failure of markets to reflect direct consumer willingness to pay through user-pays pricing also meant that the markets were not only potentially non-competitive but distorted in their capacity to reflect consumer preferences. Some distinctive welfare economic analysis of this phenomenon emerged, beginning with Steiner (1952), which showed how programme differentiation and programme duplication were a prediction of limited frequency space combined with advertiser-funded revenue raising.

Externalities and intrinsic merit concerns in areas such as political, social and cultural information, sometimes referred to as being associated with the 'persuasive and pervasive' nature of broadcasting, provided further rationales for a policy response. These can be positive and negative, so that, just as there is concern over supporting public political participation through enhancing current affairs and educative programming, so there is concern over how undue violence in programme content may lead to an actual increase in violent behaviour.

These matters have been less well analysed by economics, since investigation must transcend the limitations of providing non-market valuations. Economics has been the science of markets, with markets providing a ready metric. Transactions beyond the market had less ready analysis and, even more, less convenient and widely accepted and understood measurement (Edwards and Withers, 2001; Snowball, 2008). Broadcasting was therefore not seen as a standard market best left to competitive market forces and to generic competition regulation to redress market failure concerns.

The typical policy response was instead to regulate spectrum allocation via a government authority, to apply specialized as well as generic competition regulation to broadcasters, to impose content regulation to enhance public goods in broadcasting, and to provide for tax-funded public broadcasting also to mute profit exploitation under monopoly and to promote public good provision through appropriate programming. Murdoch (2009) referred to this as rule by 'administrative fiat'.

However, the exact balance of policy response differed substantially across countries. Also, over time, developments such as the arrival of cable and satellite analogue broadcasting allowed some changes such as user-pays elements to emerge. Timing of such developments was also highly variable, especially because introduction of new technology often was itself subject to divergent policy regimes. The recent work by Acemoglu and Robinson (2013) begins to provide a framework that can integrate these considerations.

There has been an evolution over time in the form that the individual areas of policy response have taken. Ever since the classic analysis by Coase (1959), economists have in principle favoured auctions and trading as a preferred vehicle for the allocation of property rights in the frequency spectrum, rather than 'command and control' approaches. Such market mechanism approaches are now demonstrating their relevance in practice in broadcasting policy implementation, after long delay.

DIGITAL FUTURES

The new digital technologies have often been characterized as 'convergence'. In some ways this notion revives older media discussions across the print, music, film, information technology, telecommunications and broadcasting domains. Even early television was itself seen as a convergence of radio and film, and cable television is a blending of telecommunications and television.

There is also the issue in convergence of vertical as well as horizontal linkages. With a value chain from content creation through production to wholesale supply, retail transmission, reproduction, distribution and associated complementary production in everything from sporting event rights to equipment manufacture, the gaps in the chain of substitutes and complements are complex.

The phenomenon is therefore not new, but it is important even if capable of change and evolution. Management scholarship has looked at the implications for the operation of the firm itself. Coase (1937) pointed out that firms exist to minimize the transaction costs of co-ordinating economic activity beyond what individual transactions of buyers and sellers through the market could accomplish. A question emerging with digital convergence is whether the balance may be shifting back from firms to individuals, or is the role of the firm simply changing its form?

Certainly the adjustments for print media are the most obvious as newspapers go online and develop pay-walls as they lose advertising to the new Internet media. But the corporate adjustment made here is under further challenge as individual real-time peer-to-peer sharing of news and information grows in the dispersed and low-cost 'blogosphere'. How far residual respect for editorial and journalistic skills will sustain the company model, as opposed to sole trader and individual activity, is now being tested.

Television firms have not yet experienced such dramatic change as for print, but the capacity of individuals to make their own entertainments and news, and share and access these through the Internet social media, challenges the role of the television firm – while nevertheless creating new roles for content providers, for example sports organizations, telecommunications and Internet companies and the manufacturers of mobile, wireless devices.

BROADCASTING POLICIES

Industry and policy scholarship has looked more at the government's role and less at individual firm behaviour (Galperin, 2004). On spectrum management the move to market-like mechanisms facilitated by digitization has attracted attention, and the ascendance of the convergence idea has meant extension of that logic to examination of spectrum allocation too for cognate sectors.

Much research has focused on problems of transition. How does policy 'switch off' analogue terrestrial signals and allow migration to digital while maintaining universal access to television, for instance (Adda et al., 2005; Hayashi, 2006)? But post-transition the notion of a 'level playing field' has emerged as a holy grail in some regulatory circles, meaning spectrum allocation the same way, for the same price, for the same duration for radio and television broadcasting as it is allocated for telecommunications, telephony and broadband, as anticipated in Tadayoni (2001).

As digitization frees up spectrum, reallocation of its use this way is a first step. As indicated, much attention was given to how the transition is managed, particularly where there was analogue and digital overlap required. But with analogue switched off the move to markets becomes easier, though the task of changing to common conditions for spectrum that continues to be occupied by incumbent broadcasters is a tougher political ask.

Beyond spectrum allocation, regulation of diversity and competition and of content and standards has also come under challenge under digitization. Instead of dealing with these matters through the 'command and control' mechanism of conditions attaching to licensing, the convergence logic suggests that a wider common regulation is required for all content providers irrespective of licence-holding status dependent on the spectrum allocation only.

A notion of 'content service providers' has emerged that would cover all significant enterprises providing content in the public domain. The dilemma is that operationally such content providers would realistically need to be substantial and influential entities with corporate substance. But, as discussed, the nature of the firm participating in broadcasting activity may be changing and moving in a different direction. Incumbent broadcasters and large Internet and telecommunications firms may well be manageably incorporated in this way under common regulation. But many have not been regulated in this manner in the past and so will resist – and the growth of the individual peer-to-peer broadcasters and communicators provides a small business and individual sector that will be outside.

In the area of property rights there has been concern at how the digitization era with its cheap and unlimited copying capacity may allow enhanced access at the expense of the incentive for creativity (Scotchmer, 2004). Excludability has in some dimensions been enhanced, as with user-pays arrangements, but at the same time the capability to make perfect and multiple copies has also been much enhanced. Technology itself is seen as being a vehicle for providing some further answers, as under digital rights management regimes (Cowie and Kapur, 2006), as the law that protects content creators seeks new ways of enforcing copyright.

Given the property rights regime, and inside the regulatory net, the large firms are increasingly operating as commercial entities with growing media product portfolios seeking to overcome declining individual media returns and to spread costs and risks across the media market. In this, what was once a major divide across the Atlantic as between European and American approaches is itself becoming convergent in business and perhaps in policy, as forecast by Picard (1997).

While, in principle, convergence erases difference across platforms and this can be reinforced by new 'technologically neutral' regulatory regimes thus allowing new competition, issues of market power still remain and indeed may be enhanced. The concerns in this area over monopolization of content, bundling of distribution and denial of access do not disappear even if they take new forms. Just as industry policy makers were obliged in an earlier era to puzzle over the emergence of the conglomerate and what it meant for monopoly manufacturing, like issues have arisen for broadcasting policy in recent times (Cave, 1997).

The market economists' prescription here has been to complement the use of market instruments for spectrum allocation or orbital slots with other market-like instruments. Thus, in the cultural matter of local content, the economists' prescription has been to move from compulsion on such requirements as a share of programming to direct funding of such content for broadcasting, as long has been the practice in film.

In the case of concern over monopoly the prescription most offered by economists is to utilize the general competition regime applied *ex post*, this is to say, allow firms the incentive to innovate and compete, but to have sanctions to deter anti-competitive conduct – provide guidelines to assist interpretation and this will be reinforced by the accumulation of case law. Certainly, *ex ante* regulation of the industry structure has as much buttressed monopoly in practice as it has prevented it. Public broadcasting however can still complement general competition law as a benchmark competitor based on public funding and as a provider of public good services, especially as a not-for-profit editorial intermediary (Candel, 2010).

This latter editorial function is a litmus test of how broadcasting will change. Instead of editors choosing to draw to attention what is important, consumers may increasingly choose for themselves what they want their attention drawn to or indeed use the broadcasting media increasingly as a resource. To an educated and inquiring population this may be an advance for public information and knowledge. A move from news and entertainment to knowledge may be possible – but it all depends on the consumer in the new digital world. Certainly the old media played its part in bringing down the Berlin Wall, and the new media is testing the powers of modern authoritarian regimes. How pervasive such benefits are compared to the less high-minded developments such as the abuse of rights and privacy in the pursuit of meretricious consumer sensationalism seen in the *News of the World* scandals in the UK in 2011 remains to be determined.

REFERENCES

Acemoglu, D. and J. Robinson (2013), 'Economics versus politics: difficulties of policy advice', NBER Working Paper w18921, March.
Adda, J., M. Ottavani, G. Demange and E. Auriol (2005), 'The transition to digital television', *Economic Policy*, **20** (4), 161–209.
Brown, A. and R.J. Picard (eds) (2005), *Digital Terrestrial Television in Europe*, Mahwah, NJ: Lawrence Erlbaum.
Candel, R. (2010), 'Digitalising terrestrial broadcasting: public policy and public service issues', *Communication, Politics and Culture*, **43** (2), 99–117.
Cave, M. (1997), 'Regulating digital television in a converging world', *Telecommunications Policy*, **21**, 575–96.
Coase, R.H. (1937), 'The nature of the firm', *Economica*, **4** (16), 386–405.
Coase, R.H. (1959), 'The Federal Communications Commission', *Journal of Law and Economics*, **2**, 1–40.
Convergence Review (2010), *Final Report*, Canberra: Department of Broadband, Communications and the Digital Economy.
Cowie, C. and S. Kapur (2006), 'The management of digital rights in pay TV', in M. Cave and K. Nakamura (eds), *Digital Broadcasting: Policy and Practice in the Americas, Europe and Japan*, Cheltenham, UK and Northampton, MA, USA: Edward Elgar Publishing.
Doyle, G. (ed.) (2006), *The Economics of the Mass Media*, Cheltenham, UK and Northampton, MA, USA: Edward Elgar Publishing.
Edwards, L. and G. Withers (2001), 'The budget, the election and the voter', *Australian Social Monitor*, **4** (1), June, 9–14.
Galperin, H. (2004), *New Television, Old Politics: The Transition to Digital Television in the United States and Britain*, Oxford: Oxford University Press.
Hayashi, K. (2006), 'Legal and economic issues of digital terrestrial television (DTTV) from an industrial perspective', in M. Cave and K. Nakamura (2006), *Digital Broadcasting: Policy and Practice in the Americas, Europe and Japan*, Cheltenham, UK and Northampton, MA, USA: Edward Elgar Publishing, pp. 139–61.
Inglis, K. (2006), *Whose ABC? The Australian Broadcasting Corporation, 1983–2006*, Melbourne: Black.
Murdoch, J. (2009), 'The absence of trust', MacTaggart Lecture, Edinburgh International Television Festival, available at: http://image.guardian.co.uk/sys-files/Media/documents/2009/08/28/JamesMurdoch MacTaggartLecture.pdf.
Noll, R.G., M.J. Peck and J.J. McGowan (1973), *Economic Aspects of Television Regulation*, Washington, DC: Brookings Institution.
Picard, R.G. (1997), 'Comparative aspects of media economics and its development in Europe and in the USA', in J. Heinrich and G.D. Kopper (eds), *Media Economics in Europe*, Berlin: VISTAS, pp. 15–23.
Scotchmer, S. (2004), 'The political economy of intellectual property treaties"', *Journal of Law, Economics and Organisation*, **20** (2), 415–37.
Snowball, J.D. (2008), *Measuring the Value of Culture: Methodology and Examples in Cultural Economics*, Berlin: Springer.
Steiner, P.O. (1952), 'Program patterns and preferences and the workability of competition in radio broadcasting', *Quarterly Journal of Economics*, **66** (2), 194–223.
Tadayoni, R. (2001), 'Digital terrestrial broadcasting: innovation and development or a tragedy for incumbents?', *Communications and Strategies*, **42** (2), 89–129.

FURTHER READING

For traditional analogue broadcasting economics see Noll et al. (1973) and for digital broadcasting economics see Brown and Picard (2005). Doyle (2006) provides a compendium in this field. A thorough policy review of policy settings for the new digital age is the Convergence Review (2010).

37. Games and entertainment software

John Banks and Stuart Cunningham

Games and the broader interactive entertainment industry are the major 'born global/ born digital' creative industry. The videogame industry (formally referred to as interactive entertainment) is the economic sector that develops, markets and sells videogames to millions of people worldwide. There are over 11 countries with revenues of over $1 billion. This number was expected to grow 9.1 per cent annually to $48.9 in 2011 and $68 billion in 2012, making it the fastest-growing component of the international media sector (Scanlon, 2007; Caron, 2008).

A 2009 PricewaterhouseCoopers report details that, from 2005 to 2009, the global revenues of videogames grew annually at 16 per cent. This growth rate is more than five times faster than the total media and entertainment industries (PricewaterhouseCoopers, 2009, in De Prato et al., 2012: 222). The September 2011 ITU-T Technology Watch Report (Adolph, 2011) details how revenues of the videogames industry surpassed those of the US movie industry in 2005 and those of the US music industry in 2007. In the UK, the videogames industry had surpassed the music industry in terms of revenues by 2008. The report states that, 'According to Gartner research, the global videogame industry – software, hardware and online gaming – will grow from $74 billion (estimate for 2011) to $112 billion in sales by 2015' and that, 'In comparison, movie theaters around the world reported a combined total revenue of $31.8 billion in 2010' (Adolph, 2011: 17).

The videogames industry is characterized not just by impressive growth but also by much more rapid innovation cycles (for example platform and device innovations) than the more stately history of technology in film, TV and publishing. Videogames are an outstanding example of creative content and use driving technological innovation and take-up, not the other way round, as is usually understood by innovation research and business strategy. As Miles and Green (2008) argue, many of the innovation processes and sources that characterize creative industries such as games development are not captured by the traditional science-and-technology-driven R&D indicators that dominate the manufacturing and high-tech sectors. Miles and Green identify innovations in organizational forms and business models, including co-production approaches that involve consumers in the product and experience development process (Banks, 2013; Cunningham, 2013). The demand from producers and users alike for more engrossing visuals, sound and action has driven successive upgrades of processing speed. Consistent and burgeoning demand has assisted device makers during ICT downturns. Games are not simply restricted to the domain of entertainment but increasingly have wide application in many industries, including defence, health and education. The term 'gamification' (the use of game design and mechanics in non-game contexts) and the niche category of serious games each register this process and illustrate the embedding of interactive entertainment across the broader economy. And, despite the impressive growth rates, industry data probably have not fully captured significant recent industry transformations, with a shift from traditional dedicated gaming devices such as consoles to smartphones and tablets, from big game design companies to smaller studios and independent developers releasing their games online in app stores.

ACADEMIC TRADITIONS: CRITICAL POLITICAL ECONOMY

Much of the literature reported on in this chapter is from the field of media, cultural and communication studies. These are the disciplines in which the authors specialize. However, by reviewing these fields' use of political economy, and drawing on Schumpeterian and evolutionary (dynamic) approaches, including economic sociology, the chapter offers a complementary perspective to that of mainstream economics. This can be seen, for example, in the correlation between our observations and those of economists such as Shankar and Bayus (2002), whose analysis of network effects in gaming mirrors our emphasis on a firm's customer base being a critical strategic asset.

Academic traditions analysing the games industry can be grouped into two broad analytical approaches: critical political economy and rich variations on the cultures of use approach. This sharp appraisal of the videogames industry by Toby Miller crystallizes many of the themes in the critical political economy approach:

> The global gaming industry is essentially a rather banal repetition of Hollywood history: domination by firms that buy up or destroy small businesses and centralise power in the metropole; decimation of little bedroom concerns in favour of giant conglomerates; a working mythology of consumer power; and massive underwriting by the State . . . We need to follow the money, follow the labour, follow the high-tech trash. (Miller, 2008: 232)

Certainly, there are oligopolistic structures in the global games industry, as there tend to be in most highly capitalized industries. However, the degree of change in the dominant players, and the recent industry shake-up with the entry of new players, driven by platform changes and consumer take-up, might suggest a more porous power structure than seen in many other industries. Key features of the political economy approach which we will touch on include: critique of large conglomerates per se; critique of labour conditions in the industry; and critique of the notion of the 'digital sublime' that is considered to dominate much of the field of games studies. A further, non-US centric, political economy tradition also foregrounds the need to consider the issues faced by local production industries in the context of this global industry. Within this optic on the international system, critical considerations centred on globally dominant firms, their hegemonic control and the labour conditions they promote give way to questions of business strategy, as the viability of the firm's place in the global system is paramount.

A regular feature of the global games system is buyouts of small firms which show promise by dominant publishers. Kerr notes that:

> The structure of the games industry in the UK and Ireland has been shaped by the increasing globalization of the games industry during the past two decades. During this time successful game development studios in the UK have been targeted for acquisition by major international publishers and investors. This has had a significant impact upon the development and creation of new projects as well as in some cases their location. (Kerr, 2012: 124)

But the reality is more complex, as such buyouts may also be the difference between survival and liquidation for companies which are vulnerable to the 'creative destruction' wrought in an industry with such rapid innovation cycles. This can be exacerbated by many local developers relying on a work-for-hire (or fee-for-service) business model. A recent report noted, for example, that the UK's most successful independent studios derive two-thirds of their gross revenues from work for hire (Games Investor Consulting, 2008: 13).

In some cases we see reasonably sustainable independent games development alongside of and separate from the axis of developers making games for the conglomerate publishers. In Sweden, for example, there is the success of indie developer Mojang releasing games such as *Minecraft*, which has sold over 4 million copies. Mojang distributed the game itself and used an innovative business model by making the game available for purchase at various alpha and beta stages during the development cycle (Sandqvist, 2012: 145).

Despite the high-profile success stories such as Mojang's *Minecraft*, Rovio's *Angry Birds*, *Call of Duty: Black Ops II* and Halfbrick's *Fruit Ninja*, from a macro-perspective videogames could be understood as a rather problematic and volatile industry. Arguably considerable investment in terms of capital, government grants, labour, education and other resources is tied up in companies that often fail to return profits. It is relatively easy to identify the high-profile success stories with large internationally successful blockbuster titles such as *Call of Duty: Black Ops II*. Nevertheless many companies strive for success and fail. The videogames industry is high-risk, and the likelihood of economic and commercial success is quite small (Sandqvist, 2012: 149).

ACADEMIC TRADITIONS: CULTURES OF USE

Critical political economy considers games from a top-down macro-perspective. Inevitably, it downplays the experiences of games players, whose reported depth of engagement with their engrossing interactivity tends to be 'seen through' as part of what Vincent Mosco (2004) calls the 'digital sublime'. The sublime promises of the digital are simply the latest ruse of a capitalist system working for the benefit of conglomerate games publishers. However, it is difficult to avoid the observation that video gameplay offers experiences quite different to other mainstream entertainment. Their heightened interactivity blurs the distinction between active producer and passive consumer, and their stimulation of virtual communities of networked players, contributed greatly to the phenomenal growth rates of the industry since its beginnings in the 1970s.

The experiences and cultures of players are taken up in diverse studies which constitute the second broad analytical approach. User-engagement approaches were initially structured around the ludology/narratology debate, with studies such as those of T.L. Taylor (2006, 2012), Henry Jenkins (n.d.) and Thomas Malaby (2007) recently enriching deeply our understanding of the cultures of gameplay. Game studies is, of course, even younger than the industry, and emerged from the attempt to understand games in relation to established media. In the late 1990s, ludologists claimed that their degree of interactivity and simulation distinguished videogames. Narratologists, on the other hand, including theorists such as Henry Jenkins (n.d.), emphasize the continuing significance of narrative as a cultural practice that shapes both the production and the consumption of videogames.

But the 'cultures of use' approach is broader than the early disputes between narratologists and ludologists; indeed the field has moved on from these debates. It includes rich ethnographic studies of both player and developer cultures. Bogost (2006), for example, argues for a proceduralist approach that foregrounds the distinctive significance and prominence of software design and engineering as shaping players' experience. This is an approach that focuses on the *materiality* of videogames as assemblages of artefacts, which includes technologies such as game engines and software tools. But critics such as Sicart (2011) argue that a proceduralist approach suggests that the meaning of a game can be deduced from the formal properties of the rules. This diminishes the role of the player in co-creating meaning and pleasure from the game.

Others in the field, such as DeKoven (2002) and Taylor (2006), place the emphasis not so much on technological constructs (though these remain significant in their work) but rather on the culture developed by the players. Malaby (2007) argues that videogames provide an opportunity to experiment with the importance of play beyond rules: 'a game is a semibounded and socially legitimate domain of contrived contingency that generates interpretable outcomes'. He insists on the variable importance of games: 'what is at stake in them can range from very little to the entirety of one's material, social and cultural capital' (Malaby, 2007: 96).

To illustrate this latter point, consider Castronova's (2005) study of the massive multiplayer online game (MMOG) *EverQuest*. Castronova (2005) claimed that, at the time, Norrath – the fictional setting of the *EverQuest* universe – had a GNP per capita between that of Russia and Bulgaria, and higher than countries including China and India. On other platforms, individual virtual assets have sold for substantial sums, perhaps most notably the 'Crystal Palace' in *Project Entropia*, which sold for $330 000 in 2009 (Brennan, 2009). In the 'real' world, there is an industry based on 'levelling up' and 'gold farming' mostly in low-income countries. Dibbell (2007) estimated that, at the time, there were thousands of such businesses in China, employing 'an estimated 100,000 workers, who produce the bulk of all the goods in what has become a $1.8billion worldwide trade in virtual items'.

INDUSTRY RESTRUCTURING, THE GLOBAL DOWNTURN, NEW PLATFORMS AND NEW BUSINESS MODELS

But the industry has moved on rapidly. Major platform shifts, new business models and runaway innovation started before the global financial crisis and continue through a period of slowdown in the world economy. There has been major consolidation at the console production end of the games industry, with fewer, more expensive, blockbuster titles, a hollowing out of the mid-range games market, and a rapid growth and proliferation of casual gaming and mobile applications with unprecedentedly low production costs and barriers to entry.

Major shake-outs and consolidation in the wake of the global downturn have been accompanied by rapidly evolving platform shifts from console and PC products towards small hand-held devices (for example Nintendo DS) with their cartridge software, and now with web-based, mobile and tablet applications becoming increasingly dominant. In 2013 it is expected that the sales value of PC-based videogames will decline to $4 billion, or 6 per cent of the overall videogames market value (De Prato et al., 2012: 233). These are strong examples of the speeding up of the innovation cycle and the interdependency of technology push and user-demand pull. More decisive even than technology push is business model innovation, with Apple's iTunes/App Store solving the micro-payment conundrum, leading to a radical disintermediation of the value chain of significant parts of the industry.

Much future growth lies in games going online and mobile. PricewaterhouseCoopers (2009, 2011) estimated these markets would grow to about $15 billion in 2013 (around 40 per cent of the total), whereas IDATE (2008, 2011) predicted growth from €9 billion to €17 billion for the same time period and a similar share of the global market. This growth of mobile games offers producers quite new opportunities, with smartphones combining advanced computer power, storage capacity and audiovisual capabilities with mobile broadband. These factors have contributed to the rapid growth and diffusion of mobile games.

Many companies including recent start-ups are focusing their development efforts on games for mobile platforms and smaller downloadable games. Standardization across hardware, software and distribution seems to have attracted many new developers. The prime example of this is the Apple iPhone, iOS operating system and App Store 'ecology'. The growth of online distribution also drastically cuts the need for physical logistics, introducing significant disintermediation effects: 'a range of activities in the legacy value network are potentially rendered obsolete – manufacturing boxes and disks, organization and the infrastructure of distribution, retail, sales, inventory and returns' (De Prato et al., 2012: 234). Digital distribution is proving to be especially important in the North American market. It is estimated, for example, that the online Steam distribution platform now accounts for nearly 70 per cent of all digitally distributed game revenues (Graft, 2009).

Mobile device platforms and associated distribution channels offer new opportunities for smaller, emerging and indie developers. As O'Donnell (2012: 102) states, 'iOS development can be started for free, and testing and deploying a game require the maintenance of a $99.00 per year license. Android OS development tools are free and games can be tested and deployed in the virtual "app stores" for free as well.' Essentially the barrier to entry in terms of development costs (both hardware/software costs and labour costs) is now significantly lower, whereas development costs were rising steeply until online and mobile games came to the forefront (Kerr, 2006).

However, established conglomerate games developers and publishers are also moving into mobile games with their large franchises. The success of the Electronic Arts Sports football *FIFA 13* and *Need for Speed Street Racer* apps shows this. These changes might also presage re-intermediation as Internet service providers and telecommunications companies, mobile phone manufacturers, online social networks, online retailers and game sites (for example Gameforge) become players in the games industry, while, for simpler browser-based games, developers may publish their games directly (De Prato et al., 2012: 235).

Further, it is well to note that *Call of Duty: Black Ops II*, a traditional first-person shooter game released in late 2012 for the PlayStation 3 and Xbox 360 consoles, developed by Treyarch and published by Activision, grossed over $500 million within 24 hours of going on sale to become possibly the biggest entertainment launch of all time. This is a timely reminder of the continuing economic significance of the traditional retail channel blockbuster game market. The sustainability of emerging casual gaming and mobile device market opportunities – and if they offer better long-term commercial viability than many of the earlier, high-risk business models – remains an open question.

THIRD APPROACH: ORGANIZATIONAL STUDIES DRAWING ON ECONOMIC SOCIOLOGY

What firms have survived the global shake-out and this period of runaway innovation? How are development studios reshaping and transforming themselves, their workplaces and their practices to pursue these emerging opportunities? Flowing from these industry shifts are implications for the nature of management and organizational structure of the creative firm, the changing nature of labour in the videogames industry, and the increased viability of the small firm in the industry ecology due to value-chain disintermediation. We posit that these circumstances require an approach based on organizational studies informed in particular by economic sociology.

Of course, political economy attends to organizations – but almost exclusively large, multinational hegemons. There are also important organizational foci in the cultures of use approach, for example Taylor's (2012) study of professional player leagues. Malaby's (2009) study of Linden Lab and *Second Life* offers an exemplary organizational ethnography that demonstrates the importance of understanding changing organizational structures and dynamics together with issues relating to professional identity and workplace culture. He shows how the workers of a very young but quickly growing company dealt with the perceived loss of their collaborative freedoms as a result of the rapid growth and maintenance of the runaway success of *Second Life*. Workers at Linden Lab also struggled to manage not only the virtual world they had created but also themselves in a non-hierarchical fashion. How do you regulate something that's supposed to run on its own? Can free-range principles of the game be applied to the organization that runs it?

In exploring the practices 'the Lindens' employed, Malaby (2009) uncovers what was at stake in their virtual world, what a game really is (and how people participate), and the role of the unexpected in a product like *Second Life* and an organization like Linden Lab. Malaby foregrounds that control and authority over cultural production are at stake in these workplace relations. To develop these open-ended worlds that increasingly characterize interactive entertainment, professional developers need to give up some aspects of managerial control. By embracing the contingency of these relationships with their consumers, they are making cultural production firms less hierarchical. Malaby (2009: 9) hopes that his book contributes to our being

> in a better position to understand the emerging institutions that are ever more able to shape and govern our increasingly digital lives. It explores how an organization that set out to create a deeply and complexly contingent environment is itself then remade by its creation through that

domain's emergent effects, in a constantly reiterative process, but without losing its position of greatest influence.

Economic sociology has a considerable intellectual heritage, beginning with Weber and Durkheim, and including Schumpeter, Polanyi and Talcott Parsons. Mark Granovetter's (1985) highly influential article 'Economic Action and Social Structure: The Problem of Embeddedness' revived the notion that the economy is embedded in social relations. This broad approach was developed further in actor–network theory (Callon, 1998, 2007; Latour, 2005; MacKenzie, 2006) and in institutional economics (North, 1991). In this approach, markets are seen as socio-cultural phenomena that also involve material infrastructures such as technologies. The categories of the social, cultural, technological and economic sit in non-hierarchical relations to each other. These approaches emphasize that markets and market transactions are never just economic; they are entangled with the social, cultural and technical conditions that make them possible.

The potential of this third approach is that it may help us to move past the antinomies which structure political economy (capitalist control) versus cultures of use (player/developer agency). It takes market dynamics seriously, attends to the nature of the firm as the key meso-level entity between the macro-perspective of political economy and the micro-perspective of user agency. Organizational studies based on economic sociology may also enable us to grapple with the increasingly distributed and networked characteristics of games production and consumption that may be in the process of transforming the nature both of firms and of markets in the industry. This includes the co-creativity that increasingly characterizes relationships among games consumers and developers, as gamers are no longer at the end of the value chain. Gamers are involved in the process of making, promoting and distributing games. In this sense videogames and interactive entertainment can be approached as strong examples of what we have called 'social network' markets (Potts et al., 2008).

David Stark's (2009) *The Sense of Dissonance: Accounts of Worth in Economic Life* is relevant in establishing this approach to games studies. He asks what counts as worth (a concept that includes both *value* and *values*) in business. Stark's thesis is that firms confronting rapidly changing and uncertain environments are better served by allowing multiple logics of worth and not discouraging the resulting exploration of uncertainty. This is a good description of the interactive entertainment industry with its critical inputs from creative arts, software engineering and business strategy. We are seeing new modes of value creation involving deep engagement with social media affordances, where maintaining oppositions between community and commercialism, professional and amateur, entertainment and education, information or service, don't work any more. Successful firms are exploring new ways to organize innovation processes and to manage risk. Stark (2009: 17) describes these organizational forms as heterarchic (multiple rather than linear power relations) and became interested in how the actions of agents in these firms were made possible by 'maintaining an ongoing ambiguity among the co-existing principles' (2009: xiii–xiv). This dissonance generated by the ongoing rivalries and disputes became an opportunity for organizations to identify possibilities for entrepreneurship from exploration and recombination. The frictions generated should not be avoided or shut down, as they generate a 'resourceful dissonance'.

ECONOMIC SOCIOLOGY APPLIED: THE CASE OF HALFBRICK

Halfbrick, an Australian-based games developer, provides a useful illustration. The Australian games industry has been undergoing rapid creative destruction as a result of a global shake-out dating from the global financial crisis of 2008 onward. Throughout most of its history, the industry's business models revolved around the production of titles for transnational games publishers, while the strategy of the industry's lobby, the Australian Games Developers' Association, has been to secure tax incentives and other forms of state support to undergird business sustainability. Many of the largest Australian games companies were slow to diversify into exploring new forms of interactive entertainment and went into liquidation.

Brisbane-based Halfbrick Studios, developer of the hit mobile device game *Fruit Ninja*, is one of those games companies that not only survived but grew strongly through this period. More than two years after its first release, *Fruit Ninja* was still in second place on the Top Paid iPhone Apps list. The company's success has been used as a model by the Australian government and as an indicator of opportunity for local game developers to 'reclaim their competitive advantage' (Parker, 2012). Formed in 2001, Halfbrick, like most Australian games companies, was originally a developer of licensed titles for platforms such as Game Boy Advance, Nintendo DS games hand-held console and PlayStation portable. In recent years, however, Halfbrick has transformed its business model to become an independent games developer and publisher of its own titles for mobile devices. Halfbrick's principal business model depends upon high-volume sales of games titles via app download purchases, principally from Apple's iTunes Store and merchandising sales from branded T-shirts, posters, mugs and the like (with characters or iconography from Halfbrick games). The viability of this business model depends on the radical disintermediation pioneered by Apple's App Store. The developer receives an unprecedentedly high 70 per cent return on every download.

However, it is important to keep in mind that the sustainability of the Halfbrick business model may be a brief window and the conglomerates might reassert greater control over the supply of content to the global market. The continued prominence of licensed properties such as Disney titles and the EA Sports franchises, and the marketing advantages of established mobile developers such as Zynga suggest that the dominant industry business paradigm of conglomerate control remains alive and well. The window of opportunity for smaller firms such as the Halfbricks and Mojangs may not last. As the technology platforms of these mobile platforms advance we may also see the high barriers to entry return and the end of this 'Prague spring' for independent games makers. Nevertheless, we do see here the emergence of significant new organizational cultures for managing creativity and for harnessing the capacity to adapt to rapidly changing business environments.

The authors' current research with Australian interactive entertainment companies, including Halfbrick Studios, has focused on what allowed companies to survive the sea changes this chapter has outlined (Banks, 2012). Stark's (2009) organizational model of heterarchy and 'multiple evaluative criteria' has proved helpful for understanding how Halfbrick and its confreres turned 'on a sixpence' to address almost completely different technical platforms, user dynamics, and business model and market conditions.

Halfbrick's senior managers and developers place particular value on workplace organizational culture. A common theme emerging across all levels of the company is a flat, team-based organization that devolves control as much as possible directly to the teams and encourages a work–life balance in which creativity can thrive. This comes to the fore in the organization of project teams. Halfbrick consists of some six independent teams, with each team encouraged to develop its own distinctive approach to project management. Each team also identifies and pursues its own project ideas in consultation with senior managers. CEO Shainiel Deo (2012) emphasizes that maintaining and respecting this 'team independence' are central to Halfbrick's success, which is borne out in encounters with members of teams at all levels of the organization.

Deo (2012) emphasizes his goal as building a workplace in which his staff can sustainably make quality games and have a place where they 'love to create and work'. Halfbrick has very low staff turnover – most of the developers have been with Halfbrick for a good few years, and all commented on the quality of the workplace culture and the sense of creative autonomy that they enjoyed as the most significant factors contributing to Halfbrick's success and their decision to stay with the company. Deo (2012) emphasized that this requires a certain kind of developer who can thrive and grow in this culture – developers need to be able to work collaboratively in such a decentralized and autonomous environment, and his managers 'need to be able to trust them to do this'. These characteristics were just as important factors in hiring decisions as technical skills. In a previous account, Banks (2012: 158) wrote:

> Halfbrick's developers are doing far more than designing and innovating new products: along the way, they are redesigning their firm and transforming their dominant mode of production – the project model of game development . . . This is about much more than technical affordances or design qualities as it goes to the very heart of professional developers' working lives and environments and how they imagine what it means to be professional cultural producers.

Halfbrick's organizational diversity is especially evident in 'Halfbrick Fridays'. On Halfbrick Fridays, a day is set aside from the standard development schedules for anyone in the company to brainstorm and pitch ideas and then develop them through an iterative, rapid prototyping process. The initial idea for *Fruit Ninja* emerged from a Halfbrick Friday session. Through this process developers come up with many contending and competing game ideas and then evaluate those they think may have potential for commercial viability through rounds of internal peer critique. Project teams then identify and select projects that emerge from this process for further development. Deo describes this process as both 'bottom up' and 'creatively empowering' (Banks, 2012).

Halfbrick's organizational culture may provide a strong example of what Neff (2012) calls 'venture labour' – the shifting of risk towards workers as a form of entrepreneurial labour that is required to stimulate innovation. Neff asks us to consider the challenge of also creating sustainable jobs and workplaces for such workers. We suggest that this is precisely what firms such as Halfbrick may be seeking to achieve and that these efforts provide us with a rather different perspective on the videogames and interactive entertainment industry than does a traditional political economy critique of the games publisher conglomerates. In this process of rapid and uncertain transformation confronting videogames firms and workers we suggest that it is worthwhile to pay some attention to these emerging modes of organization and workplace culture, even if they might as yet be somewhat at the margins and vulnerable to structural shifts over which they may have little control. As Stark observes, to pursue innovation in unpredictable and uncertain environments it is necessary to

> build organizations that are not only capable of learning but also capable of suspending accepted knowledge and established procedures to redraw cognitive categories and reconfigure relational boundaries – both at the level of the products and services produced by the firm and at the level of working practices and production processes within the firm. (Stark, 2009: 83)

Firms in the videogames and interactive entertainment industry may well be at the forefront of such adaptive organizational innovation.

SUMMARY AND FURTHER RESEARCH

New frameworks are needed to better understand the dynamic changes we have seen sweeping through the interactive entertainment industry over the past few years. Two broad academic traditions of analysis of the games industry present themselves: critical political economy; and cultures of use. Industry restructuring, the global downturn, new platforms and new business models have seen creative destruction in games and new opportunities for smaller firms and new practices. These opportunities are especially evident in games for mobile devices. These industry trends, we argue, bring into prominence a third emerging approach: organizational studies, which draws on economic sociology. Nevertheless, with the continued prominence of licensed properties and the marketing advantages of established mobile operators, the window of opportunity for smaller firms and independent developers pursuing these emerging opportunities and new practices may not last. Further research in this spirit should focus on gathering evidence into whether the kinds of organizational and management culture outlined in the chapter's account of applied economic sociology have managed to improve the sustainability of games firms in this most volatile of industries.

REFERENCES

Aarseth, E. (2012), 'A narrative theory of games', *Proceedings of the International Conference on the Foundations of Digital Games*, New York: Association for Computing Machinery, pp. 129–33.

Adolph, M. (2011), 'Trends in video games and gaming: ITU-T Technology Watch Report', available at: http://www.scribd.com/doc/67857275/Trends-in-Video-Games-and-Gaming (accessed 5 December 2012).

Banks, J. (2012), 'The iPhone as innovation platform: reimagining the videogames developer', in L. Hjorth, J. Burgess and I. Richardson (eds), *Studying Mobile Media: Cultural Technologies, Mobile Communication, and the iPhone*, London: Routledge, pp. 155–72.

Banks, J. (2013), *Co-Creating Videogames*, London: Bloomsbury Academic.

Bogost, I. (2006), *Unit Operations: An Approach to Videogame Criticism*, Cambridge, MA: MIT Press.

Brennan, S. (2009), 'Crystal Palace Space Station auction tops 330,000 US dollars', *Massively*, available at: http://massively.joystiq.com/2009/12/29/crystal-palace-space-station-auction-tops-330–000-us-dollars/ (accessed 20 January 2013).

Callon, M. (ed.) (1998), *The Laws of the Markets*, London: Blackwell.

Callon, M. (2007), 'What does it mean to say that economics is performative?', in D. MacKenzie, F. Muniesa and L. Siu (eds), *Do Economists Make Markets? On the Performativity of Economics*, Princeton, NJ: Princeton University Press.

Caron, F. (2008), 'Gaming expected to be a $68 billion business by 2012', *Arstechnica*, available at: http://arstechnica.com/gaming/2008/06/gaming-expected-to-be-a-68-billion-business-by-2012/ (accessed 14 March 2013).
Castronova, E. (2005), *Synthetic Worlds*, Chicago: University of Chicago Press.
Cunningham, S. (2013), *Hidden Innovation: Policy, Industry and the Creative Sector*, St Lucia: University of Queensland Press.
DeKoven, B. (2002), *The Well-Played Game: A Playful Path to Wholeness*, Lincoln, NE: Writers Club Press.
Deo, S. (2012), Interviews with the authors, Brisbane, May. Available at: http://www.slq.qld.gov.au/whats-on/calevents/livestreams/shainiel-deo (accessed 7 June 2013).
De Prato, G., S. Lindmark and J.-P. Simon (2012), 'The evolving European video game software ecosystem', in P. Zackariasson and T.L. Wilson (eds), *The Video Game Industry: Formation, Present State, and Future*, New York: Routledge.
Dibbell, J. (2007), 'The life of the Chinese gold farmer', *New York Times*, 17 June, available at: http://www.nytimes.com/2007/06/17/magazine/17lootfarmers-t.html?_r=2&pagewanted=print (accessed 20 January 2013).
Games Investor Consulting (2008), *Raise the Game: The Competitiveness of the UK's Games Development Sector and the Impact of Governmental Support Overseas*, London: NESTA.
Graft, K. (2009), 'Stardock reveals Impulse, Steam market share estimates', *Gamasutra*, available at: http://www.gamasutra.com/php-bin/news_index.php?story=26158 (accessed 5 December 2012).
Granovetter, M. (1985), 'Economic action and social structure: the problem of embeddedness', *American Journal of Sociology*, **91** (3), 481–510.
IDATE (2008), *Digiworld Yearbook 2008*, Montpellier: IDATE.
IDATE (2011), *World Video Game* Market, Montpellier: IDATE.
Jenkins, H. (n.d.), 'Game design as narrative architecture', available at: http://web.mit.edu/cms/People/henry3/games&narrative.html (accessed 5 December 2012).
Kerr, A. (2006), *The Business and Culture of Video Games: Gamework/Gameplay*, London: Sage.
Kerr, A. (2012), 'The UK and Irish game industries', in P. Zackariasson and T.L. Wilson (eds), *The Video Game Industry: Formation, Present State, and Future*, New York: Routledge.
Latour, B. (2005), *Reassembling the Social: An Introduction to Actor–Network Theory*, Oxford: Oxford University Press.
MacKenzie, D. (2006), *An Engine, Not a Camera: How Financial Models Shape Markets*, Cambridge, MA: MIT Press.
Malaby, T.M. (2007), 'Beyond play: a new approach to games', *Games and Culture*, **2** (2), 95–113.
Malaby, T.M. (2009), *Making Virtual Worlds: Linden Lab and Second Life*, Ithaca, NY: Cornell University Press.
Mäyrä, F. (2008) *An Introduction to Game Studies: Games in Culture*, London: Sage.
Miles, I. and Green, L. (2008), *Hidden Innovation in the Creative Industries: Research Report*, London: National Endowment for Science, Technology and the Arts, available at: www.nesta.org.uk/publications/reports/assets/features/hidden_innovation (accessed 14 March 2013).
Miller, T. (2008), 'Anyone for games? Via the new international division of cultural labour', in Helmut Anheier and Yudhishthir Raj Isar (eds), *The Cultural Economy*, The Cultures and Globalization Series, vol. 2, London: Sage, pp. 227–40.
Mosco, V. (2004), *The Digital Sublime: Myth, Power, and Cyberspace*, Cambridge, MA: MIT Press.
Neff, G. (2012), *Venture Labor: Work and the Burden of Risk in Innovative Industries*, Cambridge, MA: MIT Press.
North, D.C. (1991), 'Institutions', *Journal of Economic Perspectives*, **5** (1), 97–112.
O'Donnell, C. (2012), 'The North American game industry', in P. Zackariasson and T.L. Wilson (eds), *The Video Game Industry: Formation, Present State, and Future*, New York: Routledge.
Parker, L. (2012), 'Australian govt pledges $20m to home-grown game development', *Gamespot.au*, 14 November, available at: http://au.gamespot.com/news/australian-govt-pledges-20m-to-home-grown-game-development-6400005.
Potts, J., S. Cunningham, J. Hartley and P. Ormerod (2008), 'Social network markets: a new definition of creative industries', *Journal of Cultural Economics*, **32** (3), 167–85.
PricewaterhouseCoopers (2009), *Global Entertainment and Media Outlook 2009–2013*, New York: PricewaterhouseCoopers.
PricewaterhouseCoopers (2011), *Global Entertainment and Media Outlook 2010–2015*, New York: PricewaterhouseCoopers.
Sandqvist, U. (2012), 'The development of the Swedish game industry: a true success story?', in P. Zackariasson and T.L. Wilson (eds), *The Video Game Industry: Formation, Present State, and Future*, New York: Routledge.
Scanlon, J. (2007), 'The videogames industry outlook: $31.6 billion and growing', *Business Week.com*, 13

August, available at: http://www.businessweek.com/stories/2007–08–13/the-video-game-industry-outlook-31-dot-6-billion-and-growingbusinessweek-business-news-stock-market-and-financial-advice (accessed 14 March 2013).

Shankar, V., and B.L. Bayus (2002), 'Network effects and competition: an empirical analysis of the home video game industry', *Strategic Management Journal*, **24** (4), 375–84.

Sicart, M. (2011), 'Against procedurality', *Game Studies*, **11** (3), available at: http://www.gamestudies.org/1103/articles/sicart_ap (accessed 5 December 2012).

Stark, D. (2009), *The Sense of Dissonance: Accounts of Worth in Economic Life*, Princeton, NJ: Princeton University Press.

Taylor, T.L. (2006), *Play between Worlds: Exploring Online Game Culture*, Cambridge, MA: MIT Press.

Taylor, T.L. (2012), *Raising the Stakes: E-Sports and the Professionalization of Computer Gaming*, Cambridge, MA: MIT Press.

Zackariasson, P. and T. Wilson (eds) (2012), *The Video Game Industry: Formation, Present State and Future*, Abingdon: Routledge.

FURTHER READING

Frans Mäyrä's (2008) *An Introduction to Game Studies: Games in Culture* provides a summary of early work in game studies, whilst Aarseth's (2012) 'A Narrative Theory of Games' provides a contemporary perspective on the ludology versus narratology debate. Zackariasson and Wilson's (2012) volume *The Video Game Industry: Formation, Present State and Future* contains both histories and contemporary case studies on a number of worldwide game industries, and provides good coverage of topical debates in the field. Both Stark's (2009) *The Sense of Dissonance* and Neff's (2012) *Venture Labor* should be consulted for approaches to creative firms in contemporary business conditions.

Index